LONDON MATHEMATICAL SOCIETY LECTURE NOTE SERIES

Managing Editor: Professor Endre Süli, Mathematical Institute, University of Oxford,
Woodstock Road, Oxford OX2 6GG, United Kingdom

The titles below are available from booksellers, or f
www.cambridge.org/mathematics

London Mathematical Society Lecture Note Series: 458

Integrable Systems and Algebraic Geometry

A Celebration of Emma Previato's 65th Birthday

Volume 1

Edited by

RON DONAGI
University of Pennsylvania

TONY SHASKA
Oakland University, Michigan

CAMBRIDGE
UNIVERSITY PRESS

University Printing House, Cambridge CB2 8BS, United Kingdom

One Liberty Plaza, 20th Floor, New York, NY 10006, USA

477 Williamstown Road, Port Melbourne, VIC 3207, Australia

314–321, 3rd Floor, Plot 3, Splendor Forum, Jasola District Centre, New Delhi – 110025, India

79 Anson Road, #06–04/06, Singapore 079906

Cambridge University Press is part of the University of Cambridge.

It furthers the University's mission by disseminating knowledge in the pursuit of education, learning, and research at the highest international levels of excellence.

www.cambridge.org
Information on this title: www.cambridge.org/9781108715744
DOI: 10.1017/9781108773287

First published 2020

Printed and bound in Great Britain by Clays Ltd, Elcograf S.p.A.

A catalogue record for this publication is available from the British Library.

ISBN – 2 Volume Set 978-1-108-78549-5 Paperback
ISBN – Volume 1 978-1-108-71574-4 Paperback
ISBN – Volume 2 978-1-108-71577-5 Paperback

Contents of Volume 1

Contents of Volume 2

Integrable Systems

A Celebration of Emma Previato's 65th Birthday

Ron Donagi and Tony Shaska

1 A Word from the Editors

These two volumes celebrate and honor Emma Previato, on the occasion of her 65th birthday. The present volume consists of 16 articles on and around the subject of integrable systems, one of the two main areas where Emma Previato has made many major contributions. The companion volume focuses on Emma's other major research area, algebraic geometry. The articles were contributed by Emma's coauthors, colleagues, students, and other researchers who have been influenced by Emma's work over the years. They present a very attractive mix of expository articles, historical surveys and cutting edge research.

Emma Previato is a mathematical pioneer, working in her two chosen areas, algebraic geometry and integrable systems. She has been among the first women to do research, in both areas. And her work in both areas has been deep and influential. Emma received a Bachelor's degree from the University of Padua in Italy, and a PhD from Harvard University under the direction of David Mumford in 1983. Her thesis was on hyperelliptic curves and solitons. The work on hyperelliptic curves has evolved and expanded into Emma's life-long interest in algebraic geometry. The work on solitons has led to her ongoing research on integrable systems, which is the subject of the present volume.

Emma Previato has been a faculty member at the Department of Mathematics at Boston University since 1983. She has published nearly a hundred research articles, edited six books, and directed seven PhD dissertations. Her broader impact extends through her renowned teaching and her extensive mentoring activities. She runs AFRAMATH, an annual outreach symposium, and works tirelessly on several on- and off-campus mentoring programs. She has also founded and has been leading the activities of the Boston University chapters of MAA and AWM. She serves on numerous advisory boards.

1

The subject of integrable systems has a long and rich history. It brings together ideas and techniques from analysis, geometry, algebra, and several branches of physics. Emma Previato's work has made important contributions in all of these directions. We review some of her accomplishments in section 2 below.

We tried to collect in this volume a broad range of articles covering most of these areas. We summarize these contributions in section 3.

The editors and many of the authors have enjoyed years of fruitful interactions with Emma Previato. We all join in wishing her many more years of health, productivity, and great mathematics.

2 Emma Previato's Contributions

Emma Previato works in different areas, using methods from algebra, algebraic geometry, mechanics, differential geometry, analysis, and differential equations. The bulk of her research belongs to integrable equations. She is noted for often finding unexpected connections between integrability and many other areas, often including various branches of algebraic geometry.

Early Activity

As an undergraduate at the University of Padua, Italy, Emma wrote a dissertation on group lattices, followed by six journal publications [6, 5, 4, 3, 2, 1]. With methods from algebra, initiated by Dedekind in the 19th century, this area's goal is to relate the group structure to the lattice of subgroups, and provide classifications for certain properties: an excellent overview is the article by Freese [7], a review of the definitive treatise by R. Schmidt, where results from all of Emma's papers are used to give one example, a lattice criterion for a finitely generated group to be solvable.

PhD Thesis and Main Area

Emma's thesis [8], submitted at Harvard in 1983 under the supervision of D.B. Mumford [9], is still her most cited paper. Her thesis advisor was among the pioneers of this beautiful area, integrable equations, which grew and unified disparate parts of mathematics over the next twenty years, and is still very active. Emma's original tool for producing exact solutions to large classes of nonlinear PDEs, the Riemann theta function, remained one of her main interests.

Theta Functions

She later pursued more theoretical aspects of special functions, such as Prym theta functions [14, 12, 11, 10, 13] also surprisingly related to numerical results in conformal field theory, the Schottky problem [15], and Thetanulls [16].

Algebraically Completely Integrable Systems

The area of integrable PDEs is surprisingly related to algebraically completely integrable Hamiltonian systems, or ACIS, in the sense that algebro-geometric (aka finite-gap) solutions of integrable hierarchies linearize on Abelian varieties, which can be organized into angle variables for an ACIS over a suitable base, typically a subset of the moduli space of curves whose Jacobian is the fiber [11, 17]. Thanks to this discovery, the area integrates with classical geometric invariant theory, surface theory, and other traditional studies of algebraic geometry. With the appearance of the moduli spaces of vector bundles and Higgs bundles over a curve, at the hands of N. Hitchin in the 1980s, large families of ACIS were added to the examples, as well as theoretical algebro-geometric techniques. In [10, 18, 20, 19], Emma took up the challenge of generalizing the connection between ACIS and integrable hierarchies to curves beyond hyperelliptic. In [21], the families of curves are organized as divisors in surfaces.

Higher Rank and Higher-Dimensional Spectra

On the PDE side, the challenges were of two types. When the ring of functions on the (affine) spectral curve can be interpreted as differential operators with a higher-dimensional space of common eigenfunctions, the fiber of the integrable system is no longer a Jacobian: it degenerates to a moduli space of higher-rank vector bundles, possibly with some auxiliary structures [22]. Neither the PDEs nor the integrable systems have been made explicit in higher rank in general. Some cases, however, are worked out in [26, 25, 24, 23, 27]. The other challenge is to increase the dimension of the spectral variety, for example from curve to surface. Despite much work, this problem too has arguably no explicit solution in general. An attempt to set up a general theory over a multi-dimensional version of the formal Universal Grassmann Manifold of Sato which hosts all linear flows of solutions of integrable hierarchies, is given in [28], and more concrete special settings are mentioned below, under the heading of "Differential Algebra".

Special Solutions: Coverings of Curves

An important aspect of theta functions is their reducibility, a property whose investigation goes back to Weierstrass and his student S. Kowalevski. Given their special role in integrability, reducible theta functions are invaluable for applied mathematicians to approximate solutions, or even derive exact expressions and periods in terms of elliptic functions. To the algebro-geometric theory of Elliptic Solitons, initiated by I.M. Krichever and developed by A. Treibich and his thesis supervisor J.-L. Verdier, Emma contributed [31, 32, 35, 30, 34, 29, 12, 33], while [37, 36] generalize the reduction to hyperelliptic curves or Abelian subvarieties. More general aspects of elliptic (sub)covers are taken up in [38].

Another type of special solution is the one obtained by self-similarity [39]; the challenge here is to find an explicit relationship between the PDE flows and the deformation in moduli that obeys Painlevé-type equations: this is one reason why Emma's work has turned to a special function which is associated to Riemann's theta function but only exists on Jacobians: the sigma function (cf. the eponymous section below).

Poncelet and Billiards: Generalizing ACIS

Classical theorems of projective geometry can be generalized to ACIS [41, 40], while the challenge of matching them with integrable hierarchies is still ongoing [42].

Generalizing ACIS: Hitchin Systems

Explicit Hamiltonians for the Hitchin system are only available in theory: they are given explicit algebraic expression in [43] (cf. also [44], which led to work on the geometry of the moduli space of bundles [45]). An explicit integration in terms of special functions leads to the problem of non-commutative theta functions [46].

Differential Algebra

Differential Algebra is younger than Algebraic Geometry, but it has many features in common. Mumford gives credit to J.L. Burchnall and T.W. Chaundy for the first spectral curve, the Spectrum of a commutative ring of differential operators [47]. This is arguably the reason behind algebro-geometric solutions to integrable hierarchies. On the differential-algebra setting, Emma published [48, 49], connecting geometric properties of the curve with differential resultants, a major topic of elimination theory which is currently being worked out

[51, 50] and naturally leads to the higher-rank solutions: their Grassmannian aspects are taken up in [54, 53, 55, 56, 52] the higher-dimensional spectral varieties arc addressed in [57]. Other aspects of differential algebra are connected to integrability in [58] (the action of an Abelian vector field on the meromorphic functions of an Abelian variety) and [59] (a p-adic analog); in [60], the deformations act on modular forms.

The Sigma Function

Klein extended the definition of the (genus-one) Weierstrass sigma function to hyperelliptic curves and curves of genus three. H.F. Baker developed an in-depth theory of PDEs satisfied by the hyperelliptic sigma function, which plays a key role in recent work on integrable hierarchies (KdV-type, e.g.). Beginning in the 1990s, this theory of Kleinian sigma functions was revisited, originally by V.M. Buchstaber, V.Z. Enolskii, and D.V. Leykin, much extended in scope, eventually to be developed for "telescopic" curves (a condition on the Weierstrass semigroup at a point). Previato goes beyond the telescopic case in [62, 61], while she investigates the higher-genus analog of classical theorems in [69, 68, 70, 67, 66, 63, 64, 65] and their connections with integrability in [71] and [72], which gives the first algebro-geometric solutions to a dispersionless integrable hierarchy. It is not a coincidence that its integrable flow on the Universal Grassmann Manifold 'cut across' the Jacobian flows of traditional hierarchies, and this is where the two variables of the sigma function (the Jacobian, and the modular ones) should unite to explain the mystery of the Painlevé equations.

Algebraic Coding Theory

Emma's primary contribution to this area is through mentoring undergraduate and graduate thesis or funded-research projects. In fact, this research strand began at the prompting of students in computer science who asked her to give a course on curves over fields of prime characteristic, which she ran for years as a vertically-integrated seminar. Together with her PhD student Drue Coles, she published research papers pursuing Trygve Johnsen's innovative idea of error-correction for Goppa codes implemented via vector bundles [74, 75, 73], then she pursued overviews and extensions of Goppa codes to surfaces [76].

Other

Emma edited or co-edited four books [78, 80, 79, 77]. In addition to book and journal publication, Emma published reviews (BAMS, SIAM), entries in mathematical dictionaries or encyclopaedias, teaching manuals and online research or teaching materials; she also published on the topic of mentoring in the STEAM disciplines.

3 Articles in this Volume

Gesztesy and Nichols consider a particular class of integral operators $T_{\gamma,\delta}$ in $L^2(\mathbb{R}^n)$, $n \in \mathbb{N}$, $n \geq 2$, with integral kernels bounded. These integral operators (and their matrix-valued analogs) naturally arise in the study of multi-dimensional Schrödinger and Dirac-type operators and they describe an application to the case of massless Dirac-type operators.

H. Knörrer is using quaternions to give explicit formulas for the global symmetries of the three dimensional Kepler problem. The regularizations of the Kepler problem that are based on the Hopf map and on stereographic projections, respectively, are interpreted in terms of these symmetries.

Zoladek gives two proofs of the Jacobi identity for the Poisson bracket on a symplectic manifold.

Luen-Chau Li presents an expository account of the work done in the last few years in understanding a matrix Lax equation which arises in the study of scalar hyperbolic conservation laws with spectrally negative pure-jump Markov initial data. He begins with its extension to general $N \times N$ matrices, which is Liouville integrable on generic coadjoint orbits of a matrix Lie group. In the probabilistically interesting case in which the Lax operator is the generator of a pure-jump Markov process, the spectral curve is generically a fully reducible nodal curve. In this case, the equation is not Liouville integrable, but we can show that the flow is still conjugate to a straight line motion, and the equation is exactly solvable. En route, he establishes a dictionary between an open, dense set of lower triangular generator matrices and algebro-geometric data which plays an important role in the analysis.

F. Calogero and F. Payandeh study solvable dynamical systems in the plane with polynomial interactions. They present a few examples of algebraically solvable dynamical systems characterized by 2 coupled Ordinary Differential Equations. These findings are obtained via a new twist of a recent technique to identify dynamical systems solvable by algebraic operations, themselves explicitly identified as corresponding to the time evolutions of the zeros of polynomials the coefficients of which evolve according to algebraically solvable (systems of) evolution equations.

V. Dragovich, M. Radnovic, and R.F. Ranomenjanahary present recent results about double reflection and incircular nets. The building blocks are pencils of quadrics, related billiards, and quad graphs.

Alessandro Arsie and Paolo Lorenzoni present a survey of the work done by the authors in the last few years developing the theory of bi-flat F-manifolds and exploring their relationships with integrable hierarchies (dispersionless and dispersive), with Painlevé transcendents, and with complex reflection groups.

Pol Vanhaecke studies some algebraic-geometrical aspects of the periodic 6-particle Kac-van Moerbeke system. This system is known to be algebraically integrable, having the affine part of a hyperelliptic Jacobian of a genus two curve as the generic fiber of its momentum map. Particular attention goes to the divisor needed to complete this fiber into an Abelian variety: it consists of six copies of the curve, intersecting according to a pattern which is determined in the paper. The author also compares this divisor to the divisor which appears in some natural singular compactification of the fiber.

T. Brown and N. Ercolani focus on discrete Painlevé equations and connections between combinatorics and integrable systems. Two discrete dynamical systems are discussed and analyzed whose trajectories encode significant explicit information about a number of problems in combinatorial probability, including graphical enumeration on Riemann surfaces and random walks in random environments. The authors show that the two models are integrable and their analysis uncovers the geometric sources of this integrability and uses this to conceptually explain the rigorous existence and structure of elegant closed form expressions for the associated probability distributions. Connections to asymptotic results are also described. The work brings together ideas from a variety of fields including dynamical systems theory, probability theory, classical analogues of quantum spin systems, addition laws on elliptic curves, and links between randomness and symmetry.

T. Kappeler and P. Topalov study the Arnold-Liouville theorem for integrable PDEs. They present an infinite dimensional version of the Arnold-Liouville theorem.

A. Chern, F. Knoeppel, F. Pedit, and U. Pinkall study commuting Hamiltonian flows of curves in real space forms. They provide a geometric point of view of the Hamiltonian flows.

Franco Magri writes on the Kowalewski Top from the viewpoint of bihamiltonian geometry. The paper is a commentary of one section of the celebrated paper by Sophie Kowalewski on the motion of a rigid body with a fixed point. Its purpose is to show that the results of Kowalewski may be recovered by using the separability conditions obtained by Tullio Levi-Civita in 1904.

Steven Rayan and Jacek Szmigielski study Peakons and Hitchin systems. They review the Calogero-Françoise integrable system, which is a generalization of the Camassa-Holm system. We express solutions as (twisted) Higgs bundles, in the sense of Hitchin, over the projective line. We use this point of view to (a) establish a general answer to the question of linearization of isospectral flow and (b) demonstrate, in the case of two particles, the dynamical meaning of the theta divisor of the spectral curve in terms of mechanical collisions. They also outline the solution to the inverse problem for CF flows using Stieltjes' continued fractions.

Spalding and Veselov study the tropical version of Markov dynamics on the Cayley cubic. They prove that this action is semi-conjugated to the standard action of $SL_2(\mathbb{Z})$ on a torus and therefore ergodic with the Lyapunov exponent and entropy given by the logarithm of the spectral radius of the corresponding matrix.

G.S. Mauleshova and A.E.Mironov focus on one-point commuting difference operators. They study a new class of rank one commuting difference operators containing a shift operator with only positive degrees. We obtain equations which are equivalent to the commutativity conditions in the case of hyperelliptic spectral curves. Using these equations we construct explicit examples of operators with polynomial and trigonometric coefficients.

References

[1] Emma Previato, *Gruppi in cui la relazione di Dedekind è transitiva*, Rend. Sem. Mat. Univ. Padova **54** (1975), 215–229 (1976). MR0466319

[2] _____, *Una caratterizzazione dei sottogruppi di Dedekind di un gruppo finito*, Atti Accad. Naz. Lincei Rend. Cl. Sci. Fis. Mat. Natur. (8) **59** (1975), no. 6, 643–650 (1976). MR0480738

[3] _____, *Sui sottogruppi di Dedekind nei gruppi infiniti*, Atti Accad. Naz. Lincei Rend. Cl. Sci. Fis. Mat. Natur. (8) **60** (1976), no. 4, 388–394. MR0491981

[4] _____, *Groups in whose dual lattice the Dedekind relation is transitive*, Rend. Sem. Mat. Univ. Padova **58** (1977), 287–308 (1978). MR543147

[5] _____, *A lattice-theoretic characterization of finitely generated solvable groups*, Istit. Veneto Sci. Lett. Arti Atti Cl. Sci. Mat. Natur. **136** (1977/78), 7–11. MR548255

[6] _____, *Some families of simple groups whose lattices are complemented*, Boll. Un. Mat. Ital. B (6) **1** (1982), no. 3, 1003–1014. MR683488

[7] R. Freese, *Subgroup lattices of groups by roland schmidt*, Review, 1994.

[8] Emma Previato, *Hyperelliptic quasiperiodic and soliton solutions of the nonlinear Schrödinger equation*, Duke Math. J. **52** (1985), no. 2, 329–377. MR792178

[9] _____, *HYPERELLIPTIC CURVES AND SOLITONS*, ProQuest LLC, Ann Arbor, MI, 1983, Thesis (PhD)–Harvard University. MR2632885

[10] Letterio Gatto and Emma Previato, *A remark on Griffiths' cohomological interpretation of Lax equations: higher-genus case*, Atti Accad. Sci. Torino Cl. Sci. Fis. Mat. Natur. **126** (1992), no. 3-4, 63–70. MR1231817

[11] Emma Previato, *Geometry of the modified KdV equation*, Geometric and quantum aspects of integrable systems (Scheveningen, 1992), 1993, pp. 43–65. MR1253760

[12] Emma Previato and Jean-Louis Verdier, *Boussinesq elliptic solitons: the cyclic case*, Proceedings of the Indo-French Conference on Geometry (Bombay, 1989), 1993, pp. 173–185. MR1274502

[13] Bert van Geemen and Emma Previato, *Prym varieties and the Verlinde formula*, Math. Ann. **294** (1992), no. 4, 741–754. MR1190454

[14] _____, *Heisenberg action and Verlinde formulas*, Integrable systems (Luminy, 1991), 1993, pp. 61–80. MR1279817

[15] Christian Pauly and Emma Previato, *Singularities of 2Θ-divisors in the Jacobian*, Bull. Soc. Math. France **129** (2001), no. 3, 449–485. MR1881203

[16] E. Previato, T. Shaska, and G. S. Wijesiri, *Thetanulls of cyclic curves of small genus*, Albanian J. Math. **1** (2007), no. 4, 253–270. MR2367218

[17] Emma Previato, *A particle-system model of the sine-Gordon hierarchy*, Phys. D **18** (1986), no. 1-3, 312–314, Solitons and coherent structures (Santa Barbara, Calif., 1985). MR838338

[18] M. R. Adams, J. Harnad, and E. Previato, *Isospectral Hamiltonian flows in finite and infinite dimensions. I. Generalized Moser systems and moment maps into loop algebras*, Comm. Math. Phys. **117** (1988), no. 3, 451–500. MR953833

[19] Emma Previato, *Flows on r-gonal Jacobians*, The legacy of Sonya Kovalevskaya (Cambridge, Mass., and Amherst, Mass., 1985), 1987, pp. 153–180. MR881461

[20] _____, *Generalized Weierstrass ℘-functions and KP flows in affine space*, Comment. Math. Helv. **62** (1987), no. 2, 292–310. MR896099

[21] Silvio Greco and Emma Previato, *Spectral curves and ruled surfaces: projective models*, The Curves Seminar at Queen's, Vol. VIII (Kingston, ON, 1990/1991), 1991, pp. Exp. F, 33. MR1143110

[22] Emma Previato and George Wilson, *Vector bundles over curves and solutions of the KP equations*, Theta functions—Bowdoin 1987, Part 1 (Brunswick, ME, 1987), 1989, pp. 553–569. MR1013152

[23] Geoff Latham and Emma Previato, *Higher rank Darboux transformations*, Singular limits of dispersive waves (Lyon, 1991), 1994, pp. 117–134. MR1321199

[24] Geoff A. Latham and Emma Previato, *Darboux transformations for higher-rank Kadomtsev-Petviashvili and Krichever-Novikov equations*, Acta Appl. Math. **39** (1995), no. 1-3, 405–433, KdV '95 (Amsterdam, 1995). MR1329574

[25] _____, *KP solutions generated from KdV by "rank 2" transference*, Phys. D **94** (1996), no. 3, 95–102. MR1392449

[26] E. Previato, *Burchnall-Chaundy bundles*, Algebraic geometry (Catania, 1993/Barcelona, 1994), 1998, pp. 377–383. MR1651105

[27] Emma Previato and George Wilson, *Differential operators and rank 2 bundles over elliptic curves*, Compositio Math. **81** (1992), no. 1, 107–119. MR1145609

[28] Min Ho Lee and Emma Previato, *Grassmannians of higher local fields and multivariable tau functions*, The ubiquitous heat kernel, 2006, pp. 311–319. MR2218024

[29] E. Colombo, G. P. Pirola, and E. Previato, *Density of elliptic solitons*, J. Reine Angew. Math. **451** (1994), 161–169. MR1277298

[30] J. C. Eilbeck, V. Z. Enolskii, and E. Previato, *Varieties of elliptic solitons*, J. Phys. A **34** (2001), no. 11, 2215–2227, Kowalevski Workshop on Mathematical Methods of Regular Dynamics (Leeds, 2000). MR1831289

[31] J. Chris Eilbeck, Victor Z. Enolski, and Emma Previato, *Spectral curves of operators with elliptic coefficients*, SIGMA Symmetry Integrability Geom. Methods Appl. **3** (2007), Paper 045, 17. MR2299846

[32] E. Previato, *Jacobi varieties with several polarizations and PDE's*, Regul. Chaotic Dyn. **10** (2005), no. 4, 531–543. MR2191376

[33] Emma Previato, *The Calogero-Moser-Krichever system and elliptic Boussinesq solitons*, Hamiltonian systems, transformation groups and spectral transform methods (Montreal, PQ, 1989), 1990, pp. 57–67. MR1110372

[34] _____, *Monodromy of Boussinesq elliptic operators*, Acta Appl. Math. **36** (1994), no. 1-2, 49–55. MR1303855

[35] _____, *Reduction theory, elliptic solitons and integrable systems*, The Kowalevski property (Leeds, 2000), 2002, pp. 247–270. MR1916786

[36] Ron Y. Donagi and Emma Previato, *Abelian solitons*, Math. Comput. Simulation **55** (2001), no. 4-6, 407–418, Nonlinear waves: computation and theory (Athens, GA, 1999). MR1821670

[37] È. Previato and V. Z. Ènol' skiĭ, *Ultra-elliptic solitons*, Uspekhi Mat. Nauk **62** (2007), no. 4(376), 173–174. MR2358755

[38] Robert D. M. Accola and Emma Previato, *Covers of tori: genus two*, Lett. Math. Phys. **76** (2006), no. 2-3, 135–161. MR2235401

[39] G. N. Benes and E. Previato, *Differential algebra of the Painlevé property*, J. Phys. A **43** (2010), no. 43, 434006, 14. MR2727780

[40] Emma Previato, *Poncelet's theorem in space*, Proc. Amer. Math. Soc. **127** (1999), no. 9, 2547–2556. MR1662198

[41] _____, *Some integrable billiards*, SPT 2002: Symmetry and perturbation theory (Cala Gonone), 2002, pp. 181–195. MR1976669

[42] Yuji Kodama, Shigeki Matsutani, and Emma Previato, *Quasi-periodic and periodic solutions of the Toda lattice via the hyperelliptic sigma function*, Ann. Inst. Fourier (Grenoble) **63** (2013), no. 2, 655–688. MR3112844

[43] Bert van Geemen and Emma Previato, *On the Hitchin system*, Duke Math. J. **85** (1996), no. 3, 659–683. MR1422361

[44] E. Previato, *Dualities on $T^*SU_X(2, O_X)$*, Moduli spaces and vector bundles, 2009, pp. 367–387. MR2537074

[45] W. M. Oxbury, C. Pauly, and E. Previato, *Subvarieties of $SU_C(2)$ and 2θ-divisors in the Jacobian*, Trans. Amer. Math. Soc. **350** (1998), no. 9, 3587–3614. MR1467474

[46] Emma Previato, *Theta functions, old and new*, The ubiquitous heat kernel, 2006, pp. 347–367. MR2218026

[47] _____, *Seventy years of spectral curves: 1923–1993*, Integrable systems and quantum groups (Montecatini Terme, 1993), 1996, pp. 419–481. MR1397276

[48] Jean-Luc Brylinski and Emma Previato, *Koszul complexes, differential operators, and the Weil-Tate reciprocity law*, J. Algebra **230** (2000), no. 1, 89–100. MR1774759

[49] Emma Previato, *Another algebraic proof of Weil's reciprocity*, Atti Accad. Naz. Lincei Cl. Sci. Fis. Mat. Natur. Rend. Lincei (9) Mat. Appl. **2** (1991), no. 2, 167–171. MR1120136

[50] Alex Kasman and Emma Previato, *Commutative partial differential operators*, Phys. D **152/153** (2001), 66–77, Advances in nonlinear mathematics and science. MR1837898

[51] _____, *Factorization and resultants of partial differential operators*, Math. Comput. Sci. **4** (2010), no. 2-3, 169–184. MR2775986

[52] Maurice J. Dupré, James F. Glazebrook, and Emma Previato, *A Banach algebra version of the Sato Grassmannian and commutative rings of differential operators*, Acta Appl. Math. **92** (2006), no. 3, 241–267. MR2266488

[53] _____, *Curvature of universal bundles of Banach algebras*, Topics in operator theory. Volume 1. Operators, matrices and analytic functions, 2010, pp. 195–222. MR2723277

[54] _____, *Differential algebras with Banach-algebra coefficients I: from C*-algebras to the K-theory of the spectral curve*, Complex Anal. Oper. Theory **7** (2013), no. 4, 739–763. MR3079828

[55] _____, *Differential algebras with Banach-algebra coefficients II. The operator cross-ratio tau-function and the Schwarzian derivative*, Complex Anal. Oper. Theory **7** (2013), no. 6, 1713–1734. MR3129889

[56] Emma Previato and Mauro Spera, *Isometric embeddings of infinite-dimensional Grassmannians*, Regul. Chaotic Dyn. **16** (2011), no. 3-4, 356–373. MR2810984

[57] Emma Previato, *Multivariable Burchnall-Chaundy theory*, Philos. Trans. R. Soc. Lond. Ser. A Math. Phys. Eng. Sci. **366** (2008), no. 1867, 1155–1177. MR2377688

[58] _____, *Lines on abelian varieties*, Probability, geometry and integrable systems, 2008, pp. 321–344. MR2407603

[59] Alexandru Buium and Emma Previato, *Arithmetic Euler top*, J. Number Theory **173** (2017), 37–63. MR3581908

[60] Eleanor Farrington and Emma Previato, *Symbolic computation for Rankin-Cohen differential algebras: a case study*, Math. Comput. Sci. **11** (2017), no. 3-4, 401–415. MR3690055

[61] Jiryo Komeda, Shigeki Matsutani, and Emma Previato, *The sigma function for Weierstrass semigoups $\langle 3, 7, 8 \rangle$ and $\langle 6, 13, 14, 15, 16 \rangle$*, Internat. J. Math. **24** (2013), no. 11, 1350085, 58. MR3143604

[62] _____, *The Riemann constant for a non-symmetric Weierstrass semigroup*, Arch. Math. (Basel) **107** (2016), no. 5, 499–509. MR3562378

[63] J. C. Eilbeck, V. Z. Enolski, S. Matsutani, Y. Ônishi, and E. Previato, *Abelian functions for trigonal curves of genus three*, Int. Math. Res. Not. IMRN (2008), no. 1, Art. ID rnm 140, 38. MR2417791

[64] _____, *Addition formulae over the Jacobian pre-image of hyperelliptic Wirtinger varieties*, J. Reine Angew. Math. **619** (2008), 37–48. MR2414946

[65] J. C. Eilbeck, V. Z. Enolskii, and E. Previato, *On a generalized Frobenius-Stickelberger addition formula*, Lett. Math. Phys. **63** (2003), no. 1, 5–17. MR1967532

[66] Shigeki Matsutani and Emma Previato, *Jacobi inversion on strata of the Jacobian of the C_{rs} curve $y^r = f(x)$*, J. Math. Soc. Japan **60** (2008), no. 4, 1009–1044. MR2467868

[67] _____, *A generalized Kiepert formula for C_{ab} curves*, Israel J. Math. **171** (2009), 305–323. MR2520112

[68] _____, *Jacobi inversion on strata of the Jacobian of the C_{rs} curve $y^r = f(x)$, II*, J. Math. Soc. Japan **66** (2014), no. 2, 647–692. MR3201830

[69] _____, *The al function of a cyclic trigonal curve of genus three*, Collect. Math. **66** (2015), no. 3, 311–349. MR3384012

[70] E. Previato, *Sigma function and dispersionless hierarchies*, XXIX Workshop on Geometric Methods in Physics, 2010, pp. 140–156. MR2767999

[71] Shigeki Matsutani and Emma Previato, *From Euler's elastica to the mKdV hierarchy, through the Faber polynomials*, J. Math. Phys. **57** (2016), no. 8, 081519, 12. MR3541543

[72] _____, *A class of solutions of the dispersionless KP equation*, Phys. Lett. A **373** (2009), no. 34, 3001–3004. MR2559804

[73] Drue Coles and Emma Previato, *Goppa codes and Tschirnhausen modules*, Advances in coding theory and cryptography, 2007, pp. 81–100. MR2440171

[74] _____, *Decoding by rank-2 bundles over plane quartics*, J. Symbolic Comput. **45** (2010), no. 7, 757–772. MR2645976

[75] Emma Previato, *Vector bundles in error-correcting for geometric Goppa codes*, Algebraic aspects of digital communications, 2009, pp. 42–80. MR2605297

[76] Brenda Leticia De La Rosa Navarro, Mustapha Lahyane, and Emma Previato, *Vector bundles with a view toward coding theory*, Algebra for secure and reliable communication modeling, 2015, pp. 159–171. MR3380380

[77] R. Donagi, B. Dubrovin, E. Frenkel, and E. Previato, *Integrable systems and quantum groups*, Lecture Notes in Mathematics, vol. 1620, Springer-Verlag, Berlin; Centro Internazionale Matematico Estivo (C.I.M.E.), Florence, 1996, Lectures given at the First 1993 C.I.M.E. Session held in Montecatini Terme, June 14–22, 1993, Edited by M. Francaviglia and S. Greco, Fondazione CIME/CIME Foundation Subseries. MR1397272

[78] David A. Ellwood and Emma Previato (eds.), *Grassmannians, moduli spaces and vector bundles*, Clay Mathematics Proceedings, vol. 14, American Mathematical Society, Providence, RI; Clay Mathematics Institute, Cambridge, MA, 2011, Papers from the Clay Mathematics Institute (CMI) Workshop on Moduli Spaces of Vector Bundles, with a View Towards Coherent Sheaves held in Cambridge, MA, October 6–11, 2006. MR2809924

[79] Emma Previato (ed.), *Advances in algebraic geometry motivated by physics*, Contemporary Mathematics, vol. 276, American Mathematical Society, Providence, RI, 2001. MR1837106

[80] _____ (ed.), *Dictionary of applied math for engineers and scientists*, Comprehensive Dictionary of Mathematics, CRC Press, Boca Raton, FL, 2003. MR1966695

[81] _____, *Soliton equations and their algebro-geometric solutions. Vol. I [book review of mr1992536]*, Bull. Amer. Math. Soc. (N.S.) **45** (2008), no. 3, 459–467. MR3077138

[82] E. Previato, *Poncelet's porism and projective fibrations*, Higher genus curves in mathematical physics and arithmetic geometry, 2018, pp. 157–169. MR3782465

[83] _____, *Complex algebraic geometry applied to integrable dynamics: concrete examples and open problems*, Geometric methods in physics XXXV, 2018, pp. 269–280. MR3803645

[84] Alexandru Buium and Emma Previato, *The Euler top and canonical lifts*, J. Number Theory **190** (2018), 156–168. MR3805451

[85] E. Previato, *Curves in isomonodromy and isospectral deformations: Painlevé VI as a case study*, Algebraic curves and their applications, 2019, pp. 247–265. MR3916744

[86] Jiryo Komeda, Shigeki Matsutani, and Emma Previato, *The sigma function for trigonal cyclic curves*, Lett. Math. Phys. **109** (2019), no. 2, 423–447. MR3917350

Ron Donagi
Mathematics Department,
UPenn, Philadelphia, PA, 19104-6395

Tony Shaska
Department of Mathematics and Statistics,
Oakland University, Rochester, MI. 48309

1

Trace Ideal Properties of a Class of Integral Operators

Fritz Gesztesy and Roger Nichols

Dedicated with admiration to Emma Previato – Geometer Extraordinaire

Abstract. We consider a particular class of integral operators $T_{\gamma,\delta}$ in $L^2(\mathbb{R}^n)$, $n \in \mathbb{N}$, $n \geqslant 2$, with integral kernels $T_{\gamma,\delta}(\cdot, \cdot)$ bounded (Lebesgue) a.e. by

$$|T_{\gamma,\delta}(x, y)| \leqslant C \langle x \rangle^{-\delta} |x - y|^{2\gamma - n} \langle y \rangle^{-\delta}, \quad x, y \in \mathbb{R}^n, \ x \neq y,$$

for fixed $C \in (0, \infty)$, $0 < 2\gamma < n$, $\delta > \gamma$, and prove that

$$T_{\gamma,\delta} \in \mathcal{B}_p(L^2(\mathbb{R}^n)) \text{ for } p > n/(2\gamma), \ p \geqslant 2.$$

(Here $\langle x \rangle := (1 + |x|^2)^{1/2}$, $x \in \mathbb{R}^n$, and $\mathcal{B}_p$ abbreviates the ℓ^p-based trace ideal.) These integral operators (and their matrix-valued analogs) naturally arise in the study of multi-dimensional Schrödinger and Dirac-type operators and we describe an application to the case of massless Dirac-type operators.

1 Introduction

The principal aim of this paper is to derive trace ideal properties of a class of (matrix-valued) integral operators that naturally arise in the context of multi-dimensional Schrödinger and Dirac-type operators. More precisely, we will focus on integral operators $T_{\gamma,\delta}$ in $L^2(\mathbb{R}^n)$, $n \in \mathbb{N}$, $n \geqslant 2$, which in the scalar context are associated with integral kernels $T_{\gamma,\delta}(\cdot, \cdot)$ that are bounded (Lebesgue) a.e. by

$$|T_{\gamma,\delta}(x, y)| \leqslant C \langle x \rangle^{-\delta} |x - y|^{2\gamma - n} \langle y \rangle^{-\delta}, \quad x, y \in \mathbb{R}^n, \ x \neq y, \tag{1}$$

for fixed $C \in (0, \infty)$, $0 < 2\gamma < n$, $\delta > \gamma$. We then prove in Theorem 3.1 that

$$T_{\gamma,\delta} \in \mathcal{B}_p(L^2(\mathbb{R}^n)) \text{ for } p > n/(2\gamma), \ p \geqslant 2. \tag{2}$$

2010 *Mathematics Subject Classification.* Primary: 46B70, 47B10, 47G10, 47L20; Secondary: 35Q40, 81Q10.

Key words and phrases. Trace ideals, interpolation theory, massless Dirac-type operators.

This result is then applied to prove Theorem 3.4 which derives a uniform trace ideal norm bound with respect to the spectral parameter on the resolvent of massless Dirac-type operators if the Dirac-type resolvent is viewed as a map between suitable weighted $L^2(\mathbb{R}^n)$ spaces. Such estimates are well-known to imply limiting absorption principles for the Dirac-type operator in question which, in turn, have strong spectral implications (such as the absence of any singular continuous spectrum, etc.). Moreover, the fact that trace ideal bounds are involved can now be used to infer continuity properties of underlying spectral shift functions which yields further applications to the Witten index for a particular class of non-Fredholm operators as discussed, for instance, in [3]–[9], [12], [24] (see also Remark 3.5). We note that two-dimensional massless Dirac-type operators are also known to be relevant in the context of graphene, one more reason to study the massless case.

In Section 2 we collect a fair amount of background material, some of which is crucial for our main Section 3. In particular, we focus on integral operators with integral kernels closely related to the right-hand side of (1) and survey some of the pertinent literature in this context, including a fundamental criterion by Nirenberg and Walker [22] (we include a detailed proof of the latter), a result on absolute kernels, some well-known Schur tests, and a trace norm estimate due to Demuth, Stollmann, Stolz, and van Casteren [10] (again we supply the proof of the latter). Section 2 also recalls a version of the Sobolev inequality and a fundamental trace ideal interpolation result. Our principal Section 3 then proves the inclusion (2) in Theorem 3.1 and demonstrates its applicability to the case of massless Dirac-type operators in Theorem 3.4. Finally, Appendix A collects some useful results on pointwise domination of linear operators and its consequences in connection with boundedness, compactness, and Hilbert–Schmidt properties. We include a discussion of block matrix operator situations necessitated by the study of Dirac-type operators.

We conclude this introduction with some comments on the notation employed in this paper: Let $\mathcal{H}$ be a separable complex Hilbert space, $(\cdot, \cdot)_{\mathcal{H}}$ the scalar product in $\mathcal{H}$ (linear in the second argument), and $I_{\mathcal{H}}$ the identity operator in $\mathcal{H}$.

Next, if T is a linear operator mapping (a subspace of) a Hilbert space into another, then $\mathrm{dom}(T)$ and $\mathrm{ker}(T)$ denote the domain and kernel (i.e., null space) of T. The spectrum, point spectrum (the set of eigenvalues), the essential spectrum of a closed linear operator in $\mathcal{H}$ will be denoted by $\sigma(\cdot)$, $\sigma_p(\cdot)$, $\sigma_{ess}(\cdot)$, respectively. Similarly, the absolutely continuous and singularly continuous spectrum of a self-adjoint operator in $\mathcal{H}$ are denoted by $\sigma_{ac}(\cdot)$ and $\sigma_{sc}(\cdot)$.

The Banach spaces of bounded and compact linear operators on a separable complex Hilbert space $\mathcal{H}$ are denoted by $\mathcal{B}(\mathcal{H})$ and $\mathcal{B}_\infty(\mathcal{H})$, respectively; the corresponding ℓ^p-based trace ideals will be denoted by $\mathcal{B}_p(\mathcal{H})$, their norms are abbreviated by $\| \cdot \|_{\mathcal{B}_p(\mathcal{H})}$, $p \in [1, \infty)$. Moreover, $\operatorname{tr}_{\mathcal{H}}(A)$ denotes the corresponding trace of a trace class operator $A \in \mathcal{B}_1(\mathcal{H})$.

If $p \in [1, \infty) \cup \{\infty\}$, then $p' \in [1, \infty) \cup \{\infty\}$ denotes its conjugate index, that is, $p' := (1 - 1/p)^{-1}$. If Lebesgue measure is understood, we simply write $L^p(M)$, $M \subseteq \mathbb{R}^n$ measurable, $n \in \mathbb{N}$, instead of the more elaborate notation $L^p(M; d^n x)$. For $x = (x_1, \ldots, x_n) \in \mathbb{R}^n$, $n \in \mathbb{N}$, we abbreviate $\langle x \rangle := (1 + |x|^2)^{1/2}$.

Finally, $\lfloor \cdot \rfloor$ denotes the floor function on $\mathbb{R}$, that is, $\lfloor x \rfloor$ characterizes the largest integer less than or equal to $x \in \mathbb{R}$.

2 Some Background Material

This preparatory section is primarily devoted to various results of integral operators, but we also briefly recall Sobolev's inequality and some interpolation results for trace ideal operators.

We start by recalling the following version of Sobolev's inequality (see, e.g., [26, Corollary I.14]), to be employed in the proof of Theorem 3.1.

Theorem 2.1 *Let $n \in \mathbb{N}$, $r, s \in (1, \infty)$, $0 < \lambda < n$, $r^{-1} + s^{-1} + \lambda n^{-1} = 2$, $f \in L^r(\mathbb{R}^n)$, $h \in L^s(\mathbb{R}^n)$. Then, there exists $C_{r,s,\lambda,n} \in (0, \infty)$ such that*

$$\int_{\mathbb{R}^n \times \mathbb{R}^n} d^n x \, d^n y \, \frac{|f(x)||h(y)|}{|x - y|^\lambda} \leqslant C_{r,s,\lambda,n} \|f\|_{L^r(\mathbb{R}^n)} \|h\|_{L^s(\mathbb{R}^n)}. \tag{3}$$

We continue this section with a special case of a very interesting result of Nirenberg and Walker [22] also to be employed in the proof of Theorem 3.1. For convenience of the reader we offer a detailed proof.

Theorem 2.2 *Let $n \in \mathbb{N}$, $c, d \in \mathbb{R}$, $c + d > 0$, and consider*

$$K_{c,d}(x, y) = |x|^{-c} |x - y|^{(c+d)-n} |y|^{-d}, \quad x, y \in \mathbb{R}^n, \ x \neq x'. \tag{4}$$

Then the integral operator $K_{c,d}$ in $L^2(\mathbb{R}^n)$ with integral kernel $K_{c,d}(\cdot, \cdot)$ in (4) is bounded if and only if

$$c < n/2 \quad \text{and} \quad d < n/2. \tag{5}$$

Proof. To prove the necessity of the conditions (5), assume $K \in \mathcal{B}(L^2(\mathbb{R}^n))$. Then

$$\int_{\mathbb{R}^n} d^n y \, K_{c,d}(\,\cdot\,, y) f(y) \in L^2(\mathbb{R}^n), \quad f \in L^2(\mathbb{R}^n). \tag{6}$$

In particular, choosing $f = \chi_{\overline{B_n(0;1)}}$, the characteristic function of the closed unit ball in $\mathbb{R}^n$,

$$\overline{B_n(0;1)} = \{x \in \mathbb{R}^n \mid |x| \leq 1\}, \tag{7}$$

which has finite Lebesgue measure, $\left|\overline{B_n(0;1)}\right| = \pi^{n/2}/\Gamma((n/2)+1)$, one obtains

$$\int_{\overline{B_n(0;1)}} d^n y \, K_{c,d}(\,\cdot\,, y) \in L^2(\mathbb{R}^n). \tag{8}$$

Since $K_{c,d}(\,\cdot\,,\,\cdot\,)$ is symmetric in x and y, one also infers

$$\int_{\overline{B_n(0;1)}} d^n x \, K_{c,d}(x,\,\cdot\,) \in L^2(\mathbb{R}^n). \tag{9}$$

In summary, if $K_{c,d} \in \mathcal{B}(L^2(\mathbb{R}^n))$, then

$$\int_{\overline{B_n(0;1)}} d^n y \, K_{c,d}(\,\cdot\,, y) \in L^2(\mathbb{R}^n) \quad \text{and} \quad \int_{\overline{B_n(0;1)}} d^n x \, K_{c,d}(x,\,\cdot\,) \in L^2(\mathbb{R}^n). \tag{10}$$

To investigate the behavior of

$$\int_{\overline{B_n(0;1)}} \frac{d^n y}{|x|^c |x - y|^{n-c-d} |y|^d}, \quad x \in \mathbb{R}^n \setminus \{0\}, \tag{11}$$

as $|x| \to \infty$, one writes

$$\int_{\overline{B_n(0;1)}} \frac{d^n y}{|x|^c |x - y|^{n-c-d} |y|^d} = \frac{1}{|x|^{n-d}} \int_{\overline{B_n(0;1)}} \frac{d^n y}{\left|\dfrac{x}{|x|} - \dfrac{y}{|x|}\right|^{n-c-d} |y|^d},$$
$$x \in \mathbb{R}^n \setminus \{0\}. \tag{12}$$

If $x, y \in \mathbb{R}^n$ with $|x| \geq 2$ and $|y| \leq 1$, then the elementary estimates

$$\frac{1}{2} \leq 1 - \frac{1}{|x|} \leq \left|\frac{x}{|x|} - \frac{y}{|x|}\right| \leq 1 + \frac{1}{|x|} \leq \frac{3}{2} \tag{13}$$

imply

$$C_1 \leq \left|\frac{x}{|x|} - \frac{y}{|x|}\right|^{c+d-n} \leq C_2, \quad x, y \in \mathbb{R}^n, \ |x| \geq 2, \ |y| \leq 1, \tag{14}$$

for some constants $C_1, C_2 \in (0, \infty)$. In particular, the finiteness of the integral in (11) implies the finiteness of the integral

$$\int_{B_n(0;1)} \frac{d^n y}{|y|^d}. \tag{15}$$

By Lebesgue's dominated convergence theorem,

$$\lim_{|x|\to\infty} \int_{B_n(0;1)} \frac{d^n y}{\left|\frac{x}{|x|} - \frac{y}{|x|}\right|^{n-c-d} |y|^d}$$

$$= \int_{B_n(0;1)} d^n y \lim_{|x|\to\infty} \frac{1}{\left|\frac{x}{|x|} - \frac{y}{|x|}\right|^{n-c-d} |y|^d}$$

$$= \int_{B_n(0;1)} \frac{d^n y}{|y|^d} =: I_{1,n,d}, \tag{16}$$

since (13) implies

$$\lim_{|x|\to\infty} \left| \frac{x}{|x|} - \frac{y}{|x|} \right| = 1, \quad y \in \mathbb{R}^n. \tag{17}$$

Therefore, by (12) and (16),

$$\int_{B_n(0;1)} \frac{d^n y}{|x|^c |x-y|^{n-c-d} |y|^d} \sim I_{1,n,d} \cdot \frac{1}{|x|^{n-d}} \quad \text{as } |x| \to \infty, \tag{18}$$

and, similarly,

$$\int_{B_n(0;1)} \frac{d^n x}{|x|^c |x-y|^{n-c-d} |y|^d} \sim I_{1,n,c} \cdot \frac{1}{|y|^{n-c}} \quad \text{as } |y| \to \infty. \tag{19}$$

In light of (18) and (19), the containments in (10) hold only if $(n-d)2 > n$ and $(n-c)2 > n$, that is, only if $c < n/2$ and $d < n/2$.

To prove sufficiency of the conditions in (5), assume $c < n/2$ and $d < n/2$. It suffices to prove the claim $K_{c,d} \in \mathcal{B}(L^2(\mathbb{R}^n))$ in the special case where $c, d \in [0, \infty)$. The claim for general c and d then follows from this special case. Indeed, if c were negative, for example, then d would be positive, and the elementary inequality

$$\frac{|x|}{|x-y|} \leqslant 1 + \frac{|y|}{|x-y|}, \quad x, y \in \mathbb{R}^n, x \neq y, \tag{20}$$

implies

$$K_{c,d}(x,y) = \left(\frac{|x|}{|x-y|}\right)^{-c} \frac{1}{|x-y|^{n-d}|y|^d}$$

$$\leqslant M_{-c}\left(1 + \frac{|y|^{-c}}{|x-y|^{-c}}\right) \frac{1}{|x-y|^{n-d}|y|^d}$$

$$= \frac{M_{-c}}{|x-y|^{n-d}|y|^d} + \frac{M_{-c}}{|x-y|^{n-(c+d)}|y|^{c+d}}, \; x, y \in \mathbb{R}^n \backslash \{0\}, \; x \neq y, \tag{21}$$

where for each $\alpha \in [0, \infty)$, $M_\alpha \in (0, \infty)$ is a constant such that

$$(1+t)^\alpha \leqslant M_\alpha(1+t^\alpha), \quad t \in [0, \infty). \tag{22}$$

Note that the existence of M_α is guaranteed by the fact that, for each fixed $\alpha \in [0, \infty)$, the function

$$\phi_\alpha(t) = \frac{(1+t)^\alpha}{1+t^\alpha}, \quad t \in [0, \infty), \tag{23}$$

is continuous on $[0, \infty)$ and has a finite limit as $t \to \infty$. The special case under consideration (viz., $c, d \in [0, \infty)$) then implies that the right-hand side of (21) is the sum of the kernels of two integral operators in $\mathcal{B}(L^2(\mathbb{R}^n))$, so that $K_{c,d}(\cdot, \cdot)$ generates a bounded operator on $L^2(\mathbb{R}^n)$. Therefore, for the remainder of this proof, we assume $0 \leqslant c < n/2$ and $0 \leqslant d < n/2$.

By the arithmetic-geometric mean inequality,

$$|x| \geqslant \prod_{j=1}^n |x_j|^{1/n}, \quad x = (x_1, \ldots, x_n) \in \mathbb{R}^n, \tag{24}$$

implying

$$K_{c,d}(x,y) \leqslant \prod_{j=1}^n \frac{1}{|x_j|^{c/n}|x_j - y_j|^{1-(c+d)/n}|y_j|^{d/n}}, \tag{25}$$

$$x = (x_j)_{j=1}^n, \; y = (y_j)_{j=1}^n \in \mathbb{R}^n, \; x_j \neq y_j, \; x_j \neq 0, \; y_j \neq 0, \; 1 \leqslant j \leqslant n.$$

Therefore, by Lemma A.4, it suffices to show that the integral operator $J_{c,d}$ with integral kernel

$$J_{c,d}(s,t) = \frac{1}{|s|^{c/n}|s-t|^{1-(c+d)/n}|t|^{d/n}}, \quad s, t \in \mathbb{R}\backslash\{0\}, \; s \neq t, \tag{26}$$

belongs to $\mathcal{B}(L^2(\mathbb{R}))$.

The function $J_{c,d}(\cdot, \cdot)$ defined by (26) is homogeneous of degree (-1). In addition,

$$\int_0^\infty ds\, J(s,1)\, s^{-1/2} = \int_0^\infty \frac{ds}{s^{(1/2)+(c/n)}|s-1|^{1-(c+d)/n}} < \infty, \tag{27}$$

as the integrand on the right-hand side in (27) behaves like a constant times $s^{-[(1/2)+(c/n)]}$ (resp., $|s-1|^{(c+d)/n-1}$) as $s \to 0+$ (resp., $s \to 1$) and decays like a constant times $s^{(d/n)-(3/2)}$ as $s \to \infty$. Therefore, by Lemma A.5, the restriction of $J_{c,d}(\cdot, \cdot)$ to $(0, \infty) \times (0, \infty)$ generates an integral operator $J_{c,d,+} \in \mathcal{B}(L^2((0, \infty)))$. In particular,

$$
\int_{\mathbb{R}} dx \left| \int_{\mathbb{R}} dy\, J_{c,d}(x, y) f(y) \right|^2
$$
$$
\leqslant \int_0^\infty dx \left(\int_0^\infty dy\, J_{c,d}(x, y) |f(-y)| \right)^2
$$
$$
+ \int_0^\infty dx \left(\int_0^\infty dy\, J_{c,d}(x, y) |f(-y)| \right)^2
$$
$$
+ \int_0^\infty dx \left(\int_0^\infty dy\, J_{c,d}(x, y) |f(y)| \right)^2
$$
$$
+ \int_0^\infty dx \left(\int_0^\infty dy\, J_{c,d}(x, y) |f(y)| \right)^2
$$
$$
\leqslant 4 \|J_{c,d,+}\|_{\mathcal{B}(L^2((0,\infty)))} \|f\|_{L^2(\mathbb{R})}^2, \quad f \in L^2(\mathbb{R}), \tag{28}
$$

which proves $J_{c,d} \in \mathcal{B}(L^2(\mathbb{R}))$. To obtain (28), we have employed the elementary fact

$$
\max \left\{ \|f|_{(0,\infty)}\|_{L^2((0,\infty))}, \|f|_{(-\infty,0)}\|_{L^2((-\infty,0))} \right\} \leqslant \|f\|_{L^2(\mathbb{R})}, \quad f \in L^2(\mathbb{R}). \tag{29}
$$

$\square$

The following result is discussed in connection with the notion of absolute kernels in [18, p. 271].

Theorem 2.3 *Let $n \in \mathbb{N}$ and $\beta \in (0, n)$. Suppose $p_0, q_0, p, q \in (1, \infty)$ with*

$$
\frac{1}{p'} + \frac{1}{p_0} < 1, \quad \frac{1}{q} + \frac{1}{q_0} < 1, \quad \frac{1}{p} = \frac{1}{q} + \frac{1}{p_0} + \frac{1}{q_0} - \frac{\beta}{n}. \tag{30}
$$

If

$$
a \in L^{p_0}(\mathbb{R}^n) \quad \text{and} \quad b \in L^{q_0}(\mathbb{R}^n), \tag{31}
$$

then the kernel

$$
k(x, y) = a(x)|x - y|^{\beta-n} b(y) \text{ for a.e. } x, y \in \mathbb{R}^n \tag{32}
$$

generates a bounded integral operator $K \in \mathcal{B}(L^q(\mathbb{R}^n), L^p(\mathbb{R}^n))$.

While Theorem 2.3 permits a variety of functions a and b, it does not apply to the kernel $K_{c,d}$ in (4) due to the integrability requirements in (31).

Theorem 2.2 gives necessary and sufficient conditions for boundedness of the integral operator $K_{c,d}$. In general, there are no known practical necessary and sufficient conditions for the boundedness of an integral operator. However, there are various sufficient conditions which allow one to infer boundedness of an integral operator from appropriate bounds on the integral kernel itself. Illustrative examples are the well-known *Schur criteria* or *Schur tests* to which we briefly turn next for the sake of completeness.

The following well-known version of the Schur test (cf., e.g., [16, Theorem 5.2]) provides a sufficient condition for boundedness between L^2-spaces in terms of pointwise bounds on the integral kernel when integrated against a pair of measurable trial functions.

Theorem 2.4 (Schur test–first version) *Let* $(X, \mathcal{M}, d\mu)$ *and* $(Y, \mathcal{N}, d\nu)$ *be* σ-*finite measure spaces and* $k : X \times Y \to [0, \infty)$ *a measurable function. If* $\phi : X \to (0, \infty)$ *and* $\psi : Y \to (0, \infty)$ *are measurable and if* $\alpha, \beta \in (0, \infty)$ *are such that*

$$\int_Y d\nu(y) \, k(x, y)\psi(y) \leqslant \alpha\phi(x) \text{ for a.e. } x \in X \tag{33}$$

and

$$\int_X d\mu(x) \, k(x, y)\phi(x) \leqslant \alpha\psi(y) \text{ for a.e. } y \in Y, \tag{34}$$

then k *is the integral kernel of a bounded integral operator*

$$K \in \mathcal{B}\big(L^2(Y; d\nu), L^2(X; d\mu)\big) \tag{35}$$

and

$$\|K\|_{\mathcal{B}(L^2(Y;d\nu), L^2(X;d\mu))} \leqslant (\alpha\beta)^{1/2}. \tag{36}$$

Proof. If $f \in L^2(Y; d\nu)$, then using the Cauchy–Schwarz inequality one obtains

$$\int_X d\mu(x) \left(\int_Y d\nu(y) \, k(x, y)|f(y)| \right)$$

$$= \int_X d\mu(x) \left(\int_Y d\nu(y) \, k(x, y)^{1/2}\psi(y)^{1/2} \left[\frac{k(x, y)}{\psi(y)} \right]^{1/2} |f(y)| \right)^2$$

$$\leqslant \int_X d\mu(x) \left(\int_Y d\nu(y \, k(x, y)\psi(y)) \right) \left(\int_Y d\nu(y') \frac{k(x, y')}{\psi(y')} |f(y')|^2 \right)$$

$$\leqslant \int_X d\mu(x) \, \alpha\phi(x) \left(\int_Y d\nu(y) \frac{k(x, y)}{\psi(y)} |f(y)|^2 \right)$$

$$= \alpha \int_Y dv(y) \frac{|f(y)|^2}{\psi(y)} \left(\int_X d\mu(x) \, k(x, y)\phi(x) \right)$$

$$\leqslant \alpha \int_Y dv(y) \frac{|f(y)|^2}{\psi(y)} \beta \psi(y)$$

$$= \alpha\beta \int_Y dv(y) |f(y)|^2. \tag{37}$$

□

Example 2.5 (Abel kernel) *The integral operator K in $L^2((0, 1))$ generated by the kernel*

$$k(x, y) = \begin{cases} 0, & x \leqslant y, \\ (x - y)^{-1/2}, & y < x, \end{cases} \tag{38}$$

belongs to $\mathcal{B}(L^2((0, 1)))$. In fact, the Schur test applies with $\psi(x) = \phi(x) = 1$ for a.e. $x \in (0, 1)$ and $\alpha = \beta = 2$.

The proof of the following L^p-based version of the Schur test relies on Hölder's inequality (cf., e.g., [29, Satz 6.9]).

Theorem 2.6 (Schur test–second version) *Let $p, p' \in (1, \infty)$ with $p^{-1} + (p')^{-1} = 1$ and let $(X, \mathcal{M}, d\mu)$ and $(Y, \mathcal{N}, dv)$ be σ-finite measure spaces. Suppose $k : X \times Y \to \mathbb{C}$ is a measurable function and that there exist measurable functions $k_1, k_2 : X \times Y \to [0, \infty)$ such that*

$$|k(x, y)| \leqslant k_1(x, y)k_2(x, y) \text{ for a.e. } (x, y) \in X \times Y, \tag{39}$$

and

$$\|k_1(x, \cdot)\|_{L^{p'}(Y;dv)} \leqslant C_1, \quad \|k_2(\cdot, y)\|_{L^p(X;d\mu)} \leqslant C_2, \tag{40}$$

for μ-a.e. $x \in X$ and v-a.e. $y \in Y$ for some constants $C_1, C_2 \in (0, \infty)$. Then k is the integral kernel of a bounded integral operator

$$K \in \mathcal{B}\big(L^p(Y; dv), L^p(X; d\mu)\big) \tag{41}$$

and

$$\|K\|_{\mathcal{B}(L^p(Y;dv), L^p(X;d\mu))} \leqslant C_1 C_2. \tag{42}$$

While Theorems 2.4 and 2.6 provide useful sufficient conditions for an integral operator to be bounded over an L^p-space, they do not yield information about possible compactness or trace ideal properties of the integral operator. (For compactness properties, see, e.g., [16, § 13, 14], [18, § 11], [20, Ch. 2],

[28, Sect. 6.3], [32, Ch. V].) In particular, neither Theorem 2.4 nor Theorem 2.6 implies the trace ideal property 2. As an example of a result which provides sufficient conditions for an integral operator to belong to the trace class, we mention the following result on general integral operators due to [10] and provide its short proof.

Theorem 2.7 *Let $(X, \mathcal{A}, \mu)$ be a σ-finite measure space and suppose that $A(\cdot, \cdot)$, $B(\cdot, \cdot) : X \times X \to \mathbb{C}$ are measurable such that*

$$A(\cdot, x), B(x, \cdot) \in L^2(X; d\mu) \text{ for a.e. } x \in X,$$

$$\int_X d\mu(y) \, \|A(\cdot, y)\|_{L^2(X;d\mu)} \|B(y, \cdot)\|_{L^2(X;d\mu)} < \infty. \tag{43}$$

Then there exists a trace class operator $AB : L^2(X; d\mu) \to L^2(X; d\mu)$ with integral kernel

$$AB(x, y) = \int_X d\mu(t) \, A(x, t) B(t, y), \tag{44}$$

such that

$$\|AB\|_{\mathcal{B}_1(L^2(X;d\mu))} \leqslant \|A(\cdot, y)\|_{L^2(X;d\mu)} \|B(y, \cdot)\|_{L^2(X;d\mu)}. \tag{45}$$

Proof. Introducing $g(y) := \|B(y, \cdot)\|_{L^2(X;d\mu)}$, $h(y) := \|A(\cdot, y)\|_{L^2(X;d\mu)}$, and employing the (unusual) convention $g(x)^{-1} = 0$ if $g(x) = 0$, we denote by M_f the maximally defined operator of multiplication by f in the space $L^2(X; d\mu)$. Then

$$AB = A M_{h^{-1}} M_{(hg)^{1/2}} M_{(hg)^{1/2}} M_{g^{-1}} B, \tag{46}$$

and $A M_{h^{-1}} M_{(hg)^{1/2}}$ and $M_{(hg)^{1/2}} M_{g^{-1}} B$ are seen to be Hilbert–Schmidt operators. For instance,

$$\|A M_{h^{-1}} M_{(hg)^{1/2}}\|^2_{\mathcal{B}_2(L^2(X;d\mu))}$$
$$= \int_X d\mu(x) \int_X d\mu(y) \, \big| A(x, y) h(y)^{-1} (hg)(x)^{1/2} \big|^2$$
$$= \int_X d\mu(x) \int_X d\mu(y) \, \big| A(x, y) h(y)^{-1/2} (g)(x)^{1/2} \big|^2$$
$$= \int_X d\mu(y) \, g(y) h(y) < \infty. \tag{47}$$

$\square$

One easily verifies that inequality (45) becomes an equality for rank-one operators A, B in $L^2(\mathbb{R}^n)$ generated by Lebesgue-a.e. nonnegative functions.

While Theorem 2.7 provides a positive result for a certain class of integral operators, it is limited in scope to the trace class. Therefore, Theorem 2.7 is not well-suited for application to (2), owing to the condition $p > n/(2\gamma)$

in (2). To circumvent these difficulties, we shall employ interpolation methods in Section 3 to prove (2). In particular, we will make use of the following trace ideal interpolation result, see, for instance, [14, Theorem III.13.1], [31, Theorem 0.2.6] (see also [13], [15, Theorem III.5.1]) in the proof of (2) (cf. Theorem 3.1).

Theorem 2.8 *Let $p_j \in [1, \infty) \cup \{\infty\}$, $\Sigma = \{\zeta \in \mathbb{C} \,|\, \mathrm{Re}(\zeta) \in (\xi_1, \xi_2)\}$, $\xi_j \in \mathbb{R}$, $\xi_1 < \xi_2$, $j = 1, 2$. Suppose that $A(\zeta) \in \mathcal{B}(\mathcal{H})$, $\zeta \in \overline{\Sigma}$ and that $A(\cdot)$ is analytic on Σ, continuous up to $\partial \Sigma$, and that $\|A(\cdot)\|_{\mathcal{B}(\mathcal{H})}$ is bounded on $\overline{\Sigma}$. Assume that for some $C_j \in (0, \infty)$,*

$$\sup_{\eta \in \mathbb{R}} \|A(\xi_j + i\eta)\|_{\mathcal{B}_{p_j}(\mathcal{H})} \leqslant C_j, \quad j = 1, 2. \tag{48}$$

Then

$$A(\zeta) \in \mathcal{B}_{p(\mathrm{Re}(\zeta))}(\mathcal{H}), \quad \frac{1}{p(\mathrm{Re}(\zeta))} = \frac{1}{p_1} + \frac{\mathrm{Re}(\zeta) - \xi_1}{\xi_2 - \xi_1}\left[\frac{1}{p_2} - \frac{1}{p_1}\right], \quad \zeta \in \overline{\Sigma}, \tag{49}$$

and

$$\|A(\zeta)\|_{\mathcal{B}_{p(\mathrm{Re}(\zeta))}(\mathcal{H})} \leqslant C_1^{(\xi_2 - \mathrm{Re}(\zeta))/(\xi_2 - \xi_1)} C_2^{(\mathrm{Re}(\zeta) - \xi_1)/(\xi_2 - \xi_1)}, \quad \zeta \in \overline{\Sigma}. \tag{50}$$

In case $p_j = \infty$, $\mathcal{B}_\infty(\mathcal{H})$ can be replaced by $\mathcal{B}(\mathcal{H})$.

In the next section, we shall employ Theorem 2.8 to interpolate between the $\mathcal{B}(L^2(\mathbb{R}^n))$ and $\mathcal{B}_p(L^2(\mathbb{R}^n))$ properties for a family of integral operators $T_{\gamma, \delta}$ in $L^2(\mathbb{R}^n)$, $n \geqslant 2$, with kernels bounded in absolute value according to (1), for appropriate values of the parameters γ, δ.

3 Interpolation and Trace Ideal Properties of a Class of Integral Operators

In this section we combine Theorems 2.1, 2.2, 2.8, and an interpolation procedure to prove Theorem 3.1 below. The latter asserts a trace ideal containment for integral operators in $L^2(\mathbb{R}^n)$, $n \geqslant 2$, with kernels bounded in absolute value by a constant times a function of the form $\langle x \rangle^{-\delta} |x - y|^{2\gamma - n} \langle y \rangle^{-\delta}$, $x, y \in \mathbb{R}^n$, $x \neq y$, for appropriate values of the parameters γ, δ. Theorem 3.4 then provides an application of Theorem 3.1 to the case of n-dimensional massless Dirac-type operators.

A combination of Theorems 2.1, 2.2, and 2.8 yields the following general result.

Theorem 3.1 *Let $n \in \mathbb{N}$, $n \geqslant 2$, $0 < 2\gamma < n$, $\delta > \gamma$, and suppose that $T_{\gamma, \delta}$ is an integral operator in $L^2(\mathbb{R}^n)$ whose integral kernel $T_{\gamma, \delta}(\cdot, \cdot)$ satisfies the estimate*

$$|T_{\gamma,\delta}(x,y)| \leqslant C\langle x\rangle^{-\delta}|x-y|^{2\gamma-n}\langle y\rangle^{-\delta}, \quad x,y\in\mathbb{R}^n, \ x\neq y \qquad (51)$$

for some $C\in(0,\infty)$. Then,

$$T_{\gamma,\delta}\in\mathcal{B}_p\big(L^2(\mathbb{R}^n)\big), \quad p>n/(2\gamma), \ p\geqslant 2, \qquad (52)$$

and

$$\|T_{\gamma,\delta}\|_{\mathcal{B}_{n/(2\gamma-\varepsilon)}(L^2(\mathbb{R}^n))}$$

$$\leqslant \sup_{\eta\in\mathbb{R}}\big[\|T_{\gamma,\delta}(-2\gamma+\varepsilon+i\eta)\|_{\mathcal{B}(L^2(\mathbb{R}^n))}\big]^{2[-2\gamma+(n/2)+\varepsilon]/n}$$

$$\times \sup_{\eta\in\mathbb{R}}\big[\|T_{\gamma,\delta}(-2\gamma+(n/2)+\varepsilon+i\eta)\|_{\mathcal{B}_2(L^2(\mathbb{R}^n))}\big]^{2(2\gamma-\varepsilon)/n} \qquad (53)$$

for $0<\varepsilon$ sufficiently small.

Proof. Following the idea behind Yafaev's proof of [31, Lemma 0.13.4], we introduce the analytic family of integral operators $T_{\gamma,\delta}(\cdot)$ in $L^2(\mathbb{R}^n)$ generated by the integral kernel

$$T_{\gamma,\delta}(\zeta;x,y)=T_{\gamma,\delta}(x,y)\langle x\rangle^{-(\zeta/2)}|x-y|^\zeta\langle y\rangle^{-(\zeta/2)}, \quad x,y\in\mathbb{R}^n, \ x\neq y, \qquad (54)$$

noting $T_{\gamma,\delta}(0)=T_{\gamma,\delta}$. By Theorems 2.2 and A.2 (i) (for $N=1$),

$$T_{\gamma,\delta}(\zeta)\in\mathcal{B}\big(L^2(\mathbb{R}^n)\big), \quad 0<\mathrm{Re}(\zeta)+2\gamma<n, \ \delta\geqslant\gamma. \qquad (55)$$

To check the Hilbert–Schmidt property of $T_{\gamma,\delta}(\cdot)$ one estimates for the square of $|T_{\gamma,\delta}(\cdot;\cdot,\cdot)|$,

$$|T_{\gamma,\delta}(\zeta;x,y)|^2 \leqslant \langle x\rangle^{-2\delta-\mathrm{Re}(\zeta)}|x-y|^{2\mathrm{Re}(\zeta)+4\gamma-2n}\langle x\rangle^{-2\delta-\mathrm{Re}(\zeta)}, \qquad (56)$$
$$x,y\in\mathbb{R}^n, \ x\neq y,$$

and hence one can apply Theorem 2.1 upon identifying $\lambda=2n-4\gamma-2\mathrm{Re}(z)$, $r=s=n/[\mathrm{Re}(\zeta)+2\gamma]$, and $f=h=\langle\cdot\rangle^{-[2\delta+\mathrm{Re}(\zeta)]}$, to verify that $0<\lambda<n$ translates into $n/2<\mathrm{Re}(\zeta)+2\gamma<n$, and $f\in L^r(\mathbb{R}^n)$ holds with $r\in(1,2)$ if $\delta>\gamma$. Hence,

$$T_{\gamma,\delta}(\zeta)\in\mathcal{B}_2\big(L^2(\mathbb{R}^n)\big), \quad n/2<\mathrm{Re}(\zeta)+2\gamma<n, \ \delta>\gamma. \qquad (57)$$

It remains to interpolate between the $\mathcal{B}\big(L^2(\mathbb{R}^n)\big)$ and $\mathcal{B}_2\big(L^2(\mathbb{R}^n)\big)$ properties, employing Theorem 2.8 as follows. Choosing $0<\varepsilon$ sufficiently small, one identifies $\xi_1=-2\gamma+\varepsilon$, $\xi_2=-2\gamma+(n/2)+\varepsilon$, $p_1=\infty$, $p_2=2$, and hence obtains

$$p(\mathrm{Re}(\zeta))=n/[\mathrm{Re}(\zeta)+2\gamma-\varepsilon], \qquad (58)$$

in particular, $p(0)>n/(2\gamma)$ (and of course, $p(0)\geqslant 2$). Since ε may be taken arbitrarily small, (52) follows from (58) and (53) is a direct consequence of (50). $\qquad\square$

While subordination in general only applies to $\mathcal{B}_p$-ideals with p even (see the discussion in [27, p. 24 and Addendum E]), the use of complex interpolation in Theorem 3.1 (and the focus on bounded and Hilbert–Schmidt operators) permits one to avoid this restriction.

Theorem 3.1 represents the principal result of this paper and to the best of our knowledge it appears to be new.

The singularity structure on the diagonal of the integral kernels $K_{c,d}$ introduced in (4) naturally matches the one of multi-dimensional Schrödinger and Dirac-type operators as we will indicate next.

As a brief preparation we first record the asymptotic behavior of Hankel functions of the first kind with index $v \geqslant 0$ (cf. e.g., [1, Sect. 9.1]), $H_v^{(1)}(\cdot)$, as the latter are crucial in the context of constant coefficient (i.e., free, or non-interacting) Schrödinger and Dirac-type operators, a natural first step in studying Schrödinger and Dirac-type operators with nontrivial interaction terms (i.e., potentials). Employing, for instance, [1, p. 360, 364], one obtains

$$H_0^{(1)}(\zeta) \underset{\zeta \to 0}{=} (2i/\pi)\ln(\zeta) + O\big(|\ln(\zeta)||\zeta|^2\big), \tag{59}$$

$$H_v^{(1)}(\zeta) \underset{\zeta \to 0}{=} -(i/\pi)2^v \Gamma(v)\zeta^{-v}$$

$$+ \begin{cases} O\big(|\zeta|^{\min(v,-v+2)}\big), & v \notin \mathbb{N}, \\ O\big(|\ln(\zeta)||\zeta|^v\big) + O\big(\zeta^{-v+2}\big), & v \in \mathbb{N}, \end{cases} \tag{60}$$

$$\mathrm{Re}(v) > 0,$$

$$H_v^{(1)}(\zeta) \underset{\zeta \to \infty}{=} (2/\pi)^{1/2}\zeta^{-1/2}e^{i\zeta-(v\pi/2)-(\pi/4)}, \quad v \geqslant 0, \ \mathrm{Im}(\zeta) \geqslant 0. \tag{61}$$

Starting with the Laplacian in $L^2(\mathbb{R}^n)$,

$$h_0 = -\Delta, \quad \mathrm{dom}(h_0) = H^2(\mathbb{R}^n), \tag{62}$$

the Green's function of h_0, denoted by $g_0(z; \cdot, \cdot)$, is then of the form,

$$g_0(z; x, y) := (h_0 - zI)^{-1}(x, y)$$

$$= \begin{cases} (i/4)\big(2\pi z^{-1/2}|x-y|\big)^{(2-n)/2} H_{(n-2)/2}^{(1)}\big(z^{1/2}|x-y|\big), & n \geqslant 2, \ z \in \mathbb{C}\backslash\{0\}, \\ \dfrac{1}{(n-2)\omega_{n-1}}|x-y|^{2-n}, & n \geqslant 3, \ z = 0, \end{cases}$$

$$z \in \mathbb{C}\backslash[0, \infty), \ \mathrm{Im}\big(z^{1/2}\big) > 0, \ x, y \in \mathbb{R}^n, \ x \neq y, \tag{63}$$

where $\omega_{n-1} = 2\pi^{n/2}/\Gamma(n/2)$ ($\Gamma(\cdot)$ the Gamma function, cf., e.g., [1, Sect. 6.1]) represents the area of the unit sphere S^{n-1} in $\mathbb{R}^n$.

As $z \to 0$, $g_0(z; \cdot, \cdot)$ is continuous on the off-diagonal for $n \geqslant 3$,

$$\lim_{z \to 0} g_0(z; x, y) = g_0(0; x, y) = \frac{1}{(n-2)\omega_{n-1}} |x - y|^{2-n}, \tag{64}$$

$$x, y \in \mathbb{R}^n, \ x \neq y, \ n \in \mathbb{N}, \ n \geqslant 3,$$

but blows up for $n = 2$ as

$$g_0(z; x, y) \underset{z \to 0}{=} -\frac{1}{2\pi} \ln\left(z^{1/2} |x - y|/2\right)\left[1 + O\left(z|x - y|^2\right)\right] + \frac{1}{2\pi} \psi(1) \tag{65}$$

$$+ O\left(|z||x - y|^2\right), \quad x, y \in \mathbb{R}^2, \ x \neq y.$$

Here $\psi(w) = \Gamma'(w)/\Gamma(w)$ denotes the digamma function (cf., e.g., [1], Sect. 6.3]). This briefly illustrates the relevance of the diagonal singularity structure $|x - y|^{(c+d)-n}$ in $K_{c,d}$ in (4).

To describe an application to massless Dirac operators we need additional preparations. To rigorously define the free massless n-dimensional Dirac operators to be studied in the sequel, we now introduce the following set of basic hypotheses assumed for the remainder of this section.

Hypothesis 3.2 *Let $n \in \mathbb{N}$, $n \geqslant 2$.*

(*i*) *Set $N = 2^{\lfloor (n+1)/2 \rfloor}$ and let α_j, $1 \leqslant j \leqslant n$, $\alpha_{n+1} := \beta$, denote $n+1$ anti-commuting Hermitian $N \times N$ matrices with squares equal to I_N, that is,*

$$\alpha_j^* = \alpha_j, \quad \alpha_j \alpha_k + \alpha_k \alpha_j = 2\delta_{j,k} I_N, \quad 1 \leqslant j, k \leqslant n+1. \tag{66}$$

Here I_N denotes the $N \times N$ identity matrix.

(*ii*) *Introduce in $[L^2(\mathbb{R}^n)]^N$ the free massless Dirac operator*

$$H_0 = \alpha \cdot (-i\nabla) = \sum_{j=1}^{n} \alpha_j(-i\partial_j), \quad \mathrm{dom}(H_0) = [W^{1,2}(\mathbb{R}^n)]^N, \tag{67}$$

where $\partial_j = \partial/\partial x_j$, $1 \leqslant j \leqslant n$.

(*iii*) *Next, consider the self-adjoint matrix-valued potential*
$V = \{V_{\ell,m}\}_{1 \leqslant \ell, m \leqslant N}$ satisfying for some fixed $\rho > 1$, $C \in (0, \infty)$,

$$V \in \left[L^\infty(\mathbb{R}^n)\right]^{N \times N}, \quad |V_{\ell,m}(x)| \leqslant C\langle x \rangle^{-\rho} \text{ for a.e. } x \in \mathbb{R}^n, 1 \leqslant \ell, m \leqslant N. \tag{68}$$

Under these assumptions on V, the massless Dirac operator H in $[L^2(\mathbb{R}^n)]^N$ is defined via

$$H = H_0 + V, \quad \mathrm{dom}(H) = \mathrm{dom}(H_0) = [W^{1,2}(\mathbb{R}^n)]^N. \tag{69}$$

Here we employed the short-hand notation

$$[L^2(\mathbb{R}^n)]^N = L^2(\mathbb{R}^n; \mathbb{C}^N), \quad [W^{1,2}(\mathbb{R}^n)]^N = W^{1,2}(\mathbb{R}^n; \mathbb{C}^N), \text{ etc.} \quad (70)$$

Then H_0 and H are self-adjoint in $[L^2(\mathbb{R}^n)]^N$, with essential spectrum covering the entire real line,

$$\sigma_{ess}(H) = \sigma_{ess}(H_0) = \sigma(H_0) = \mathbb{R}, \quad (71)$$

a consequence of relative compactness of V with respect to H_0. In addition,

$$\sigma_{ac}(H_0) = \mathbb{R}, \quad \sigma_p(H_0) = \sigma_{sc}(H_0) = \emptyset. \quad (72)$$

With the exception of the comment following (75) and one more in connection with spectral shift functions in Remark 3.5, we will now drop the self-adjointness hypothesis on the $N \times N$ matrix V and still define a closed operator H in $[L^2(\mathbb{R}^n)]^N$ as in (69).

Turning to the Green's matrix of the massless free Dirac operator H_0 we assume

$$z \in \mathbb{C}_+, \ x, y \in \mathbb{R}^n, \ x \neq y, \ n \in \mathbb{N}, \ n \geqslant 2, \quad (73)$$

and compute for the Green's function $G_0(z; \cdot, \cdot)$ of H_0,

$$
\begin{aligned}
G_0(z; x, y) &:= (H_0 - zI)^{-1}(x, y) \\
&= i4^{-1}(2\pi)^{(2-n)/2}|x-y|^{2-n}z\,[z|x-y|]^{(n-2)/2}H^{(1)}_{(n-2)/2}(z|x-y|)I_N \\
&\quad - 4^{-1}(2\pi)^{(2-n)/2}|x-y|^{1-n}[z|x-y|]^{n/2}H^{(1)}_{n/2}(z|x-y|)\,\alpha \cdot \frac{(x-y)}{|x-y|}.
\end{aligned}
$$
$$(74)$$

The Green's function $G_0(z; \cdot, \cdot)$ of H_0 continuously extends to $z \in \overline{\mathbb{C}_+}$. In addition, in the massless case $m = 0$, the limit $z \to 0$ exists,

$$
\begin{aligned}
\lim_{\substack{z \to 0, \\ z \in \overline{\mathbb{C}_+}\backslash\{0\}}} G_0(z; x, y) &:= G_0(0; x, y) \\
&= i2^{-1}\pi^{-n/2}\Gamma(n/2)\,\alpha \cdot \frac{(x-y)}{|x-y|^n}, \quad x, y \in \mathbb{R}^n, \ x \neq y, \ n \in \mathbb{N}, \ n \geqslant 2,
\end{aligned}
$$
$$(75)$$

and no blow up occurs for all $n \in \mathbb{N}, n \geqslant 2$. This observation is consistent with the sufficient condition for the Dirac operator $H = H_0 + V$ (in dimensions $n \in \mathbb{N}, n \geqslant 2$), with V an appropriate self-adjoint $N \times N$ matrix-valued potential, having no eigenvalues, as derived in [19, Theorems 2.1, 2.3].

Returning to our analysis of the resolvent of H_0, the asymptotic behavior (59)–(61) implies for some $c_n \in (0, \infty)$,

$$\|G_0(0; x, y)\|_{\mathcal{B}(\mathbb{C}^N)} \leqslant c_n|x-y|^{1-n}, \quad x, y \in \mathbb{R}^n, \ x \neq y, \ n \in \mathbb{N}, \ n \geqslant 2, \quad (76)$$

and for given $R \geqslant 1$,

$$\|G_0(z; x, y)\|_{B(\mathbb{C}^N)} \leqslant c_{n,R}(z) e^{-\mathrm{Im}(z)|x-y|} \begin{cases} |x-y|^{1-n}, & |x-y| \leqslant 1, \ x \neq y, \\ 1, & 1 \leqslant |x-y| \leqslant R, \\ |x-y|^{(1-n)/2}, & |x-y| \geqslant R, \end{cases}$$

$$z \in \overline{\mathbb{C}_+}, \ x, y \in \mathbb{R}^n, \ x \neq y, \ n \in \mathbb{N}, \ n \geqslant 2, \tag{77}$$

for some $c_{n,R}(\cdot) \in (0, \infty)$ continuous and locally bounded on $\overline{\mathbb{C}_+}$.

For future purposes we now rewrite $G_0(z; \cdot, \cdot)$ as follows:

$$G_0(z; x, y)$$
$$= i 4^{-1} (2\pi)^{(2-n)/2} |x-y|^{2-n} z \, [z|x-y|]^{(n-2)/2} H^{(1)}_{(n-2)/2}(z|x-y|) I_N$$
$$- 4^{-1} (2\pi)^{(2-n)/2} |x-y|^{1-n} [z|x-y|]^{n/2} H^{(1)}_{n/2}(z|x-y|) \alpha \cdot \frac{(x-y)}{|x-y|}$$
$$= |x-y|^{1-n} f_n(z, x-y), \tag{78}$$
$$z \in \overline{\mathbb{C}_+}, \ x, y \in \mathbb{R}^n, \ x \neq y, \ n \in \mathbb{N}, \ n \geqslant 2,$$

where f_n is continuous and locally bounded on $\overline{\mathbb{C}_+} \times \mathbb{R}^n$, in addition,

$$\|f_n(z, x)\|_{B(\mathbb{C}^N)} \leqslant c_n(z) e^{-\mathrm{Im}(z)|x|} \begin{cases} 1, & 0 \leqslant |x| \leqslant 1, \\ |x|^{(n-1)/2}, & |x| \geqslant 1, \end{cases} \tag{79}$$
$$z \in \overline{\mathbb{C}_+}, \ x, y \in \mathbb{R}^n,$$

for some constant $c_n(\cdot) \in (0, \infty)$ continuous and locally bounded on $\overline{\mathbb{C}_+}$. In particular, decomposing $G_0(z; \cdot, \cdot)$ into

$$G_0(z; x, y) = G_0(z; x, y) \chi_{[0,1]}(|x-y|) + G_0(z; x, y) \chi_{[1,\infty)}(|x-y|)$$
$$:= G_{0,<}(z; x-y) + G_{0,>}(z; x-y), \tag{80}$$
$$z \in \overline{\mathbb{C}_+}, \ x, y \in \mathbb{R}^n, \ x \neq y, \ n \in \mathbb{N}, \ n \geqslant 2,$$

one verifies that

$$|G_{0,>}(z; x-y)_{j,k}| \leqslant \begin{cases} C_n |x-y|^{-(n-1)}, & z = 0, \\ C_n(z)|x-y|^{-(n-1)/2}, & z \in \overline{\mathbb{C}_+}, \end{cases} \tag{81}$$
$$x, y \in \mathbb{R}^n, \ |x-y| \geqslant 1, \ 1 \leqslant j, k \leqslant N,$$

for some constants $C_n, C_n(\cdot) \in (0, \infty)$, in particular,

$$G_{0,>}(z; \cdot) \in [L^\infty(\mathbb{R}^n)]^{N \times N}, \quad z \in \overline{\mathbb{C}_+}, \tag{82}$$

and that $G_{0,>}(\cdot; \cdot)$ is continuous on $\overline{\mathbb{C}_+} \times \mathbb{R}^n$.

Starting our analysis of integral operators connected to the resolvent of H_0 we first note that Theorem 2.2 implies the following fact.

Theorem 3.3 *Let* $n \in \mathbb{N}$, $n \geqslant 2$. *Then the integral operator* $R_0(\delta)$ *in* $[L^2(\mathbb{R}^n)]^N$ *with integral kernel* $R_0(\delta; \cdot, \cdot)$ *bounded entrywise by*

$$|R_0(\delta; \cdot, \cdot)_{j,k}| \leqslant C\langle\cdot\rangle^{-\delta}|G_0(0; \cdot, \cdot)_{j,k}|\langle\cdot\rangle^{-\delta}, \quad \delta \geqslant 1/2, \ 1 \leqslant j, k \leqslant N, \tag{83}$$

for some $C \in (0, \infty)$, *is bounded,*

$$R_0(\delta) \in \mathcal{B}([L^2(\mathbb{R}^n)]^N). \tag{84}$$

In a similar fashion, the integral operator $R_0(z, \delta)$ *in* $[L^2(\mathbb{R}^n)]^N$, *with integral kernel* $R_0(z, \delta; \cdot, \cdot)$ *bounded entrywise by*

$$|R_0(z, \delta; \cdot, \cdot)_{j,k}| \leqslant C\langle\cdot\rangle^{-\delta}|G_0(z; \cdot, \cdot)_{j,k}|\langle\cdot\rangle^{-\delta},$$
$$\delta \geqslant (n+1)//4, \ z \in \overline{\mathbb{C}_+}, \ 1 \leqslant j, k \leqslant N, \tag{85}$$

for some $C \in (0, \infty)$, *is bounded,*

$$R_0(z, \delta) \in \mathcal{B}([L^2(\mathbb{R}^n)]^N), \quad z \in \overline{\mathbb{C}_+}. \tag{86}$$

Proof. The inclusion (84) is an immediate consequence of (75) and hence the estimate $|G_0(0; x, y)_{j,k}| \leqslant C|x - y|^{1-n}$, $x, y \in \mathbb{R}^n$, $x \neq y$, $1 \leqslant j, k \leqslant N$, Theorem 2.2, choosing $c = d = 1/2$ in (4), and an application of Theorem A.2 (i) and Remark A.3.

To prove the inclusion (86) we employ the estimates (59)–(61) (cf. also (77)) to obtain

$$|G_0(z; x, y)_{j,k}| \leqslant C(z)|x - y|^{1-n}\chi_{[0,1]}(|x - y|)$$
$$+ D(z)|x - y|^{(1-n)/2}\chi_{[1,\infty)}(|x - y|), \tag{87}$$
$$z \in \overline{\mathbb{C}_+}, \ x, y \in \mathbb{R}^n, \ x \neq y, \ 1 \leqslant j, k \leqslant N,$$

for some $C, D(z) \in (0, \infty)$, and apply Theorems 2.2 and A.2 (i) (cf. also Remark A.3) to both terms on the right-hand sides of (87). The part $0 \leqslant |x - y| \leqslant 1$ leads to $\delta \geqslant 1/2$, whereas the part $|x - y| \geqslant 1$ yields $\delta \geqslant (n + 1)/4$, implying (86). $\qquad\square$

Combining Theorems 2.1, 2.2, 2.8, and 3.1 then yields the second principal result of this section, an application to massless Dirac-type operators.

Theorem 3.4 *Let* $n \in \mathbb{N}$, $n \geqslant 2$. *Then the integral operator* $R_0(\delta)$ *in* $[L^2(\mathbb{R}^n)]^N$ *with integral kernel* $R_0(\delta; \cdot, \cdot)$ *permitting the entrywise bound*

$$|R_0(\delta; \cdot, \cdot)_{j,k}| \leqslant C\langle\cdot\rangle^{-\delta}|G_0(0; \cdot, \cdot)_{j,k}|\langle\cdot\rangle^{-\delta}, \quad \delta > 1/2, \ 1 \leqslant j, k \leqslant N, \tag{88}$$

for some $C \in (0, \infty)$, satisfies

$$R_0(\delta) \in \mathcal{B}_p\big([L^2(\mathbb{R}^n)]^N\big), \quad p > n. \tag{89}$$

In a similar fashion, the integral operator $R_0(z, \delta)$ in $[L^2(\mathbb{R}^n)]^N$ with integral kernel $R_0(z, \delta; \cdot, \cdot)$ permitting the entrywise bound

$$|R_0(z, \delta; \cdot, \cdot)_{j,k}| \leqslant C\langle \cdot \rangle^{-\delta} |G_0(z; \cdot, \cdot)_{j,k}| \langle \cdot \rangle^{-\delta},$$

$$z \in \overline{\mathbb{C}_+}, \ \delta > (n+1)/4, \ 1 \leqslant j, k \leqslant N, \tag{90}$$

for some $C \in (0, \infty)$, satisfies

$$R_0(z, \delta) \in \mathcal{B}_p\big([L^2(\mathbb{R}^n)]^N\big), \quad p > n, \ z \in \overline{\mathbb{C}_+}. \tag{91}$$

Proof. We will apply the fact (96).

The inclusion (89) is immediate from (75) (employing the elementary estimate $|G_0(0; x, y)_{j,k}| \leqslant C|x - y|^{1-n}$, $x, y \in \mathbb{R}^n$, $x \neq y$, $1 \leqslant j, k \leqslant N$) and Theorem 3.1 (with $\gamma = 1/2$).

To prove the inclusion (91) we again employ the estimate (87). An application of Theorem 3.1 to both terms in (87), then yields for the part where $0 \leqslant |x - y| \leqslant 1$ that $\gamma = 1/2$ and hence $\delta > 1/2$ and $p > n$. Similarly, for the part where $|x - y| \geqslant 1$ one infers $\gamma = (n+1)/4$ and hence $\delta > (n+1)/4$ and $p > 2n/(n+1)$, $p \geqslant 2$, and thus one concludes $\delta > (n+1)/4$ and $p > n$. $\qquad\qquad\square$

Remark 3.5 To put Theorem 3.4 a bit into perspective we note that inclusions of the type (89), even in the far weaker situation with $\mathcal{B}_p\big([L^2(\mathbb{R}^n)]^N\big)$ replaced by $\mathcal{B}\big([L^2(\mathbb{R}^n)]^N\big)$, imply a global limiting absorption principle with strong spectral implications (such as, the absence of any singular spectrum) for the underlying Dirac-type operators, H_0 and $H = H_0 + cV$, for sufficiently small coupling constants $c \in \mathbb{C}$. (For details in this limiting absorption context context we refer to [4], [5], [25, Sects. XIII.7, XIII.8], [30, Ch. 4], [31, Chs. 1, 2, 6] and the detailed bibliography cited therein). The actual $\mathcal{B}_p\big([L^2(\mathbb{R}^n)]^N\big)$ result in Theorem 3.4 permits one to go a step further and derive continuity properties of the spectral shift function (cf., e.g., [30, Ch. 8], [31, Ch. 9]) between the pair of self-adjoint operators (H, H_0) (here we again assume the $N \times N$ matrix-valued potential V to be self-adjoint), which in turn permits a discussion of the Witten index of class of non-Fredholm model operators as discussed in [3]–[9], [12], [24], with additional material in preparation. $\qquad\qquad\diamond$

We conclude this section by noting once more that massless Dirac operators, particularly, in two dimensions, are known to be of relevance in applications to graphene. This fact, and particularly the prominent role massless Dirac-type operators play in connection with the Witten index of certain classes of non-Fredholm operators, explains our interest in them.

Appendix A. Some Remarks on Block Matrix Operators

In this appendix we collect some useful (and well-known) material on pointwise domination of linear operators in connection with boundedness, compactness, and the Hilbert–Schmidt property, with particular emphasis on the block matrix operator situation (required in the context of Dirac-type operators).

Definition A.1 *Let $(M; \mathcal{M}; \mu)$ be a σ-finite, separable measure space, μ a nonnegative, measure with $0 < \mu(M) \leqslant \infty$, and consider the linear operators A, B defined on $L^2(M; d\mu)$. Then B pointwise dominates A*

$$\text{if for all } f \in L^2(M; d\mu), \ |(Af)(\cdot)| \leqslant (B|f|)(\cdot) \ \mu\text{-a.e. on } M. \tag{92}$$

For a linear block operator matrix $T = \{T_{j,k}\}_{1 \leqslant j,k \leqslant N}$, $N \in \mathbb{N}$, in the Hilbert space $[L^2(M; d\mu)]^N$ (where $[L^2(M; d\mu)]^N = L^2(M; d\mu; \mathbb{C}^N)$), we recall that $T \in \mathcal{B}_2([L^2(M; d\mu)]^N)$ if and only if $T_{j,k} \in \mathcal{B}_2(L^2(M; d\mu))$, $1 \leqslant j, k \leqslant N$. Moreover, we recall that (cf. e.g., [2, Theorem 11.3.6])

$$\|T\|^2_{\mathcal{B}_2(L^2(M;d\mu)^N)} = \int_{M \times M} d\mu(x)\, d\mu(y)\, \|T(x, y)\|^2_{\mathcal{B}_2(\mathbb{C}^N)}$$

$$= \int_{M \times M} d\mu(x)\, d\mu(y) \sum_{j,k=1}^{N} |T_{j,k}(x, y)|^2$$

$$= \sum_{j,k=1}^{N} \int_{M \times M} d\mu(x)\, d\mu(y)\, |T_{j,k}(x, y)|^2$$

$$= \sum_{j,k=1}^{N} \|T_{j,k}\|^2_{\mathcal{B}_2(L^2(M;d\mu))}, \tag{93}$$

where, in obvious notation, $T(\cdot, \cdot)$ denotes the $N \times N$ matrix-valued integral kernel of T in $[L^2(M; d\mu)]^N$, and $T_{j,k}(\cdot, \cdot)$ represents the integral kernel of $T_{j,k}$ in $L^2(M; d\mu)$, $1 \leqslant j, k \leqslant N$.

In addition, employing the fact that for any $N \times N$ matrix $D \in \mathbb{C}^{N \times N}$,

$$\|D\|_{\mathcal{B}(\mathbb{C}^N)} \leqslant \|D\|_{\mathcal{B}_2(\mathbb{C}^N)} \leqslant N^{1/2} \|D\|_{\mathcal{B}(\mathbb{C}^N)}, \tag{94}$$

one also obtains

$$\|T\|^2_{\mathcal{B}_2(L^2(M;d\mu)^N)} \leqslant N \int_{M \times M} d\mu(x)\, d\mu(y)\, \|T(x, y)\|^2_{\mathcal{B}(\mathbb{C}^N)}. \tag{95}$$

More generally, for $\mathcal{H}$ a complex separable Hilbert space and $T = \{T_{j,k}\}_{1 \leqslant j,k \leqslant N}$, $N \in \mathbb{N}$, a block operator matrix in $\mathcal{H}^N$, one confirms that

$$T \in \mathcal{B}(\mathcal{H}^N) \ (\text{resp., } T \in \mathcal{B}_p(\mathcal{H}^N), \ p \in [1, \infty) \cup \{\infty\})$$

if and only if   (96)

for each $1 \leqslant j, k \leqslant N$, $T_{j,k} \in \mathcal{B}(\mathcal{H}^N)$ (resp., $T_{j,k} \in \mathcal{B}_p(\mathcal{H}^N)$, $p \in [1, \infty) \cup \{\infty\}$).

In other words, for membership of T in $\mathcal{B}(\mathcal{H}^N)$ or $\mathcal{B}_p(\mathcal{H}^N)$, $p \in [1, \infty) \cup \{\infty\}$, it suffices to focus on each of its matrix elements $T_{j,k}$, $1 \leqslant j, k \leqslant N$. (For necessity of the last line in (96) it suffices to multiply T from the left and right by $N \times N$ diagonal matrices with $I_{\mathcal{H}}$ on the jth and kth position, respectively, and zeros otherwise, to isolate $T_{j,k}$ and appeal to the ideal property. For sufficiency, it suffices to write T as a sum of N^2 terms with $T_{j,k}$ at the j, kth position and zeros otherwise.)

The next result is useful in connection with Section 3.

Theorem A.2 *Let $N \in \mathbb{N}$ and suppose that T_1, T_2 are linear $N \times N$ block operator matrices defined on $[L^2(M; d\mu)]^N$, such that for each $1 \leqslant j, k \leqslant N$, $T_{2,j,k}$ pointwise dominates $T_{1,j,k}$. Then the following items (i)–(iii) hold:*

(i) If $T_2 \in \mathcal{B}([L^2(M; d\mu)]^N)$ then $T_1 \in \mathcal{B}([L^2(M; d\mu)]^N)$ and

$$\|T_1\|_{\mathcal{B}([L^2(M;d\mu)]^N)} \leqslant \|T_2\|_{\mathcal{B}([L^2(M;d\mu)]^N)}. \tag{97}$$

(ii) If $T_2 \in \mathcal{B}_\infty([L^2(M; d\mu)]^N)$ then $T_1 \in \mathcal{B}_\infty([L^2(M; d\mu)]^N)$ and

$$\|T_1\|_{\mathcal{B}([L^2(M;d\mu)]^N)} \leqslant \|T_2\|_{\mathcal{B}([L^2(M;d\mu)]^N)}. \tag{98}$$

(iii) If $T_2 \in \mathcal{B}_2([L^2(M; d\mu)]^N)$ then $T_1 \in \mathcal{B}_2([L^2(M; d\mu)]^N)$ and

$$\|T_1\|_{\mathcal{B}_2([L^2(M;d\mu)]^N)} \leqslant \|T_2\|_{\mathcal{B}_2([L^2(M;d\mu)]^N)}. \tag{99}$$

Proof. For item *(ii)* we refer to [11] and [23] (see also [21]) combined with (96) as we will not use it in this paper. While the proofs of items *(i)* and *(iii)* are obviously well-known, we briefly recall them here as we will be using these facts in Section 3. Starting with item *(i)*, we introduce the notation $f = (f_1, \ldots, f_N) \in [L^2(M; d\mu)]^N$ and $|f| = (|f_1|, \ldots, |f_N|) \in [L^2(M; d\mu)]^N$ and compute,

$$\|T_1 f\|^2_{[L^2(M;d\mu)]^N} = \sum_{j=1}^N \|(T_1 f)_j\|^2_{L^2(M;d\mu)} = \sum_{j=1}^N ((T_1 f)_j, (T_1 f)_j)_{L^2(M;d\mu)}$$

$$= \sum_{j=1}^N \left| \sum_{k,\ell=1}^N (T_{1,j,k} f_k, T_{1,j,\ell} f_\ell)_{L^2(M;d\mu)} \right|$$

$$\leqslant \sum_{j=1}^{N} \sum_{k,\ell=1}^{N} |(T_{1,j,k}f_k, T_{1,j,\ell}f_\ell)_{L^2(M;d\mu)}|$$

$$\leqslant \sum_{j=1}^{N} \sum_{k,\ell=1}^{N} (|T_{1,j,k}f_k|, |T_{1,j,\ell}f_\ell|)_{L^2(M;d\mu)}$$

$$\leqslant \sum_{j=1}^{N} \sum_{k,\ell=1}^{N} (T_{2,j,k}|f_k|, T_{2,j,\ell}|f_\ell|)_{L^2(M;d\mu)}$$

$$= \sum_{j=1}^{N} ((T_2|f|)_j, (T_2|f|)_j)_{L^2(M;d\mu)}$$

$$= \|T_2|f|\|^2_{[L^2(M;d\mu)]^N}$$

$$\leqslant \|T_2\|^2_{\mathcal{B}(L^2(M;d\mu)^N)} \||f|\|^2_{[L^2(M;d\mu)]^N}$$

$$= \|T_2\|^2_{\mathcal{B}(L^2(M;d\mu)^N)} \|f\|^2_{[L^2(M;d\mu)]^N}, \tag{100}$$

implying item (i). For item (iii) we recall from [27, Theorem 2.13] that $T_{1,j,k} \in \mathcal{B}_2(L^2(M;d\mu))$, $1 \leqslant j,k \leqslant N$, and $\|T_{1,j,k}\|_{\mathcal{B}_2(L^2(M;d\mu))} \leqslant \|T_{2,j,k}\|_{\mathcal{B}_2(L^2(M;d\mu))}$, $1 \leqslant j,k \leqslant N$, and hence by (93),

$$\|T_1\|^2_{\mathcal{B}_2([L^2(M;d\mu)]^N)} = \sum_{j,k=1}^{N} \|T_{1,j,k}\|^2_{\mathcal{B}_2(L^2(M;d\mu))} \leqslant \sum_{j,k=1}^{N} \|T_{2,j,k}\|^2_{\mathcal{B}_2(L^2(M;d\mu))}$$

$$= \|T_2\|^2_{\mathcal{B}_2([L^2(M;d\mu)]^N)}. \tag{101}$$

$\square$

Remark A.3 We note that the subordination assumption $|(Af)(\cdot)| \leqslant (B|f|)(\cdot)$ μ-a.e. on M, if A and B are integral operators in $\mathcal{H}$ with integral kernels $A(\cdot,\cdot)$ and $B(\cdot,\cdot)$, respectively, is implied by the condition $|A(\cdot,\cdot)| \leqslant B(\cdot,\cdot)$ $\mu \otimes \mu$-a.e. on $M \times M$ since

$$|(Af)(x)| = \left| \int_M d\mu(y)\, A(x,y)f(y)) \right| \leqslant \int_M d\mu(y)\, |A(x,y)||f(y)|$$

$$\leqslant \int_M d\mu(y)\, B(x,y)|f(y)| = (B|f|)(x) \text{ for a.e. } x \in M. \tag{102}$$

$\diamond$

Next, we state the following result.

Lemma A.4 *Let* $n \in \mathbb{N}$ *and suppose that* $K : \mathbb{R}^n \times \mathbb{R}^n \to [0,\infty)$ *satisfies*

$$0 \leqslant K(x,y) \leqslant \prod_{j=1}^{n} K_j(x_j, y_j), \quad x = (x_j)_{j=1}^{n}, y = (y_j)_{j=1}^{n} \in \mathbb{R}^n, \tag{103}$$

for functions $K_j : \mathbb{R} \times \mathbb{R} \rightarrow [0, \infty)$. If $K_j(\cdot, \cdot)$ is the kernel of a bounded integral operator $K_j \in \mathcal{B}(L^2(\mathbb{R}))$ for each $1 \leqslant j \leqslant n$, then $K(\cdot, \cdot)$ is the kernel of a bounded integral operator $K \in \mathcal{B}(L^2(\mathbb{R}^n))$, and

$$\|K\|_{\mathcal{B}(L^2(\mathbb{R}^n))} \leqslant \prod_{j=1}^{n} \|K_j\|_{\mathcal{B}(L^2(\mathbb{R}))}. \tag{104}$$

Proof. We proceed by induction on n. The claim is evident in the case $n = 1$. Let $n \in \mathbb{N}$, and suppose the claim is true for $n - 1 \in \mathbb{N}$. In order to establish the claim for n, we compute for $f \in L^2(\mathbb{R}^n)$:

$$\int_{\mathbb{R}^n} d^n x \left| \int_{\mathbb{R}^n} d^n y \, K(x, y) f(y) \right|^2$$

$$\leqslant \int_{\mathbb{R}} dx_1 \cdots \int_{\mathbb{R}} dx_n$$

$$\times \left(\int_{\mathbb{R}} dy_1 \cdots \int_{\mathbb{R}} dy_n \prod_{j=1}^{n} K(x_j, y_j) |f(y_1, \ldots, y_{n-1}, y_n)| \right)^2$$

$$\leqslant \|K_1\|^2_{\mathcal{B}(L^2(\mathbb{R}))} \cdots \|K_{n-1}\|^2_{\mathcal{B}(L^2(\mathbb{R}))}$$

$$\times \int_{\mathbb{R}} dx_n \left\| \int_{\mathbb{R}} dy_n \, K_n(x_n, y_n) |f(y_1, \ldots, y_{n-1}, y_n)| \right\|^2_{L^2(\mathbb{R}^{n-1}; dy_1 \cdots dy_{n-1})}, \tag{105}$$

and

$$\int_{\mathbb{R}} dx_n \left\| \int_{\mathbb{R}} dy_n \, K_n(x_n, y_n) |f(y_1, \ldots, y_n)| \right\|^2_{L^2(\mathbb{R}^{n-1}; dy_1 \cdots dy_{n-1})}$$

$$= \int_{\mathbb{R}} dx_n \left[\int_{\mathbb{R}} dy_1 \cdots \int_{\mathbb{R}} dy_{n-1} \right.$$

$$\times \left. \left(\int_{\mathbb{R}} dy_n \, K_n(x_n, y_n) |f(y_1, \ldots, y_{n-1}, y_n)| \right)^2 \right]$$

$$= \int_{\mathbb{R}} dy_1 \cdots \int_{\mathbb{R}} dy_{n-1} \left[\int_{\mathbb{R}} dx_n \right.$$

$$\times \left. \left(\int_{\mathbb{R}} dy_n \, K_n(x_n, y_n) |f(y_1, \ldots, y_{n-1}, y_n)| \right)^2 \right]$$

$$\leqslant \|K_n\|^2_{\mathcal{B}(L^2(\mathbb{R}))} \int_{\mathbb{R}} dy_1 \cdots \int_{\mathbb{R}} dy_{n-1} \left(\int_{\mathbb{R}} dy_n \, |f(y_1, \ldots, y_{n-1}, y_n)|^2 \right)$$

$$= \|K_n\|^2_{\mathcal{B}(L^2(\mathbb{R}))} \int_{\mathbb{R}^n} dy_1 \cdots dy_n \, |f(y_1, \ldots, y_{n-1}, y_n)|^2$$

$$= \|K_n\|^2_{\mathcal{B}(L^2(\mathbb{R}))} \|f\|^2_{L^2(\mathbb{R}^n)}. \tag{106}$$

To obtain the inequality in (106), we used the boundedness property of K_n in the form

$$\int_{\mathbb{R}} dx_n \left(\int_{\mathbb{R}} dy_n \, K_n(x_n, y_n) | f(y_1, \dots, y_{n-1}, y_n) | \right)^2 \tag{107}$$

$$\leqslant \|K_n\|^2_{\mathcal{B}(L^2(\mathbb{R}))} \int_{\mathbb{R}} dy_n \, |f(y_1, \dots, y_{n-1}, y_n)|^2 \text{ for a.e. } (y_j)_{j=1}^{n-1} \in \mathbb{R}^{n-1}. \tag{108}$$

The claim and the estimate in (104) now follow upon combining (105) and (106). $\qquad\qquad\square$

We conclude with one more fact from [17, Theorem 319]:

Lemma A.5 *Let* $p \in (1, \infty)$. *If* $K : \mathbb{R} \times \mathbb{R} \to [0, \infty)$ *is homogeneous of degree* (-1) *and the* (*necessarily identical*) *quantities*

$$\int_0^\infty ds \, K(s, 1) \, s^{-1/p'} \quad \text{and} \quad \int_0^\infty dt \, K(1, t) \, t^{-1/p} \tag{109}$$

are equal to some number $C \in (0, \infty)$, *then the integral operator* K *with kernel* $K(\,\cdot\,, \,\cdot\,)$ *belongs to* $\mathcal{B}(L^p((0, \infty)))$ *and*

$$\|K\|_{\mathcal{B}(L^p((0,\infty)))} \leqslant C. \tag{110}$$

Acknowledgments. We are indebted to Alan Carey, Jens Kaad, Galina Levitina, Denis Potapov, Fedor Sukochev, and Dima Zanin for helpful discussions and to the referee for a very careful reading of our manuscript.

References

[1] M. Abramowitz and I. A. Stegun, *Handbook of Mathematical Functions*, Dover, New York, 1972.

[2] M. Sh. Birman and M. Solomyak, *Spectral Theory of Selfadjoint Operators in Hilbert Space*, Mathematics and its Applications, Reidel, 1987.

[3] A. Carey, F. Gesztesy, H. Grosse, G. Levitina, D. Potapov, F. Sukochev, and D. Zanin, *Trace formulas for a class of non-Fredholm operators: A review*, Rev. Math. Phys. **28**, no. 10, (2016), 1630002 (55 pages).

[4] A. Carey, F. Gesztesy, J. Kaad, G. Levitina, R. Nichols, D. Potapov, and F. Sukochev, *On the global limiting absorption principle for massless Dirac operators*, Ann. H. Poincaré **19**, 1993–2019 (2018).

[5] A. Carey, F. Gesztesy, G. Levitina, R. Nichols, D. Potapov, F. Sukochev, and D. Zanin, *On the limiting absorption principle for massless Dirac operators and the computation of spectral shift functions*, in preparation.

[6] A. Carey, F. Gesztesy, G. Levitina, D. Potapov, F. Sukochev, and D. Zanin, *On index theory for non-Fredholm operators: a* $(1 + 1)$*-dimensional example*, Math. Nachrichten **289**, 575–609 (2016).

[7] A. Carey, F. Gesztesy, G. Levitina, D. Potapov, F. Sukochev, and D. Zanin, *Trace formulas for a* $(1 + 1)$-*dimensional model operator*, preprint, 2014.

[8] A. Carey, F. Gesztesy, G. Levitina, and F. Sukochev, *On the index of a non-Fredholm model operator*, Operators and Matrices **10**, 881–914 (2016).

[9] A. Carey, F. Gesztesy, D. Potapov, F. Sukochev, and Y. Tomilov, *On the Witten index in terms of spectral shift functions*, J. Analyse Math. **132**, 1–61 (2017).

[10] M. Demuth, P. Stollmann, G. Stolz, and J. van Casteren, *Trace norm estimates for products of integral operators and diffusion semigroups*, Integral Eq. Operator Th. **23**, 145–153 (1995).

[11] P. G. Dodds and D. H. Fremlin, *Compact operators in Banach lattices*, Israel J. Math. **34**, 287–320 (1979).

[12] F. Gesztesy, Y. Latushkin, K. A. Makarov, F. Sukochev, and Y. Tomilov, *The index formula and the spectral shift function for relatively trace class perturbations*, Adv. Math. **227**, 319–420 (2011).

[13] F. Gesztesy, Y. Latushkin, F. Sukochev, and Y. Tomilov, *Some operator bounds employing complex interpolation revisited*, in *Operator Semigroups Meet Complex Analysis, Harmonic Analysis and Mathematical Physics*, W. Arendt, R. Chill and Yu. Tomilov (eds.), Operator Theory: Advances and Applications, Vol. 250, Birkhäuser–Springer, 2015, pp. 213–239.

[14] I. C. Gohberg and M. G. Krein, *Introduction to the Theory of Linear Nonselfadjoint Operators*, Translations of Mathematical Monographs, Vol. 18, Amer. Math. Soc., Providence, RI, 1969.

[15] I. C. Gohberg and M. G. Krein, *Theory and Applications of Volterra Operators in Hilbert Space*, Translations of Mathematical Monographs, Vol. 24, Amer. Math. Soc., Providence, RI, 1970.

[16] P. R. Halmos and V. S. Sunder, *Bounded Integral Operators on* L^2 *Spaces*, Springer, Berlin, 1978.

[17] G. H. Hardy, J. E. Littlewood, and G. Pólya, *Inequalities*, Cambridge University Press, New York, 1959.

[18] K. Jörgens, *Linear Integral Operators*, transl. by G. F. Roach, Pitman, Boston, 1982.

[19] H. Kalf, T. Okaji, and O. Yamada, *The Dirac operator with mass* $m_0 \geqslant 0$: *Non-existence of zero modes and of threshold eigenvalues*, Doc. Math. **20**, 37–64 (2015); Addendum, Doc. Math. **22**, 1–3 (2017).

[20] M. A. Krasnoselskii, P. P. Zabreiko, E. I. Pustylnik, P. E. Sobolevskii, *Integral Operators in Spaces of Summable Functions*, transl. by T. Ando, Noordhoff Intl. Publ., Leyden, 1976.

[21] H. Leinfelder, *A remark on a paper of Loren D. Pitt*, Bayreuter Math. Schr. **11**, 57–66 (1982).

[22] L. Nirenberg and H. F. Walker, *The null spaces of elliptic partial differential operators on* $\mathbb{R}^n$, J. Math. Anal. Appl. **42**, 271–301 (1973).

[23] L. D. Pitt, *A compactness condition for linear operators in function spaces*, J. Operator Th. **1**, 49–54 (1979).

[24] A. Pushnitski, *The spectral flow, the Fredholm index, and the spectral shift function*, in *Spectral Theory of Differential Operators: M. Sh. Birman 80th Anniversary Collection*, T. Suslina and D. Yafaev (eds.), AMS Translations, Ser. 2, Advances in the Mathematical Sciences, Vol. 225, Amer. Math. Soc., Providence, RI, 2008, pp. 141–155.

[25] M. Reed and B. Simon, *Methods of Modern Mathematical Physics. IV: Analysis of Operators*, Academic Press, New York, 1978.

[26] B. Simon, *Quantum Mechanics for Hamiltonians Defined as Quadratic Forms*, Princeton University Press, Princeton, NJ, 1971.

[27] B. Simon, *Trace Ideals and Their Applications*, Mathematical Surveys and Monographs, Vol. 120, 2nd ed., Amer. Math. Soc., Providence, RI, 2005.

[28] J. Weidmann, *Linear Operators in Hilbert Spaces*, Graduate Texts in Mathematics, Vol. 68, Springer, New York, 1980.

[29] J. Weidmann, *Lineare Operatoren in Hilberträumen. Teil I: Grundlagen*, Teubner, Stuttgart, 2000.

[30] D. R. Yafaev, *Mathematical Scattering Theory. General Theory*, Amer. Math. Soc., Providence, RI, 1992.

[31] D. R. Yafaev, *Mathematical Scattering Theory. Analytic Theory*, Math. Surveys and Monographs, Vol. 158, Amer. Math. Soc., Providence, RI, 2010.

[32] P. P. Zabreiko, A. I. Koshelev, M. A. Krasnoselskii, S. G. Mikhlin, L. S. Rakovshchik, and V. Ya. Stetsenko, *Integral Equations – A Reference Text*, transl. by T. O. Shaposhnikova, R. S. Anderssen, S. G. Mikhlin, Noordhoff Intl. Publ., Leyden, 1975.

Fritz Gesztesy
Department of Mathematics,
Baylor University,
One Bear Place #97328, Waco,
TX 76798-7328, USA
E-mail address: Fritz_Gesztesy@baylor.edu
URL: http://www.baylor.edu/math/index.php?id=935340

Roger Nichols
Mathematics Department,
The University of Tennessee at Chattanooga,
415 EMCS Building, Dept. 6956, 615 McCallie Ave,
Chattanooga, TN 37403, USA
E-mail address: Roger-Nichols@utc.edu
URL: http://www.utc.edu/faculty/roger-nichols/index.php

2

Explicit Symmetries of the Kepler Hamiltonian

Horst Knörrer

To Emma Previato on the occasion of her 65th birthday

Abstract. Using quaternions, we give explicit formulas for the global symmetries of the three dimensional Kepler problem. The regularizations of the Kepler problem that are based on the Hopf map and on stereographic projections, respectively, are interpreted in terms of these symmetries.

1 Introduction

Consider the motion of a point mass in $\mathbb{R}^3$ under the influence of a gravitational field created by a point mass that is fixed at the origin. The "Kepler problem" in classical mechanics is the problem to determine the position $q(t) \in \mathbb{R}^3$ of the movable mass as a function of time t.

Using Newton's second law "force = mass $\times$ acceleration" and Newton's law on gravitation, that the force on the movable point mass is directed towards the origin and is proportional to the inverse square of its distance $\|q(t)\|$ from the origin, one immediately derives the second order differential equation

$$\ddot{q} = -\mu \frac{q}{\|q\|^3}. \tag{1}$$

The constant $\mu > 0$ depends on the gravitational constant and the two masses involved. Introducing the "momentum" $p = \dot{q}$, one sees that (1) is a Hamiltonian system with respect to the standard symplectic form $\Sigma_{j=1}^{3} dp_j \wedge dq_j$ and the "Kepler Hamiltonian"

$$H(q, p) = \frac{1}{2} \|p\|^2 - \frac{\mu}{\|q\|}. \tag{2}$$

Based on astronomical observations, Johannes Kepler (1571–1630) had stated the three "Kepler laws" about planetary motion (we recall them below). Isaac Newton (1643–1727) used the Kepler laws to guess and derive what we

38

now call "Newton's second law" and the "inverse square law", as they were stated above. At the time this was called the "direct problem". The book [5] contains, among others, a detailed description of Newton's way to attack this problem. More or less as an afterthought Newton then solved the "inverse problem", namely, that – in modern language – (1) implies Kepler's laws. In fact, the first edition of Newton's *Principia* does not contain any proof that Kepler's first law follows from Newton's laws; the second edition (which appeared 22 years later) has only a two line sketch. See [30] and, for the controversies about it, [2] §6 and the introduction to [9].

To recall, Kepler's laws are:

Kepler's first law: *Let $q(t)$ be a maximal solution of* (1). *Its orbit is either an ellipse which has one focal point at the origin, a branch of a hyperbola which has one focal point at the origin, a parabola whose focal point is the origin, or a ray emanating from the origin.*

Kepler's second law (Equal areas in equal times): *The area swept out by the vector joining the origin to the point $q(t)$ in a given time is proportional to that time.*

Kepler's third law: *If the motion is elliptical, then the squares of the periods of the planets are proportional to the cubes of their semimajor axes.*

From a modern point of view, Kepler's second law is just a reformulation of the law of conservation of angular momentum $L = q \times \dot{q}$; and the third law follows from conservation of angular momentum and the formula for the area of an ellipse. See for example [1] 8.E. The treatment of the Kepler problem in modern textbooks on mechanics always starts with the observation that the angular momentum L and the Hamiltonian H are conserved quantities. The case $L = 0$ is easy to deal with. For the case $L \neq 0$ there are several strategies to derive Kepler's first law. We sketch some of them:

- Rewrite (1) in polar coordinates, also express the conserved quantities "energy" and "angular momentum" in terms of polar coordinates, and use this to determine a first order differential equation for the angle in terms of the radius, which can be integrated explicitly. See [1] 8.E.
- Deduce from (1) that

$$\frac{d}{dt}\left(\mu \frac{q}{\|q\|}\right) = -L \times \ddot{q} = \frac{d}{dt}\left(L \times \dot{q}\right)$$

integrate this equation and take the inner product with L on both sides of the resulting equation. One gets an equation which can be viewed as an algebraic version of the "gardener's construction" of a conic section. See [11] 3.1.

- Observe that the "Laplace Lenz Runge vector"

$$A = \tfrac{1}{\mu}\left(\|\dot{q}\|^2 - \tfrac{\mu}{\|q\|}\right)q - \tfrac{1}{\mu}(q \cdot \dot{q})\,\dot{q} \tag{3}$$

is a conserved quantity, i.e., that it is independent of time; and use this fact to derive the "directrix description" of a conic section. See [8] (3.9). Below, we will comment further on the Laplace Lenz Runge vector.

- Look at the "hodograph", i.e. the curve traced out by the momentum vector $p(t) = \dot{q}(t)$. On can use (1) to show that this curve has constant curvature, and hence is a circle. Rotate this circle by $\tfrac{\pi}{2}$ around the origin. The Kepler orbit can then be identified as a polar reciprocal (with respect to the origin) of the rotated hodograph, and hence is a conic section. See [11] 8.6.4.

  A purely geometric proof of Kepler's first law using the hodograph was given by Hamilton, Kelvin and Tait, Maxwell, Fano, and Feynman independently. See [9], [15]. This argument is in parts very close to Newton's original argument.

- Introduce the "elliptic anomaly" s, which is a reparametrization of time characterized by

$$\frac{ds}{dt} = \frac{\sqrt{2|H|}}{\|q\|} \tag{4}$$

and rewrite (1) in terms of this new variable s. If one also uses the fact that the Laplace Lenz Runge vector is a conserved quantity, one gets to a quick proof of Kepler's first law in Cartesian coordinates. See [7] 2.2.1.

If, in the case of an elliptical orbit, one chooses the Cartesian coordinate system to have its origin at the center of the ellipse, its x–direction to be the direction of the major axis, and its y–direction to be the direction of the minor axis, then

$$q(s) = \left(a \cos(s - s_0),\; b \sin(s - s_0),\; 0\right) \tag{5}$$

with an integration constant s_0 and $2a$, $2b$ the length of the major resp. minor axis.

  With all these approaches, one gets Kepler's first law about the shape of the orbit, but not yet an explicit solution of (1), i.e. a formula for q as a function of t. This problem leads to the "Kepler equation", whose solution yields (in the case of elliptical orbits) a Fourier series representation of $q(t)$. The coefficients of this Fourier series are values of Bessel functions. See [35], 17.2. The folklore, that Bessel invented "his" functions with the purpose of solving the Kepler equation seems to be not quite correct; see [6] § 3.

  The Kepler problem has four obvious conserved quantities, namely energy H and the three components of angular momentum L. The discussion above shows that there are more conserved quantities: For example, the three components of A, the Laplace Lenz Runge vector[1] (3). A posteriori, i.e. when Newton's first law is established, one can see A (in the case of elliptical orbits and with the normalization chosen here) as the vector pointing in the direction of the major axis of the ellipse whose length is the eccentricity of the ellipse. For this reason it is sometimes also called "eccentricity vector", see [14] and [7], 2.2. After its probably first discovery by J.Hermann (a disciple of the Bernoullis), it was rediscovered many times, among others by P.S.Laplace and W.R.Hamilton. See [13], [14].

In his calculation of the spectrum of the hydrogen atom, W.Pauli [29] introduced a quantum mechanical observable whose classical version would be the Laplace Lenz Runge vector. For the classical problem, Pauli quoted the paper [22] of W.Lenz, who in turn attributes this vector to C.Runge [32], who however does not claim any originality. See [13]. Pauli's calculation uses the commutation relations between the quantum mechanical observables that commute with the Hamiltonian for the hydrogen atom, and in particular the fact that these observables form a Lie algebra isomorphic to $sO(4)$. See also [16], Section 6.

The analogue in classical mechanics of the commutator in quantum mechanics is the Poisson bracket for functions. As the angular momentum and the Laplace Lenz Runge vector are conserved quantities for (1), their components Poisson commute with the Kepler Hamiltonian (2). Consequently, for every fixed energy E, the Hamiltonian vectorfields associated to the components of angular momentum and of the Laplace Lenz Runge vector are tangent to the level set

$$H^{-1}(E) = \{(q, p) \in \mathbb{R}^3 \times \mathbb{R}^3 \mid H(q, p) = E\}$$

It is easy to verify that their $\mathbb{R}$–linear span is closed under the Lie bracket, and that it forms a Lie algebra which is isomorphic to $sO(4)$ when $E < 0$, isomorphic to the Lie algebra of the group $E^+(3)$ of rigid motions of $\mathbb{R}^3$ when $E = 0$, and isomorphic to $sl(2, \mathbb{C})$ when $E > 0$. This strongly suggests that, in the case of negative energy, the above mentioned level set can be compactified to a manifold with an action of the group $SU(2) \times SU(2)$ (the universal cover of $SO(4)$) that commutes with the Kepler flow. This is indeed the case. Of course integrating the Hamiltonian vectorfields associated to the components of angular momentum and of the Laplace Lenz Runge vector gives this group action. To my knowledge, this has only been done implicitly; see [3] §4, [31]. The main purpose of this paper is to write down an explicit formula for this action. See Theorem 3.1 and Lemma 3.2.

Both in quantum mechanics and in classical mechanics, the a priori surprising fact that the Kepler problem has a bigger symmetry group than

---

[1] In fact, all other conserved quantities are functions of the ones mentioned above.

the group $SO(3)$, which one would naively expect, led to the discovery of connections to other physical problems. In particularly V.Fock [10] in 1935 used the Hopf map to relate the spectrum of the hydrogen atom to the spectrum of the quantum mechanical harmonic oscillator in four dimensions, giving an even clearer understanding of the degeneracies of the eigenvalues. In classical mechanics, the problem of treating Hamiltonians close to, but not equal to the Kepler Hamiltonian (as they arise in the treatment of the three body problem) led to different ways of "regularizing" the Kepler flow. As already pointed out by Levi Civita (see [11] 8.40, 8.41), the reparametrization by the elliptic anomaly (4) is useful for this purpose, because all orbits have the same period 2π (see (5)), and even the collision orbits where the movable mass crashes into the origin are embedded into periodic orbits. The reparametrization by the elliptic anomaly corresponds to multiplying the Hamiltonian (2) by an appropriate function. See (6) below. P.Kuustanheimo and E.Stiefel [21] in 1965 used the Hopf map to relate the classical Kepler problem to classical harmonic oscillators on $\mathbb{C}^2$. J.Moser [27] in 1970 used stereographic projection in the momentum variable to relate the regularized Kepler flow to the geodesic flow on the three dimensional sphere[2]. For a given negative energy E, this construction embeds the level set $H^{-1}(E)$ as a dense subset into the unit tangent bundle of the three sphere[3]. On this unit tangent bundle, the $SO(4)$ symmetry is obvious. Other, may be more pedagogical descriptions of this construction may be found in [26] or [11] 8.3.

The explicit formula for the $SU(2) \times SU(2)$ symmetry of the Kepler problem presented in this paper is in terms of fractional linear transformations on the complex plane for the two dimensional subproblem, and in terms of a generalization of fractional linear transformations on the set of pure quaternions in the three dimensional situation. In view of the construction of Györgyi/Moser this is not surprising, since stereographic projection intertwines rotational symmetry and fractional linear transformations (see [36] III.8.b in the two dimensional situation and Lemma 3.3 below in the three dimensional situation). As a biproduct, the explicit description of the symmetry given here puts the somewhat complicated looking relation of [20] between the Kuustanheimo–Stiefel regularization and the Györgyi/Moser regularization [4] into a simple context: The total space of the Kuustanheimo–Stiefel regularization is isomorphic to a dense subset of $SU(2) \times SU(2)$, and

---

[2] A closely related construction was given by G.Györgyi [17].

[3] The Ligon–Schaaf regularization [24] that treats $\{(q, p) \in \mathbb{R}^3 \times \mathbb{R}^3 \mid H(q, p) < 0\}$ simultanuously may viewed as a generalization of the Györgyi/Moser construction. See [18], [25].

[4] Kummer's construction uses the fact that the cotangent bundle of the three sphere can be viewed as a coadjoint orbit of $SU(2, 2)$, related to the action of this group on the total space of the Kustaanheimo–Stiefel regularization.

the maps to the unit tangent bundle of the three sphere and to the level set $H^{-1}(E)$ are just the quotient maps by the isotropy group of a point. See the discussion after Lemma 3.2 and after Lemma 3.3.

As pointed out above, the main content of this paper are explicit formulas for the symmetries of the Kepler problem and the group theoretical understanding of some of its regularizations that derive from these formulas. We first present these formulas in two dimensions, using the identification of $\mathbb{R}^2$ with $\mathbb{C}$. As pointed out, the formulas are in terms of fractional linear transformations. Complex coordinates for the two dimensional Kepler problem have been used for a long time; see, for example [23] I.5. Nevertheless the formulas of Theorem 2.1 for the symmetries in the two dimensional case seem to be new.

In three dimensions we identify $\mathbb{R}^3$ with the space $\mathbb{H}_{\text{pure}}$ of pure quaternions, describe the symmetry groups in terms of quaternions, and write the action of the symmetry groups on $\mathbb{H}_{\text{pure}} \times \mathbb{H}_{\text{pure}}$ in terms of "quaternionic fractional linear transformations". For the case of negative energy this is done in Section 3, where we state and prove all the results mentioned above. These results have analogues in the case of positive energy. In that case, the symmetry group is $SL(2, \mathbb{C})$, and the analog of the regularization of Györgyi/Moser was constructed by Belbruno [4]/Osipov [28], using stereographic projection from the unit tangent bundle of a hyperboloid. The group action on $\mathbb{H}_{\text{pure}} \times \mathbb{H}_{\text{pure}}$ in this situation is described in Theorem 4.1, and Lemma 4.2 gives the group theoretic interpretation of the Belbruno/Osipov regularization. Finally, in Theorem 5.1, we also treat the case of zero energy, where the symmetry group is $E^+(3)$.

Throughout the paper we work with the Hamiltonian that describes the Kepler problem using the elliptic anomaly as independent variable. If $E \neq 0$, the rescaled problem is the Hamiltonian flow of the Hamiltonian $\tilde{H}(q, p) = \frac{1}{\sqrt{2|E|}} \|q\| (\|p\|^2 - 2E) - \frac{2\mu}{\sqrt{2|E|}}$ on the hypersurface $\{(q, p) \mid \tilde{H}(q, p) = 0\}$. See [8], II.5.10b. Scaling q by the factor $\sqrt{2|E|}$ and p by the factor $\frac{1}{2|E|}$ allows to reduce the discussion to the cases $2|E| = 1$ or $E = 0$. That is, we consider the Hamiltonian flows of

$$K_{\pm}(q, p) = \|q\| (\|p\|^2 \pm 1) - 2\mu \qquad \text{and}$$
$$K_0(q, p) = \|q\| \|p\|^2 - 2\mu \tag{6}$$

on the hypersurfaces $\{(q, p) \mid K_{\pm}(q, p) = 0\}$ and $\{(q, p) \mid K_0(q, p) = 0\}$, respectively. The case of negative energy (that is, bounded orbits) is described by K_+, the case of positive energy by K_-. Explicitly, the dynamical systems associated to the Hamiltonians (6) are

$$\dot{q} = 2\|q\| p \qquad \dot{p} = -\frac{1}{\|q\|^2} (K(q, p) + 2\mu)q \tag{7}$$

where $K = K_-, K_+, K_0$.

2 Two Dimensions

For an invertible complex 2×2 matrix $A = \begin{pmatrix} a & b \\ c & d \end{pmatrix}$ and $(q, p) \in \mathbb{C} \times \mathbb{C}$ set

$$A \cdot (q, p) = \left((\bar{c}\bar{p} + \bar{d})^2 \cdot q \, , \, \frac{ap + b}{cp + d} \right)$$

whenever it is defined. Observe that the second component is the standard action of $GL(2, \mathbb{C})$ on the complex plane by fractional linear transformations. If $\det A = 1$, the factor $(\bar{c}\bar{p} + \bar{d})^2$ is the complex conjugate of the inverse of the derivative of the fractional linear transformation $p \mapsto \frac{ap+b}{cp+d}$. Therefore, for $A_1, A_2 \in SL(2, \mathbb{C})$

$$(A_1 A_2) \cdot (q, p) = A_1 \cdot \left(A_2 \cdot (q, p) \right)$$

whenever defined. Also, the action is symplectic with respect to the standard symplectic form $\operatorname{Re}(d\bar{p} \wedge dq)$.

Theorem 2.1 *Let* $(q, p) \in \mathbb{C} \times \mathbb{C}$. *Then*

$$K_+ \left(A \cdot (q, p) \right) = K_+(q, p) \text{ for all } A \in SU(2)$$
$$K_- \left(A \cdot (q, p) \right) = K_-(q, p) \text{ for all } A \in SU(1, 1)$$

$$K_0 \left(A \cdot (q, p) \right) = K_0(q, p) \text{ for all } A \in P = \left\{ \begin{pmatrix} a & 0 \\ c & a^{-1} \end{pmatrix} \,\middle|\, a, c \in \mathbb{C}, \, a \neq 0 \right\}$$

whenever defined.

Proof. Let $A = \begin{pmatrix} a & b \\ c & d \end{pmatrix} \in GL(2, \mathbb{C})$. Then

$$K_\pm \left(A \cdot (q, p) \right) + 2\mu = |\bar{c}\bar{p} + \bar{d}|^2 \, |q| \left(\left| \frac{ap + b}{cp + d} \right|^2 \pm 1 \right)$$

$$= |q| \left(|ap + b|^2 \pm |cp + d|^2 \right)$$

$$= |q| \left(|p_1|^2 \pm |p_2|^2 \right)$$

where

$$\begin{pmatrix} p_1 \\ p_2 \end{pmatrix} = A \cdot \begin{pmatrix} p \\ 1 \end{pmatrix}$$

If $A \in SU(2)$ then $|p_1|^2 + |p_2|^2 = |p|^2 + 1$, and if $A \in SU(1, 1)$ then $|p_1|^2 - |p_2|^2 = |p|^2 - 1$. The proof for K_0 is similar. $\square$

3 Three Dimensions, Negative Energy

Recall that the quaternions are the four dimensional skew field over $\mathbb{R}$

$$\mathbb{H} = \{x_0 \cdot 1 + x_1 \cdot \imath + x_2 \cdot j + x_3 \cdot k \mid x_0, \cdots, x_3 \in \mathbb{R}\}$$

with the associative multiplication characterized by

$$1 \cdot u = u \cdot 1 = u \quad \text{for all } u \in \mathbb{H}$$
$$\imath^2 = j^2 = k^2 = -1$$
$$\imath j = -j\imath = k, \ jk = -kj = \imath, \ k\imath = -\imath k = j$$

For a quaternion

$$u = x_0 \cdot 1 + x_1 \cdot \imath + x_2 \cdot j + x_3 \cdot k$$

we denote by

$$u^* = x_0 \cdot 1 - x_1 \cdot \imath - x_2 \cdot j - x_3 \cdot k \qquad \text{the conjugate quaternion}$$

$$|u| = \sqrt{uu^*} = \sqrt{x_0^2 + x_1^2 + x_2^2 + x_3^2} \qquad \text{its norm}$$

$$\mathrm{Re}\,(u) = x_0 = \frac{1}{2}(u + u^*) \qquad \text{its real part}$$

$$\mathcal{P}\,(u) = x_1 \cdot \imath + x_2 \cdot j + x_3 \cdot k = \frac{1}{2}(u - u^*) \qquad \text{its pure part}$$

$\mathbb{R}$ is considered as the subfield $\{x_0 \cdot 1 \mid x_0 \in \mathbb{R}\}$ of $\mathbb{H}$. The multiplicative inverse of a non zero quaternion u is $u^{-1} = \frac{1}{|u|^2}u^*$. It is easy to check that $(uv)^* = v^*u^*$ for $u, v \in \mathbb{H}$.

We identify $\mathbb{R}^3$ with the space of pure quaternions

$$\mathbb{H}_{\mathrm{pure}} = \{u \in \mathbb{H} \mid \mathrm{Re}\,(u) = 0\} = \{u \in \mathbb{H} \mid u = \mathcal{P}(u)\}$$
$$= \{u \in \mathbb{H} \mid u^2 \in \mathbb{R}, \ u^2 \le 0\}$$

by mapping $q = (q_1, q_2, q_3) \in \mathbb{R}^3$ to $q_1 \cdot \imath + q_2 \cdot j + q_3 \cdot k \in \mathbb{H}_{\mathrm{pure}}$. The multiplicative group of quaternions of norm one

$$\mathbb{H}_{\mathrm{unit}} = \{u \in \mathbb{H} \mid |u| = 1\}$$

is isomorphic to $SU(2)$. The product

$$G_+ = \mathbb{H}_{\mathrm{unit}} \times \mathbb{H}_{\mathrm{unit}}$$

(with componentwise multiplication) is a double cover of $SO(4)$, see for example [19], sec 3.

For $(\alpha_1, \alpha_2) \in G_+$ and $(q, p) \in \mathbb{H}_{\mathrm{pure}} \times \mathbb{H}_{\mathrm{pure}}$ set (whenever defined)

$$(\alpha_1, \alpha_2) \bullet (q, p) = \left((bp + a)\,q\,(bp + a)^*, \ (ap + b)(bp + a)^{-1}\right)$$

where $a = \frac{1}{2}(\alpha_1 + \alpha_2), \ b = \frac{1}{2}(-\alpha_1 + \alpha_2)$.

Theorem 3.1 *Let $(\alpha_1, \alpha_2) \in G_+$. Then for all $(q, p) \in \mathbb{H}_{\text{pure}} \times \mathbb{H}_{\text{pure}}$.*

(i) $(\alpha_1, \alpha_2) \bullet (q, p) \in \mathbb{H}_{\text{pure}} \times \mathbb{H}_{\text{pure}}$.

(ii) $(\alpha_1' \alpha_1, \alpha_2' \alpha_2) \bullet (q, p) = (\alpha_1', \alpha_2') \bullet \big((\alpha_1, \alpha_2) \bullet (q, p)\big)$ *for all* $(\alpha_1', \alpha_2') \in G_+$.

(iii) $K_+\big((\alpha_1, \alpha_2) \bullet (q, p)\big) = K_+(q, p)$.

Furthermore the map $(q, p) \mapsto (\alpha_1, \alpha_2) \bullet (q, p)$ is symplectic with respect to the symplectic form $\operatorname{Re} dq^* \wedge dp$.

The statements of the theorem apply whenever all expressions are defined. For a similar description of this symmetry see [12] 2.5.

Proof. Again set $a = \frac{1}{2}(\alpha_1 + \alpha_2)$, $b = \frac{1}{2}(-\alpha_1 + \alpha_2)$. By construction, $|a|^2 + |b|^2 = 1$ and $ab^* = -ba^*, a^*b = -b^*a$.

(i) Since $q \in \mathbb{H}_{\text{pure}}$ we have $q^* = -q$. Therefore

$$\big((bp + a)\, q\, (bp + a)^*\big)^* = (bp + a)\, q^*\, (bp + a)^* = -(bp + a)\, q\, (bp + a)^*$$

so that the first component of $(\alpha_1, \alpha_2) \bullet (q, p)$ lies in $\mathbb{H}_{\text{pure}}$. Similarly

$$\begin{aligned}
\big((ap + b)(bp + a)^*\big)^* &= (bp + a)(ap + b)^* = (bp + a)(p^*a^* + b^*) \\
&= b|p|^2 a^* + bpb^* + ap^*a^* + ab^* \\
&= -a|p|^2 b^* - bp^*b^* - apa^* - ba^* \\
&= -(ap + b)(p^*b^* + a^*) = -(ap + b)(bp + a)^*
\end{aligned}$$

so that

$$\begin{aligned}
\big((ap + b)(bp + a)^{-1}\big)^* &= \frac{1}{|bp + a|^2}\big((ap + b)(bp + a)^*\big)^* \\
&= -(ap + b)(bp + a)^{-1}
\end{aligned}$$

and $(ap + b)(bp + a)^{-1} \in \mathbb{H}_{\text{pure}}$.

(ii) By construction, the second component of $(\alpha_1, \alpha_2) \bullet (q, p)$ is

$$\begin{aligned}
(ap + b)(bp + a)^{-1} &= \big[\alpha_2(p + 1) + \alpha_1(p - 1)\big]\big[\alpha_2(p + 1) - \alpha_1(p - 1)\big]^{-1} \\
&= \frac{1 + \alpha_1 \frac{p-1}{p+1}\alpha_2^{-1}}{1 - \alpha_1 \frac{p-1}{p+1}\alpha_2^{-1}}
\end{aligned}$$

$$\tag{8}$$

Consequently

$$(ap + b)(bp + a)^{-1} + 1 = \frac{2}{1 - \alpha_1 \frac{p-1}{p+1} \alpha_2^{-1}}$$

$$(ap + b)(bp + a)^{-1} - 1 = \frac{2\,\alpha_1 \frac{p-1}{p+1} \alpha_2^{-1}}{1 - \alpha_1 \frac{p-1}{p+1} \alpha_2^{-1}}$$

Therefore, by (8), the second component of $(\alpha_1', \alpha_2') \bullet \big((\alpha_1, \alpha_2) \bullet (q, p)\big)$ is

$$\left[\alpha_2' \frac{1}{1 - \alpha_1 \frac{p-1}{p+1} \alpha_2^{-1}} + \alpha_1' \frac{\alpha_1 \frac{p-1}{p+1} \alpha_2^{-1}}{1 - \alpha_1 \frac{p-1}{p+1} \alpha_2^{-1}} \right]$$

$$\times \left[\alpha_2' \frac{1}{1 - \alpha_1 \frac{p-1}{p+1} \alpha_2^{-1}} - \alpha_1' \frac{\alpha_1 \frac{p-1}{p+1} \alpha_2^{-1}}{1 - \alpha_1 \frac{p-1}{p+1} \alpha_2^{-1}} \right]^{-1}$$

$$= \left[\alpha_2' + \alpha_1' \alpha_1 \frac{p-1}{p+1} \alpha_2^{-1} \right] \left[\alpha_2' - \alpha_1' \alpha_1 \frac{p-1}{p+1} \alpha_2^{-1} \right]^{-1}$$

$$= \frac{1 + \alpha_1' \alpha_1 \frac{p-1}{p+1} (\alpha_2' \alpha_2)^{-1}}{1 - \alpha_1' \alpha_1 \frac{p-1}{p+1} (\alpha_2' \alpha_2)^{-1}}$$

which, again by (8), is the second component of $(\alpha_1' \alpha_1,\ \alpha_2' \alpha_2) \bullet (q, p)$.

The derivative of the map $p \mapsto (ap + b)(bp + a)^{-1}$ at the point p is the linear map

$$u \longmapsto \frac{d}{dt}\big(a(p + tu) + b\big)\big(b(p + tu) + a\big)^{-1}\Big|_{t=0}$$

$$= au(bp + a)^{-1} - (ap + b)(bp + a)^{-1} bu\,(bp + a)^{-1}$$

$$= \big[a - (ap + b)(bp + a)^{-1} b\big] u\,(bp + a)^{-1}$$

Now

$$|bp + a|^2 \big[a - (ap + b)(bp + a)^{-1} b\big]$$

$$= (|b|^2 |p|^2 + |a|^2 + bpa^* + ap^* b^*)\,a - (ap + b)(bp + a)^* b$$

$$= (|b|^2 |p|^2 + |a|^2)a + |a|^2 bp + apa^* b - a|p|^2 |b|^2 - bp^* |b|^2$$

$$\quad - apa^* b - ba^* b$$

$$= |a|^2 a + |a|^2 bp + bp|b|^2 + bb^* a$$

$$= bp + a$$

So the derivative is the linear map $u \longmapsto \frac{1}{|bp+a|^2}(bp + a)\,u\,(bp + a)^{-1}$ on $\mathbb{H}_{\text{pure}}$. Its inverse is the map $u \longmapsto (bp + a)^*\,u\,(bp + a)$; and the adjoint of the inverse (with respect to the bilinear form $< u, v > = \text{Re } uv^*$) is the map $u \longmapsto (bp + a)\,u\,(bp + a)^*$. Thus the first component of $(q, p) \mapsto (\alpha_1, \alpha_2) \bullet (q, p)$

is the adjoint of the derivative of the second component. This proves (ii) and also the fact that the map is symplectic.

(iii) $K_+\big((\alpha_1, \alpha_2) \bullet (q, p)\big) + 2\mu = |bp + a|^2 |q|^2 \left(\dfrac{|ap + b|^2}{|bp + a|^2} + 1 \right)$

$$= |q|^2 \big(|ap + b|^2 + |bp + a|^2\big)$$
$$= |q|^2 \big(|a|^2 |p|^2 + |b|^2 + apb^* + bp^*a^*$$
$$+ |b|^2 |p|^2 + |a|^2 + ap^*b^* + bpa^*\big)$$
$$= |q|^2 \big(|p|^2 + 1\big) = K_+(q, p) + 2\mu$$

$\square$

The Kustaanheimo–Stiefel map

For simplicity we assume from now on that $2\mu = 1$. Then the point $(\iota, 0)$ lies on the energy hypersurface $K_+^{-1}(0)$. We consider the map

$$\varphi_+ : \{(\alpha_1, \alpha_2) \in G_+ \,|\, \alpha_2 \neq -\alpha_1\} \longrightarrow K_+^{-1}(0)$$
$$(\alpha_1, \alpha_2) \longmapsto (\alpha_1, \alpha_2) \bullet (\iota, 0) = (a\iota a^*, ba^{-1})$$

where, as before, $a = \frac{1}{2}(\alpha_1 + \alpha_2)$, $b = \frac{1}{2}(-\alpha_1 + \alpha_2)$.

Under this parametrization, the angular momentum is $L = \mathcal{P}(q \cdot p) = \frac{1}{4}(\alpha_1 \iota \alpha_1^* - \alpha_2 \iota \alpha_2^*)$, and the eccentricity vector is $-\frac{q}{|q|} + 2\mathcal{P}(p \cdot L) = -\frac{1}{2}(\alpha_1 \iota \alpha_1^* + \alpha_2 \iota \alpha_2^*)$. By direct calculation one shows

Lemma 3.2 φ_+ *maps the flow* $t \mapsto (\alpha_1 e^{\iota t}, \alpha_2 e^{-\iota t})$ *on* G_+ *to the regularized Kepler flow, that is the flow of the dynamical system* (7).

For their regularization of the Kepler problem, Kustaanheimo–Stiefel [21] use the map that sends $(\mathbf{z}, \mathbf{w}) \in (\mathbb{C}^2 \setminus \{0\}) \times \mathbb{C}^2$ to $(q, p) \in \mathbb{R}^3$ given by

$$q = (\mathbf{z} \cdot \sigma^3 \mathbf{z}^*, \ \mathbf{z} \cdot \sigma^1 \mathbf{z}^*, \ \mathbf{z} \cdot \sigma^2 \mathbf{z}^*),$$
$$p = \frac{1}{\|\mathbf{z}\|^2} \big(\mathrm{Im}(\mathbf{z} \cdot \sigma^3 \mathbf{w}^*), \ \mathrm{Im}(\mathbf{z} \cdot \sigma^1 \mathbf{w}^*), \ \mathrm{Im}(\mathbf{z} \cdot \sigma^2 \mathbf{w}^*) \big)$$

where

$$\sigma^1 = \begin{pmatrix} 0 & 1 \\ 1 & 0 \end{pmatrix}, \quad \sigma^2 = \begin{pmatrix} 0 & -\iota \\ \iota & 0 \end{pmatrix}, \quad \sigma^3 = \begin{pmatrix} 1 & 0 \\ 0 & -1 \end{pmatrix},$$

are the Pauli matrices. See [8] II.4.12 or [7] I.5.3. The first component of this map is the Hopf map

$$h(\mathbf{z}) = \big(|z_1|^2 - |z_2|^2, \ 2\,\mathrm{Re}\,(z_1 \bar{z}_2), \ 2\,\mathrm{Im}\,(z_1 \bar{z}_2) \big)$$

Its second component is $\frac{1}{2\|\mathbf{z}\|^2}\frac{d}{dt}h(\mathbf{z}+\imath t\mathbf{w})\big|_{t=0}$. The Kustaanheimo–Stiefel map is the restriction of the map above to the set

$$I = \left\{(\mathbf{z}, \mathbf{w}) \in (\mathbb{C}^2 \setminus \{0\}) \times \mathbb{C}^2 \mid \mathrm{Re}\,(z_1\bar{w}_1 + z_2\bar{w}_2) = 0\right\}$$

In the case of negative energy, one further restricts to

$$I_+ = \left\{(\mathbf{z}, \mathbf{w}) \in I \mid \|\mathbf{z}\|^2 + \|\mathbf{w}\|^2 = 1\right\}$$

We identify $\mathbb{C}^2$ with $\mathbb{H}$ by mapping $(z_1, z_2) \in \mathbb{C}^2$ to $z_1 + k z_2 \in \mathbb{H}$. Under this identification I corresponds to

$$\tilde{I} = \left\{(u, v) \in (\mathbb{H} \setminus \{0\}) \times \mathbb{H} \mid u^*v + v^*u = 0\right\}$$

and I_+ to

$$\tilde{I}_+ = \left\{(u, v) \in I \mid |u|^2 + |v|^2 = 1\right\}$$

As before, we identify phase space $\mathbb{R}^3 \times \mathbb{R}^3$ with $\mathbb{H}_{\mathrm{pure}} \times \mathbb{H}_{\mathrm{pure}}$. With these identifications, the Kustaanheimo Stiefel map is

$$\mathcal{KS} : \tilde{I} \longrightarrow \mathbb{H}_{\mathrm{pure}} \times \mathbb{H}_{\mathrm{pure}}$$
$$(u, v) \longmapsto (u\,\imath\,u^*, -v\,u^{-1})$$

Indeed, if $u = z_1 + kz_2$, $b = w_1 + kw_2$, then

$$u\,\imath\,u^* = (z_1 + kz_2)\,\imath\,(\bar{z}_1 - \bar{z}_2 k) = z_1\imath\bar{z}_1 + kz_2\imath\bar{z}_1 - z_1\,\imath\,\bar{z}_2 k - kz_2\,\imath\,\bar{z}_2 k$$
$$= |z_1|^2\imath + jz_2\bar{z}_1 + z_1\bar{z}_2 j - j|z_2|^2 k$$
$$= \left(|z_1|^2 - |z_2|^2\right)\imath + 2\,\mathrm{Re}(z_1\bar{z}_2)\,j + 2\,\mathrm{Im}(z_1\bar{z}_2)\,k$$

corresponds to $h(\mathbf{z})$, and $\frac{1}{2\|\mathbf{z}\|^2}\frac{d}{dt}h(\mathbf{z}+\imath t\mathbf{w})\big|_{t=0}$ corresponds to

$$\frac{1}{2\|u\|^2}\frac{d}{dt}(u + tv\imath)\,\imath\,(u + tv\imath)^* = \frac{1}{2\|u\|^2}(-vu^* + uv^*)$$
$$= -\frac{1}{\|u\|^2}vu^* = -vu^{-1}$$

For similar descriptions of the Kustaanheimo–Stiefel map, see [33] and [34]. Summarizing the discussion above, we see that the restriction of the Kustaanheimo–Stiefel map to $\tilde{I}_+$ is the same as the composition of φ_+ with the diffeomorphism

$$\tilde{I}_+ \longrightarrow \{(\alpha_1, \alpha_2) \in G_+ \mid \alpha_2 \neq -\alpha_1\}$$
$$(u, v) \longmapsto (u - v, u + v)$$

(Then $a = \frac{1}{2}(\alpha_1 + \alpha_2) = u$, $b = \frac{1}{2}(-\alpha_1 + \alpha_2) = -v$.). Lemma 3.2 gives the explicit solution to the harmonic oscillator in the Kustaanheimo–Stiefel regularization.

The method of stereographic projection

Let $S^3 = \{(x_0, x_1, x_2, x_3) \in \mathbb{R}^4 \mid x_0^2 + \cdots + x_3^2 = 1\}$ be the three dimensional sphere. We denote by $N = (1, 0, 0, 0)$ the "north pole". Stereographic projection is the map from $S^3 \setminus \{N\}$ to $\mathbb{R}^3$ that sends x to $s(x) = \frac{1}{1-x_0}(x_1, x_2, x_3)$. Its inverse is the map

$$p = (p_1, p_2, p_3) \longmapsto B(p) = \frac{1}{\|p\|^2 + 1}(\|p\|^2 - 1, 2p_1, 2p_2, 2p_3)$$

For $x \in S^3$ we denote by $T_x S^3 = \{y \in \mathbb{R}^4 \mid x \cdot y = 0\}$ the tangent space to S^3 in the point x. The derivative of the stereographic projection in x is the map

$$y \longmapsto \frac{1}{1 - x_0}(y_1, y_2, y_3) + \frac{y_0}{(1 - x_0)^2}(x_1, x_2, x_3)$$

If $p = s(x)$, the adjoint of the derivative is the map $\mathbb{R}^3 \longrightarrow T_x S^3$, $q \longmapsto \frac{1}{2}A(q, p)$ where

$$A(q, p) = 2 < p, q > (1, -p_1, -p_2, -p_3) + (\|p\|^2 + 1)(0, q_1, q_2, q_3)$$

The inverse of the adjoint is

$$y \longmapsto s_a(x, y) = (1 - x_0)(y_1, y_2, y_3) + y_0(x_1, x_2, x_3)$$

The stereographic projection is the starting point for the "Moser regularization" and for the Ligon–Schaaf map[5]. Let $T_1 S^3 = \{(x, y) \mid x \in S^3, y \in T_x S^3, \|y\| = 1\}$ be the unit tangent bundle of S^3. The Moser map is the bijection

$$T_1 S^3 \setminus T_N S^3 \longrightarrow \mathbb{R}^3 \times \mathbb{R}^3$$

$$(x, y) \longmapsto \left(\frac{1}{2}s_a(x, y), s(x)\right)$$

We identify $\mathbb{R}^4$ with $\mathbb{H}$ by mapping (x_0, x_1, x_2, x_3) to $x_0 \cdot 1 + x_1 \cdot \imath + x_2 \cdot j + x_3 \cdot k$. Then N corresponds to $1 \in \mathbb{H}$ and $T_1 S^3$ to

$$TS = \{(x, y) \in \mathbb{H} \times \mathbb{H} \mid |x| = |y| = 1, \operatorname{Re} xy^* = 0\}$$

In this language the Moser map is

$$\mathcal{M} : TS \setminus \{(1, y) \mid \operatorname{Re} y = 0\} \longrightarrow \mathbb{H}_{\text{pure}} \times \mathbb{H}_{\text{pure}}$$

$$(x, y) \longmapsto \left(\frac{1}{4}(y - y^*) + \frac{1}{4}[xy^* - x^*y], \frac{1+x}{1-x}\right)$$

---

[5] Observe that on the energy hypersurface $K_+^{-1}(0) = \left\{(q, p \mid \|q\| = \frac{1}{\|p\|^2 + 1}\right\}$

$A(q, p) = \left(2 < p, q >; (\|p\|^2 + 1)q - 2 < p, q > p\right) = \left(2 < p, q >; \frac{q}{\|q\|} - 2 < p, q > p\right)$

and $B(p) = \left(\frac{\|p\|^2 - 1}{\|p\| + 1}; \frac{2p}{\|p\| + 1}\right) = \left(\|q\|(\|p\|^2 - 1); 2\|q\|p\right) = \left(2\|q\|\|p\|^2 - 1; 2\|q\|p\right)$.

Recall that A and B are the building blocks of the Ligon–Schaaf map [24], see [8] II.3.4.

since

$$2 \left(1 - \mathrm{Re}(x)\right) \mathcal{P}(y) + 2 \mathrm{Re}(y)\mathcal{P}(x) - \left(1 - \frac{1}{2}(x + x^*)\right)(y - y^*)$$
$$+ \frac{1}{2}(x - x^*)(y + y^*)$$
$$= (y - y^*) + [xy^* - x^*y]$$

and $\frac{1}{1-\mathrm{Re}(x)}\mathcal{P}(x) = \frac{x-x^*}{2-x-x^*} = \frac{x^2-1}{2x-x^2-1} = \frac{1+x}{1-x}$.

The group G_+ acts on TS by $\left((\alpha_1, \alpha_2), (x, y)\right) \longmapsto (\alpha_1 x \alpha_2^*, \alpha_1 y \alpha_2^*)$. We show that the Moser map is equivariant with respect to the actions of G_+.

Lemma 3.3 *Let* $(\alpha_1, \alpha_2) \in G_+$. *Then for all* $(x, y) \in TS$

$$\mathcal{M}(\alpha_1 x \alpha_2^*, \alpha_1 y \alpha_2^*) = (\alpha_1, \alpha_2) \bullet \mathcal{M}(x, y)$$

Proof. The first components of the map $(q, p) \longmapsto (\alpha_1, \alpha_2) \bullet (q, p)$ is the adjoint of the inverse of the derivative of the second component, and the second component of the map $(x, y) \longmapsto (\alpha_1 x \alpha_2^*, \alpha_1 y \alpha_2^*)$ is the adjoint of the inverse of the derivative of the first. Also, the first component of $\mathcal{M}$ is half the inverse of the derivative of the stereographic projection. Therefore it suffices to show that, for all $(x, y) \in TS$, the second component of $\mathcal{M}(\alpha_1 x \alpha_2^*, \alpha_1 y \alpha_2^*)$ is equal to the second component of $(\alpha_1, \alpha_2) \bullet \mathcal{M}(x, y)$.

Set $a = \frac{1}{2}(\alpha_1 + \alpha_2)$, $b = \frac{1}{2}(-\alpha_1 + \alpha_2)$. Then the second component of $(\alpha_1, \alpha_2) \bullet \mathcal{M}(x, y)$ is

$$\left(a\frac{1+x}{1-x} + b\right)\left(b\frac{1+x}{1-x} + a\right)^{-1}$$
$$= \left(a(1+x) + b(1-x)\right)\left(b(1+x) + a(1-x)\right)^{-1}$$
$$= (\alpha_2 + \alpha_1 x)(\alpha_2 - \alpha_1 x)^{-1}$$
$$= \frac{1 + \alpha_1 x \alpha_2^*}{1 - \alpha_1 x \alpha_2^*}$$

which is the second component of $\mathcal{M}(\alpha_1 x \alpha_2^*, \alpha_1 y \alpha_2^*)$. $\qquad\square$

Since $\mathcal{M}(-1, \iota) = (\iota, 0)$, it follows that

$$\varphi_+(\alpha_1, \alpha_2) = \mathcal{M}\left(-\alpha_1\alpha_2^*, \alpha_1 \iota \alpha_2^*\right) \qquad \text{for all } (\alpha_1, \alpha_2) \in G_+$$

If, as before, one interprets φ_+ as the Kustaanheimo–Stiefel map, then this formula is the relation between the Kustaanheimo–Stiefel regularization and the Moser regularization described in [20].

4 Three Dimensions, Positive Energy

The quaternions can be viewed as the subalgebra of the algebra $\mathrm{Mat}(2 \times 2, \mathbb{C})$ of complex 2×2 matrices. In fact, $\mathrm{Mat}(2 \times 2, \mathbb{C}) \cong \mathbb{H} \otimes_{\mathbb{R}} \mathbb{C} \cong \mathbb{H} \times \mathbb{H}$ where the multiplication on $\mathbb{H} \times \mathbb{H}$ is given by

$$(\alpha, \beta) * (\alpha', \beta') = (\alpha\alpha' - \beta\beta', \ \alpha\beta' + \beta\alpha')$$

In this description, the group $SL(2, \mathbb{C})$ corresponds to

$$G_- = \{(\alpha, \beta) \in \mathbb{H} \times \mathbb{H} \mid |\alpha|^2 - |\beta|^2 = 1, \ \alpha\beta^* + \beta\alpha^* = 0\}$$

For $(\alpha, \beta) \in G_-$ and $(q, p) \in \mathbb{H}_{\text{pure}} \times \mathbb{H}_{\text{pure}}$ we set

$$(\alpha, \beta) \circ (q, p) = \left((\beta p + \alpha) q (\beta p + \alpha)^*, \ (\alpha p - \beta)(\beta p + \alpha)^{-1}\right)$$

Theorem 4.1 *Let $(\alpha, \beta) \in G_-$. Then for all $(q, p) \in \mathbb{H}_{\text{pure}} \times \mathbb{H}_{\text{pure}}$.*

(i) $(\alpha, \beta) \circ (q, p) \in \mathbb{H}_{\text{pure}} \times \mathbb{H}_{\text{pure}}$.
(ii) $\left((\alpha', \beta') * (\alpha, \beta)\right) \circ (q, p) = (\alpha', \beta') \circ \left((\alpha, \beta) \circ (q, p)\right)$ *for all*
   $(\alpha', \beta') \in G_-$.
(iii) $K_-\left((\alpha, \beta) \circ (q, p)\right) = K_-(q, p)$.

Furthermore the map $(q, p) \mapsto (\alpha, \beta) \circ (q, p)$ is symplectic with respect to the symplectic form $\mathrm{Re}\, dq^* \wedge dp$.

The proof is similar to the case of negative energy.

We consider the map

$$\varphi_- : \{(\alpha, \beta) \in G_- \mid \beta \neq 0\} \longrightarrow K_-^{-1}(0)$$

$$(\alpha, \beta) \longmapsto (\beta \iota \beta^*, \alpha\beta^{-1}) = \lim_{t \to \infty} (\alpha, \beta) \circ \left(\frac{\iota}{t^2 - 1}, t\iota\right)$$

φ_- maps the flow $t \mapsto (\alpha, \beta) * (\cosh t, \iota \sinh t)$ on G_- to the regularized Kepler flow.

In the case of positive energy, one restricts the Kustaanheimo–Stiefel map to $I_- = \{(\mathbf{z}, \mathbf{w}) \in I \mid \|\mathbf{z}\|^2 - \|\mathbf{w}\|^2 = 1\}$. Under the identification of $\mathbb{C}^2 \times \mathbb{C}^2$ with $\mathbb{H} \times \mathbb{H}$ this set corresponds to $\tilde{I}_- = \{(u, v) \in I \mid |u|^2 - |v|^2 = -1\}$. $\tilde{I}_-$ can parametrized by G_- through the map $(\alpha, \beta) \longmapsto (-\beta, \alpha)$. The composition of this isomorphism with the Kustaanheimo Stiefel map $\mathcal{K}S$ coincides with φ_-.

Let $H_+^3 = \{(x_0, x_1, x_2, x_3) \in \mathbb{R}^4 \mid x_0^2 - x_1^2 - x_2^2 - x_3^2 = 1, \ x_0 > 0\}$ be the upper sheet of the three dimensional hyperboloid. Further denote by

$$T_1 H_+^3 = \{(x, y) \mid x \in H_+^3, \ y \in \mathbb{R}^4$$

$$x_0 y_0 - x_1 y_1 - x_2 y_2 - x_3 y_3 = 0, \ y_0^2 - y_1^2 - y_2^2 - y_3^2 = -1\}$$

the "unit tangent bundle" of H_+^3. The Belbruno map is

$$\{(x, y) \in T_1 H_+^3 \mid x \neq N\} \longrightarrow \mathbb{R}^3 \times \mathbb{R}^3$$

$$(x, y) \longmapsto \left(\frac{1 - x_0}{2}(y_1, y_2, y_3) + \frac{y_0}{2}(x_1, x_2, x_3), \; \frac{1}{1 - x_0}(x_1, x_2, x_3) \right)$$

The second component is the stereographic projection from the three dimensional hyperboloid.

Here, we identify $\mathbb{R}^4$ with $\mathbb{R} \times \mathbb{H}_{\text{pure}}$ by mapping (x_0, x_1, x_2, x_3) to $(x_0, x_1 \cdot \imath + x_2 \cdot j + x_3 \cdot k)$. Under this identification[6] the unit tangent bundle of the hyperboloid corresponds to the set TH of pairs $\big((x_0, \mathbf{x}), (y_0, \mathbf{y})\big) \in (\mathbb{R} \times \mathbb{H}_{\text{pure}})^2$ for which

$$(x_0, \mathbf{x}) * (x_0, -\mathbf{x}) = (1, 0), \quad (y_0, \mathbf{y}) * (y_0, -\mathbf{y}) = (-1, 0)$$

$$(x_0, \mathbf{x}) * (y_0, -\mathbf{y}) - (y_0, \mathbf{y}) * (x_0, -\mathbf{x}) = (0, 0)$$

In this language the Belbruno map is

$$\mathcal{B}: \{\big((x_0, \mathbf{x}), (y_0, \mathbf{y})\big) \in TH \mid (x_0, \mathbf{x}) \neq (1, 0)\} \longrightarrow \mathbb{H}_{\text{pure}} \times \mathbb{H}_{\text{pure}}$$

$$\big((x_0, \mathbf{x}), (y_0, \mathbf{y})\big) \longmapsto \left(\frac{1 - x_0}{2}\mathbf{y} + \frac{y_0}{2}\mathbf{x}, \; \frac{1}{1 - x_0}\mathbf{x} \right)$$

The group G_- acts on $\mathbb{R} \times \mathbb{H}_{\text{pure}}$ by

$$\big((\alpha, \beta), (x_0, \mathbf{x})\big) \mapsto (\alpha, \beta) * (x_0, \mathbf{x}) * (\alpha^*, -\beta^*)$$

Lemma 4.2 *Let* $(\alpha, \beta) \in G_-$. *Then for all* $\big((x_0, \mathbf{x}), (y_0, \mathbf{y})\big) \in TH$

$$\mathcal{B}\big((\alpha, \beta) * (x_0, \mathbf{x}) * (\alpha^*, -\beta^*), \; (\alpha, \beta) * (y_0, \mathbf{y}) * (\alpha^*, -\beta^*)\big)$$

$$= (\alpha, \beta) \circ \mathcal{B}\big((x_0, \mathbf{x}), (y_0, \mathbf{y})\big)$$

The proof is again similar to the case of negative energy. Also, for $(\alpha, \beta) \in G_-$

$$\varphi_-(\alpha, \beta) = -\mathcal{B}\big((\alpha, \beta) * (1, 0) * (\alpha^*, -\beta^*), \; (\alpha, \beta) * (0, \imath) * (\alpha^*, -\beta^*)\big)$$

5 Three Dimensions, Zero Energy

In this situation the symmetry group is

$$G_0 = \left\{ \begin{pmatrix} \alpha \\ c \end{pmatrix} \; \middle| \; \alpha \in \mathbb{H}_{\text{unit}}, \; c \in \mathbb{H}, \; \alpha^* c + c^* \alpha = 0 \right\}$$

---

[6] The isomorphism between $\mathbb{H} \times \mathbb{H} \cong \mathbb{H} \otimes_{\mathbb{R}} \mathbb{C}$ and $\text{Mat}(2 \times 2, \mathbb{C})$, mentioned above, maps $\mathbb{R} \times \mathbb{H}_{\text{pure}}$ to the space of Hermitian matrices.

with the multiplication

$$\begin{pmatrix} \alpha' \\ c' \end{pmatrix} * \begin{pmatrix} \alpha \\ c \end{pmatrix} = \begin{pmatrix} \alpha'\alpha \\ \alpha'c + c'\alpha \end{pmatrix}$$

For $\begin{pmatrix} \alpha \\ c \end{pmatrix} \in G_0$ and $(q, p) \in \mathbb{H}_{\text{pure}} \times \mathbb{H}_{\text{pure}}$ set (whenever defined)

$$\begin{pmatrix} \alpha \\ c \end{pmatrix} \bullet (q, p) = \left((cp + \alpha)\, q\, (cp + \alpha)^* \,,\; \alpha p (cp + \alpha)^{-1} \right)$$

Theorem 5.1 *Let* $\begin{pmatrix} \alpha \\ c \end{pmatrix} \in G_0$. *Then for all* $(q, p) \in \mathbb{H}_{\text{pure}} \times \mathbb{H}_{\text{pure}}$.

(i) $\begin{pmatrix} \alpha \\ c \end{pmatrix} \bullet (q, p) \in \mathbb{H}_{\text{pure}} \times \mathbb{H}_{\text{pure}}$.

(ii) $\left[\begin{pmatrix} \alpha' \\ c' \end{pmatrix} * \begin{pmatrix} \alpha \\ c \end{pmatrix} \right] \bullet (q, p) = \begin{pmatrix} \alpha' \\ c' \end{pmatrix} \bullet \left[\begin{pmatrix} \alpha \\ c \end{pmatrix} \bullet (q, p) \right]$ *for all* $\begin{pmatrix} \alpha' \\ c' \end{pmatrix} \in G_0$.

(iii) $K_0 \left(\begin{pmatrix} \alpha \\ c \end{pmatrix} \bullet (q, p) \right) = K_0(q, p)$.

Furthermore the map $(q, p) \mapsto \begin{pmatrix} \alpha \\ c \end{pmatrix} \bullet (q, p)$ *is symplectic with respect to the symplectic form* $\operatorname{Re} dq^* \wedge dp$.

The proof is similar to the case of negative energy.

Again, we consider the map

$$\varphi_0 : \{(\alpha, c) \in G_0 \mid c \neq 0\} \longrightarrow K_0^{-1}(0)$$

$$\begin{pmatrix} \alpha \\ c \end{pmatrix} \longmapsto (c \iota c^*, \alpha c^{-1}) = \lim_{t \to \infty} \begin{pmatrix} \alpha \\ c \end{pmatrix} \bullet \left(\frac{\iota}{t}, t\iota \right)$$

φ_0 maps the flow $t \mapsto \begin{pmatrix} \alpha \\ c \end{pmatrix} * \begin{pmatrix} 1 \\ -\iota t \end{pmatrix} = \begin{pmatrix} \alpha \\ c - t\alpha\, \iota \end{pmatrix}$ on G_0 to the regularized Kepler flow.

Also observe that φ_0 is the composition of the maps

$$G_0 \longrightarrow \mathbb{H}_{pure} \times \mathbb{H}_{pure} \,;\quad \begin{pmatrix} \alpha \\ c \end{pmatrix} \longmapsto (c\alpha^{-1}, \alpha\iota\,\alpha^*) \quad \text{and}$$

$$(\mathbb{H}_{pure} \setminus \{0\}) \times \mathbb{H}_{pure} \longrightarrow \mathbb{H}_{pure} \times \mathbb{H}_{pure} \,;\quad (x, y) \longmapsto \left(-xyx, \frac{1}{x} \right)$$

This gives the regularization described in [4], Theorem 3, or [28].

References

[1] V.I.Arnol'd: Mathematical Methods of Classical Mechanics. Springer Verlag 1978
[2] V.I.Arnol'd: Huygens & Barrrow, Newton & Hooke. Birkhäuser 1990

[3] H.Bacry, H.Ruegg, J.Souriau: Dynamical groups and spherical potentials in classical mechanics. Commun. Math. Phys. **3**, 323–333 (1966)

[4] J.Belbruno: Two Body motion under the inverse square central force and equivalent geodesic flows. Celestial Mechanics **15**, 467–476 (1977)

[5] J.Brackenridge: The Key to Newton's Dynamics: The Kepler Problem and the Principia. University of California Press 1995

[6] P.Colwell: Solving Kepler's Equation. Willmann–Bell 1993

[7] B.Cordani: The Kepler Problem. Birkhäuser Verlag 2003

[8] R.Cushman, L.Bates: Global Aspects of Classical Integrable Systems. Birkhäuser Verlag 1997

[9] D.Derbes: Reinventing the wheel: Hodographic solutions to the Kepler problems. Am. J. Phys **69**, 481–489 (2001)

[10] V.Fock: Zur Theorie des Wasserstoffatoms. Z. Phys **98**, 145–154 (1935)

[11] H.Geiges: The Geometry of Celestial Mechanics. Cambridge University Press 2016

[12] P.Girard: Quaternions, Clifford Algebras and Relativistic Physics. Birkhäuser Verlag 2007

[13] H.Goldstein: Prehistory of the "Runge–Lenz" vector. Am. J. Physics **43**, 737–738 (1975)

[14] H.Goldstein: More on the prehistory of the "Runge–Lenz" vector. Am. J. Physics **44**, 1123–1124 (1976)

[15] D,.Goodstein, J.Goodstein: Feynman's Lost Lecture. Random House 1996

[16] V.Guillemin, Shl.Sternberg: Variations on a Theme of Kepler. AMS Colloquium Publications **42**, 1990

[17] G.Györgyi: Kepler's equation, Fock variables, Bacry's generators and Dirac brackets. Nuov. Cim. **53 A**, 717–736 (1968)

[18] G.Heckman, T.de Laat: On the regularization of the Kepler problem. J. of Symplectic Geometry **10**, 463–473 (2012)

[19] M.Koecher, R.Remmert: Hamilton's Quaternions. In: H.-D. Ebbinghaus et al.: Numbers. Springer Verlag 1991, pp. 189–220

[20] M.Kummer: On the regularization of the Kepler problem. Commun. Math. Phys. **84**, 133–152 (1982)

[21] P.Kustaanheimo, E.Stiefel: Perturbation theory of Kepler motion based on spinor regularization. J. Reine Angew. Math. **218**, 609–636 (1965)

[22] W.Lenz: Über den Bewegungsverlauf und die Quantenzustände der gestörten Keplerbewegung. Zeitschrift fr Physik A 24 **24**, 197–207 (1924)

[23] T.Levi–Civita: Fragen der Klassischen und Relativistischen Mechanik. Springer Verlag 1924

[24] T.Ligon, M. Schaaf: On the global symmetry of the classical Kepler Problem. Reports on Mathematical Physics **9**, 281–300 (1976)

[25] Ch.Marle: A property of conformally Hamiltonian vector fields; application to the Kepler problem. Journal of Geometric Mechanics **4**, 181–206 (2012)

[26] J.Milnor: On the Geometry of the Kepler problem. American Mathematical Monthly **90**, 353–365 (1983)

[27] J.Moser: Regularization of Kepler's problem and the averaging method on a manifold. Comm. Pure Appl. Math. **23**, 609–636 (1970)

[28] Y.Osipov: The Kepler problem and geodesic flows in spaces of constant curvature. Celestial Mechanics **16**, 191–208 (1977)

[29] W.Pauli: Über das Wasserstoffspektrum vom Standpunkt der neuen Quanten-
      mechanik. Zeitschrift fr Physik **36**, 336–363 (1926)
[30] B.Pourciau: Reading the Master: Newton and the Birth of Celestial Mechanics.
      American Mathematical Monthly **104**, 1–19 (1997)
[31] H.Rogers: Symmetry transformations of the classical Kepler problem. J.
      Math.Phys. **14**, 1125–1129 (1973)
[32] C.Runge: Vektoranalysis. Hirzel (Leipzig) 1919
[33] M.Vivarelli: The KS–transformation in hypercomplex form. Celestial Mechanics
      29, 45–50 (1983)
[34] J.Waldvogel: Quaternions and the perturbed Kepler problem. Celestial Mech.
      Dynam. Astronom. **95**, 201–212 (2006)
[35] G.N.Watson: A Treatise on the Theory of Bessel Functions. Cambridge University
      Press 1922
[36] H.Weyl: The Theory of Groups and Quantum Mechanics. Dover 1931

Horst Knörrer
Department of Mathematics, ETH Zürich, CH 8092 Zürich, Switzerland
E-mail address: knoerrer@math.ethz.ch

3

A Note on the Commutator of Hamiltonian Vector Fields

Henryk Żołądek

Abstract. We present two proofs of the Jacobi identity for the Poisson bracket on a symplectic manifold.

1 Introduction

A differential $2-$form ω on a manifold M is **symplectic** if:

1. is non-degenerate (if $\omega(u, \cdot) \equiv 0$ on $T_x M$ then the vector $u \in T_x M$ is zero);
2. is closed ($d\omega = 0$).

It is standard that property 1 allows to associate with any smooth function H on M the Hamiltonian vector field X_H and to define the **Poisson bracket** $\{F, G\}$ of two functions F and G on M. We have

$$\{F, G\} = X_G F, \tag{1}$$

where the **Hamiltonian vector field** X_H is defined by

$$i_{X_H}\omega = -dH \tag{2}$$

and $(i_Y\omega)(v) = \omega(Y(x), v)$ for a vector field Y and $v \in T_x M$.

Also it is known that the commutator of two Hamiltonian vector fields is the Hamiltonian vector field generated by the Poisson bracket of these functions,

$$[X_F, X_G] = X_{\{G,F\}}, \tag{3}$$

and that this is equivalent to the **Jacobi identity** of the Poisson bracket.

2000 *Mathematics Subject Classification.* Primary 53D05; Secondary 37J05

Key words and phrases. Poisson bracket, Jacobi identity, Cartan formula

Supported by Polish NCN OPUS Grant No 2017/25/B/BST1/00931

Indeed,

$$X_{\{G,F\}}H = \{H, \{G, F\}\} \quad \text{(Jac. id.)}$$
$$= \{\{H, G\}, F\} + \{G, \{H, F\}\}$$
$$= X_F X_G H - X_G X_F H$$
$$= [X_F, X_G]H.$$

But it is less widespread that the Jacobi identity is directly related with property 2 of the symplectic form.

In V. Arnold's book [2], which is widely read and frequently cited, the proof of the Jacobi identity (in Section 40) uses Darboux theorem. The latter states that near any point in $x_0 \in M$ the form ω can be reduced to the following 'constant' form

$$\Omega = \sum \mathrm{d}p_i \wedge \mathrm{d}q_i, \tag{4}$$

i.e., $h^*\omega = \Omega$ for a germ $h : (\mathbb{R}^{2n}, 0) \longmapsto (M, x_0)$ of local diffeomorphism. On the other hand, the geometric proof of the Darboux theorem in Section 43 of [2] uses the commutativity of the phase flows generated by vector fields X_{p_i} and X_{q_i}, where the functions p_i and q_i are defined inductively. Moreover, nowhere in his book is the closeness of ω used in this context.

Note also that the life of Poisson geometers is much easier. Recall that a **Poisson structure** on a manifold is a Poisson bracket on the space of functions on it which is antisymmetric and obeys the Jacobi identity. There is no non-degeneracy condition.

A correct and complete approach to the latter subject can be found in the book [1] of R. Abraham and J. Marsden. Firstly, the Jacobi identity follows from the property $\mathrm{d}\omega = 0$ via the known Cartan formula for evaluation of $\mathrm{d}\omega$ on three vector fields. Next, there exists a direct (although rather formal) proof of the Darboux theorem, which does not involve Hamiltonian vector fields; so, Arnold's argument can work.

The purpose of this note (aimed mainly at young researchers) is to present (following mainly [1]) the two proofs of the Jacobi identity for the Poisson bracket in compact form. In the next section we use the Cartan formula and in Section 3 we discuss the Darboux theorem.

2 The Cartan Formula

The **Cartan formula** in the case of $2-$forms follows:

$$\mathrm{d}\omega\,(X, Y, Z) = X\omega(Y, Z) - Y\omega(X, Z) + Z\omega(X, Y)$$
$$- \omega\,([X, Y], Z) + \omega\,([X, Z], Y) - \omega\,([Y, Z], X), \tag{5}$$

where X, Y, X are vector fields.

We have $\{F, G\} = dF(X_G) = -\left(i_{X_F}\omega\right)(X_G) = -\omega(X_F, X_G)$ and hence

$$\{H, \{F, G\}\} = X_H \omega(X_F, X_G). \tag{6}$$

Moreover,

$$\omega([X_H, X_F], X_G) =$$
$$-\left(i_{X_G}\omega\right)([X_H, X_F]) =$$
$$dG([X_H, X_F]) =$$
$$[X_H, X_F]G =$$
$$X_H(X_F G) - X_F(X_H G) =$$
$$X_H \omega(X_F, X_G) - X_F \omega(X_H, X_G). \tag{7}$$

We use these formulas to prove the **Jacobi identity**. We want to show that the quantity

$$\Phi = \{H, \{F, G\}\} - \{\{H, F\}, G\} - \{F, \{H, G\}\} \tag{8}$$

vanishes, i.e., that the operation $K \longmapsto \{H, K\}$ is a derivation in a (noncommutative) algebra. By Eq. (6) we have

$$\Phi = X_H \omega(X_F, X_G) + X_G \omega(X_H, X_F) - X_F \omega(X_H, X_G)$$
$$= X_H \omega(X_F, X_G) - X_G \omega(X_F, X_H) + X_F \omega(X_G, X_H);$$

with $X = X_H$, $Y = X_F$ and $Z = X_G$ it is the first part of the right-hand side of Eq. (5). Using $d\omega = 0$ we obtain

$$\Phi = \omega([X_H, X_F], X_G) - \omega([X_H, X_G], X_F) + \omega([X_F, X_G], X_H). \tag{9}$$

Now we use Eqs. (7) and we find that the right-hand side of Eq. (9) equals 2Φ. Therefore $\Phi = 0$.

The same transformations can be found in [1, Prop. 3.3.17] and in [3, Prop. 1.1.7].

The general Cartan formula for a $k-$ form α and vector fields X_j, $j = 1, \ldots, k+1$, takes the form

$$d\alpha(X_1, \ldots, X_{k+1}) = \sum (-1)^{j+1} X_j \alpha(X_1, \ldots, X_{k+1})$$
$$+ \sum_{i<j} (-1)^{i+j} \alpha\left([X_i, X_j], X_1, \ldots, X_{k+1}\right)$$

and is proved by induction, with use of the **homotopy formula** (sometimes called also the Cartan formula)

$$\mathcal{L}_X \alpha = i_X d\alpha + d i_X \alpha, \tag{10}$$

where $\mathcal{L}_X \alpha = \frac{d}{dt}|_{t=0} \left(g_X^t\right)^* \alpha$ is the Lie derivative and $\left\{g_X^t\right\}$ is the phase flow generated by a vector field X; we refer to Arnold's book [2] for its geometric proof. Note also that

$$X\alpha\,(X_1,\ldots,X_k) = \mathcal{L}_X\left\{\alpha\,(X_1,\ldots,X_k)\right\}$$
$$= (\mathcal{L}_X\alpha)\,(X_1,\ldots,X_k) + \sum \alpha\left(X_1,\ldots,\mathcal{L}_X X_j,\ldots,X_k\right).$$

For $k=0$ it is obvious. For $k=1$ we have

$$\mathrm{d}\alpha\,(X,Y) = (i_X\mathrm{d}\alpha)\,(Y) = (\mathcal{L}_X\alpha)\,(Y) - \mathrm{d}i_X\alpha(Y)$$
$$= [\mathcal{L}_X\alpha(Y) - \alpha\,(\mathcal{L}_X Y)] - Y i_X\alpha$$
$$= X\alpha(Y) - \alpha\,([X,Y]) - Y\alpha(X).$$

Then for $k=2$ we get

$$\mathrm{d}\omega\,(X,Y,Z) = i_X\mathrm{d}\omega\,(Y,Z) = (\mathcal{L}_X\omega)\,(Y,Z) - \mathrm{d}i_X\omega\,(Y,Z)$$
$$= \{X\omega(Y,Z) - \omega\,([X,Y],Z) - \omega\,(Y,[X,Z])\}$$
$$- \{Y i_X\omega\,(Z) - Z i_X\omega\,(Y) - i_X\omega\,([Y,Z])\},$$

where $Y i_X\omega(Z) = Y\omega(X,Z)$, $Z i_X\omega(Y) = Z\omega(X,Y)$ and $i_X\omega\,([Y,Z]) = \omega\,(X,[Y,Z])$.

3 The Darboux Theorem

We begin with the proof of this theorem; it is a homotopy proof (used e.g. in [5] and sometimes called the Moser's deformation method).

We can assume that we have a nondegenerate and closed 2−form ω in a neighborhood of 0 in $\mathbb{R}^{2n}$. Let ω_0 be a constant 2−form such that $\omega_0 = \omega(0)$ and let $\omega_1 = \omega - \omega_0$. Consider the 1−parameter family of closed differential forms

$$\omega_t = \omega_0 + t\omega_1, \quad 0 \le t \le 1.$$

We look for a family $\{f_t\}$ of diffeomorphisms such that

$$f_t^*\omega_0 = \omega_t \tag{11}$$

and $f_t(0) = 0$. Then $h = f_1^{-1}$ transforms ω to the constant form ω_0 which can be reduced to Ω by a construction of the corresponding symplectic basis.

Let $V_t(x) = \frac{\mathrm{d}}{\mathrm{d}t}f_t(x)$ be the corresponding (non-autonomous in general) vector field and let α be a 1−form such that $\omega_1 = \mathrm{d}\alpha$ (the Poincaré lemma) and $\alpha(0) = 0$.

Differentiating the both sides of Eq. (11) with respect to t we find $\mathcal{L}_{V_t}\omega_0 = \omega_1$ and using the homotopy formula (with $\mathrm{d}\omega_0 = 0$) we get

$$\mathrm{d}i_{V_t}\omega_0 = \mathrm{d}\alpha.$$

By the non-degeneracy of ω_0 the equation $i_{V_t}\omega_0 = \alpha$ is solvable with respect to the vector field V_t such that $V_t(0) = 0$. In fact, is an autonomous vector field, $V_t = V$, and generates the flow $f_t = g_V^t$ such that $g_V^t(0) = 0$.

Practically the same arguments can be found in [1, Th. 3.2.2] and in [4, Th. 1.2]

Before passing from the Darboux theorem to the Jacobi formula we will check how the Hamiltonian vector field varies when the symplectic form changes under diffeomorphisms.

We use the notation

$$\omega_x(u, v) = \langle \omega(x); u, v \rangle, \quad u, v \in T_x M.$$

So, the definition of the Hamiltonian vector field takes the form

$$\langle \omega(x); X_H(x), v \rangle = -\langle dH(x), v \rangle, \quad v \in T_x M.$$

Let $g : N \longmapsto M$ be a diffeomorphism. For $x \in N$ and $y = g(x)$ we have the derivative map $Dg(x) : T_x N \longmapsto T_y M$. Thus for $u, v \in T_x N$ we have

$$\langle g^* \omega(x); u, v \rangle = \langle \omega(g(x)); Dg(x)u, Dg(x)v \rangle.$$

Next, for $w \in T_x N$ we have

$$-\langle dg^* H(x), w \rangle$$
$$= -\langle dH(g(x)), Dg(x)w \rangle$$
$$= \langle \omega(y); X_H^{(\omega)}(y), Dg(x)w \rangle$$
$$= \langle \omega(y); Dg(x) \cdot (Dg)^{-1}(x) X_H^{(\omega)}(y), Dg(x)w \rangle$$
$$= \langle g^* \omega(x); (Ad_g)_* X_H^{(\omega)}, w \rangle,$$

where by $X_H^{(\omega)}$ we denote the Hamiltonian vector field defined by means of the form ω. We get the identity

$$X_{g^* H}^{(g^* \omega)} = (Ad_g)_* X_H^{(\omega)} = (Dg)^{-1} X_H^{(\omega)} \circ g,$$

which means that with the change of the symplectic structure the corresponding Hamiltonian vector fields change naturally.

In particular, we have

$$\{g^* F, g^* H\}^{(g^* \omega)} = X_{g^* H}^{(g^* \omega)}(g^* F) = g^* X_H^{(\omega)}(F) = g^* \{F, G\}^{(\omega)}.$$

Now we can recall Arnold's proof [2] of the Jacobi identity with a constant symplectic form (like Ω). Consider the quantity (8). It is a linear combination of second order derivatives of the three functions (with combinations of products of first order derivatives of the other functions as coefficients). Let us look for the part containing the second order derivatives of F :

$$\{H, \{F, G\}\} - \{\{H, F\}, G\} = \{\{F, H\}, G\} - \{\{F, G\}, H\}$$
$$= (X_G X_H - X_H X_G) F = [X_G, X_H] F.$$

But the differential operator $[X_G, X_H]$ is of first order and hence the latter expression vanishes.

References

[1] R. Abraham and J. Marsden, "Foundations of Mechanics", The Benjaming/Cummings Publ. Comp., London, 1978.

[2] V. I. Arnold, "Mathematical Methods of Classical Mechanics", Springer–Verlag, New York, 1989 [Russian: Nauka, Moskva, 1974].

[3] J.-P. Dufour and N. T. Zhung, "Poisson Structures and Their Normal Forms", Birkhaüser, Basel, 2000.

[4] H. Hofer and E. Zehnder, "Symplectic Invariants and Hamiltonian Dynamics", Birkhaüser, Basel, 2010.

[5] J. Moser, *On the volume elements on a manifold*, Trans. Amer. Math. Soc. **120** (1965), 286–294.

Henryk Żoła̧dek
Institute of Mathematics,
University of Warsaw,
ul. Banacha 2, 02-097 Warsaw, Poland
E-mail address: zoladek@mimuw.edu.pl

4

Nodal Curves and a Class of Solutions of the Lax Equation for Shock Clustering and Burgers Turbulence

Luen-Chau Li

To Emma Previato on the occasion of her 65th birthday

Abstract. In this paper, we present an expository account of the work done in the last few years in understanding a matrix Lax equation which arises in the study of scalar hyperbolic conservation laws with spectrally negative pure-jump Markov initial data. We begin with its extension to general $N \times N$ matrices, which is Liouville integrable on generic coadjoint orbits of a matrix Lie group. In the probabilistically interesting case in which the Lax operator is the generator of a pure-jump Markov process, the spectral curve is generically a fully reducible nodal curve. In this case, the equation is not Liouville integrable, but we can show that the flow is still conjugate to a straight line motion, and the equation is exactly solvable. En route, we establish a dictionary between an open, dense set of lower triangular generator matrices and algebro-geometric data which plays an important role in our analysis.

1 Introduction

From many points of view, it is useful to study evolution equations with random initial conditions. A case in point is the well-known (inviscid) Burgers equation:

$$u_t + \left(\frac{1}{2} u^2 \right)_x = 0, \tag{1}$$

where such studies are often motivated by various areas of application, including cosmology, fluid mechanics, statistical physics, and statistics (see, for example, [B, SZ, GMS, G, W] and the references therein.) Indeed, the study of Burgers equation with random initial conditions or with an additional random forcing term on the right hand side is known as Burgers turbulence [FB, Woy]. Among the random initial conditions that have been considered, we cite Brownian motion [SAF, S, Ber], white noise [B, AE, G, FM], Levy process (with no positive jumps) on a half-line [CD, Ber]. So far, the methods used have been quite ad hoc, with different techniques applied to different initial conditions. Amongst such works, there are two in particular [G, Ber] which seem to indicate that for the specific initial stochastic process in x under study, there is some sort of integrability in the probability law of the solution $u(x, t)$ for $t > 0$. (By integrability here, we simply mean that we can write down explicit formulas.) So it is natural to ask if such miracles can be extended to the unique entropy solution of the general scalar hyperbolic conservation law:

$$u_t + f(u)_x = 0, \quad u(x, 0) = u_0(x), \quad x \in \mathbb{R}, \quad t > 0, \tag{2}$$

corresponding to a strictly convex, C^1 flux function f [MS] . In contrast to other authors in their work on the Burgers case, Menon and Srinivasan in [MS] were trying to find a broad class of stochastic processes whose structure in x is preserved by the unique entropy solution of (2). In this context, an important contribution in [MS] is the following closure theorem: if $u_0(x)$ is a spectrally negative strong Markov process in x, then $u(x, t)$ is Markov and spectrally negative for each $t > 0$. Moreover, under the additional assumption that $u_0(x)$ is Feller and that the evolution in (2) preserves the Feller property, the formal arguments in [MS] shows that the evolution of the generator $\mathcal{A}$ of the process is described by means of a zero curvature equation:

$$\partial_t \mathcal{A} - \partial_x \mathcal{B} = [\mathcal{A}, \mathcal{B}]. \tag{3}$$

In the special case when $u_0(x)$ is stationary, we have $\partial_x \mathcal{B} = 0$ and hence the zero curvature equation reduces to the Lax equation

$$\partial_t \mathcal{A} = [\mathcal{A}, \mathcal{B}], \tag{4}$$

where $\mathcal{A}$ and $\mathcal{B}$ are operators defined by

$$\mathcal{A}\varphi(u) = b(u, t)\varphi'(u) + \int_{-\infty}^{u} n(u, v, t)(\varphi(v) - \varphi(u)) \, dv,$$

$$\mathcal{B}\varphi(u) = -f'(u)b(u, t)\varphi'(u) - \int_{-\infty}^{u} [f]_{u,v}(\varphi(v) - \varphi(u)) \, n(u, v, t) \, dv \tag{5}$$

for $\varphi \in C_c^{\infty}(\mathbb{R})$. Here, $b(u, t)$ is the drift, $n(u, v, t) \, dv$ is the jump measure, and $[f]_{u,v}$ is just the shorthand for the shock speed connecting states u and v:

$$[f]_{u,v} = \frac{f(v) - f(u)}{v - u}. \tag{6}$$

Thus at least formally, the work in [MS] establishes an integrable structure[1] for the problem, which is quite remarkable. In this connection, we must mention the recent work of Kaspar and Rezakhanlou [KR], in which they established rigorously that for a class of monotone pure-jump initial conditions, $(u(x,t))_{x \in \mathbb{R}}$ is a Feller process for each $t > 0$, and the evolution of its generator $\mathcal{A}$ is indeed governed by the Lax equation in (4).

Prior to the work in [KR], Menon [M] also considered the class of monotone pure-jump Markov processes in x, but in the simpler situation in which u can only take values in a fixed, finite set

$$\{-\infty < u_1 < u_2 < \cdots < u_N < \infty\} \tag{7}$$

on the line. In this case, the sample paths of the processes are piecewise constant paths. Since $b \equiv 0$ and

$$n(u_j, v)\, dv = \sum_{k<j} a_{jk}\delta(u_k - v)\, dv, \quad a_{jk} \geq 0, \ j = 1, \ldots, N, \tag{8}$$

it follows from the expressions for $\mathcal{A}$ and $\mathcal{B}$ above that

$$\mathcal{A}\varphi(u_j) = \sum_{k<j} a_{jk}\varphi(u_k) - \varphi(u_j)\sum_{k<j} a_{jk}, \tag{9}$$

$$\mathcal{B}\varphi(u_j) = \sum_{k<j} a_{jk} f_{jk}\varphi(u_k) - \varphi(u_j)\sum_{k<j} f_{jk}a_{jk}, \tag{10}$$

$j = 1, \ldots, N$. Thus the operator $\mathcal{A}$ gives rise to a lower triangular generator matrix A with row sums zero such that $a_{jk} \geq 0$ for $k < j$. Similarly, the operator $\mathcal{B}$ gives rise to a lower triangular matrix B with row sums zero such that $b_{jk} = f_{jk}a_{jk}$ for $k < j$, where

$$f_{jk} = -\frac{f(u_j) - f(u_k)}{u_j - u_k}, \ k < j. \tag{11}$$

Hence the equation $\partial_t \mathcal{A} = [\mathcal{A}, \mathcal{B}]$ leads to the finite dimensional system

$$\dot{A} = [A, B], \tag{12}$$

where A is the generator of a Markov process which has only a finite number of states (a.k.a. Markov chain). From the standard theory of Markov chains [N], we recall that for $1 \leq j < i \leq N$, the (i, j) entry of the generator matrix A is the rate of going from state u_i to state u_j, (note the typo in [L3] where i and j should be switched), while $a_i := -a_{ii}$ is the rate of leaving state u_i.

In [M], the author in fact introduced an extension of (12) above for general matrices $L \in \mathfrak{g} = gl(N, \mathbb{R})$. To exhibit the structure of this equation, we introduce the following Lie subalgebras of $\mathfrak{g}$:

[1] Here the term integrable structure simply refers to the zero-curvature equation/Lax pair, which is the compatibility condition of two linear problems. It does not carry the connotation that such equations are completely integrable.

$$\mathfrak{m} = \left\{ X = (x_{ij}) \in \mathfrak{g} \mid \sum_{j=1}^{N} x_{ij} = 0, i = 1, \ldots, N \right\}, \tag{13}$$

$\mathfrak{d}$ = diagonal subalgebra of $\mathfrak{g}$.

Then we have the (vector space) direct-sum decomposition

$$\mathfrak{g} = \mathfrak{m} \oplus \mathfrak{d} \tag{14}$$

with associated projection maps $\Pi_{\mathfrak{m}}$ and $\Pi_{\mathfrak{d}}$. If we now introduce the real symmetric matrix $F = (f_{jk}) \in \mathfrak{g}$, where

$$f_{kj} = f_{jk} = -\frac{f(u_j) - f(u_k)}{u_j - u_k}, \ k < j, f_{jj} = -f'(u_j), \tag{15}$$

then the equation alluded to above can be written in the form

$$\dot{L} = [L, \Pi_{\mathfrak{m}}(F \circ L)], \quad L \in \mathfrak{g}, \tag{16}$$

where $F \circ X = (f_{ij}x_{ij})$ is the Hadamard product and it is easy to verify that (12) is a special case of this more general equation, i.e., $B = \Pi_{\mathfrak{m}}(F \circ A)$ in (12) with $A \in \mathfrak{q} \cap \mathfrak{b}_-$. Here, $\mathfrak{b}_-$ is the Lie subalgebra of $\mathfrak{g}$ consisting of lower triangular matrices, and $\mathfrak{q}$ is the cone in $\mathfrak{m}$ given by

$$\mathfrak{q} = \{ X = (x_{ij}) \in \mathfrak{m} \mid x_{ij} \geq 0, i \neq j \}. \tag{17}$$

Note that associated with the splitting in (14) is the classical r-matrix $R = \Pi_{\mathfrak{m}} - \Pi_{\mathfrak{d}}$ and its associated Lie algebra $\mathfrak{g}_R = (\mathfrak{g}, [\cdot, \cdot]_R)$ and (16) is Hamiltonian in the $(-)$-Lie-Poisson structure on $\mathfrak{g}_R^* \simeq \mathfrak{g}$ [M]. In [L1, L2], the author shows that (16) is Liouville integrable on the generic coadjoint orbits of the Lie group G_R (which integrates the Lie algebra $\mathfrak{g}_R$) and solves the equation explicitly using Riemann theta functions. However, in [L3], we showed that $\mathfrak{q} \cap \mathfrak{b}_-$ is not invariant under the $Ad^*_{G_R}$-action. For this reason, the equation (12) in the probabilistically interesting case is not even Hamiltonian on the $Ad^*_{G_R}$-orbits. Nevertheless, our work in [L3] showed that the flow generated by (12) is still conjugate to a straight line motion, and the equation is exactly solvable. The proof of these statements is achieved by an explicit change of variable which maps an open, dense subset of $\mathfrak{q} \cap \mathfrak{b}_-$ to an associated set of algebro-geometric data consisting of a fully reducible nodal curve C (the spectral curve of (12)) and a set of complementary variables $(\lambda_{ij}(A))_{2 \leq i < j \leq N}$ whose logarithms move linearly under the flow. In this way, we therefore have a class of solutions of the full equation (4).

The goal of this work is to give an expository account of [L1, L2, L3], our emphasis here is on the progression of ideas and we explain our method of analysis. By putting the material in the above references together, it is our hope that the reader can come to appreciate the similarities as well as the differences between the generic case and the probabilistically interesting case. The paper is organized as follows. In Section 2, we begin by giving some background material. In Section 3, we discuss the Liouville integrability of (16) on generic coadjoint orbits of G_R, where the spectral curve is smooth. In Section 4, we consider (12), which describes the evolution of the infinitesimal generator matrix of the pure-jump Markov process. Here the spectral curve is generically a fully reducible nodal curve, and we study several normalized eigenvectors of the matrix loop $A_u(h)^T = (hu - A)^T$ on the normalization C^ν of C. (Here h is a spectral parameter, and $u = \text{diag}(u_1, \ldots, u_N)$.) In Section 5, we exhibit two (related) ways to linearize the flow. In Section 6, we discuss the correspondence between an open, dense subset of $\mathfrak{q} \cap \mathfrak{b}_-$ and algebro-geometric data, thereby completing the linearization picture. In Section 7, the last section, we present two ways to explicitly solve the Cauchy problem: one way is by means of the reconstruction formula for A in Section 6, and the other is based on Riemann-Hilbert factorization problems which we solve via another reconstruction formula in the same section.

2 Preliminaries

Let G be the identity component of $GL(N, \mathbb{R})$ and let D be the subgroup of G consisting of diagonal matrices with positive diagonal entries. Consider the group action $\Phi : G \times \mathbb{R}^N \longrightarrow \mathbb{R}^N$, $(g, x) \mapsto gx$. Then the isotropy subgroup of $e = (1, 1, \ldots, 1)^T \in \mathbb{R}^N$ is the Markov group

$$M = \{g \in G \mid \Phi(g, e) = e\}. \tag{18}$$

The elements of M are invertible stochastic matrices. (Here we are using the word stochastic matrix in a slightly weaker sense in that we do not require its entries to be nonnegative.) As a matter of fact, M is also isomorphic to the identity component $Aff^0(N - 1, \mathbb{R})$ of the affine group of $\mathbb{R}^{N-1}$ [P], [L1] consisting of maps $T : \mathbb{R}^{N-1} \longrightarrow \mathbb{R}^{N-1}$ of the form $T(x) = Cx + b$, where $C \in GL^0(N - 1, \mathbb{R})$, the identity component of $GL(N - 1, \mathbb{R})$ and $b \in \mathbb{R}^{N-1}$.

Clearly, the Lie groups G, D and M integrate the Lie algebras $\mathfrak{g}$, $\mathfrak{d}$ and $\mathfrak{m}$ respectively. In addition to these Lie groups, we will also need to introduce the Lie subgroup B_- which integrates $\mathfrak{b}_-$ consisting of invertible lower triangular matrices.

The Lie group G_R which integrates $\mathfrak{g}_R$ admits the following description. As a submanifold of G,

$$G_R = \{g \in G \mid (g^{-1}e, e_j) > 0, j = 1, \ldots, N\}, \tag{19}$$

where $e_1, \ldots, e_N$ are the vectors in the canonical basis of $\mathbb{R}^N$. In [L1], it was shown that if $g \in G_R$, then we have the unique factorization

$$g = \mathbf{m}(g)\mathbf{d}(g)^{-1}, \quad \mathbf{m}(g) \in M, \mathbf{d}(g) \in D. \tag{20}$$

Moreover, if for $g, h \in G_R$, we define

$$g \cdot h = \mathbf{m}(g) h \mathbf{d}(g)^{-1}, \tag{21}$$

then a direct calculation (see, for example, [DLT] for the method) shows that $(G_R, \cdot)$ is a Lie group with Lie algebra $\mathfrak{g}_R$ and the coadjoint action of G_R on $\mathfrak{g}_R^* \simeq \mathfrak{g}$ is given by

$$
\begin{aligned}
Ad^*_{G_R}(g^{-1})L &= \Pi_{\mathfrak{d}^\perp} \mathbf{m}(g) L \mathbf{m}(g)^{-1} + \Pi_{\mathfrak{m}^\perp} \mathbf{d}(g) L \mathbf{d}(g)^{-1} \\
&= \Pi_{\mathfrak{d}^\perp} \mathbf{m}(g) L \mathbf{m}(g)^{-1} + \Pi_{\mathfrak{m}^\perp} L,
\end{aligned}
\tag{22}
$$

where explicitly,

$$
\Pi_{\mathfrak{m}^\perp} X = e \left(\sum_{j=1}^N x_{jj} e_j \right)^T , \quad \Pi_{\mathfrak{d}^\perp} X = X - e \left(\sum_{j=1}^N x_{jj} e_j \right)^T . \tag{23}
$$

Proposition 2.1 *(a) The equation (16) is Hamiltonian on the coadjoint orbits of the Lie group G_R with Hamiltonian $H = \frac{1}{2} tr(L(F \circ L))$.*
(b) The set $\mathfrak{m}$ is preserved by the coadjoint action of G_R, but $\mathfrak{b}_-$ is not. Consequently, $\mathfrak{b}_- \cap \mathfrak{m}$ is not preserved by the coadjoint action of G_R.

Remark 2.2 It is due to part (b) of the above proposition that the flow of (12) is not Hamiltonian on the coadjoint orbits of G_R.

Next, we let LG be the group of loops $g : S^1 \longrightarrow GL(N, \mathbb{C})$ satisfying the reality condition $\overline{g(z)} = g(\bar{z})$. Then the Lie algebra of LG is the loop algebra $L\mathfrak{g}$ consisting of smooth loops $X : S^1 \longrightarrow gl(N, \mathbb{C})$ satisfying the reality condition $\overline{X(h)} = X(\bar{h})$ (cf. [M]). We will denote by LG_+ (resp. LG_-) the Lie subgroup of LG consisting of loops g which extend analytically to the interior (resp. exterior) of the unit circle and we let $L\mathfrak{g}_+$ (resp. $L\mathfrak{g}_-$) be its corresponding Lie algebra. Then

$$\widetilde{M} = \{ g \in LG_+ \mid g(0) \in M \} \tag{24}$$

and

$$\widetilde{D} = \{ g \in LG_- \mid g(\infty) \in D \} \tag{25}$$

are Lie subgroups of LG_+ and LG_- which integrate

$$
\begin{aligned}
\widetilde{\mathfrak{m}} &= \{ X \in L\mathfrak{g}_+ \mid X_0 \in \mathfrak{m} \}, \\
\widetilde{\mathfrak{d}} &= \{ X \in L\mathfrak{g}_- \mid X_0 \in \mathfrak{d} \},
\end{aligned}
\tag{26}
$$

respectively. As extension of (14), we have the splitting

$$L\mathfrak{g} = \widetilde{\mathfrak{m}} \oplus \widetilde{\mathfrak{d}} \tag{27}$$

with associated projection maps $\Pi_{\widetilde{\mathfrak{m}}}$ and $\Pi_{\widetilde{\mathfrak{d}}}$.

Now let

$$\tilde{G} = \{g \in LG \mid g = g_+ g_-^{-1}, g_+ \in \tilde{M}, g_- \in \tilde{D}\} \tag{28}$$

and endow it with a Lie group structure with the multiplication

$$g * h = g_+ h g_-^{-1}, g, h \in \tilde{G}. \tag{29}$$

Then $(\tilde{G}, *)$ integrates $\tilde{\mathfrak{g}} = \tilde{\mathfrak{m}} \ominus \tilde{\mathfrak{d}}$, the Lie algebra anti-direct sum of $\tilde{\mathfrak{m}}$ and $\tilde{\mathfrak{d}}$. Using the pairing

$$(X, Y) = \oint_{|h|=1} \mathrm{tr}\,(X(h)Y(h)) \frac{dh}{2\pi i h}, \tag{30}$$

we identify the algebraic dual $\tilde{\mathfrak{g}}^*$ with $\tilde{\mathfrak{g}}$. In the next result, we will make the stronger assumption that f can be extended to an entire function on $\mathbb{C}$.

Proposition 2.3 *(a) If $L_u(h) = hu - L$, then the equation (16) is equivalent to*

$$\dot{L}_u(h) = [L_u(h), \Pi_{\mathfrak{m}}(F \circ L) + hf(u)]$$
$$= [L_u(h), (\Pi_{\tilde{\mathfrak{m}}}\, g(L_u))(h)], \tag{31}$$

where $g(L_u)(h) = f(L_u(h)h^{-1})h$.

(b) The above equation is Hamiltonian on the coadjoint orbits of $(\tilde{G}, *)$ through

$$\mathcal{T}_u = \{L_u \mid L_u(h) = hu - L, L \in \mathfrak{g}\} \subset \tilde{\mathfrak{g}} \simeq \tilde{\mathfrak{g}}^* \tag{32}$$

with Hamiltonian $H_{\mathcal{F}}(L_u) = \oint_{|h|=1} \mathrm{tr}\,(h\mathcal{F}(L_u(h)h^{-1})) \frac{dh}{2\pi i} = -\frac{1}{2}\mathrm{tr}\,(L(F \circ L))$, where $\mathcal{F}(x) = \int_0^x f(s)\,ds$. Moreover, we can identify the $\mathrm{Ad}_{\tilde{G}}^*$ orbit through $L_u \in \mathcal{T}_u$ with the $\mathrm{Ad}_{G_R}^*$ orbit through $L \in \mathfrak{g}$.

The equation in the above proposition was solved by Riemann-Hilbert factorization problems in [L1]. For the corresponding equation in the probabilistically interesting case:

$$\dot{A}_u(h) = [A_u(h), (\Pi_{\tilde{\mathfrak{m}}}\, g(A_u))(h)], A_u(h, 0) = hu - A_0, A_0 \in \mathfrak{q} \cap \mathfrak{b}_-, \tag{33}$$

a new Lie group comes into play due to the fact that $A \in \mathfrak{b}_- \cap \mathfrak{m}$, but otherwise, the factorization problem to be solved [L3] bears resemblance to the one in [L1] from a group-theoretic perspective. However, the algebraic geometries involved are quite different.

3 Complete Integrability in the Generic Case

We begin with a simple proposition in which we identify a collection of coadjoint orbit invariants of the group G_R. As the reader will see, this turns out to be the complete set. However, the proof of this fact is by no means simple.

Proposition 3.1 *The coordinate functions x_{ii} defined by $x_{ii}(L) = \ell_{ii}$ for $L \in \mathfrak{g}$, $i = 1, \ldots, N$ are invariant under the coadjoint action of G_R.*

In order to define the generic elements in $\mathfrak{g}$ and the generic coadjoint orbits of G_R, we introduce the function

$$\mathcal{S}(L^0, L) = \det \begin{pmatrix} L^0 \otimes I_N - I_N \otimes L^T & -e \otimes I_N \\ I_N \otimes e^T & 0 \end{pmatrix} \tag{34}$$

for $L^0, L \in \mathfrak{g}$. Here, I_N is the $N \times N$ identity matrix and $\otimes$ is the symbol for the Kronecker product. To simplify notation, we set $\mathcal{S}(L) = \mathcal{S}(L, L)$. We refer the reader to the sketch of proof of Theorem 3.6 below where he/she can see how this function was discovered in [L1].

Proposition 3.2 *If $\Lambda^0 = \operatorname{diag}(\lambda_1^0, \ldots, \lambda_N^0)$ and $\Lambda = \operatorname{diag}(\lambda_1, \ldots, \lambda_N)$ are regular elements in $\mathfrak{d}$, that is, $\lambda_i^0 - \lambda_j^0 \neq 0$ and $\lambda_i - \lambda_j \neq 0$ for all $i \neq j$, then*

$$\mathcal{S}(\Lambda^0, \Lambda) = (-1)^{\frac{1}{2}N(N-1)} \prod_{i=2}^{N} \prod_{j=1}^{i-1} (\lambda_i^0 - \lambda_j^0)(\lambda_i - \lambda_j). \tag{35}$$

Thus $\mathcal{S}(L^0, L)$ is not identically zero, and indeed, $\mathcal{S}(L^0, L)$ is a semi-invariant of the coadjoint action of G_R, as the following result shows.

Proposition 3.3 *For all $g^0, g \in G_R$, we have*

$$\mathcal{S}(Ad_{G_R}^*((g^0)^{-1})L^0, Ad_{G_R}^*(g^{-1})L) = \det(\mathbf{m}(g^0)) \det(\mathbf{m}(g)) \mathcal{S}(L^0, L) \tag{36}$$

and hence

$$\mathcal{S}(Ad_{G_R}^*(g^{-1})L) = (\det(\mathbf{m}(g)))^2 \mathcal{S}(L). \tag{37}$$

As a consequence, the subset of $\mathfrak{g}$ defined by

$$\mathcal{G} = \{L \in \mathfrak{g} \mid \mathcal{S}(L) \neq 0\} \tag{38}$$

is an open, dense subset of $\mathfrak{g}$ invariant under the coadjoint action of G_R.

Definition 3.4 *A matrix $L \in \mathfrak{g}$ is generic iff $L \in \mathcal{G}$. An orbit $\mathcal{O}_{L^0}$ of the coadjoint action of G_R through $L^0 \in \mathfrak{g}$ is generic iff L^0 is a generic matrix.*

Proposition 3.5 $S(L^0, L) = (-1)^{\frac{1}{2}N(N-1)}\phi(L^0)\phi(L)$, *where*

$$\phi(L^0) = \det(e \ L^0 e \ (L^0)^2 e \ \ldots \ (L^0)^{N-1} e),$$
$$\phi(L) = \det(e \ Le \ L^2 e \ \ldots \ L^{N-1} e). \tag{39}$$

From this proposition, it is clear that the connected components

$$\mathcal{G}^+ = \{L \in \mathfrak{g} \mid \phi(L) > 0\},$$
$$\mathcal{G}^- = \{L \in \mathfrak{g} \mid \phi(L) < 0\} \tag{40}$$

of $\mathcal{G}$ are invariant under the coadjoint action of G_R. For this reason, we can restrict our attention to orbits in $\mathcal{G}^+$ in the next theorem as the analysis in the other component is similar.

Theorem 3.6 *If* $L^0 \in \mathcal{G}^+$, *then*

$$\mathcal{O}_{L^0} = \{L \in \mathcal{G}^+ \mid \ell_{ii} = \ell_{ii}^0, i = 1, \ldots, N\}. \tag{41}$$

Hence $\mathcal{O}_{L^0}$ *is of dimension* $N(N-1)$.

Proof Sketch. We will give a brief sketch of the proof to give an indication of how the function $S(L^0, L)$ was discovered. First of all, it is clear from Proposition 3.1 that $\mathcal{O}_{L^0}$ is a subset of the level set. Conversely, suppose L is an element of the level set, we want to show that there exists $m \in M$ such that $L = \Pi_{\partial\perp} mL^0 m^{-1} + \Pi_{\mathfrak{m}\perp} L^0$. Put $p = m^{-1}$, it is not hard to show that the above equation is equivalent to $L^0 p - pL - ea^T = 0$ where both p and $a \in \mathbb{R}^N$ are unknowns to be determined. Using the vec operator (recall that vec $(X) = (x_{11} \ldots x_{1l} \ldots, x_{k1} \ldots x_{kl})^T$ for a $k \times l$ matrix X), the above equation together with the equation $pe - e = 0$ can be formulated as the linear system

$$\begin{pmatrix} L^0 \otimes I_N - I_N \otimes L^T & -e \otimes I_N \\ I_N \otimes e^T & 0 \end{pmatrix} \begin{pmatrix} \text{vec}(p) \\ a \end{pmatrix} = \begin{pmatrix} 0 \\ e \end{pmatrix} \tag{42}$$

of $N^2 + N$ linear equations in $N^2 + N$ unknowns with the additional stipulation that p is invertible. The rest of the proof then consists of getting an explicit expression for $\det p$ from which we can check that indeed p is invertible. It is perhaps interesting to note that after a lengthy argument, we can show that $\det p = \frac{\phi(L^0)}{\phi(L)}$, where $\phi(L^0)$ and $\phi(L)$ are defined in Proposition 3.5 above. $\quad\square$

To establish the Liouville integrability of the system on generic coadjoint orbits of G_R, we introduce the characteristic polynomial of $L_u(h)$:

$$\det(L_u(h) - zI) = \sum_{r=0}^{N} \sum_{k=0}^{r} I_{rk}(L) h^k z^{N-r}, \tag{43}$$

which is invariant under the Hamiltonian flow and therefore provides us with a collection of conserved quantities. Clearly, the functions I_{rr} for $r = 0, \ldots, N$ are constants. To count the number of genuine integrals from the characteristic polynomial of $L_u(h)$ and to show that the Hamiltonian $H(L)$ is related to the I_{rk}'s, we have the following result.

Proposition 3.7 *(a) The functions $I_{r,r-1}$, $r = 1, \ldots, N$ are invariant under the coadjoint action of G_R.*

(b) The Hamiltonian $H(L) = \frac{1}{2} \operatorname{tr}(L(F \circ L))$ is generated by $I_{r,r-2}(L)$, $2 \leq r \leq N$ and $I_{r',r'-1}(L)$, $1 \leq r' \leq N$. More precisely, if we let

$$A_k(x) = \prod_{i=1, i \neq k}^{N} (x - u_i), \quad k = 1, \ldots, N, \tag{44}$$

then

$$H(L) = (-1)^N \sum_{r=1}^{N-1} \left(\sum_{k=1}^{N} \frac{f(u_k)}{A_k(u_k)} u_k^{N-1-r} \right) I_{r+1,r-1}(L) + c(L), \tag{45}$$

where $c(L)$ is quadratic in $I_{r',r'-1}(L)$, $1 \leq r' \leq N$.

(c) The functions I_{rk}, $k = 0, \ldots, r - 2$, $r = 1, \ldots, N$ provide a collection of $\frac{1}{2}N(N-1)$ conserved quantities in involution on the generic coadjoint orbits of G_R.

To construct the (putative) angles, introduce the homogenization of $I(L; h, z)$:

$$\widetilde{I}(L; \zeta, h, z) = \det(hu - \zeta L - zI). \tag{46}$$

Then we have the spectral curve

$$C(L) = \{[\zeta : h : z] \in \mathbb{CP}^2 \mid \widetilde{I}(L; \zeta, h, u) = 0\}, \tag{47}$$

which is smooth for L in a Zariski open set $\mathcal{U} \subset \mathfrak{g}$ and the genus of $C(L)$ is given by the standard formula $g = g(C(L)) = \frac{1}{2}(N-1)(N-2)$. For $P = [\zeta : h : z] \in C(L)$, let $\mathcal{M}(L, P) = hu - \zeta L - zI$, then $f(P) = \ker \mathcal{M}(L, P)^T$ defines the holomorphic eigenvector map $f : C(L) \longrightarrow \mathbb{CP}^{N-1}$. Let

$$H = \{[z_1 : \cdots : z_N] \in \mathbb{CP}^{N-1} \mid z_1 + \cdots + z_N = 0\}, \tag{48}$$

then the divisor $D(L) = f^* H$ is well-defined and is of degree $\frac{N(N-1)}{2} = g + N - 1$. Now we impose an additional assumption on matrices $L \in \mathcal{G} \cap \mathcal{U}$: (GA) the spectrum $\sigma(L)$ is simple, i.e., L has distinct eigenvalues $\lambda_1, \ldots, \lambda_N$ (a priori, the eigenvalues can be labelled as smooth functions only locally), then the points $P_+^i = [1 : 0 : -\lambda_i]$, $i = 1, \ldots, N$ where $h = 0$ are distinct and we have $(h) = P_+ - P_-$, $P_\pm = \sum_{i=1}^{N} P_\pm^i$.

Under this additional assumption, we introduce the holomorphic 1-forms

$$\omega_{r+1,k-1} = \frac{h^{k-1} z^{N-r-1}}{I_z(h,z)} \, dh, \ 1 \le k \le r-1, r = 2, \ldots, N-1 \qquad (49)$$

and the meromorphic 1-forms (with poles at the points of P_+)

$$\omega_{r+1,-1} = \frac{z^{N-r-1}}{h I_z(h,z)} \, dh, \ r = 1, \ldots, N-1, \qquad (50)$$

and we define

$$\phi_{rk}(L) = \int_{D_0(L)}^{D(L)} \omega_{r+1,k-1}, \ 1 \le k \le r-1, r = 2, \ldots, N-1,$$

$$\phi_{r0}(L) = \int_{D_0(L)}^{D(L)} \omega_{r+1,-1}, \ r = 1, \ldots, N-1, \qquad (51)$$

where $D_0(L) = (g + N - 1)P_-^1$. Note that in the definition of $\phi_{r0}(L)$ above, the path of integration going from the point P_-^1 to the points in $D(L)$ have to avoid the points in P_+. The variables introduced in (51) are defined modulo the lattice Λ_L in the space $\mathbb{C}^{g+N-1}$ generated by the columns of the $(g + N - 1) \times (2g + N)$ matrix

$$A_L = \begin{pmatrix} A_1 & 0 \\ A_2 & A_3 \end{pmatrix} \qquad (52)$$

where

(i) A_1 is the $g \times 2g$ period matrix corresponding to $\{\omega_{rk} \mid 0 \le k \le r-3, r = 3, \ldots, N\}$ and a canonical basis $\delta_1, \ldots, \delta_{2g}$ of $H_1(C(L), \mathbb{Z})$,

(ii) A_2 is the $(N-1) \times 2g$ matrix whose (i, j) entry is $\oint_{\delta_j} \omega_{i+1,-1}$,

(iii) A_3 is the $(N-1) \times N$ matrix whose (i, j) entry is $\oint_{\alpha_j} \omega_{i+1,-1}$, where α_j is a small closed contour enclosing the pole $P_+^j = [1 : 0 : -\lambda_j]$ of $\omega_{i+1,-1}$.

Theorem 3.8 (a) $\{\phi_{rk}, I_{r'k'}\}_R(L) = -\delta_{kk'}\delta_{r',r+1}, \ 1 \le k \le r-1,$
$r = 2, \ldots, N-1, 0 \le k' \le r-2, r' = 2, \ldots, N.$
(b) $\{\phi_{r0}, I_{r'k'}\}_R(L) = -\delta_{k'0}\delta_{r',r+1}, \ r = 1, \ldots, N-1, 0 \le k' \le r-2,$
$r' = 2, \ldots, N.$

From the above Poisson bracket relations and the expression of the Hamiltonian $H = \frac{1}{2}\text{tr}(L(F \circ L))$ in Proposition 3.7 above, the following corollaries follow.

Corollary 3.9 (a) $\{\phi_{rk}, H\}_R(L) = (-1)^{N-1} \left(\sum_{j=1}^{N} \frac{f(u_j)}{A_j(u_j)} u_j^{N-1-r} \right) \delta_{k,r-1},$
$1 \le k \le r-1, r = 2, \ldots, N-1.$
(b) $\{\phi_{r0}, H\}_R(L) = (-1)^{N-1} \left(\sum_{j=1}^{N} \frac{f(u_j)}{A_j(u_j)} u_j^{N-1-r} \right) \delta_{r-1,0}, \ r = 1, \ldots,$
$N - 1.$

Corollary 3.10 *The functions I_{rk}, $k = 0, \ldots, r - 1$, $r = 1, \ldots, N$ are functionally independent on the open, dense subset of $\mathcal{G} \cap \mathcal{U}$ satisfying (GA).*

Theorem 3.11 *(a) The Hamiltonian system generated by $H(L) = \frac{1}{2} tr\,(L(F \circ L))$ is Liouville integrable on the generic coadjoint orbits of G_R.*

(b) For initial data $L_0 \in \mathcal{G} \cap \mathcal{U}$ satisfying the additional assumption (GA), the factorization problem

$$e^{thf\left(u - \frac{L_0}{h}\right)} = g_+(h, t)g_-(h, t)^{-1}, \quad h \in \mathbb{CP}^1 \setminus \{0, \infty\},$$

$$g_+(\cdot, t) \mid S^1 \in \widetilde{M}, \quad g_-(\cdot, t) \mid S^1 \in \widetilde{D} \tag{53}$$

can be solved explicitly for $g_\pm(h, t)$ by means of Riemann theta functions associated with the Riemann surface of the spectral curve $C(L_0)$ and the solution of (31) is given by

$$L_u(h, t) = g_\pm(h, t)^{-1} L_u(h, 0) g_\pm(h, t). \tag{54}$$

4 Nodal Curves and Normalized Eigenvectors

For the rest of the paper, we will study the Lax equation (12), which is not Hamiltonian on the $Ad^*_{G_R}$-orbits.

We begin by taking $A \in \mathfrak{b}_-$ which is not necessarily in $\mathfrak{m}$. We will make the following genericity assumption on A:

(GA1-a) $a_{ii} \neq 0$, $i = 1, \ldots N$, $a_{ii} \neq a_{jj}$, $i \neq j$ for $A \in \mathfrak{b}_- \setminus \mathfrak{m}$,

(GA1-b) $a_{ii} < 0$, $i = 2, \ldots, N$, $a_{ii} \neq a_{jj}$, $i \neq j$ for $A \in \mathfrak{b}_- \cap \mathfrak{m}$.

In this case, the spectral curve C is the projective curve whose affine part is given by the equation

$$I(A; h, z) := \det\,(A_u(h) - zI) = \prod_{i=1}^{N}(z_i(h) - z) = 0, \ z_i(h) = hu_i - a_{ii}, \tag{55}$$

where $a_{11} = 0$ if A is also in $\mathfrak{m}$. Hence C is fully reducible with components C_i, $i = 1, \ldots, N$ (each with multiplicity 1), where

$$C_i = \{[\zeta : h : z] \mid hu_i - \zeta a_{ii} - z = 0\}, \quad i = 1, \ldots, N. \tag{56}$$

By (GA1), it follows that the points $Q^i_+ = [1 : 0 : -a_{ii}]$, $i = 1, \ldots, N$ where $h = 0$ are distinct. The points $Q_i = [0 : 1 : u_i] \in C_i$, $i = 1, \ldots, N$ are the points at infinity of the curve C. Thus the divisor of the coordinate function h of the curve C is given by

$$(h) = Q_+ - Q_-, \tag{57}$$

where $Q^i_- = Q_i$, $i = 1, \ldots, N$. As usual, we will identify $\mathbb{CP}^1 \setminus \{\infty\}$ with $\mathbb{C}$.

If $1 \leq i < j \leq N$, we have $C_i \cap C_j = \{P_{ij}\}$, where

$$P_{ij} = \left[1 : b_{ij} : \frac{a_{ii}u_j - a_{jj}u_i}{u_i - u_j} \right], \quad b_{ij} = \frac{a_{ii} - a_{jj}}{u_i - u_j}. \tag{58}$$

We will make the additional assumption that
(GA2) the P_{ij}'s are all distinct.

Clearly, each P_{ij} is an ordinary double point for C and therefore C is a *nodal curve*. Since each irreducible component C_i ($\simeq \mathbb{CP}^1$) is smooth, the normalization of C is

$$C^v = \sqcup_{i=1}^{N} C_i, \tag{59}$$

and the canonical normalization map is given by

$$v : C^v \longrightarrow C = \cup_{i=1}^{N} C_i, ([\zeta : h : hu_i - a_{ii}], i) \mapsto [\zeta : h : hu_i - a_{ii}]. \tag{60}$$

Clearly, $v^{-1}(P_{ij}) = \{(P_{ij}, i), (P_{ij}, j)\}$, $1 \leq i < j \leq N$. Away from these points, the map v is one-to-one onto $C \setminus C_{\text{sing}}$, where C_{sing} is the set of singular points of C. We will use the notation

$$P_{ij}^+ = (P_{ij}, i), \quad P_{ij}^- = (P_{ij}, i), 1 \leq i < j \leq N \tag{61}$$

to denote the node-branches. Since

$$\delta = \text{Card}(C_{\text{sing}}) = \text{Card}(\{P_{ij}\}_{i<j}) = \frac{N(N-1)}{2}, \tag{62}$$

$$\gamma = \text{number of irreducible components of } C = N,$$

it follows that the arithmetic genus of C is given by [ACG]

$$g = \delta - \gamma + 1 = \frac{(N-1)(N-1)}{2}. \tag{63}$$

We will work with the eigenvectors of $A_u(h)^T$. For this purpose, let $(e_1, \ldots, e_N)$ be the standard ordered basis of $\mathbb{C}^N$ and introduce the subspaces

$$V_i = \mathbb{C}e_1 \oplus \cdots \oplus \mathbb{C}e_i, \quad i = 1, \ldots, N. \tag{64}$$

Then we have the standard flag

$$\{0\} \subset V_1 \subset V_2 \subset \cdots \subset V_N = \mathbb{C}^N. \tag{65}$$

Since $A_u(h)^T$ is upper triangular, it stabilizes this flag, i.e., $A_u(h)^T V_i \subset V_i$ for each i. In particular, it is easy to see that the eigenspace of $A_u(h)^T$ corresponding to the eigenvalue $z = z_i(h)$ is contained in the subspace V_i, $i = 1, \ldots N$. We will introduce several normalized eigenvectors of $A_u(h)^T$ on the normalization C^v which serve different purposes. We will start with $\hat{v}$, normalized by the condition

$$(e, \hat{v}(P)) = 1 \tag{66}$$

for a general point $P \in C^\nu$, where e is the vector with all components equal to 1 introduced in Section 2. In order to present our next result, we have to introduce some notation. First, for any positive integer $r < N$, let $Q_{r,N}$ denote the set of sequences $I = (i_1, \ldots, i_r)$ where each component is an integer satisfying $1 \le i_\mu \le N$ and such that $1 \le i_1 < \cdots < i_r \le N$. If X is an $N \times N$ matrix, and $I, J \in Q_{r,N}$, we let $X[I|J]$ denote the submatrix of X whose (a, b) entry is x_{i_a, j_b} for $a = 1, \ldots, r, b = 1, \ldots, r$.

Proposition 4.1 *The normalized eigenvector $\widehat{v}$ is such that $\widehat{v}\,|_{C_1} = e_1$ and for each $2 \le i \le N$, we have $\widehat{v}_{i+1}\,|_{C_i} = \cdots = \widehat{v}_N\,|_{C_i} = 0$ (by convention, this is vacuous if $i = N$), and*

$$\widehat{v}_i\,|_{C_i} = (-1)^{i+1} \frac{det(A_u(h)^T - z_i(h)I)[1, \ldots, i-1 \mid 1, \ldots, i-1]}{\widetilde{\Delta}_i(h)},$$

$$\widehat{v}_k\,|_{C_i} = (-1)^{k+1} \frac{det(A_u(h)^T - z_i(h)I)[1, \ldots, i-1 \mid 1, \ldots, \widehat{k}, \ldots, i]}{\widetilde{\Delta}_i(h)},$$

$$(67)$$

for $k < i$, where $\widetilde{\Delta}_i(h)$ is the determinant of the upper Hessenberg matrix

$$
\begin{pmatrix}
1 & 1 & \cdots & 1 & 1 \\
(u_1 - u_i)(h - b_{i1}) & -a_{21} & \cdots & -a_{i-1,1} & -a_{i1} \\
0 & (u_2 - u_i)(h - b_{i2}) & \cdots & -a_{i-1,2} & -a_{i2} \\
\vdots & \vdots & \vdots & \vdots & \vdots \\
0 & 0 & \cdots & (u_{i-1} - u_i)(h - b_{i,i-1}) & -a_{i,i-1}
\end{pmatrix}.
$$

$$(68)$$

Consequently, for $2 \le i \le N$,

$$\widehat{v}_i\,|_{C_i} = (-1)^{i+1} \frac{\prod_{\mu=1}^{i-1}(u_\mu - u_i)(h - b_{\mu i})}{\widetilde{\Delta}_i(h)},$$

$$\widehat{v}_{i-1}\,|_{C_i} = (-1)^{i+1} \frac{a_{i,i-1}\prod_{\mu=1}^{i-2}(u_\mu - u_i)(h - b_{\mu i})}{\widetilde{\Delta}_i(h)},$$

$$\widehat{v}_k\,|_{C_i} = (-1)^{k+1}$$

$$\times \frac{\prod_{\mu=1}^{k-1}(u_\mu - u_i)(h - b_{\mu i})\, det(A_u(h)^T - z_i(h)I)[k, \ldots, i-1 \mid k+1, \ldots, i]}{\widetilde{\Delta}_i(h)}$$

$$= (-1)^{i+1} \frac{\prod_{\mu=1}^{k-1}(u_\mu - u_i)(h - b_{\mu i})\left[a_{ik}\prod_{\mu=k+1}^{i-1}(u_\mu - u_i)(h - b_{\mu i}) + l.o.t.\right]}{\widetilde{\Delta}_i(h)},$$

$$(69)$$

for $k < i - 1$.

From (68), the leading coefficient of $\widetilde{\Delta}_i(h)$ is given by $(-1)^{i+1} \prod_{\mu=1}^{i-1} (u_\mu - u_i)$. We will henceforth denote by $\Delta_i(h)$ the monic polynomial defined by

$$\widetilde{\Delta}_i(h) = (-1)^{i+1} \prod_{\mu=1}^{i-1} (u_\mu - u_i)\Delta_i(h), \quad 2 \le i \le N. \tag{70}$$

We will also make the following additional genericity assumption on A:
(GA3) $\frac{\Delta_j(b_{ij})}{\Delta_i(b_{ij})} \in \mathbb{R}^* = \mathbb{R} \setminus \{0\}$ for all $2 \le i < j \le N$, and $\Delta_i(b_{1i}) \ne 0$
for $i = 2, \dots, N$.

Corollary 4.2 *For* $1 \le i < j \le N$, *we have*

$$(\widehat{v}\,|_{C_i})(P_{ij}^+) = (\widehat{v}\,|_{C_j})(P_{ij}^-). \tag{71}$$

Therefore, we can also regard $\widehat{v}$ as a rational map $C \longrightarrow \mathbb{C}^N$ continuous at the nodes of C and has poles at the zeros of the polynomials $\Delta_i(h)$, regarded as a function on C_i, $i = 2, \dots, N$.

For each irreducible component C_i of the spectral curve C, let $D_i \in \mathrm{Div}\,(C_i)$ denote the divisor of zeros of the monic polynomial $\Delta_i(h)$ or equivalently, the divisor of poles of $\widehat{v}\,|_{C_i}$, $i = 2, \dots, N$. As $\Delta_i(h)$ is a polynomial of degree $i - 1$, it follows that $\deg D_i = i - 1$, $i = 2, \dots, N$. Therefore,

$$D(A) = D_2 + \cdots + D_N \in \mathrm{Div}\,(C), \quad \deg D(A) = g + N - 1. \tag{72}$$

The divisor $D(A)$ will play an important role in the linearization of the flow and its explicit integration. In order to state our next corollary, we introduce the polynomials $B_{ik}(h) = \prod_{\mu=1}^{k-1}(h - b_{\mu i})$ for $k \le i$.

Corollary 4.3 *For $A \in \mathfrak{b}_-$, the eigenvector $\widehat{v}$ has the following structure.*

(a) If $D_i \in \mathrm{Div}\,(C_i)$ is as defined above, then the divisor of zeros and poles of the nontrivial components of $\widehat{v}$ are given by

$$(\widehat{v}_k\,|_{C_i})_\infty = D_i, \ k \le i,$$
$$(\widehat{v}_i\,|_{C_i})_0 = P_{1i} + \cdots + P_{i-1,i},$$
$$(\widehat{v}_k\,|_{C_i})_0 = P_{1i} + \cdots + P_{k-1,i}$$
$$\qquad + \text{an effective divisor of degree } i - k - 1, k < i \tag{73}$$

so that we can write

$$\widehat{v}\,|_{C_i} = \sum_{k=1}^{i-1} \frac{B_{ik}(h)}{\Delta_i(h)} \left(\sum_{\lambda=0}^{i-k-1} C_{k\lambda}^{(i)} h^\lambda \right) e_k + \frac{B_{ii}(h)}{\Delta_i(h)} e_i. \tag{74}$$

(b) In the notation of part (a), we have

$$C_{k,i-k-1}^{(i)} = \frac{a_{ik}}{u_k - u_i}, k < i, \ i = 2, \dots, N. \tag{75}$$

We next restrict to the case where $A \in \mathfrak{b}_- \cap \mathfrak{m}$.

Proposition 4.4 *If $A \in \mathfrak{b}_- \cap \mathfrak{m}$, then*

$$\tilde{\Delta}_i(h) = h\tilde{\Delta}_i^o(h), \quad i = 2, \ldots, N, \tag{76}$$

where $\tilde{\Delta}_2^o(h) = -(u_1 - u_2)$ and for $i \geq 3$, $\tilde{\Delta}_i^o(h)$ is the polynomial of degree $i - 2$ given by

$$\tilde{\Delta}_i^o(h) = \det \begin{pmatrix} 1 & 1 & \cdots & 1 & 1 \\ u_1 - u_i & u_2 - u_i & \cdots & u_{i-1} - u_i & 0 \\ 0 & (u_2 - u_i)(h - b_{i2}) & \cdots & a_{i-1,2} & -a_{i2} \\ \vdots & \vdots & \vdots & \vdots & \vdots \\ 0 & 0 & \cdots & (u_{i-1} - u_i)(h - b_{i,i-1}) & -a_{i,i-1} \end{pmatrix}. \tag{77}$$

Hence

$$D_i = Q_+^i + D_i' \tag{78}$$

where $\deg D_i' = i - 2$, $i = 3, \ldots, N$.

We will set

$$D'(A) = \sum_{i=3}^{N} D_i'. \tag{79}$$

Clearly, $\deg D'(A) = g$.

Remark 4.5 (a) If we had considered in an analogous way the divisor associated with the spectral problem $A_u(h)v = zv$ instead, then we would have $v \mid_{C_N} = e_N$, and the fact that $A \in \mathfrak{m}$ would come in play only in the last step when we restrict the problem to C_1. Indeed, since $Ae = 0$, the divisor associated with $v \mid_{C_1}$ would depend on the divisors associated with $v \mid_{C_i}$'s ($i < N$) in a highly nontrivial way.
(b) In spite of the above remark, the divisor defined analogously by the spectral problem $A_u(h)v = zv$ can still be used in defining the linearization map. However, it is a subtle fact that this divisor does not work in the construction of additional linearizing variables in the case when $A \in \mathfrak{b}_- \setminus \mathfrak{m}$.

For $A \in \mathfrak{b}_- \cap \mathfrak{m}$, we will make the genericity assumption (GA4): for $i = 3, \ldots, N$,

$$\tilde{\Delta}_i^o(0) = \det \begin{pmatrix} 1 & 1 & \cdots & 1 & 1 \\ u_1 - u_i & u_2 - u_i & \cdots & u_{i-1} - u_i & 0 \\ 0 & a_{ii} - a_{22} & \cdots -a_{i-1,2} & -a_{i2} \\ \vdots & \vdots & \vdots & \vdots & \vdots \\ 0 & 0 & \cdots & a_{ii} - a_{i-1,i-1} & -a_{i,i-1} \end{pmatrix} \neq 0. \tag{80}$$

With this assumption, it follows that none of the points of D_i' can be the point Q_+^j, $j = 3, \ldots, N$.

The other two normalized eigenvectors which we will find useful are defined as follows:

$$\phi \mid_{C_i} := B_{ii}(h) \frac{\widehat{v} \mid_{C_i}}{\widehat{v}_i \mid_{C_i}} = \Delta_i(h) \widehat{v} \mid_{C_i},$$

$$\psi \mid_{C_i} := \frac{\phi \mid_{C_i}}{\Delta_i(b_{1i})}, \quad i = 1, \ldots, N, \tag{81}$$

where we use the convention that $B_{11}(h) \equiv 1$, $\Delta_1(h) \equiv 1$ and $b_{11} = 1$.

From this definition, it is clear that $\phi \mid_{C_i}$ is characterized by the normalization condition

$$\phi_i \mid_{C_i} (h) = B_{ii}(h), \quad i = 1, \ldots, N. \tag{82}$$

On the other hand we obtain the relation

$$\Delta_i(h) = (e, \phi \mid_{C_i} (h)), \quad i = 2, \ldots, N, \tag{83}$$

which is useful in computing the evolution of $D(A)$ under the flow. It should be pointed out that these two normalized eigenvectors may take different values at $P_{ij}^{\pm}$. For example, we have

$$\psi \mid_{C_j} (P_{ij}^-) = \frac{\Delta_j(b_{ij})}{\Delta_j(b_{1j})} \cdot \frac{\Delta_i(b_{1i})}{\Delta_i(b_{ij})} \cdot \psi \mid_{C_i} (P_{ij}+), \quad 1 \le i < j \le N \tag{84}$$

and (GA3) guarantees that

$$\lambda_{ij}(A) := \frac{\Delta_j(b_{ij})}{\Delta_j(b_{1j})} \cdot \frac{\Delta_i(b_{1i})}{\Delta_i(b_{ij})} \in \mathbb{R}^*. \tag{85}$$

As the reader will see in Section 6, we will call C and $(\lambda_{ij}(A))_{2 \le i < j \le N}$ the *algebro-geometric data* associated to A satisfying the genericity assumptions. This correspondence is the change of variable map which is used in linearizing the flow. Note that we can define a line bundle $\widetilde{E}$ on C^v by letting $\widetilde{E}_P$ be the one-dimensional eigenspace of $A_u(h(P))^T$ with eigenvalue $z(P)$ for $P \in C^v$. At each node P_{ij}, $1 \le i < j \le N$, we can regard (84) as defining the isomorphism

$$\widetilde{E}_{P_{ij}^+} \longrightarrow \widetilde{E}_{P_{ij}^-}, \quad c\psi \mid_{C_i} (P_{ij}^+) \mapsto c\psi \mid_{C_j} (P_{ij}^-). \tag{86}$$

In this way, we can interpret $\lambda_{ij}(A)$ as the gluing data of $\widetilde{E}_{P_{ij}^+}$ and $\widetilde{E}_{P_{ij}^-}$ at P_{ij}. Hence we obtain a line bundle E over C by identifying the fibers $\widetilde{E}_{P_{ij}^+}$, $\widetilde{E}_{P_{ij}^-}$ via (86).

5 Linearization of the Flow

There are two ways to linearize the flows $A(t)$ as defined by the equation

$$\dot{A} = [A, \Pi_{\mathrm{m}}(F \circ A)] \tag{87}$$

under the genericity assumptions on $A_0 \in \mathfrak{q} \cap \mathfrak{b}_-$ set forth in the previous section.

To obtain formulas similar to those in [L1, L2], we have to introduce the dualizing shealf [ACG] ω_C of the nodal curve C which is the analog of the sheaf of holomorphic 1-forms in the smooth case. By definition, for each open set $U \subset C$, $\omega_C(U)$ is the space of rational 1-forms η on $v^{-1}(U) \subset C^v$ having at most simple poles at the pair of points $\{P_{ij}^+, P_{ij}^-\}$ lying over each node $P_{ij} \in U$, satisfying

$$\operatorname{Res}_{P_{ij}^+} \eta + \operatorname{Res}_{P_{ij}^-} \eta = 0. \tag{88}$$

For the nodal curve C, the dimension of its space of global sections $H^0(C, \omega_C)$ is given by g. Indeed, a basis which serves our purpose is given in the following.

Proposition 5.1 *A basis of $H^0(C, \omega_C)$ is given by the regular differentials*

$$\omega_{rk} = \frac{h^k z^{N-r}}{I_z(L; h, z)} \, dh, \ 0 \le k \le r - 3, \ r = 3, \dots, N. \tag{89}$$

Remark 5.2 Note that the formal expressions for the ω_{rk}'s in the above proposition are exactly the same as the ones in the smooth case. The meaning of the corresponding objects, of course, are entirely different here.

To define the variables which move linearly under the flow, set

$$D_0'(A) = \sum_{i=3}^{N} (i - 2) Q_+^i \tag{90}$$

and put

$$\phi_{rk}(A) = \int_{D_0'(A)}^{D'(A)} \omega_{r+1,k-1}, 1 \le k \le r - 1, r = 2, \dots, N - 1. \tag{91}$$

These variables are defined modulo the lattice Λ generated by integral linear combinations of the columns of the $g \times (g + N - 1)$ matrix whose entries are

$$\oint_{\gamma_{ij}^+} \omega_{r+1,k-1}, 1 \le i < j \le N, \tag{92}$$

for $1 \le k \le r - 1, r = 2, \dots, N - 1$ where γ_{ij}^+ are small closed contours enclosing P_{ij}^+. To be more precise, we can arrange the forms in the order

$$\omega_{30}; \omega_{40}, \omega_{41}; \cdots \omega_{r+1,0}, \dots, \omega_{r+1,r-2}; \cdots; \omega_{N0}, \dots, \omega_{N,N-2}. \tag{93}$$

In a similar way, we can arrange the points P_{ij}^+ in the order

$$P_{12}^+, \ldots, P_{1N}^+; \cdots : P_{i,i+1}^+, \ldots, P_{iN}^+; \cdots ; \cdots ; P_{N-1,N}^+. \qquad (94)$$

Thus we have

Proposition 5.3 *The map* $A \mapsto (\phi_{rk}(A))_{1\leq k\leq r-1, 2\leq r\leq N-1}$ *is well-defined from the open subset of* $\mathfrak{q} \cap \mathfrak{b}_-$ *satisfying (GA1-b), (GA2), (GA3) and (GA4) into* $\mathbb{C}^g / \Lambda \simeq (\mathbb{C}^*)^g$.

Theorem 5.4 *For* $A(0) \in \mathfrak{q} \cap \mathfrak{b}_-$ *satisfying the genericity assumptions (GA1-b), GA2), (GA3) and (GA4), we have*

$$\frac{d}{dt}\phi_{rk}(A(t)) = (-1)^{N-1} \left(\sum_{j=1}^{N} \frac{f(u_j)}{A_j(u_j)} u_j^{N-1-r} \right) \delta_{k,r-1}, \qquad (95)$$

for $1 \leq k \leq r-1, r = 2, \ldots, N-1$, *where*

$$A_j(x) = \prod_{\substack{i=1 \\ i\neq j}}^{N}(x - u_i), \quad j = 1, \ldots, N. \qquad (96)$$

The second way to linearize the dynamics is by means of the variables $\lambda_{ij}(A)$ introduced in the last section.

Theorem 5.5 *For* $A(0) \in \mathfrak{q} \cap \mathfrak{b}_-$ *satisfying the genericity assumptions (GA1-b), (GA2), (GA3) and (GA4), we have*

$$\frac{d}{dt} \log \lambda_{ij}(A(t)) = f(u_1)(b_{1i} - b_{1j}) + f(u_i)(b_{ij} - b_{1i}) + f(u_j)(b_{1j} - b_{ij}) \qquad (97)$$

for $2 \leq i < j \leq N$.

The two ways to linearize the flows are in fact connected to each other. Since $\omega_{r+1,k-1} |_{C_i}$ is a meromorphic 1-form on $C_i \simeq \mathbb{CP}^1$ and indeed we have the representation

$$\omega_{r+1,k-1} |_{C_i} = \sum_{\mu>i} \frac{\operatorname{Res}_{P_{i\mu}^+}\omega_{r+1,k-1}}{h - b_{i\mu}} \, dh + \sum_{\mu<i} \frac{\operatorname{Res}_{P_{\mu i}^-}\omega_{r+1,k-1}}{h - b_{\mu i}} \, dh \qquad (98)$$

for each i, a careful calculation of $\phi_{rk}(A)$ yields the following result.

Proposition 5.6 *For* $A \in \mathfrak{q} \cap \mathfrak{b}_-$ *satisfying the genericity assumptions, we have*

$$\phi_{rk}(A) = -\sum_{i=2}^{N-1} \sum_{j=i+1}^{N} \left(\operatorname{Res}_{P_{ij}^+}\omega_{r+1,k-1} \right) \log \lambda_{ij}(A) + constant \qquad (99)$$

for $1 \leq k \leq r-1, r = 2, \ldots, N-1$.

Thus we can regard the second set of variables $(\lambda_{ij}(A))_{2 \leq i < j \leq N}$ as building blocks of the $\phi_{rk}(A)$'s. The flow $A(t)$ on the open dense subset of $\mathfrak{q} \cap \mathfrak{b}_-$ as defined by the genericity assumptions is indeed conjugate to a straight line motion on $(\mathbb{C}^*)^g$. However, we can reach this conclusion only after showing that the map that takes A to the spectral curve C and $(\lambda_{ij}(A))_{2 \leq i < j \leq N}$ is a change of variable. This is what we turn to in the next section.

6 A Dictionary Between $\mathfrak{q} \cap \mathfrak{b}_-$ and Algebro-geometric Data

We will denote by $(\mathfrak{q} \cap \mathfrak{b}_-)'$ the set of all A in $\mathfrak{q} \cap \mathfrak{b}_-$ satisfying the genericity assumptions. For $A \in (\mathfrak{q} \cap \mathfrak{b}_-)'$, we can associate with it the spectral curve C : $\prod_{i=1}^N (hu_i - \zeta a_{ii} - z) = 0$ as well as $(\lambda_{ij}(A))_{2 \leq i < j \leq N}$. We will call C with its irreducible components C_i, $i = 1, \ldots, N$ and $(\lambda_{ij}(A))_{2 \leq i < j \leq N}$ the *algebro-geometric data* associated to A. Our goal here is to study the correspondence

$$(\mathfrak{q} \cap \mathfrak{b}_-)' \ni A \mapsto \text{ the spectral curve } C \text{ with components } C_1, \ldots, C_N,$$
$$(\lambda_{ij}(A))_{2 \leq i < j \leq N}.$$

$$(100)$$

This naturally splits into two parts, the forward problem and the inverse problem.

6.1 The Forward Problem

Our main result can be summarized in the following theorem.

Theorem 6.1 *The matrix* $A \in (\mathfrak{q} \cap \mathfrak{b}_-)'$ *is uniquely determined by the curve* C *with irreducible components* $C_1, \ldots, C_N$ *together with* $(\lambda_{ij}(A))_{2 \leq i < j \leq N}$. *That is, the map*

$$(\mathfrak{q} \cap \mathfrak{b}_-)' \ni A \mapsto \text{ the spectral curve } C \text{ with components } C_1, \ldots, C_N,$$
$$(\lambda_{ij}(A))_{2 \leq i < j \leq N}$$

$$(101)$$

is one-to-one.

We will give a sketch of the main steps in the proof.

Step 1. Show that $\Delta_i(h)$ are uniquely determined by the algebro-geometric data, $3 \leq i \leq N$.

Recall that $\Delta_i(h) = h\Delta_i^o(h)$. So we define

$$\lambda_{ij}^o(A) = \frac{\Delta_j^o(b_{ij})}{\Delta_j^o(b_{1j})} \cdot \frac{\Delta_i^o(b_{1i})}{\Delta_i^o(b_{ij})}, \quad 2 \leq i < j \leq N. \qquad (102)$$

Also, we introduce the polynomials

$$P_j(h) = \frac{\Delta_j^o(h)}{\Delta_j^o(b_{1j})}, \quad j = 2, \ldots, N. \tag{103}$$

Then

$$P_j(b_{ij}) = \begin{cases} \lambda_{ij}^o(A) P_i(b_{ij}), & i = 2, \ldots, j-1 \\ 1 & i = 1. \end{cases} \tag{104}$$

By Lagrange interpolation, we can express $P_j(h)$ in terms of $P_i(b_{ij})$, $i = 2, \ldots j - 1$. Then we continue to iterate! Once we have $P_j(h)$, we then divide by its leading coefficient to obtain $\Delta_j^o(h)$.

Step 2. Show that $\widehat{v}$ is uniquely determined by the algebro-geometric data.

In view of the relation

$$\psi \mid_{C_i} = \frac{\Delta_i(h)}{\Delta_i(b_{1i})} \widehat{v} \mid_{C_i}, \tag{105}$$

and Step 1, it suffices to show that the eigenvector ψ is uniquely determined by the algebro-geometric data. Now ψ satisfies the conditions

$$\psi(P_{ij}^-) = \lambda_{ij}(A)\psi(P_{ij}^+), \quad 1 \le i < j \le N, \tag{106}$$

where $\lambda_{1j}(A) = 1$. In order to describe what we can get from (106), we need to introduce some notations. First, we let

$$\pi_{jk}(h) = \frac{B_{jk}(h)}{\Delta_j(b_{1j})}, \quad \ell_{ij}^{(k)}(h) = \prod_{\substack{\mu=k \\ \mu \ne i}}^{j-1} \frac{h - b_{\mu j}}{b_{ij} - b_{\mu j}}$$

$$\ell_{ij}(h) \equiv \ell_{ij}^{(1)}(h). \tag{107}$$

For $1 \le k < j \le N$, and any positive integer $r \le j - k$, we denote by $Q_{r+1}^{(k,j)}$ the set of integer sequences $\alpha = (\alpha_1, \ldots, \alpha_{r+1})$ of length $r + 1$ such that

$$k = \alpha_1 < \cdots < \alpha_{r+1} = j. \tag{108}$$

We start by equating the k-th component of both sides of (106) for $k \le i \le j$. By using (74) and (105) above, we can derive the following recursion formula after some calculations:

$$\psi_k \mid_{C_j} (h) = \pi_{jk}(h) \sum_{i_1=k}^{j-1} \lambda_{i_1 j}(A) \frac{(\psi_k \mid_{C_{i_1}})(b_{i_1 j})}{\pi_{jk}(b_{i_1 j})} \ell_{i_1 j}^{(k)}(h)$$

$$= B_{jk}(h) \sum_{i_1=k}^{j-1} \lambda_{i_1 j}(A) \frac{(\psi_k \mid_{C_{i_1}})(b_{i_1 j})}{B_{jk}(b_{i_1 j})} \ell_{i_1 j}^{(k)}(h). \tag{109}$$

Iterating, it follows after further simplification of the resulting formula that

$$\psi_k \mid_{C_j} (h)\Delta_k(b_{1k})$$

$$= \sum_{r=1}^{j-k} \sum_{\alpha \in \mathcal{Q}_{r+1}^{(k,j)}} \prod_{i=1}^{r} \lambda_{\alpha_i,\alpha_{i+1}}(A)B_{kk}(b_{k\alpha_2}) \prod_{i=2}^{r} \ell_{\alpha_{i-1},\alpha_i}(b_{\alpha_i,\alpha_{i+1}})\ell_{\alpha_r,j}(h).$$

$$(110)$$

Step 3. Show that A is uniquely determined by the algebro-geometric data.

The diagonal entries of A are uniquely determined by the curve C with its irreducible components $C_1, \ldots, C_N$. For the off-diagonal entries in the lower triangular part, it follows from the formula for $\psi_k \mid C_j(h)$ obtained in Step 2 above and (75) that

$$\frac{a_{jk}}{u_k - u_j}$$

$$= \frac{\Delta_j(b_{1j})}{\Delta_k(b_{1k})} \sum_{r=1}^{j-k} \sum_{\alpha \in \mathcal{Q}_{r+1}^{(k,j)}} \prod_{i=1}^{r} \lambda_{\alpha_i \alpha_{i+1}}(A) \prod_{i=2}^{r} \ell_{\alpha_{i-1},\alpha_i}(b_{\alpha_i,\alpha_{i+1}}) \frac{B_{kk}(b_{k\alpha_2})}{A_{\alpha_r,j}(b_{\alpha_r,j})},$$

$$(111)$$

where

$$A_{ij}(h) = \prod_{\substack{\mu=1 \\ \mu \neq i}}^{j-1} (h - b_{\mu j}), \quad i < j. \tag{112}$$

6.2 The Inverse Problem

The formulas that we obtained in solving the forward problem allow us to express the genericity assumptions (GA3) and (GA4) in terms of algebro-geometric data. This is the origin of our definition below. In order to state this definition, recall that given a curve $C : \prod_{i=1}^{N}(hu_i - \zeta a_{ii} - z) = 0$ with irreducible components $C_i : hu_i - \zeta a_{ii} - z = 0, i = 1, \ldots, N$ satisfying the genericity assumptions (GA1-b) and (GA2), we can define for $1 \leq i < j \leq N$ the quantities

$$b_{ij} = \frac{a_{ii} - a_{jj}}{u_i - u_j}, \quad \ell_{ij}(h) = \prod_{\substack{\mu=1 \\ \mu \neq i}}^{j-1} \frac{h - b_{\mu j}}{b_{ij} - b_{\mu j}}, \quad A_{ij}(h) = \prod_{\substack{\mu=1 \\ \mu \neq i}}^{j-1} (h - b_{\mu j}). \tag{113}$$

We can also define

$$B_{kk}(h) = \prod_{\mu=1}^{k-1} (h - b_{\mu k}), \quad k = 1, \ldots, N, \tag{114}$$

with the convention that $B_{11}(h) \equiv 1$. If in addition, we are given $(\lambda_{ij})_{2 \leq i < j \leq N} \in (\mathbb{R}^*)^g$, we define Δ_{1j}, $j = 2, \ldots N$, by

$$\frac{1}{\Delta_{1j}} = \sum_{r=1}^{j-1} \sum_{\alpha \in Q_{r+1}^{(1,j)}} \frac{1}{b_{1\alpha_2}} \prod_{\mu=1}^{r} \lambda_{\alpha_\mu, \alpha_{\mu+1}} \prod_{\mu=2}^{r} \ell_{\alpha_{\mu-1}, \alpha_\mu}(b_{\alpha_\mu, \alpha_{\mu+1}}) \frac{1}{A_{\alpha_r, j}(b_{\alpha_r, j})}, \tag{115}$$

provided the quantities on the right hand side is nonzero, and where $\lambda_{1j} = 1$, $j = 2, \ldots, N$. (We will henceforth use this convention).

Definition 6.2 *A curve* $C : \prod_{i=1}^{N} (hu_i - \zeta a_{ii} - z) = 0$ *with irreducible components* $C_i : hu_i - \zeta a_{ii} - z = 0$, $i = 1, \ldots, N$ *together with* $(\lambda_{ij})_{2 \leq i < j \leq N} \in (\mathbb{R}^*)^g$ *is called an admissible set of data iff it has the following properties:*

(a) $a_{11} = 0$, $a_{ii} < 0$ *for* $i = 2, \ldots N$, $a_{ii} \neq a_{jj}, i \neq j$;

(b) *the points* $P_{ij} = \left[1 : \frac{a_{ii} - a_{jj}}{u_i - u_j} : \frac{a_{ii}u_j - a_{jj}u_i}{u_i - u_j} \right]$, $1 \leq i < j \leq N$ *are distinct;*

(c) *for* $j = 3, \ldots, N$,

$$\sum_{r=1}^{j-1} \sum_{\alpha \in Q_{r+1}^{(1,j)}} \frac{1}{b_{1\alpha_2}} \prod_{\mu=1}^{r} \lambda_{\alpha_\mu, \alpha_{\mu+1}} \prod_{\mu=2}^{r} \ell_{\alpha_{\mu-1}, \alpha_\mu}(b_{\alpha_\mu, \alpha_{\mu+1}}) \frac{1}{A_{\alpha_r, j}(b_{\alpha_r, j})} \neq 0, \tag{116}$$

(d) *for* $j = 3, \ldots, N$,

$$\sum_{r=1}^{j-1} \sum_{\alpha \in Q_{r+1}^{(1,j)}} \frac{1}{b_{1\alpha_2}} \prod_{\mu=1}^{r} \lambda_{\alpha_\mu, \alpha_{\mu+1}} \prod_{\mu=2}^{r} \ell_{\alpha_{\mu-1}, \alpha_\mu}(b_{\alpha_\mu, \alpha_{\mu+1}}) \ell_{\alpha_r, j}(0) \neq 0, \tag{117}$$

(e) *for all* $1 \leq k < j \leq N$,

$$\frac{\Delta_{1j}}{\Delta_{1k}} \sum_{r=1}^{j-k} \sum_{\alpha \in Q_{r+1}^{(k,j)}} \prod_{\mu=1}^{r} \lambda_{\alpha_\mu \alpha_{\mu+1}} \prod_{\mu=2}^{r} \ell_{\alpha_{\mu-1}, \alpha_\mu}(b_{\alpha_\mu, \alpha_{\mu+1}}) \frac{B_{kk}(b_{k\alpha_2})}{A_{\alpha_r, j}(b_{\alpha_r, j})} \leq 0, \tag{118}$$

with the definition in (115).

(f) *for all* $2 \leq i < j \leq N$,

$$
\frac{\displaystyle\sum_{r=1}^{j-1} \sum_{\alpha \in Q_{r+1}^{(1,j)}} \frac{1}{b_{1\alpha_2}} \prod_{\mu=1}^{r} \lambda_{\alpha_\mu, \alpha_{\mu+1}} \prod_{\mu=2}^{r} \ell_{\alpha_{\mu-1}, \alpha_\mu}(b_{\alpha_\mu, \alpha_{\mu+1}}) \ell_{\alpha_r, j}(b_{ij})}{\displaystyle\sum_{r=1}^{i-1} \sum_{\alpha \in Q_{r+1}^{(1,i)}} \frac{1}{b_{1\alpha_2}} \prod_{\mu=1}^{r} \lambda_{\alpha_\mu, \alpha_{\mu+1}} \prod_{\mu=2}^{r} \ell_{\alpha_{\mu-1}, \alpha_\mu}(b_{\alpha_\mu, \alpha_{\mu+1}}) \ell_{\alpha_r, i}(b_{ij})} \neq 0,
$$

$$\tag{119}$$

and for all $i = 3, \ldots, N$,

$$
\sum_{r=1}^{i-1} \sum_{\alpha \in Q_{r+1}^{(1,i)}} \frac{1}{b_{1\alpha_2}} \prod_{\mu=1}^{r} \lambda_{\alpha_\mu, \alpha_{\mu+1}} \prod_{\mu=2}^{r} \ell_{\alpha_{\mu-1}, \alpha_\mu}(b_{\alpha_\mu, \alpha_{\mu+1}}) \ell_{\alpha_r, i}(b_{1i}) \neq 0.
$$

$$\tag{120}$$

We will denote the collection of all admissible sets of data by $\mathcal{D}$. Our theorem is the following.

Theorem 6.3 *The map*

$$(\mathfrak{q} \cap \mathfrak{b}_-)' \ni A \mapsto \textit{the spectral curve } C \textit{ with components } C_1, \ldots, C_N,$$

$$(\lambda_{ij}(A))_{2 \leq i < j \leq N}$$

$$\tag{121}$$

is a bijection onto $\mathcal{D}$.

In the forward problem, we have shown that the map is one-to-one, thus the task here is to show that the map is onto. To do so, we take an admissible data from $\mathcal{D}$ consisting of $C : \prod_{i=1}^{N}(hu_i - \zeta a_{ii} - z) = 0$ and $(\lambda_{ij})_{2 \leq i < j \leq N} \in (\mathbb{R})^g$. We will construct $A \in (\mathfrak{q} \cap \mathfrak{b}_-)'$ such that the spectral curve is C and $\lambda_{ij}(A) = \lambda_{ij}$.

Again we give the main ideas of the proof. First of all, we construct Δ_{1j}, $j = 2, \ldots, N$ and we put $\Delta_{11} = 1$. Then we construct $A \in \mathfrak{b}_-$ such that the (i, i) entry is given by the number a_{ii} from the curve C_i, $i = 1, \ldots, N$ and the (j, k) entry is given by

$$
\frac{a_{jk}}{u_k - u_j} = \frac{\Delta_{1j}}{\Delta_{1k}} \sum_{r=1}^{j-k} \sum_{\alpha \in Q_{r+1}^{(k,j)}} \prod_{\mu=1}^{r} \lambda_{\alpha_\mu \alpha_{\mu+1}} \prod_{\mu=2}^{r} \ell_{\alpha_{\mu-1}, \alpha_\mu}(b_{\alpha_\mu, \alpha_{\mu+1}}) \frac{B_{kk}(b_{k\alpha_2})}{A_{\alpha_r, j}(b_{\alpha_r, j})}
$$

$$\tag{122}$$

for all $1 \leq k < j \leq N$.

Next, set $\psi \mid_{C_1} (h) = e_1$. For $2 \leq j \leq N$, put $\psi_j \mid_{C_j} (h) = \frac{B_{jj}(h)}{\Delta_{1j}}$. Then for $1 \leq k < j$, construct $\psi_k \mid_{C_j}$ by the expression

$$\psi_k \mid_{C_j} (h) \Delta_{1k}$$

$$= \sum_{r=1}^{j-k} \sum_{\alpha \in Q_{r+1}^{(k,j)}} \prod_{\mu=1}^{r} \lambda_{\alpha_\mu,\alpha_{\mu+1}} B_{kk}(b_{k\alpha_2}) \prod_{\mu=2}^{r} \ell_{\alpha_{\mu-1},\alpha_\mu}(b_{\alpha_\mu,\alpha_{\mu+1}}) \ell_{\alpha_r,j}(h) \tag{123}$$

and we define

$$\psi \mid_{C_j} (h) = \sum_{k=1}^{j} \psi_k \mid_{C_j} (h) e_k, \quad j = 2, \ldots, N. \tag{124}$$

These constructions constitute Step 1. In Step 2, we check by a direct but lengthy calculation that $(A_u(h)^T - z_j(h)I)\psi \mid_{C_j} (h) = 0$. In Step 3, we then show that $A \in \mathfrak{m}$ by summing each row of A. In Step 4, to show that A is such that $\lambda_{ij}(A) = \lambda_{ij}$ for all $2 \leq i < j \leq N$, it is sufficient to verify that $\psi_k \mid_{C_j} (P_{ij}) = \lambda_{ij}\psi_k \mid_{C_i} (P_{ij})$ for $k \leq i$. Finally, in Step 5, we check that A satisfies the genericity assumptions so that $A \in (\mathfrak{q} \cap \mathfrak{b}_-)'$.

7 Solving the Cauchy Problem

We begin by recalling the following result in [M].

Proposition 7.1 ([M]) *If $A_0 \in \mathfrak{q} \cap \mathfrak{b}_-$, the Cauchy problem*

$$\dot{A} = [A, \Pi_\mathfrak{m}(F \circ A)], \quad A(0) = A_0 \tag{125}$$

has a unique global solution $A(t) \in \mathfrak{q} \cap \mathfrak{b}_-, t \geq 0$.

It is interesting to note that the convexity of f is responsible for $A(t) \in \mathfrak{q}$ in the above proposition [M].

We now present our two methods to explicitly solve the Cauchy problem.

Method 1. Using the reconstruction formula for a_{jk} and the evolution of algebro-geometric data

Theorem 7.2 *Let $A_0 \in \mathfrak{q} \cap \mathfrak{b}_-$ satisfies the genericity assumptions, and let $a_{ii}(0) = a_{ii}, i = 1, \ldots, N$ be the diagaonal entries of A_0 with $a_{11}(0) = 0$. Let $(\lambda_{ij}(0))_{2 \leq i < j \leq N}$ denote the other part of the algebro-geometric data corresponding to A_0, If $A(t)$ is the unique global solution to (125), then the off-diagonal entries of $A(t)$ in the lower triangular part of the matrix are given by (with the convention that $\Delta_k(b_{1k}, t) = 1$ when $k = 1$)*

$$a_{jk}(t)$$

$$= \frac{\Delta_j(b_{1j}, t)}{\Delta_k(b_{1k}, t)} (u_k - u_j) \sum_{r=1}^{j-k} \sum_{\alpha \in Q_{r+1}^{(k,j)}} \prod_{\mu=1}^{r} \lambda_{\alpha_\mu \alpha_{\mu+1}}(t) \prod_{\mu=2}^{r} \ell_{\alpha_{\mu-1},\alpha_\mu}(b_{\alpha_\mu,\alpha_{\mu+1}})$$

$$\cdot \frac{B_{kk}(b_{k\alpha_2})}{A_{\alpha_r,j}(b_{\alpha_r,j})}, \tag{126}$$

where

$$\frac{\Delta_j(b_{1j}, t)}{\Delta_k(b_{1k}, t)}$$

$$= \frac{\displaystyle\sum_{r=1}^{k-1} \sum_{\alpha \in Q_{r+1}^{(1,k)}} \frac{1}{b_{1\alpha_2}} \prod_{\mu=1}^{r} \lambda_{\alpha_\mu,\alpha_{\mu+1}}(t) \prod_{\mu=2}^{r} \ell_{\alpha_{\mu-1},\alpha_\mu}(b_{\alpha_\mu,\alpha_{\mu+1}}) \frac{1}{A_{\alpha_r,k}(b_{\alpha_r,k})}}{\displaystyle\sum_{r=1}^{j-1} \sum_{\alpha \in Q_{r+1}^{(1,j)}} \frac{1}{b_{1\alpha_2}} \prod_{\mu=1}^{r} \lambda_{\alpha_\mu,\alpha_{\mu+1}}(t) \prod_{\mu=2}^{r} \ell_{\alpha_{\mu-1},\alpha_\mu}(b_{\alpha_\mu,\alpha_{\mu+1}}) \frac{1}{A_{\alpha_r,j}(b_{\alpha_r,j})}}. \tag{127}$$

In these formulas, for $1 \le k < j \le N$, $\alpha \in Q_{r+1}^{(k,j)}$, *we have*

$$\prod_{\mu=1}^{r} \lambda_{\alpha_\mu,\alpha_{\mu+1}}(t) = \left(\prod_{\mu=1}^{r} \lambda_{\alpha_\mu,\alpha_{\mu+1}}(0) \right) \exp\left\{ \left[b_{1k}(f(u_1) - f(u_k)) \right. \right.$$

$$+ \sum_{\mu=1}^{r} b_{\alpha_\mu,\alpha_{\mu+1}}(f(u_{\alpha_\mu}) - f(u_{\alpha_{\mu+1}}))$$

$$\left. \left. - b_{1j}(f(u_1) - f(u_j)) \right] t \right\}. \tag{128}$$

Method 2. Solving a Riemann-Hilbert factorization problem

In this method, we will (for simplicity) make the stronger assumption that f can be extended to an entire function on $\mathbb{C}$.

Let $\mathfrak{b}_-^{\mathbb{C}}$ be the complexification of $\mathfrak{b}_-$ and let $B_-^{\mathbb{C}}$ be its Lie group consisting of complex $N \times N$ invertible lower triangular matrices. We begin by giving an existence result on the factorization problem and show how the solution of the factorization problem enables us to explicitly solve the Cauchy problem.

Theorem 7.3 *For* $t \ge 0$, *there exist unique holomorphic matrix-valued functions*

$$g_+(\cdot, t) : \mathbb{CP}^1 \setminus \{\infty\} \longrightarrow B_-^{\mathbb{C}},$$

$$g_-(\cdot, t) : \mathbb{CP}^1 \setminus \{0\} \longrightarrow B_-^{\mathbb{C}}, \tag{129}$$

which are smooth in t, satisfy the conditions

$$\text{(a)} \quad g_+(\cdot, t) \mid S^1 \in \widetilde{M}, \; g_+(0, t) \in B_- \cap M, \quad \text{(b)} \quad g_-(\cdot, t) \mid S^1 \in \widetilde{D},$$
$$(130)$$

and solve the factorization problem

$$e^{thf\left(u - \frac{A_0}{h}\right)} = g_+(h, t)g_-(h, t)^{-1}, \; h \in \mathbb{CP}^1 \setminus \{0, \infty\} \quad (131)$$

for $A_0 \in \mathfrak{q} \cap \mathfrak{b}_-$

Moreover, the solution of (33) with initial data $A_u(h, 0) = hu - A_0$ is given by

$$A_u(h, t) = g_{\pm}(h, t)^{-1} A_u(h, 0) g_{\pm}(h, t). \quad (132)$$

In particular, $A(t) = g_+(0, t)^{-1} A(0) g_+(0, t)$.

We now give a sketch of the main steps in the explicit solution of the factorization problem in the above theorem for $A_0 \in (\mathfrak{q} \cap \mathfrak{b}_-)'$. So let C be the spectral curve given by $\det(hu - A_0 - zI) = 0$, and let $\widehat{v}(P) \in \mathbb{C}^N$ satisfy the eigenvalue problem

$$A_u(h(P), 0)^T \widehat{v}(P) = z(P)\widehat{v}(P), \; (e, \widehat{v}(P)) = 1, \; P \in C^\nu. \quad (133)$$

Then

$$h(P)f\left(u - \frac{A_0}{h(P)}\right)^T \widehat{v}(P) = \mu(P)\widehat{v}(P), \quad (134)$$

where

$$\mu(P) = h(P)f\left(\frac{z(P)}{h(P)}\right). \quad (135)$$

Therefore, if we define

$$v_{\pm}(t, P) = g_{\pm}(h(P), t)^T \widehat{v}(P), \quad (136)$$

then it follows from (131), (132) and (134) above that

$$v_+(t, P) = e^{t\mu(P)}v_-(t, P), \; P \in C^\nu$$
$$A_u(h(P), t)^T v_{\pm}(t, P) = z(P)v_{\pm}(t, P), \quad (137)$$

where $A_u(h, t) := hu - A(t)$. Hence the components of $v_{\pm}(t, P)$ satisfy scalar Riemann-Hilbert problems. Now let $D_i(0)$, $i = 2, \ldots, N$ be the divisors of poles of $\widehat{v} \mid_{C_i}$. Write $D_i(0) = \sum_{j=1}^{j-1}[1 : h_{ij} : z_{ij}]$, we let $\Delta_i(h) = \prod_{j=1}^{i-1}(h - h_{ij})$. Therefore, if we introduce $\omega_{\pm}(t, \cdot)$ such that

$$\omega_+(t, \cdot)\mid_{C_i} = \frac{e^{t(hf(u_i) - f'(u_i)a_{ii})}}{\Delta_i(h)}$$

$$\omega_-(t, \cdot)\mid_{C_i} = \frac{e^{-t(\mu|_{C_i} - f(u_i)h + f'(u_i)a_{ii})}}{\Delta_i(h)}.$$ (138)

Then we have

$$\omega_+(t, \cdot)\mid_{C_i} = e^{t\mu(\cdot)}\omega_-(t, \cdot)\mid_{C_i},$$

$$(\omega_+(t, \cdot)\mid_{C_i}) \geq -D_i(0) \quad \text{on } C_i \setminus Q_i,$$

$$(\omega_-(t, \cdot)\mid_{C_i}) \geq -D_i(0) \quad \text{on } C_i \setminus Q_+^i.$$ (139)

Therefore, by comparing the factorization problems restricted to C_i, we have

$$\omega_+(t, \cdot)^{-1}v_+(t, \cdot)\mid_{C_i} = \omega_-(t, \cdot)^{-1}v_-(t, \cdot)\mid_{C_i} \quad \text{on } C_i \setminus \{Q_i, Q_+^i\}. \quad (140)$$

Therefore the vector-valued function

$$\tilde{v}(t, P)\mid_{C_i} = \begin{cases} \omega_+(t, P)^{-1}v_+(t, P)\mid_{C_i}, & P \in C_i \setminus Q_i \\ \omega_-(t, P)^{-1}v_-(t, P)\mid_{C_i}, & P \in C_i \setminus Q_+^i \end{cases}$$ (141)

is meromorphic on C_i, $i = 2, \ldots N$. Indeed, by a detailed analysis, we can show that

$$\tilde{v}(t, h)\mid_{C_i} = c_i(t)\phi(t, h)\mid_{C_i},$$ (142)

where $\phi(t, h)\mid_{C_i}$ is the eigenvector of $A_u(h, t)^T$ normalized by the condition $\phi_i(t, h)\mid_{C_i} = B_{ii}(h)$. This eigenvector, by the results in Section 5 and Section 6, can be written down explicitly. Indeed, we have $\phi_i(t, h)\mid_{C_i} = B_{ii}(h)$, while for $k < i$,

$$\phi_k(t, h)\mid_{C_i} = \frac{\Delta_i(b_{1i}, t)}{\Delta_k(b_{1k}, t)} \sum_{r=1}^{i-k} \sum_{\alpha \in Q_{r+1}^{(k,i)}} \prod_{\mu=1}^{r} \lambda_{\alpha_\mu \alpha_{\mu+1}}(t) B_{kk}(b_{k\alpha_2})$$

$$\times \prod_{\mu=2}^{r} \ell_{\alpha_{\mu-1}, \alpha_\mu}(b_{\alpha_\mu, \alpha_{\mu+1}})\ell_{\alpha_r, i}(h),$$ (143)

where the terms depending on t are given by (127) and (128). Hence

$$v_\pm(t, h)\mid_{C_i} = c_i(t)\omega_\pm(t, h)\phi(t, h)\mid_{C_i}$$ (144)

where $c_i(t)$ is to be determined. To do so, we use the fact that $g_+(0, t)^{-1} \in M$ from which we find

$$c_i(t) = e^{tf'(u_i)a_{ii}} \frac{\Delta_i^o(0)}{\Delta_i^o(0, t)}, \quad i = 1, \ldots, N.$$ (145)

Moreover, if we introduce

$$\Phi(t, h) = (\phi(t, h) \,|_{C_1} \ldots \phi(t, h) \,|_{C_N}), \quad \Phi(h) = \Phi(0, h),$$
$$E(t, h) = \operatorname{diag}\left(e^{t(hf(u_1) - f'(u_1)a_{11})}, \ldots, e^{t(hf(u_N) - f'(u_N)a_{NN})}\right), \tag{146}$$

then

$$g_+(h, t) = (\Phi(h)^T)^{-1} c(t) E(t, h) \Phi(t, h)^T$$
$$g_+(0, t) = (\Phi(0)^T)^{-1} \operatorname{diag}\left(\frac{\Delta_1^o(0)}{\Delta_1^o(0, t)}, \ldots, \frac{\Delta_N^o(0)}{\Delta_N^o(0, t)}\right) \Phi(t, 0)^T. \tag{147}$$

Remark 7.4 If $P(x, t)$ denote the transition matrix (semigroup) of the *continuous-space* Markov chain $(u(x, t))_{x \in \mathbb{R}}$, $t \geq 0$, then $P(x, t) = e^{xA(t)}$, $x \geq 0$. Hence it follows from the formula for $A(t)$ in terms of A_0 above that

$$P(x, t) = g_+(0, t)^{-1} P(x, 0) g_+(0, t). \tag{148}$$

This is a formula which is not available to us without using the factorization method. Note that in particular, the quantities

$$\mathbb{P}[u(x + x_0, t) = u_i | u(x_0, t) = u_i] = e^{xa_{ii}(0)}, \quad x \geq 0 \tag{149}$$

are independent of t, as the diagonal entries of the infinitesimal generator matrix are conserved quantities.

Acknowledgments The author gratefully acknowledges the support from the Simons Foundation through grants #278994 and #585813.

References

[ACG] E. Arbarello, M. Cornalba, P. Griffiths, Geometry of algebraic curves. Volume II. Springer, Heidelberg, 2011. xxx+963 pp.

[AE] M. Avellaneda, Weinan, E., Statistical properties of shocks in Burgers turbulence. *Commun. Math. Phys.* **172** (1995), 13–38.

[Ber] J. Bertoin, The inviscid Burgers equation with Brownian initial velocity. *Commun. Math. Phys.* **193** (1998), 397–406.

[B] J. M. Burgers, The nonlinear diffusion equation. Dordrecht, Reidel, 1974.

[CD] L. Carraro, J. Duchon, Équation de Burgers avec conditions initiales à accroissements indépendants et homogènes. *Ann. Inst. Henri Poincaré, Anal. Non Linéaire* **15** (1998), 431–458.

[ChD] M.-L. Chabanol, J. Duchon, Markovian solutions of inviscid Burgers equation. *J. Stat. Phys.* **114** (2004), 525–534.

[DLT] P. Deift, L.C. Li and C. Tomei, Matrix factorizations and integrable systems. *Comm. Pure Appl. Math.* **42** (1989), 443–521.

[FB] U. Frisch, J. Bec, "Burgulence". Turbulence: nouveaux aspects/New trends in turbulence (Les Houches, 2000), 341–383, EDP Sci., Les Ulis, 2001.

[FM]   L. Frachebourg, P. Martin, Exact statistical properties of the Burgers equation. *J. Fluid Mech.* **417** (2000), 323–349.

[G]   P. Groeneboom, Brownian motion with a parabolic drift and Airy functions. *Probab. Theory Relat. Fields* **81** (1989), 79–109.

[GMS] S. Gurbatov, A. Malakhov, A. Saichev, Nonlinear random waves and turbulence in nondispersive media: Waves, Rays and Particles. Manchester University Press, Manchester, 1991.

[KR]   D. Kaspar, F. Rezakhanlou, Scalar conservation laws with monotone pure-jump Markov initial conditions. *Probab. Theory Relat. Fields* **165** (2016), 867–899.

[L1]   L.-C. Li, A finite dimensional integrable system arising in the study of shock clustering. *Commun. Math. Phys.* **340** (2015), 1109–1142.

[L2]   L.-C. Li, Erratum to: A finite dimensional integrable system arising in the study of shock clustering. *Commun. Math. Phys.* **352** (2017), 1265–1269.

[L3]   L.-C. Li, An exact discretization of a Lax equation for shock clustering and Burgers turbulence I: dynamical aspects and exact solvability. *Commun. Math. Phys.* **361** (2018), 415–466

[LM]   P. Libermann, C.-M. Marle, Symplectic geometry and analytical mechanics. Mathematics and its applications, 35. D. Reidel Publishing Co., Dordrecht, 1987. xvi+526 pp.

[MS]   G. Menon, R. Srinivasan, Kinetic theory and Lax equations for shock clustering and Burgers turbulence. *J. Stat. Phys.* **140** (2010), 1195–1223.

[M]   G. Menon, Complete integrability of shock clustering and Burgers turbulence. *Arch. Ration. Mech. Anal.* **203** (2012), 853–882.

[MP]   G. Menon, R. Pego, Universality classes in Burgers turbulence. *Commun. Math. Phys.* **273** (2007), 177–202.

[N]   J. Norris, *Markov chains*, Cambridge series in statistical and probabilistic mathematics, 2. Cambridge University Press, Cambridge, 1998. xvi +237 pp.

[P]   D. Poole, The stochastic group. *Amer. Math. Monthly* **102** (1995), 798–801.

[SAF] Z.-S. She, E. Aurell, U. Frisch, The inviscid Burgers equation with initial data of Brownian type. *Commun. Math. Phys.* **148** (1992), 623–641.

[S]   Y. Sinai, Statistics of shocks in solutions of inviscid Burgers equation. *Commun. Math. Phys.* **148** (1992), 601–621.

[STS] M. Semenov-Tian-Shansky, What is a classical r-matrix? *Funct. Anal. Appl.* **17** (1983), 259–272.

[SZ]   S. F. Shandarin, Ya. B. Zeldovich, The large-scale structure of the universe: Turbulence, intermittency, structures in a self-gravitating medium. *Rev. Mod. Phys.* **61** (1989), 185–220.

[W]   M. Winkel, Limit clusters in the inviscid Burgers turbulence with certain random initial velocities. *J. Stat. Phys.* **107** (2002), 893–917.

[Woy] W. Woyczyński, Göttingen lectures on Burgers-KPZ turbulence, *Lecture Notes in Math.*, vol. 1700, Springer-Verlag, Berlin-Heidelberg, 1998.

Luen-Chau Li

Department of Mathematics, Pennsylvania State University, University Park, PA 16802, USA
E-mail address: luenli@math.psu.edu

5

Solvable Dynamical Systems in the Plane with Polynomial Interactions

Francesco Calogero and Farrin Payandeh

Abstract. In this paper we report a few examples of *algebraically solvable* dynamical systems characterized by 2 coupled Ordinary Differential Equations which read as follows:

$$\dot{x}_n = P^{(n)}(x_1, x_2), \quad n = 1, 2,$$

with $P^{(n)}(x_1, x_2)$ specific *polynomials* of relatively low degree in the 2 dependent variables $x_1 \equiv x_1(t)$ and $x_2 \equiv x_2(t)$. These findings are obtained via a new twist of a recent technique to identify dynamical systems *solvable by algebraic operations*, themselves explicitly identified as corresponding to the time evolutions of the *zeros* of polynomials the *coefficients* of which evolve according to *algebraically solvable* (systems of) evolution equations.

1 Introduction

It has been recently noted [1] that, if the quantities $x_n(t)$ respectively $y_m(t)$ denote the N zeros respectively the N coefficients of a *generic* time-dependent monic polynomial $p_N(z; t)$ of degree N,

$$p_N(z; t) = z^N + \sum_{m=1}^{N} \left[y_m(t) \, z^{N-m} \right] = \prod_{n=1}^{N} \left[z - x_n(t) \right], \tag{1}$$

there hold the following identities relating the time evolution of these quantities:

$$\dot{x}_n = - \left[\prod_{\ell=1, \, \ell \neq n}^{N} (x_n - x_\ell) \right]^{-1} \sum_{m=1}^{N} \left[\dot{y}_m \, (x_n)^{N-m} \right], \quad n = 1, 2, \ldots, N. \tag{2a}$$

Notation 1.1 Hereafter all quantities are *a priori* assumed to be *complex* numbers, with the following exceptions: *indices* such as n, m, take positive integer values (over ranges specified on a case-by-case basis: indeed, in most

93

of this paper the range is limited just to the 2 values 1 and 2 for n, and to 1 and 2 or 1, 2 and 3 for m), while the independent variable t ("time") is *real* and it is generally assumed to run from 0 to $+\infty$. The t-dependence of time-dependent variables such as $x_n(t)$ and $y_m(t)$ is often not *explicitly* displayed (even, inconsistently, in the same formula: of course when this is unlikely to cause misunderstandings); and superimposed dots on these variables denote of course time-differentiations, $\dot{x}_n \equiv dx_n(t)/dt$, $\dot{y}_m \equiv dy_m(t)/dt$. It is of course not excluded that *complex* numbers take *real* or *imaginary* values, as indicated below on a case-by-case basis: indeed the words "in the plane" in the title of this paper refer to the standard case in which the two coordinates $x_1(t)$ and $x_2(t)$ are interpreted as the 2 *real* coordinates of a point moving in the Cartesian $x_1 x_2$-plane, or as the 2 *complex* coordinates of 2 points moving in the *complex* plane (or, equivalently, of 2 *real* two-vectors moving in a plane: see below).　■

Analogous formulas to (2a) also exist for higher time-derivatives [2] [3], and via such formulas many new *algebraically solvable* dynamical systems have been recently identified and discussed, especially dynamical systems characterized by second-order Ordinary Differential Equations (ODEs) of Newtonian type ("accelerations equal forces"): for an overview see [4] and references therein. But in this paper our treatment is confined to systems involving *first-order* time-derivatives.

In this paper we moreover confine attention to the very simplest such systems: characterized by *first-order* Ordinary Differential Equations involving only 2 dependent variables. Let us tersely review here—in this very simple context—how this approach works.

Systems of *algebraically solvable* first-order ODEs for the zeros $x_n(t)$ are obtained from the identities (2a) by assuming that the N coefficients $y_m(t)$ satisfy themselves an *algebraically solvable* system of first-order ODEs. Note that in the very simple case with $N = 2$ the equations (2a) read simply as follows:

$$\dot{x}_n = (-1)^n \left(\frac{x_n \, \dot{y}_1 + \dot{y}_2}{x_1 - x_2} \right), \quad n = 1, 2 . \tag{2b}$$

Now assume that the system of 2 ODEs

$$\dot{y}_1 = f_1 (y_1, y_2), \quad \dot{y}_2 = f_2 (y_1, y_2), \tag{3a}$$

be *algebraically solvable* (of course, for an appropriate assignment of the 2 functions $f_1 (y_1, y_2)$ and $f_2 (y_1, y_2)$). Then the system

$$\dot{x}_n = (-1)^n \left[\frac{x_n \, f_1 (-x_1 - x_2, x_1 x_2) + f_2 (-x_1 - x_2, x_1 x_2)}{x_1 - x_2} \right], \quad n = 1, 2 \tag{3b}$$

is as well *algebraically solvable*, because it clearly corresponds to (2b) via the 2 identities

$$y_1(t) = -[x_1(t) + x_2(t)], \quad y_2(t) = x_1(t)x_2(t) \tag{4a}$$

clearly associated to the polynomial (1) with $N = 2$,

$$p_2(z;t) = z^2 + y_1(t)z + y_2 = [z - x_1(t)][z - x_2(t)] . \tag{4b}$$

Indeed the solution of its initial-values problem—to compute $x_1(t)$ and $x_2(t)$ via (2b) from the assigned initial data $x_1(0)$ and $x_2(0)$—can be achieved via the following 3 steps: *(i)* from the initial data $x_1(0)$ and $x_2(0)$ compute the corresponding initial data $y_1(0)$ and $y_2(0)$ via the simple formulas (4a) (at $t = 0$); *(ii)* compute $y_1(t)$ and $y_2(t)$ from the initial data $y_1(0)$ and $y_2(0)$ via the, assumedly *algebraically solvable*, system of evolution equations (3a) characterizing the time-evolution of these variables; *(iii)* the variables $x_1(t)$ and $x_2(t)$ are then obtained as the 2 *zeros* of the, now known, polynomial (4b) (via an *algebraic* operation, indeed one that in this case of a polynomial of *second-degree* can be performed *explicitly*: note however that this operation yields 2 *a priori* indistinguishable functions $x_n(t)$ with $n = 1, 2$; to identify which is $x_1(t)$ and which is $x_2(t)$ these solutions must be followed back—by continuity in time, from the time t to the initial time 0—to identify which one of them corresponds to the initially assigned data $x_1(0)$ respectively $x_2(0)$).

The new twist of this approach on which the findings reported in this paper are based is to assume that the two functions $f_m(y_1, y_2)$ with $m = 1, 2$—besides implying the solvability of the system (3a)—feature the additional properties to be *polynomial* in their arguments and moreover to satisfy identically—i. e., for all values of the variable x—the relation

$$x\, f_1(-2x, x^2) + f_2(-2x, x^2) = 0 , \tag{5a}$$

which clearly implies that the 2 polynomials

$$x_n f_1(-x_1 - x_2, x_1 x_2) + f_2(-x_1 - x_2, x_1 x_2) , \quad n = 1, 2 \tag{5b}$$

contain both the factor $x_1 - x_2$. Therefore this condition (5a) is sufficient to imply that the system of ODEs (3b) in fact feature a *polynomial* right-hand side:

$$\dot{x}_n = P^{(n)}(x_1, x_2) , \quad n = 1, 2 , \tag{6}$$

with $P^{(n)}(x_1, x_2)$ *polynomial* in its 2 arguments.

In the following Section 2 we discuss a rather simple example manufactured in this manner (hereafter referred to as "Example 1.2"), the equations of motion of which read as follows:

Example 1.2

$$\dot{x}_n = a + b\left[(x_n)^2 - 4x_1x_2 - (x_{n+1})^2\right], \quad n = 1, 2 \; \; mod \, [2] \,, \quad (7)$$

with a and b two arbitrary parameters.   ∎

In Section 3 and its subsections we discuss 3 other somewhat analogous models (hereafter referred to respectively as "Examples 1.3, 1.4, 1.5") obtained via a recent development of the above approach to identify *algebraically solvable* dynamical systems, in which the role of the *generic* polynomial (1) is however replaced by a polynomial featuring, for all time, *one double zero* [5]. The equations of motion characterizing these 3 dynamical systems read as follows:

Example 1.3

$$\dot{x}_1 = a + b\left[(x_1)^2 + 7x_1x_2 + (x_2)^2\right],$$

$$\dot{x}_2 = a + b\left[7\,(x_1)^2 + 4x_1x_2 - 2\,(x_2)^2\right] \; \blacksquare \;; \quad (8)$$

Example 1.4

$$\dot{x}_n = x_n\left[a - b\,(x_1)^2\,x_2\right], \quad n = 1, 2 \; \blacksquare \;; \quad (9)$$

Example 1.5

$$\dot{x}_n = x_n\,[a + bx_1\,(x_1 + 2x_2)], \quad n = 1, 2 \; \blacksquare \;; \quad (10)$$

again, in each of these 3 cases, with *a* and *b* two arbitrary parameters.

Remark 1.6   Of course in all these examples the presence of the 2 *a priori arbitrary* parameters *a* and *b* is somewhat insignificant: indeed, both can clearly be replaced by *unity* by rescaling the independent variable ($t \Rightarrow \alpha t$) and the dependent variables ($x_n \Rightarrow \beta x_n$) (with obvious appropriate assignments of the parameters α and β). Moreover all these examples with an *arbitrary nonvanishing* value of the parameter *a* can be obtained via analogous models with $a = 0$ via a simple change of the independent variable (see below Subsection 4.2). While models featuring more arbitrary parameters can be derived from these via a simple change of dependent variables (see below Subsection 4.1).   ∎

Indeed, in Section 4 and its subsections we tersely outline some variants of the examples discussed in Sections 2 and 3, thereby enlarging the class of *algebraically solvable* dynamical systems identifiable via the technique introduced in this paper. These models might be of interest in applicative contexts: indeed, dynamical systems of the type discussed in this paper play a role in an ample variety of such contexts (say, from population dynamics

Example 1 97

to chemical reaction to econometric projections, etc.: you name it). But in this paper we merely focus on the presentation of the technique that subtends the identification of this kind of *algebraically solvable* dynamical systems characterized by coupled systems of 2 first-order ODEs with polynomial right-hand sides, see (6).

Finally Section 5 mentions possible future developments of thes findings.

2 Example 1

In this Section 2 we demonstrate the *algebraically solvable* character of the dynamical system (7).

The starting point of our treatment is the dynamical system (2b) with

$$\dot{y}_1 = \alpha_0 + \alpha_1 y_2, \quad \dot{y}_2 = \beta_0 y_1 + \beta_1 (y_1)^3 , \tag{11a}$$

corresponding to (3a) with

$$f_1 (y_1, y_2) = \alpha_0 + \alpha_1 y_2, \quad f_2 (y_1, y_2) = \beta_0 y_1 + \beta_1 (y_1)^3 , \tag{11b}$$

where α_0, α_1, β_0, β_1 are 4 *a priori arbitrary* parameters.

These equations of motion clearly imply that the condition (5a) is satisfied provided

$$\alpha_0 = 2\beta_0, \quad \alpha_1 = 8\beta_1 ; \tag{12}$$

and it is as well easily seen that there then obtains the system (7) with $a = -\beta_0$, $b = \beta_1$, via the insertion of (11a) in (2b) (of course with $y_1 = - (x_1 + x_2)$ and $y_2 = x_1 x_2$: see (4a)).

On the other hand it is easily seen that the system (11a) is *explicitly solvable*: indeed the equations of motion (11a) clearly imply the *second-order* ODE (of Newtonian type: "acceleration equal force")

$$\ddot{y}_1 = \alpha_1 \left[\beta_0 y_1 + \beta_1 (y_1)^3 \right] , \tag{13a}$$

namely, via (12),

$$\ddot{y}_1 = 8\beta_1 \left[\beta_0 y_1 + \beta_1 (y_1)^3 \right] . \tag{13b}$$

This second-order ODE—which is of course integrable via two quadratures—is clearly the Newtonian equation of motion of the simplest anharmonic oscillator (although, in the *real* domain, with a force that at large distance pushes the solution away from the origin). The most direct way to demonstrate the *algebraically solvable* character of this equation of motion is to exhibit its solution which—as the interested reader will easily verify—reads, in terms (for instance) of the *first* Jacobian elliptic function (see for instance [6]), as follows:

$$y_1(t) = \mu \, \text{sn} \, (\lambda t + \rho, k) \,, \tag{14a}$$

where μ and λ are determined in terms of the parameter k as follows

$$\lambda^2 = -\frac{8\beta_0\beta_1}{1+k^2}, \quad \mu^2 = -\frac{2\beta_0 k^2}{\beta_1 \left(1+k^2\right)} \,, \tag{14b}$$

ρ is determined in terms of the initial datum $y_1(0)$ as follows,

$$y_1(0) = \mu \, \text{sn} \, (\rho, k) \,, \tag{14c}$$

and the parameter k is determined in terms of the initial data $y_1(0)$ and $y_2(0)$ as the solution of the following algebraic equation

$$\left\{[\dot{y}_1(0)]^2 - 8\beta_0\beta_1 \left[y_1(0)\right]^2 - (2\beta_1)^2 \left[y_1(0)\right]^4\right\} \left(1+k^2\right)^2 = (4\beta_0)^2 k^2 \,, \tag{14d}$$

where of course (see the first (11a))

$$\dot{y}_1(0) = 2\beta_0 + 8\beta_1 y_2(0) \,. \tag{14e}$$

And of course, once $y_1(t)$ is known, $y_2(t)$ is given directly by the first (11a).

Remark 2.1   For given assigned values of $y_1(0)$ and $y_2(0)$ (hence $\dot{y}_1(0)$, see (14e)), (14d) is a *quadratic* equation for k^2; the choice of the appropriate solution for k^2 among the 2 solutions of this elementary equation must of course be made *cum grano salis*.   ∎

3 Examples 2, 3 and 4

In the 3 subsections of this Section 3 we demonstrate the *algebraically solvable* character of the 3 dynamical systems (8), (9) and (10).

But let us first summarize some relevant findings of [5].

Let $p_3(z; t)$ be a time-dependent polynomial of *third* degree in its argument z which, for all time, features a *double pole*:

$$p_3(z; t) = z^3 + \sum_{m=1}^{3} \left[y_m(t)z^{3-m}\right] = \left[z - x_1(t)\right]^2 \left[z - x_2(t)\right] \,. \tag{15a}$$

This of course implies that its 3 *coefficients* $y_m(t)$ are expressed as follows in terms of the *double zero* $x_1(t)$ and the *zero* (of unit multiplicity) $x_2(t)$:

$$y_1 = -\left(2x_1 + x_2\right), \quad y_2 = x_1\left(x_1 + 2x_2\right), \quad y_3 = -\left(x_1\right)^2 x_2 \,; \tag{15b}$$

and correspondingly that the 3 coefficients $y_m(t)$ are, for all time, related to each other by the (single) condition implied by the simultaneous vanishing at $z = x_1(t)$ of both $p_3(z;t)$ and its z-derivative $p_{3,z}(z;t)$:

$$p_3(x_1;t) = [x_1(t)]^3 + \sum_{m=1}^{3} \left\{ y_m(t) [x_1(t)]^{3-m} \right\} = 0, \qquad (15c)$$

$$p_{3,z}(x_1;t) = 3[x_1(t)]^2 + 2y_1(t)x_1(t) + y_2(t) = 0. \qquad (15d)$$

In an analogous manner (see the treatment in Section 2, and if need be [5]) it is possible to obtain the following 3 pairs of formulas (analogous to, but of course somewhat different from, the formulas (2b)):

$$\dot{x}_1 = -\frac{2x_1 \dot{y}_1 + \dot{y}_2}{2(x_1 - x_2)}, \qquad \dot{x}_2 = \frac{(x_1 + x_2)\dot{y}_1 + \dot{y}_2}{x_1 - x_2}; \qquad (16a)$$

$$\dot{x}_1 = -\frac{(x_1)^2 \dot{y}_1 - \dot{y}_3}{2x_1(x_1 - x_2)}, \qquad \dot{x}_2 = \frac{x_1 x_2 \dot{y}_1 - \dot{y}_3}{x_1(x_1 - x_2)}; \qquad (16b)$$

$$\dot{x}_1 = \frac{x_1 \dot{y}_2 + 2\dot{y}_3}{2x_1(x_1 - x_2)}, \qquad \dot{x}_2 = -\frac{x_1 x_2 \dot{y}_2 + (x_1 + x_2)\dot{y}_3}{(x_1)^2(x_1 - x_2)}. \qquad (16c)$$

It is then clear—in close analogy to the treatment described above (see Section 1)—that each of these 3 pairs of formulas opens the way to the identification of *algebraically solvable* dynamical systems involving the 2 dependent variables $x_1(t)$ and $x_2(t)$: as separately discussed in the following 3 subsections.

3.1 Example 2

In this Subsection 3.1 we demonstrate the *algebraically solvable* character of the dynamical systems (8).

Now the starting point of our treatment is—instead of the system (2b)—the slightly different system (16a). Clearly this system is *solvable by algebraic operations* if the quantities $y_1(t)$ and $y_2(t)$ satisfy the system (3a) and this system is itself *solvable*. Then the system satisfied by the variables $x_1(t)$ and $x_2(t)$—obtained by replacing, in the right hand side of (8), $\dot{y}_1$ and $\dot{y}_2$ via the equations of motion (3a)—reads

$$\dot{x}_1 = -\frac{2x_1 f_1\left(-2x_1 - x_2, (x_1)^2 + 2x_1 x_2\right) + f_2\left(-2x_1 - x_2, (x_1)^2 + 2x_1 x_2\right)}{2(x_1 - x_2)},$$

$$\dot{x}_2 = (x_1 - x_2)^{-1}\left[(x_1 + x_2) f_1\left(-2x_1 - x_2, (x_1)^2 + 2x_1 x_2\right) \right.$$
$$\left. + f_2\left(-2x_1 - x_2, (x_1)^2 + 2x_1 x_2\right)\right], \qquad (17)$$

corresponding now to the assignment (15b) (instead of (4a)) of $y_1(t)$ and $y_2(t)$ in terms of $x_1(t)$ and $x_2(t)$.

It is now clear that the conditions on the 2 functions $f_1(y_1, y_2)$ and $f_2(y_1, y_2)$ which are sufficient to guarantee that the right-hand side of the equations of motion (17) be *polynomial* in the 2 dependent variables $x_1(t)$ and $x_2(t)$ are that these 2 functions $f_1(y_1, y_2)$ and $f_2(y_1, y_2)$ be themselves *polynomial* in their 2 variables y_1 and y_2 and moreover that there hold identically—i.e., for all values of x—the relation

$$2xf_1(-3x, 3x^2) + f_2(-3x, 3x^2) = 0 . \tag{18}$$

We now assume that the time-evolution of the 2 quantities $y_1(t)$ and $y_2(t)$ be again characterized by the equations of motion (11a)—the *solvable* character of which has been pointed out in Section 2—hence by the assignments (11b) of the two functions $f_1(y_1, y_2)$ and $f_2(y_1, y_2)$. It is then easily seen that the condition (18) entails now the relations

$$\alpha_0 = \frac{3\beta_0}{2}, \quad \alpha_1 = \frac{9\beta_1}{2} \tag{19}$$

(instead of (12)).

It is then easily seen that the corresponding dynamical system satisfied by the coordinates $x_1(t)$ and $x_2(t)$ is just the system of 2 ODEs (8) (with $a = -\beta_0/2, b = -\beta_1/2$).

There remains to report—from [5]—how to obtain from the variables $y_1(t)$ and $y_2(t)$ the variables $x_1(t)$ and $x_2(t)$. The variable $x_1(t)$ is that one of the 2 roots of the—of course *explicitly solvable*—second-degree polynomial equation in x

$$3x^2 + 2y_1(t)x + y_2(t) = 0 \tag{20}$$

(see (15d)) which, by continuity in t, corresponds at $t = 0$ to the initially assigned datum $x_1(0)$. While $x_2(t)$ is then given by the formula

$$x_2(t) = -y_1(t) - 2x_1(t) \tag{21}$$

(see the first of the 3 formulas (15b)).

3.2 Example 3

In this Subsection 3.2 we demonstrate the *algebraically solvable* character of the dynamical systems (9).

Now the starting point of our treatment is the system of 2 coupled ODEs (16b). Clearly this system is *solvable by algebraic operations* if the quantities $y_1(t)$ and $y_3(t)$ satisfy the system

$$\dot{y}_1 = f_1(y_1, y_3), \quad \dot{y}_3 = f_3(y_1, y_3) , \tag{22}$$

and this system is itself *solvable* (of course for an appropriate assignment of the 2 functions $f_1(y_1, y_3)$ and $f_3(y_1, y_3)$). Then the system satisfied by the variables $x_1(t)$ and $x_2(t)$—obtained by replacing, in the right-hand side of (16b), $\dot{y}_1$ and $\dot{y}_3$ via these equations of motion (22)—reads

$$\dot{x}_1 = \frac{-(x_1)^2 f_1\left(-2x_1 - x_2, -(x_1)^2 x_2\right) + f_3\left(-2x_1 - x_2, -(x_1)^2 x_2\right)}{2x_1(x_1 - x_2)},$$

$$\dot{x}_2 = [x_1(x_1 - x_2)]^{-1}\left[x_1 x_2 f_1\left(-2x_1 - x_2, (x_1)^2 + 2x_1 x_2\right)\right.$$

$$\left. - f_3\left(-2x_1 - x_2, (x_1)^2 + 2x_1 x_2\right)\right], \qquad (23)$$

corresponding to the assignment (15b) of $y_1(t)$ and $y_3(t)$ in terms of $x_1(t)$ and $x_2(t)$; and it is easily seen that sufficient conditions to guarantee that this become a system of 2 ODEs featuring in their right-hand sides a *polynomial* dependence on the 2 dependent variables $x_1(t)$ and $x_2(t)$ are that these 2 functions $f_1(y_1, y_2)$ and $f_3(y_1, y_2)$ be themselves *polynomial* in their 2 variables y_1 and y_3 and moreover that there hold identically—i. e., for all values of x—the relation

$$\frac{x^2 f_1\left(-3x, -x^3\right) - f_3\left(-3x, -x^3\right)}{x} = 0. \qquad (24)$$

Let us now assume that the two functions $f_1(y_1, y_3)$ and $f_3(y_1, y_3)$ read as follows:

$$f_1(y_1, y_3) = y_1(\alpha_1 + \alpha_2 y_3), \quad f_3(y_1, y_3) = y_3(\beta_1 + \beta_2 y_3), \qquad (25a)$$

so that the system (22) read

$$\dot{y}_1 = y_1(\alpha_1 + \alpha_2 y_3), \quad \dot{y}_3 = y_3(\beta_1 + \beta_2 y_3). \qquad (25b)$$

Here the 4 parameters $\alpha_1, \alpha_2, \beta_1, \beta_2$ are *a priori* arbitrary, but clearly to satisfy (24) it is necessary and sufficient that (as we hereafter assume, in this Subsection)

$$\beta_1 = 3\alpha_1, \quad \beta_2 = 3\alpha_2. \qquad (26)$$

It is then a matter of trivial algebra to verify that the corresponding system of 2 ODEs for the 2 dependent variables $x_1(t)$ and $x_2(t)$ is just (9), with $a = \alpha_1$, $b = -\alpha_2$.

It is on the other hand easily seen that the system (25b) is *explicitly solvable*: by firstly integrating by a quadrature the ODE satisfied by the dependent variable $y_3(t)$, and by then integrating the *linear* ODE satisfied by the dependent variable $y_1(t)$. There results the following neat expressions of the 2 variables $y_1(t)$ and $y_3(t)$:

$$y_1(t) = y_1(0)\varphi(t), \quad y_3(t) = y_3(0)[\varphi(t)]^3, \tag{27a}$$

$$\varphi(t) = \left\{ \left[1 - \left(\frac{b}{a}\right) y_3(0) \right] \exp(3at) - 1 \right\}^{-1/3}. \tag{27b}$$

The subsequent computation of the 2 dependent variables $x_1(t)$ and $x_2(t)$ from the 2 quantities $y_1(t)$ and $y_3(t)$ can then be easily performed: it involves the *algebraic* operation of solving a *cubic* equation (a task which can actually be performed explicitly), as the interested reader will easily ascertain (or, if need be, see [5]).

Remark 3.1   If the parameter a is *imaginary*—$a = i\omega$ (with, here and hereafter, **i** the *imaginary unit*, so that $\mathbf{i}^2 = -1$) and ω *real* and *nonvanishing*, $\omega \neq 0$—both *coefficients* $y_1(t)$ and $y_3(t)$ are clearly periodic with period $T = 2\pi/|\omega|$, see (27): actually $y_3(t)$ is clearly periodic with period $T/3$; while $y_1(t)$ is certainly periodic with period T but—depending on the value of the initial datum $y_3(0)$—it might also be periodic with period $T/3$. Hence the 2 coordinates $x_1(t)$ and $x_2(t)$ are themselves periodic with period T (or possibly a small integer multiple of T; see [7] [8]).

3.3 Example 4

In this Subsection 3.3 we demonstrate the *algebraically solvable* character of the dynamical systems (10).

Now the starting point of our treatment is system (16c). Clearly this system is *solvable by algebraic operations* if the quantities $y_2(t)$ and $y_3(t)$ satisfy the system

$$\dot{y}_2 = f_2(y_2, y_3), \quad \dot{y}_3 = f_3(y_2, y_3), \tag{28}$$

and this system is itself *solvable* (of course for an appropriate assignment of the 2 functions $f_2(y_1, y_3)$ and $f_3(y_1, y_3)$). Then the system satisfied by the variables $x_1(t)$ and $x_2(t)$—obtained by replacing, in the right-hand side of (16c), $\dot{y}_2$ and $\dot{y}_3$ via these equations of motion (28)—reads

$$\dot{x}_1 = \frac{x_1 f_2 \left(x_1(x_1 + 2x_2), -(x_1)^2 x_2 \right) + 2 f_3 \left(x_1(x_1 + 2x_2), -(x_1)^2 x_2 \right)}{2x_1(x_1 - x_2)},$$

$$\dot{x}_2 = -(x_1)^{-2}(x_1 - x_2)^{-1} \left\{ x_1 x_2 f_2 \left(x_1(x_1 + 2x_2), -(x_1)^2 x_2 \right) \right.$$
$$\left. + (x_1 + x_2) f_3 \left(x_1(x_1 + 2x_2), -(x_1)^2 x_2 \right) \right\}, \tag{29}$$

corresponding to the assignment (15b) of $y_2(t)$ and $y_3(t)$ in terms of $x_1(t)$ and $x_2(t)$; and it is easily seen that sufficient conditions to guarantee that this become a system of 2 ODEs featuring in their right-hand sides a *polynomial* dependence on the 2 dependent variables $x_1(t)$ and $x_2(t)$ are that these 2 functions $f_2(y_1, y_2)$ and $f_3(y_1, y_2)$ be themselves *polynomial* in their 2 variables y_2 and y_3 and moreover that there hold identically—i.e., for all values of x—the relation

$$\frac{xf_2\left(3(x)^2, -(x)^3\right) + 2f_3\left(3(x)^2, -(x)^3\right)}{x} = 0. \tag{30}$$

Let us now assume that the two functions $f_2(y_1, y_3)$ and $f_3(y_1, y_3)$ read as follows:

$$f_2(y_2, y_3) = y_2(\alpha_1 + \alpha_2 y_2), \quad f_3(y_2, y_3) = y_3(\beta_1 + \beta_2 y_2), \tag{31a}$$

so that the system (22) read

$$\dot{y}_2 = y_2(\alpha_1 + \alpha_2 y_2), \quad \dot{y}_3 = y_3(\beta_1 + \beta_2 y_2). \tag{31b}$$

Here the 4 parameters $\alpha_1, \alpha_2, \beta_1, \beta_2$ are *a priori* arbitrary, but clearly to satisfy (30) it is necessary and sufficient that (as we hereafter assume, in this subsection)

$$\beta_1 = \frac{3\alpha_1}{2}, \quad \beta_2 = \frac{3\alpha_2}{2}. \tag{32}$$

It is then a matter of trivial algebra to verify that the corresponding system of 2 ODEs for the 2 dependent variables $x_1(t)$ and $x_2(t)$ is just (10), with $a = \alpha_1/2$, $b = \alpha_2/2$.

It is on the other hand easily seen that the system (31b) is *explicitly solvable*: it is indeed, up to simple notational changes, identical to the system (31b) discussed in the preceding Subsection 3.2.

And the subsequent computation of the 2 dependent variables $x_1(t)$ and $x_2(t)$ from the 2 quantities $y_2(t)$ and $y_3(t)$ can as well be easily performed: it involves again the *algebraic* operation of solving a *cubic* equation (a task which can actually be performed explicitly), as the interested reader will easily ascertain (or, if need be, see [5]).

4 Variants

In this Section 4 and its subsections we tersely outline some interesting variants of the *algebraically solvable* models discussed above, which might be of interest for possible utilizations of these findings in applicative contexts.

4.1 First Variant

Each of the 4 dynamical systems identified above as *algebraically solvable*—
see (7), (8), (9), (10)—features only 2 arbitrary parameters, a and b. Systems
featuring more free parameters can of course be obtained from these via the
trivial change of dependent variables

$$x_1(t) = u_{10} + u_{11}\xi_1(t) + u_{12}\xi_2(t), \quad x_2(t) = u_{20} + u_{21}\xi_1(t) + u_{22}\xi_2(t),$$

$$\text{(33a)}$$

featuring the 6 parameters $u_{n\ell}$, $n = 1, 2$, $\ell = 0, 1, 2$. This change of variables
is easily inverted:

$$\xi_1(t) = u^{-1}\{u_{22}[x_1(t) - u_{10}] - u_{12}[x_2(t) - u_{20}]\},$$
$$\xi_2(t) = u^{-1}\{u_{21}[x_1(t) - u_{10}] - u_{11}[x_2(t) - u_{20}]\}, \quad \text{(33b)}$$

where, here and hereafter,

$$u = u_{11}u_{22} - u_{12}u_{21}. \quad \text{(33c)}$$

Clearly the properties of *algebraic solvability* are not affected, although the
relevant formulas become marginally more complicated, requiring the solution
of some (*purely algebraic*) equations. On the other hand the new systems of
2 ODEs satisfied by the new dependent variables $\xi_1(t)$ and $\xi_2(t)$ feature now
several more free parameters. For instance for Example 1.2 the equations that
replace (7) read as follows:

$$\dot{\xi}_n = A_n + B_{n1}\xi_1 + B_{n2}\xi_2 + C_{n1}(\xi_1)^2 + C_{n2}(\xi_2)^2 + C_{n3}\xi_1\xi_2, \quad \text{(34a)}$$

$$A_1 = u^{-1}\left\{(u_{22} - u_{12})(a - 4bu_{10}u_{20}) + b(u_{22} + u_{12})\left[(u_{10})^2 - (u_{20})^2\right]\right\},$$
$$\text{(34b)}$$

$$A_2 = u^{-1}\left\{(u_{21} - u_{11})(a - 4bu_{10}u_{20}) + b(u_{21} + u_{11})\left[(u_{10})^2 - (u_{20})^2\right]\right\},$$
$$\text{(34c)}$$

$$B_{1n} = 2bu^{-1}[2(u_{12} - u_{22})(u_{10}u_{2n} + u_{20}u_{1n})$$
$$+ (u_{22} + u_{12})(u_{10}u_{1n} - u_{20}u_{2n})], \quad n = 1, 2, \quad \text{(34d)}$$

$$B_{2n} = 2bu^{-1}[2(u_{11} - u_{21})(u_{10}u_{2n} + u_{20}u_{1n})$$
$$+ (u_{21} + u_{11})(u_{10}u_{1n} - u_{20}u_{2n})], \quad n = 1, 2, \quad \text{(34e)}$$

$$C_{1n} = bu^{-1}\left\{4(u_{12} - u_{22})u_{1n}u_{2n} + (u_{22} + u_{12})\left[(u_{1n})^2 - (u_{2n})^2\right]\right\}, \quad n = 1, 2,$$
$$\text{(34f)}$$

$$C_{2n} = bu^{-1}\left\{4\,(u_{11}-u_{21})\,u_{1n}u_{2n} + (u_{21}+u_{11})\left[(u_{1n})^2 - (u_{2n})^2\right]\right\}, \quad n=1,2,$$

(34g)

$$C_{13} = 2bu^{-1}[2(u_{12}-u_{22})(u_{11}u_{22}+u_{12}u_{21}) + (u_{22}+u_{12})(u_{11}u_{12}-u_{21}u_{22})],$$

(34h)

$$C_{23} = 2bu^{-1}[2(u_{11}-u_{21})(u_{11}u_{22}+u_{12}u_{21}) + (u_{21}+u_{11})(u_{11}u_{12}-u_{21}u_{22})].$$

(34i)

Note that if $a = u_{10} = u_{20} = 0$ then $A_n = B_{nm} = 0$ and the equations of motion (34a) have homogeneous right-hand sides (of degree 2) featuring only the 6 coefficients $C_{n\ell}$ with $n = 1, 2$ and $\ell = 1, 2, 3$, expressed in terms of the 5 *arbitrary* parameters b and u_{nm} with n and m taking the values 1 and 2.

It is left to the interested reader to obtain analogous formulas for Examples 1.3, 1.4, 1.5.

4.2 Second Variant

If the (*autonomous*) system of 2 coupled ODEs

$$\dot{x}_n = f_n\,(x_1, x_2), \quad n = 1, 2,$$

(35a)

features *homogeneous* functions $f_n\,(x_1, x_2)$ satisfying the scaling property

$$f_n\,(cx_1, cx_2) = c^p f_n\,(x_1, x_2), \quad n = 1, 2, \quad p \neq 1$$

(35b)

(where c is an arbitrary parameter), then by setting

$$w_n(t) = \exp\left(\frac{\alpha t}{p-1}\right) x_n\,(\tau(t)), \quad \tau(t) = \frac{\exp(\alpha t) - 1}{\alpha}, \quad n = 1, 2,$$

(36a)

one gets for the new dependent variables $w_n(t)$ the new (*autonomous!*) system

$$\dot{w}_n = \frac{\alpha}{p-1} w_n + f_n\,(w_1, w_2), \quad n = 1, 2.$$

(36b)

Then—if the original system (35a) is *algebraically solvable*—the solutions $w_n(t)$ of this system satisfy interesting properties: in particular, if $\alpha = i\omega$ is *imaginary*—with ω a *nonvanishing real* parameter and p a *real rational* number—then *all* solutions of these systems (36b) are *completely periodic* with some *rational integer* multiple of the basic period $T = 2\pi/|\omega|$ (*isochrony!*). For more details on the transformation (36a) and its implications see [8] and references therein.

Note that the 4 dynamical systems of Examples 1.2, 1.3, 1.4, 1.5 belong to the class (35) if the parameter a vanishes, $a = 0$: with $p = 2$ in the cases of

Examples 1.2 and 1.3 (see (7) and (8)), with $p = 4$ in the case of Example 1.4 and $p = 3$ in the case of Example 1.5 (see (9) and (10)); and that these properties continue to hold after the generalization described in the preceding Subsection 4.1, provided the parameters u_{10} and u_{20} vanish, $u_{10} = u_{20} = 0$ (see (33)).

4.3 Third Variant

Let us note that the dynamical systems detailed in the Examples reported above can be reformulated as describing the evolution of *real* 2-vectors $\vec{r}_n(t)$ lying in a (*real*) plane. Indeed set

$$\vec{r}_n(t) \equiv (Re\,[x_n(t)]\,,\,Im\,[x_n(t)])\,, \qquad n = 1, 2, \tag{37a}$$

$$\vec{a} \equiv (Re\,[a]\,,\,Im\,[a])\,, \qquad \vec{b} \equiv (Re\,[b]\,,\,-Im\,[b])\,. \tag{37b}$$

Note the minus sign in the definition of the second component of the 2-vector $\vec{b}$.

Then—as the diligent reader will easily verify—the version of (7) yielded by this notational change reads as follows:

$$\dot{\vec{r}}_n = \vec{a} + 2\vec{r}_n \left[\vec{b} \cdot (\vec{r}_n - 2\vec{r}_{n+1}) \right] - 2\vec{r}_{n+1} \left[\vec{b} \cdot (\vec{r}_{n+1} + 2\vec{r}_n) \right]$$
$$+ \vec{b} \left[(r_{n+1})^2 - (r_n)^2 + 4\,(\vec{r}_n \cdot \vec{r}_{n+1}) \right], \qquad n = 1, 2 \ \ mod\,[2]\,. \tag{38}$$

Here of course the dot among two vectors denotes the standard scalar product, and $(r_n)^2 \equiv \vec{r}_n \cdot \vec{r}_n$. Note the *covariant* character of these equations.

The interested reader will have no difficulty to reformulate in an analogous manner the equations of motion (8) of Example 1.3; and analogous reformulations of the equations of motion of Examples 1.4 and 1.5 are also possible (hint: before applying the same procedure as indicated above, see (37), replace b with b^3 in (9), and b with b^2 in (10)).

5 Outlook

The literature on the simple kind of dynamical systems treated in this paper is of course vast; see for instance [9], [10] and standard compilations of solvable ODEs such as [11]. But it seems to us that—in spite of their simplicity—the findings reported in this paper (including their variants mentioned in Section 4) are new.

Further applications of the approach described in this paper are of course also possible: for instance by exploiting the extension of the results of [5] to time-dependent polynomials featuring zeros of *arbitrary* multiplicity (see some progress made in this direction by Oksana Bihun's recent paper [12]; we will report additional progress in a paper now in preparation [13])); or by exploiting the extensions of the fundamental results—such as (2a)—on which the findings reported in this paper are based, from polynomials to rational functions [14].

And of course extensions of the approach of this paper to systems of higher-order ODEs (including in particular second-order ODEs of Newtonian type: "accelerations equal forces"), to PDEs, to discrete-time evolutions (see [4]) shall deserve further investigations.

6 Acknowledgements

FP would like to thank the Physics Department of the University of Rome "La Sapienza" for the hospitality from March to July 2018 (during her sabbatical), when the results reported in this paper were obtained.

References

[1] F. Calogero, "New solvable variants of the goldfish many-body problem", Studies Appl. Math. **137** (1), 123–139 (2016); DOI: 10.1111/sapm.12096.

[2] O. Bihun and F. Calogero, "Novel solvable many-body problems", J. Nonlinear Math. Phys. **23**, 190-0212 (2016). DOI: 10.1080/14029251.2016.1161260.

[3] M. Bruschi and F. Calogero, "A convenient expression of the time-derivative $z_n^{(k)}(t)$, of arbitrary order k of the zero $z_n(t)$ of a time-dependent polynomial $p_N(z;t)$ of arbitrary degree N in z, and solvable dynamical systems", J. Nonlinear Math. Phys. **23**, 474–485 (2016).

[4] F. Calogero, *Zeros of Polynomials and Solvable Nonlinear Evolution Equations*, Cambridge University Press, Cambridge, U. K., 2018 (in press, about 170 pages).

[5] O. Bihun and F. Calogero, "Time-dependent polynomials with *one double* root, and related new solvable systems of nonlinear evolution equations", Qual. Theory Dyn. Syst. (in press). doi.org/10.1007/s12346-018-0282-3; http://arxiv.org/abs/1806.07502.

[6] A. Erdélyi (editor), *Higher Transcendental Functions*, vol. 2, McGraw-Hill, New York, 1953.

[7] D. Gómez-Ullate and M. Sommacal, "Periods of the Goldfish Many-Body Problem", J. Nonlinear Math. Phys. **12**, Suppl. 1, 351–362 (2005).

[8] F. Calogero, *Isochronous systems*, Oxford University Press, 2008 (264 pages; marginally update of motion paperback version, 2012).

[9] D. D. Hua, L. Cairó and M. R. Feix, "Time independent invariants for the quadratic system", J. Phys. A: Math. Gen. **26**, 7097–7114 (1993).

[10] G. R. Nicklason, "The general phase plane solution of the 2D homogeneous system with equal malthusian terms: the quadratic case", Canad. Appl. Math. Quart. **13**, 89–106 (2005).

[11] A. D. Polyanin and V. T. Zaitsev, *Handbook of Exact Solutions for Ordinary Differential Equations*, CRC Press, Boca Raton, USA, 2018.

[12] O. Bihun, "Time-dependent polynomials with one multiple root and new solvable dynamical systems", arXiv:1808.00512v1 [math-ph] 1 Aug 2018.

[13] F. Calogero and F. Payandeh, "Polynomials with multiple zeros and solvable dynamical systems including models in the plane with polynomial interactions", J. Math. Phys. **60**, 082701 (2019); doi.org/101063.1.5082249; arXiv:1904.00496v1 [math-ph] 31 Mar 2019.

[14] F. Calogero, "Zeros of Rational Functions and Solvable Nonlinear Evolution Equations", J. Math. Phys. **59**, 072701 (2018); doi.org/10.1063/1.5033543.

Francesco Calogero
Physics Department, University of Rome "La Sapienza", Rome, Italy
INFN, Sezione di Roma 1
francesco.calogero@roma1.infn.it, francesco.calogero@uniroma1.it

Farrin Payandeh
Department of Physics, Payame Noor University (PNU), PO BOX
19395-3697 Tehran, Iran
f_payandeh@pnu.ac.ir, farrinpayandeh@yahoo.com

6

The Projection Method in Classical Mechanics

A. M. Perelomov

Abstract. The new method of explicit integration of equations of motion of systems of classical mechanics (the projection method [38], [39], [19]) is described. For Calogero, Sutherland and Toda systems the explicit solutions of equations of motion are given. These solutions are impossible to obtain by other known methods.

1 Introduction

The classical integrable systems are completely integrable Hamiltonian systems with n degrees of freedom, i.e. Hamiltonian systems that have n functionally independent integrals of motion in involution. By the Bour–Liouville–Arnold theorem ([7], [31], [2]) for these systems motion can be reduced to the motion on a torus (or on Euclidean space).[1]

Until the mid-1970s only few such systems were known. The situation changed after the discovery by Gardner, Greene, Kruskal and Miura [17] of a new method of solving nonlinear evolution equations – *the inverse scattering method*, also called *the isospectral deformation method*. The algebraic formulation of this method was given by Lax [30]. This approach was applied first to infinite-dimensional systems – the Korteweg-de Vries equation.

The main idea of the isospectral deformation method is to rewrite the equations of motion of dynamical system in the form of so called *Lax representation*

$$i\,\dot{L} = [M, L], \tag{1}$$

where the matrices L and M are Hermitian and depend on dynamical variables, "dot" means the derivative with respect to time t.

---

[1] We assume that the reader is familiar with the basic concepts of the theory of Hamiltonian systems, for example see the books [56], [2], [50], [3].

From this representation it follows that the integrals of motion of system can be considered as the eigenvalues of matrix L, whose dependence on time is such that the spectrum of this matrix does not change when the dynamical variables evolve according to the equations of motion.

The application of this method to systems of classical mechanics initiated by Flaschka [14],[15], Manakov [32], and Moser [33], [34] demonstrated the complete integrability of a variety of classical systems.

We would like to underline here that integrability of all known systems of this kind is due to some higher (hidden) symmetry. Namely, all systems of this kind discovered so far are related to Lie algebras, although often this relationship is not so simple as the one expressed by the well-known theorem of E. Noether.

However, explicit integration of the equations of motion of completely integrable systems is a much more difficult problem, because the Bour–Liouville–Arnold theorem does not give the recipe for explicit integration of the equations of motion. Nevertheless in some cases we may use for this the *projection method*, introduced first by Olshanetsky and Perelomov [38],[39].

The idea of this method is to consider another dynamical system with N degrees of freedom ($N > n$) such that the original system is the projection of this new system. The dynamical behavior of the new system should be sufficiently simple that its equations of motion can be solved in closed form. Therefore, if the projection is described by explicit formulae, this gives an explicit solution of the equations of motion of the original system.

In this paper we study a class of completely integrable systems of n particles on a line or on a circle with Hamiltonian

$$H(\mathbf{p}, \mathbf{q}) = \frac{1}{2}\mathbf{p}^2 + U(\mathbf{q}),$$

$$\mathbf{p} = (p_1, p_2, \ldots, p_n), \quad \mathbf{q} = (q_1, q_2, \ldots, q_n),$$

$$\mathbf{p}^2 = \sum_{k=1}^{n} p_k^2, \quad (\mathbf{p}, \mathbf{q}) = \sum_{k=1}^{n} p_k q_k,$$

interacting pairwise via potentials $g^2 v(q)$

$$U(\mathbf{q}) = g^2 \sum_{k<l} v(q_k - q_l).$$

Potential $v(q)$ is one of the following five types[2]

$$\text{I.} \qquad v(q) = q^{-2},$$

$$\text{II.} \qquad v(q) = \sinh^{-2} q,$$

---

[2] For more details on these systems, see [35],[41] and books [50], [11].

$$\text{III.} \qquad v(q) = \sin^{-2} q,$$

$$\text{IV.} \qquad v(q) = \wp(q), \quad \text{or}$$

$$\text{IV}'. \qquad v(q) = \text{sn}^{-2}(q, k),$$

$$\text{V.} \qquad v(q) = q^{-2} + \omega^2 q^2. \tag{2}$$

Here $\wp(q)$ is the Weierstrass function, i.e. double periodic meromorphic function with poles of second order at the points $2n_1\omega_1 + 2n_2\omega_2$, and ω_1, ω_2 are two half-periods; $\text{sn}(q, k)$ is elliptic sinus [5].[3]

Note that systems of type I–V are invariant relative to the finite group S_n – group of permutation n variables $q_j, j = 1, \ldots, n$.

In addition we shall consider Toda systems [55], i.e. systems in which only nearest neighbor particles interact according to the law

$$\text{VI.} \qquad v(q) = \exp(-2q). \tag{3}$$

Note that potentials of type I–V are singular for $q_j = q_k$, see (4). Therefore, the order of the particles cannot change during the motion, and we assume that $q_k < q_j$, for $j < k$. Thus, in cases I and II the configuration space is a cone Λ, given by the inequalities

$$q_j - q_{j+1} > 0, \quad j = 1, \ldots, n - 1$$

and by the equation $\sum q_j = 0$.

For systems of type III and IV the configuration space is a convex polyhedron Λ_a (a simplex), defined by the equations

$$q_j - q_{j+1} > 0, \quad j = 1, \ldots, n - 1, \quad q_1 - q_n < d/a, \quad \sum q_j = 0,$$

where d is the real period of the function $v(q)$.

For $n = 3$, the configuration spaces Λ and Λ_a are the interior of an angle $\pi/3$, and an equilateral triangle, respectively.

From the form of the potentials it follows that in cases I and II we are dealing with a system of n particles on a line, while in cases III and IV with a system of n particles on a circle.

Initially some exact results concerning systems I–V were obtained in 1969–1975 for the quantum case. The papers by Calogero [8], [9] and Perelomov [45], concern systems of types I and V (Calogero model), while the papers by Sutherland [53], [54] deal with systems of type III (Sutherland model).

In classical mechanics, the isospectral deformation method was first applied to the Toda lattice by Flaschka [14],[15], Manakov [32] and Moser [34], and then to systems of types I–V [33], [10], [37], [46], [47]. Various aspects are

---

[3] It is known that system of type I is completely integrable for $n = 3$ [22], $n = 4$ [45] and $n = 5$ [16].

investigated in numerous articles, in particular for systems of type I–IV and VI were obtained explicit expressions for matrices L and M (see book [50]).

For systems of type V we need to modify Lax representation [46]. Namely, instead of Lax equation (1) we should consider the equations

$$i \dot{L}^{\pm} = [M, L^{\pm}] \pm \omega L^{\pm} . \tag{4}$$

From them it follows that the quantities

$$B_k^{\pm} = k^{-1} \operatorname{tr} (L^{\pm})^k$$

are not integrals of motion, nevertheless they have a very simple time dependence

$$B_k^{\pm}(t) = B_k^{\pm}(0) \exp(\mp ik\omega t).$$

From (4) it is not difficult to obtain integrals of motion. For example, the matrices

$$L_1 = L^+ L^-, \ L_2 = L^- L^+$$

satisfy the usual Lax equation

$$i \dot{L}_j = [M, L_j], \quad j = 1, 2.$$

Therefore, the eigenvalues of these matrices or the traces of powers of the matrices are integrals of motion. The form of the matrices $L^{\pm}$ was found in [46]

$$L^{\pm} = L \pm i\omega Q, \quad Q = \operatorname{diag}(q_1, \dots, q_n).$$

2 Projection Method

Let us describe shortly the projection method, introduced first by Olshanetsky and Perelomov [38],[39] (for details see the book [50]).

The main idea of this method is the following. Suppose we have a dynamical system on the n-dimensional manifold $M = \{x\}$ which generates the flow

$$g_t : M \to M, \quad x_t = g_t x_0.$$

Let us try to find another dynamical system on the manifold $\tilde{M} = \{y\}$ of dimension $N > n$

$$\tilde{g}_t : \tilde{M} \to \tilde{M}, \quad y_t = \tilde{g}_t y_0$$

so that the following conditions hold:

(i) The dynamical behavior of the new system is sufficiently simple so that the equations of motion of this system can be integrated explicitly.

(ii) The initial system is the projection π of the larger system

$$\pi : \tilde{M} \to M$$

and this projection can be described by explicit formulae.

(iii) There exists a map $\rho : M \to \tilde{M}$ which can be described explicitly, such that $\pi \cdot \rho = \mathrm{id}$.

In this case, we may integrate the equations of motion for the Hamiltonian system on M explicitly. Indeed, we have

$$x_t = \pi \cdot y_t = \pi \cdot \tilde{g}_t \cdot y_0 = \pi \cdot \tilde{g}_t \cdot \rho \cdot x_0, \tag{5}$$

i.e.

$$x_t = \pi \cdot \tilde{g}_t \cdot \rho \cdot x_0 \tag{6}$$

See Fig. 6.1.

The main difficulty of this method is finding $\tilde{M}$, π, $\tilde{g}_t$, and ρ. We do not know how to do this in general, but we have a number of interesting and nontrivial examples of application of this method.

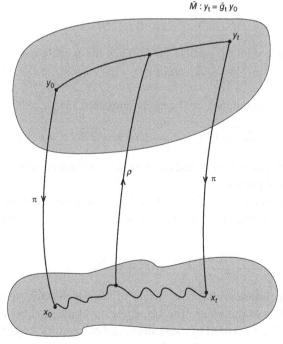

$$\tilde{M} : y_t = \tilde{g}_t \, y_0$$

$$M : x_t = g_t \, x_0, \quad x = \pi \cdot y.$$

Figure 6.1

In particular, as was shown in the papers [38], [39], systems of type I-III are projections of systems describing a free motion (geodesic flow) in certain symmetric spaces of zero, negative, and positive curvature, respectively, where the dimension of these spaces is significantly larger than the number of degrees of freedom of the original system. After the corresponding projection onto the n-dimensional space, we obtain the motion in the potential field (2).

We consider application of this method to the explicit integration of equations of motion for these systems in the rest of the paper. Note also that projection method was modified to the quantum case [43].

3 Rational Case

Systems of type I. Note first that the case of two particles is reduced to the case of one degree of freedom, i.e. to the system with Hamiltonian

$$H = \frac{1}{2}p^2 + g^2 v(q), \quad v(q) = q^{-2},\qquad(7)$$

which is invariant under transformation

$$p \to -p, \; q \to -q.\qquad(8)$$

Let the energy of the particle be E. Then the particle moves in domain $q_0 \le q < \infty$, $(q_0)^2 = g^2/E$, and $p(t) \to p^\pm$, $q(t) \sim p^\pm t$, at $t \to \pm\infty$, $p^- = -p^+$, $p^+ = \sqrt{2E}$.

The equations of motion can be easily integrated and we obtain

$$q = \sqrt{a^2 + b^2 \tau^2}, \; a = q_0, \; b = \sqrt{2E}, \; \tau = t - t_0.$$

However for the case of three and more particles it is not clear how to integrate the equations of motion.

So we give first the solution for the case of two particles by the projection method. Let us consider a Euclidean plane $\{\mathbf{x} : \mathbf{x} = (x_1, x_2)\}$ with standard metric

$$ds^2 = (dx_1^2 + dx_2^2).$$

This metric is invariant under the action of three-parametric group G – the group of motion of this plane. The Lie algebra $\mathcal{G}$ of the group G is generated by elements X_1, X_2 related to translations, and element X_3 related to rotation around origin. The commutation relations for $\mathcal{G}$ are

$$[X_1, X_2] = 0, \; [X_3, X_1] = X_2, \; [X_3, X_2] = -X_1.$$

The Lie algebra $\mathcal{G}$ admits the involution I which fixes the subalgebra $\mathcal{K}$ generated by X_3

$$I(X_1, X_2, X_3) = (-X_1, -X_2, X_3).$$

Group K generated by X_3 is a maximal compact subgroup in G. It is a group of rotations around origin. The factor space $X^0 = G/K$ is our two-dimensional plane, it is a Riemannian symmetric space of zero curvature.[4] The equation for geodesics on this space has the simple form

$$\ddot{\mathbf{x}} = 0$$

and its solution is

$$\mathbf{x} = \mathbf{a} + \mathbf{b}\,\tau, \quad \mathbf{a} \in X^0, \ \mathbf{b} \in X^0, \ \tau = t - t_0.$$

The orbits of group K are circles with center in origin and they define the natural projection to semiaxis $0 \le q < \infty$:

$$\pi\,\mathbf{x} = q = \sqrt{x_1^2 + x_2^2}.$$

So we come to the result:

The motion of system with Hamiltonian (7) is the projection π of geodesic motion on symmetric space X^0 of zero curvature.

Using this geometric approach it is not difficult to obtain a number of useful consequences. Let us give a couple of them.

Let us recall first Jacobi's sentence from his Köenigsberg "Lectures on Dynamics" (1843) [21].

The main difficulty in integrating of the given differential equations is the introduction of appropriate variables. However, we have not any rule for this and should take an inverse way. Namely, if we found a good substitution, then we should look for the problems where it may be applied with success.

For our case variables p and q are not invariant under transformation (8), they are multivalued functions of t and they are "bad" variables. But variables p^2, q^2, pq are invariant under this transformation, they are single-valued functions of t and they are "good" variables.

Corollary 3.1 *The even polynomial $P(q)$ of degree $2n$ is a polynomial of degree $2n$ in t.*

Corollary 3.2 *The rational function $R(p, q)$ invariant under transformation (8) is a rational function of t.*

---

[4] There is complete classification of symmetric spaces by Cartan, see the book [20].

Let us give, following [38], the generalization of this approach to the case of systems with arbitrary number of degrees of freedom.

Let us consider motion on geodesics (free motion) in symmetric space X_n^0 of zero curvature which is subspace of the matrix space of Hermitian $n \times n$ matrices with zero trace.

The equations of motion here have the form

$$\ddot{X} = 0,$$

with general solution

$$X(t) = A + B t, \quad A, B \in X_n^0. \tag{9}$$

Note that the matrices A and B in (9) cannot be arbitrary. In fact, for them the "angular momentum" matrix $l = i[X, \dot{X}]$ has $(n - 1)$ coinciding eigenvalues.

Now, by means of a unitary transformation $U \in SU(n)$, we transform the Hermitian matrix $X(t)$ to diagonal form

$$X(t) = U(t) \, Q(t) \, U^{-1}(t), \tag{10}$$

where

$$Q(t) = \mathrm{diag}(q_1, \ldots, q_n). \tag{11}$$

Differentiating (10) with respect to t, we obtain

$$U(t) \, L(t) \, U^{-1}(t) = B, \tag{12}$$

where

$$L = P + i[M, Q],$$
$$P = \dot{Q}, \ M = -i U^{-1} \dot{U}. \tag{13}$$

Here L and M are Hermitian $n \times n$ matrices.

Differentiating (12) with respect to t, we obtain Lax equation

$$i \dot{L} = [M, L].$$

Note that matrices L and M should satisfy the equation (13). These matrices are known [33] and matrix L has the form

$$L_{jk} = p_j \delta_{jk} + i(1 - \delta_{jk})(q_j - q_k)^{-1}.$$

Without loss of generality we may assume that $U(0) = I$. Then the matrices A and B in (9) are expressed in terms of the initial conditions by the formulae

$$A = Q(0), \quad B = L(0).$$

Thus we have obtained the final result:

The coordinates $q_j(t)$, the solutions of the equations of motion for system of type I. (2), *are the eigenvalues of the matrix*

$$Q(0) + L(0)t. \tag{14}$$

Now let us discuss the scattering process [33]. The potential $U(q)$ in (2) vanishes as $q_j - q_k \to \infty$, and thus

$$p_j(t) \to p_j^{\pm}, \quad q_j(t) \sim p_j^{\pm}t + q_j^{\pm}, \quad \text{as } t \to \pm\infty. \tag{15}$$

Thus the scattering process is determined by a canonical transformation from the variables (p_j^-, q_k^-) to the variables (p_j^+, q_k^+). It may be shown also that the quantities p_j^+ differ from the p_k^- only by a permutation

$$p_1^- = p_n^+, \; p_2^- = p_{n-1}^+, \; \dots, \; p_n^- = p_1^+$$
$$q_1^- = q_n^+, \; q_2^- = q_{n-1}^+, \; \dots, \; q_n^- = p_1^+. \tag{16}$$

Relations (14) and (15) mean that the scattering in this particular problem reduces to successive scatterings of individual pairs of particles.

Systems of type V. In this case, instead of a free motion, we consider harmonic motion in space X_n^0

$$\ddot{X} + \omega^2 X = 0, \quad X \in X_n^0.$$

The solution of this equation has the form

$$X(t) = \cos \omega t \, A + \omega^{-1} \sin \omega t \, B.$$

Representing as before this expression in the form (10) we arrive at the following statement:

Coordinates $q_j(t)$ of the system of type V *are eigenvalues of the matrix*

$$\cos \omega t \, Q(0) + \omega^{-1} \sin \omega t \, L(0). \tag{17}$$

Corollary 3.3 *A polynomial of degree k in q_j, which is invariant under permutations, is a polynomial of degree k in t ($\omega = 0$) or in $\sin \omega t$ and $\cos \omega t$ ($\omega \neq 0$).*

We mention also that the explicit solution of the equations of motion for systems of type I and V (14) and (17)), enables us to establish a simple relation between these solutions.

Suppose that $q_j(t)$ is a solution of the equations of motion for a system of type I ($\omega = 0$). Then it follows from formulas (13) and (16) that the quantities

$$\tilde{q}_j(t) = \cos \omega t \, q_j(\omega^{-1} \tan \omega t)$$

give a solution of the corresponding system of type V ($\omega \neq 0$). The converse statement is also true. A similar connection for systems of more general form was established in [48].

We note further that the quantities

$$\text{tr}\,(Q^{k_1} L^{l_1} Q^{k_2} L^{l_2} \ldots)$$

have an extremely simple time dependence. Such quantities are polynomials of degree $k = \sum k_j$ in t ($\omega = 0$) or in $\cos \omega t$ and $\sin \omega t$ ($\omega \neq 0$).

Corollary 3.4 *Coordinates $q_j(t)$ of the systems of type I and V are the roots of polynomial*

$$P_n(x, t) = \sum_{j=0}^{n} (-1)^j\, a_j(t)\, x^{n-j},$$

where function $a_j(t)$ is the polynomial of degree j in t ($\omega = 0$) or in $\sin(t)$ and $\cos(t)$ ($\omega \neq 0$).

From formulas (13) and (16) it follows that if $q(t)$ and $\tilde{q}(t)$ are solutions of equations of motions for systems I and V, then

$$\tilde{q}(t) = \cos \omega t \; q(\omega^{-1} \tan \omega t)\,.$$

This means that the more complicated case V can be obtained from the more simple case I by using the simple change of variables.

Note that this is the particular case of more general theorem [48].
Consider two classical dynamical systems characterized by the Hamiltonians

$$H = \frac{1}{2} \sum_{j=1}^{n} p_j^2 + U(q), \quad q = (q_1, \ldots, q_n)$$

and

$$\tilde{H} = \frac{1}{2} \sum_{j=1}^{n} (p_j^2 + \omega^2(t) q_j^2) + \kappa(t) U(q),$$

where the potential $U(q)$ is a homogeneous function of degree k

$$U(\lambda q) = \lambda^k U(q).$$

Let $q(t)$ and $\tilde{q}(t)$ be the solutions of equations of motion for the corresponding systems

$$\ddot{q}_j = F_j(q), \quad F_j(q) = -\partial U/\partial q_j\,, \tag{18}$$

$$\ddot{\tilde{q}}_j = \kappa(t) F_j(\tilde{q}) - \omega^2(t)\tilde{q}_j\,. \tag{19}$$

Let $\alpha_1(t)$ and $\alpha_2(t)$ be two independent solutions of equation

$$\ddot{\alpha} + \omega^2(t)\alpha = 0$$

and the function $\beta(t)$ is defined by the formula

$$\beta(t) = c\,\alpha_2(t)/\alpha_1(t) = c\int \alpha_1^{-2}(\tau)d\tau.$$

Suppose also that the function $\kappa(t)$ has the form

$$\kappa(t) = c^2\,\alpha_1^{-(k+2)}(t).$$

Theorem 3.5 ([48]) *If $q(t)$ is the solution of (18), then*

$$\tilde{q}(t) = \alpha_1(t)\,q(\beta(t))$$

is the solution of (19).

Thus, the solution of a more complicated system (19) is reduced to that of a more simple system (18).

4 Hyperbolic and Trigonometric Cases

Hyperbolic case. Let us start with the simplest case of the system with one degree of freedom with Hamiltonian

$$H = \frac{1}{2}p^2 + g^2\sinh^{-2}(q).$$

Note that H is invariant under transformation (8).

Then the particle with energy E moves in domain $q_0 \le q < \infty$, $\sinh^2(q_0) = g^2/E$. Integrating the equations of motion we obtain

$$\cosh q = a\cosh(b\,\tau), \quad a = \cosh(q_0), \quad b = \sqrt{2\,E}, \quad \tau = t - t_0. \tag{20}$$

Corollary 4.1 *Rational function of $\cosh q$, $\sinh q$ and p invariant under transformation (8) is meromorphic function of t. Function no invariant under this transformation is multivalued function.*

Let us give this solution by the projection method. Consider the geodesic motion on the upper sheet of a hyperboloid of two sheets (the metric on the hyperboloid is induced by the metric of the ambient space)

$$H^2 = \{\mathbf{x} : \mathbf{x}^2 = x_0^2 - x_1^2 - x_2^2 = 1, \ x_0 > 0\}. \tag{21}$$

The projection π is defined by

$$\pi\,x = q, \quad \cosh q = x_0.$$

The equations of the geodesic flow on the hyperboloid are easily integrated:

$$\mathbf{x}(t) = \mathbf{a}\cosh(bt) + \mathbf{b}\sinh(bt)$$

with

$$\mathbf{a}^2 = a_0^2 - a_1^2 - a_2^2 = 1, \quad \mathbf{b}^2 = -1, \quad (\mathbf{a}, \mathbf{b}) = 0.$$

This yields the expression (20) for $q(t)$.

Consider now the case of n particles. Let X_n^- be the space of negative curvature corresponding to the space X_n^0, namely the space of Hermitian positive definite $n \times n$ matrices with unit determinant. It is the homogeneous space $X_n^- = SL(n, \mathbb{C})/SU(n)$. The action of the group $G = SL(n, \mathbb{C})$ has the form

$$X \to gXg^+, \quad X \in X_n^-, \quad g \in G, \quad g^+ = \bar{g}^t.$$

Here "bar" means complex conjugation, ^{t} means transposition.

It is clear that the following curve is a geodesic line on X_n^-

$$X(t) = B \exp(2\mathcal{A}t) B^+, \quad B \in SL(n, \mathbb{C}), \quad \mathcal{A}^+ = \mathcal{A}, \ \mathrm{tr}\,\mathcal{A} = 0, \ X^+(t) = X(t).$$

Now let us take the transformation to diagonal form

$$X(t) = U(t)\exp(2Q(t))U^{-1}(t),$$

where $U(t)$ is a unitary matrix, $Q(t) = \mathrm{diag}(q_1(t), \ldots, q_n(t))$. Differentiating this equation with respect to t, we get the Lax equation (1), where

$$L = P + \frac{i}{4}[\exp(-2Q)M\exp(2Q) - \exp(2Q)M\exp(-2Q)].$$

Here $P = \dot{Q}$, $M = -iU^{-1}\dot{U}(t)$ is the "angular velocity of rotation".

Note that there are different Lax representations for system of type II.(2) For us it is convenient to use the following expression for matrix L

$$L_{jk} = p_j \delta_{jk} + i\,(1 - \delta_{jk})\coth(q_j - q_k). \tag{22}$$

And we get the final result:

The quantities $\exp 2q_j(t)$, *where* $q_j(t)$ *are solutions of equations of motion for a system of type II, are the eigenvalues of matrix*

$$X(t) = B \exp(2\mathcal{A}t) B^+, \tag{23}$$

where

$$B = \exp Q(0), \quad Q = \mathrm{diag}(q_1, \ldots, q_n), \quad \mathcal{A} = L(0). \tag{24}$$

Trigonometric case. Suppose that a particle of unit mass moves along a geodesic on a two-dimensional sphere

$$S^2 = \{\mathbf{x}: \ \mathbf{x}^2 = x_0^2 + x_1^2 + x_2^2 = 1\}.$$

We define the projection π by

$$q = \pi\mathbf{x}, \quad \cos q = x_0.$$

In this case projecting gives a system of type III with Hamiltonian

$$H = \frac{1}{2}p^2 + g^2 \sin^{-2} q, \qquad q_0 < q < \pi - q_0.$$

By integrating the equations of geodesic on the sphere we find

$$\mathbf{x}(t) = \mathbf{a}\cos\omega t + \mathbf{b}\sin\omega t$$

with

$$\mathbf{a}^2 = a_0^2 + a_1^2 + a_2^2 = 1, \quad \mathbf{b}^2 = 1, \quad (\mathbf{a}, \mathbf{b}) = 0.$$

This yields

$$\cos q(t) = a\cos(bt), \quad a = \cos q_0, \quad b = g/\sin q_0.$$

Let us go now to the general case. For this consider the Cartan-related space of positive curvature X_n^+, which is the space of the unitary matrices $X_n^+ = SU(n)$ and the motion along geodesic lines in this space. In the same way as in the preceding section, we obtain the solution of the equations of motion for the system of type III. It is given by the same formulas (22)–(24) in which we should take function cot instead coth.

Note that the relation between X_n^- and X_n^+ is the special case of duality established by E. Cartan for a wide class of symmetric spaces [20]. This duality generalizes the well-known relation between hyperbolic and spherical geometries.

5 Non-periodic Toda Lattice

The Toda lattice is a system of particles on the line with exponential interaction of nearest neighbors.

For infinitely many particles on the line such a system was first considered by Toda [55], who discovered that nonlinear waves may propagate without dissipation in this unharmonic lattice.

The case of a finite number of particles differs from it and should be treated separately. First of all, one must distinguish the non-periodic lattice of n particles on the line, where the last particle does not interact with the first one, from the periodic lattice where the last and the first particles interact as other nearest neighbors do. We give here only the simpler case of the non-periodic Toda lattice and some of its generalizations. The periodic lattice requires the more sophisticated mathematical techniques of abelian integrals and theta functions and is not considered here.

Before going to the general case, we consider the case of one degree of freedom. Let us consider a geodesic flow on a two-sheeted hyperboloid X_2^- (21). In order to obtain from this a dynamical system of the type (3) VI, we must introduce the so-called horocyclic coordinate system

$$x_0 = \cosh q + \frac{z^2}{2}\exp q, \ x_1 = \sinh q - \frac{z^2}{2}\exp q, \ x_2 = z\exp q.$$

We use this name for coordinate system because the parabola ($q = $ const, $z \in \mathbb{R}$) is the horocycle in Lobachevsky geometry. We define the projection π as

$$\pi \mathbf{x} = q, \quad \exp q = x_0 + x_1.$$

The result is:

The coordinate $q(t)$ is horocyclic projection of geodesics on the space X_2^-.

It is obvious that the angular momentum $\ell = \dot{z}\exp(2q)$ defined by the cyclic coordinate z, is conserved. Using this fact we obtain the Hamiltonian of Toda lattice,

$$H = \frac{1}{2}p^2 + g^2\exp(-2q).$$

In general case we consider the space X_n^- of positive definite symmetric matrices with determinant one. This is the homogeneous space $X_n^- = SL(n, \mathbb{R})/SO(n)$. The formulas for the metric and the geodesics, given in Sec.4, remain valid also in the present case. We need only remember that we are now dealing with real matrices.

Let Z be the subgroup in $SL(n, \mathbb{R})$ of upper triangular matrices with ones on the diagonal, and H be the subgroup of diagonal matrices. We can represent each symmetric matrix $X \in X_n^-$ uniquely in the form

$$X = z(X)h^2(X)z^t(X), \quad z(X) \in Z, \ h(X) \in H. \tag{25}$$

The pair $(h(X), z(X))$ are the so-called horospheric coordinates of the matrix X, while $h(X)$ is called the horospheric projection.

If $\Delta_j(x)$ are the lower principal minors of the matrix X and $h(x) = \text{diag}(h_1, \ldots, h_n)$, then it can be easily shown that

$$h_1^2 = \frac{1}{\Delta_{n-1}}, \ h_2^2 = \frac{\Delta_{n-1}}{\Delta_{n-2}}, \ \ldots, \ h_{n-1}^2 = \frac{\Delta_2}{\Delta_1}, \ h_n^2 = \Delta_1.$$

From this follows two statements:

1. *Under the horospheric projection (25), motion along the geodesic in the space X_n^- goes over into a motion in the potential (3).*

2. *Let* $q^0 = (q_1^0, \ldots, q_n^0)$; $p^0 = (p_1^0, \ldots, p_n^0)$ *be initial conditions for equations of motions of the Toda lattice, and let the matrix* $\mathcal{A}$ *has the form*

$$\mathcal{A}_{jk} = p_j^0 \delta_{jk} + \exp(q_{j-1}^0 - q_j^0)\delta_{j,k+1} + \exp(q_j^0 - q_{j+1}^0)\delta_{j,k-1},$$

and Δ_j *be the lower principal minors of order* j *of the matrix* $\mathcal{A}$. *Then the solutions* $q_j(t)$ *of the equations of motion have the form*

$$q_j(t) = q_j^0 + \ln \frac{\Delta_{n+1-j}}{\Delta_{n-j}}, \quad \Delta_n = \Delta_0 = 1.$$

The projection method also works for the Toda lattice of any number of particles [40], [42]. Lax representation for this case was found earlier in [14], [32].

Note that in the paper by Bogoyavlensky [6] was given the generalization of Toda lattices to systems related to root systems of simple Lie algebras. The original Toda lattice is related to Lie algebra $sl(n, \mathbb{R})$ of real traceless $n \times n$ matrices. In [40], [42], [24], see also [25], the group-theoretical methods were used for explicit integration of equations of motion for such systems.

The periodic Toda lattice is a much more complicated system. We only point out here that it is completely integrable, but unlike the nonperiodic lattice it is not asymptotically free: in fact, the motion is conditionally periodic. The equations of motion of the periodic Toda lattice were reduced to quadratures [23] and later integrated by Krichever [28] in terms of multidimensional theta functions by using algebro-geometric methods developed by Dubrovin, Matveev and Novikov, see review papers [12], [13].

6 Systems with Two Types of Particles

By a simple change of variables, proposed in [10], we can obtain a certain generalization of systems of type II. Let

$$q_j \to q_j + i\frac{\pi}{2}, \quad 0 < n_1 \le j \le n. \tag{26}$$

Then the potential

$$U(q) = g^2 \sum_{j<k} \sinh^{-2}(q_j - q_k)$$

transforms to the potential

$$U(q) = g^2 \sum_{j,k} \sinh^{-2}(q_j - q_k), \quad 1 \le j < k \le n_1; \ n_1 < j < k \le n,$$

$$- g^2 \sum_{j>n_1, k \le n_1} \cosh^{-2}(q_j - q_k). \tag{27}$$

This system contains n_1 particles of one sign and $n_2 = n - n_1$ of the opposite sign. Each pair of particles of opposite sign attract one another with a potential $-g^2 \cosh^{-2}(q_j - q_k)$. At the same time, particles of the same sign repel one another with the potential $g^2 \sinh^{-2}(q_j - q_k)$.

Projection method works also in this case and all preceding results concerning systems of type II remain valid after the substitution (26). In particular, the quantities $\exp(2q_j)$ are eigenfunctions of the matrix

$$X(t) = B \exp(2\mathcal{A}t)B^+.$$

But this formula does not answer explicitly the important question: are there bound states in a system with this potential (27)? The answer to this was given in [44].

7 Elliptic Case

1. Introduction. Elliptic Calogero system was considered first in [10]. It is a Hamiltonian system, describing the motion of n particles on the circle $S^1 = \mathbb{R}/\omega\mathbb{Z}$, $\omega \in \mathbb{R}$, interacting pairwise via potential $v(q) = \wp(q) = \wp(q|\omega, \omega')$ or, what is equivalent, $v(q) = \text{sn}^{-2}(q, k)$ (see formulas (2) IV., IV.$'$). Here $\wp(q)$ is the Weierstrass elliptic function and $\text{sn}(q, k)$ is elliptic sinus [5], [57]. In [10] Lax representation was found and n integrals of motion were obtained. In [47] it was proved that these integrals are in involution, so this system is completely integrable.

This system is very complicated and to investigate it we should use the methods of algebraic geometry [12], [18], [13]. In the paper by Krichever [29] Lax representation with spectral parameter was found and its use obtained formulae for solution of equations of motion in terms of genus n theta functions.

In the paper by Gavrilov and Perelomov [19] was found the generalization of projection method for this system. Namely, it was proved that every symmetric elliptic function in coordinates $q_1, q_2, \ldots, q_n$ is a meromorphic function of time. This makes it possible to obtain the explicit formulae for solution of equations of motion in terms of genus $(n-1)$ theta functions. Here we give only formulations of main results. The proofs and other details can be found in the original paper.

Denote by Γ_1 the elliptic curve $\mathbb{C}/\{2\omega\mathbb{Z} + 2\omega'\mathbb{Z}\}$ with period lattice generated by 2ω and $2\omega'$. The Hamiltonian H is invariant under the obvious action of the permutation group S_n, so the phase space of the complexified system is the cotangent bundle $T^*(S^n\Gamma_1)$ of the nth symmetric product $S^n\Gamma_1$.

The Lax representation with spectral parameter [29] means that the equations of motion of the system under consideration are equivalent to the matrix equation

$$i \, \dot{L}(\lambda) = [L(\lambda), M(\lambda)],$$

where $L(\lambda) = L(p, q; \lambda)$ and $M(\lambda) = M(p, q; \lambda)$ are two matrices of order n:

$$\{L(\lambda)\}_{jk} = p_j \, \delta_{jk} + i \, (1 - \delta_{jk}) \, \Phi(q_j - q_k, \lambda);$$

$$\{M(\lambda)\}_{jk} = \delta_{jk} \left(\sum_{l \neq j} \wp(q_j - q_l) - \wp(\lambda) \right) + (1 - \delta_{jk}) \, \Phi'(q_j - q_k, \lambda);$$

$$\Phi(q, \lambda) = \frac{\sigma(q - \lambda)}{\sigma(q) \sigma(\lambda)} \, \exp(\zeta(\lambda) q);$$

$$\sigma(q) = q \prod_{m,n}' \left(1 - \frac{q}{\omega_{mn}} \right) \exp \left[\frac{q}{\omega_{mn}} + \frac{1}{2} \left(\frac{q}{\omega_{mn}} \right)^2 \right],$$

and

$$\zeta(q) = \frac{\sigma'(q)}{\sigma(q)}, \quad \omega_{mn} = m\omega + n\omega'.$$

As it was shown in [29], the equations of motion can be "linearized" on the Jacobian of the spectral curve

$$\Gamma_n = \{(\lambda, \mu) : f(\lambda, \mu) \equiv \det(L(\lambda) - \mu I) = 0\}.$$

Namely, let

$$\theta(\mathbf{z} | B) = \sum_{\mathbf{n} \in \mathbb{Z}^n} e^{\pi i \langle \mathbf{n}, B\mathbf{n} \rangle + 2\pi i \langle \mathbf{n}, \mathbf{z} \rangle}, \quad \mathbf{z} \in \mathbb{Z}^n$$

be the Riemann theta function with period matrix B, where

$$B = (B_{ij}), \quad B = B^t, \quad \mathrm{Im} B > 0, \quad (x, y) = \sum_j x_j y_j, i, j = 1, \ldots, n.$$

If B is the period matrix of the curve Γ_n, then for suitable constant vectors $U, V, W \in \mathbb{C}^n$ and for a fixed parameter $t \in \mathbb{C}$, the equation

$$\theta(Uq + Vt + W) = 0, \quad q \in \mathbb{C} \tag{28}$$

has exactly n solutions $q = q_j(t)$ on the Jacobian Jac (Γ_n) of the curve Γ_n. The functions $q_j(t)$ provide solutions of the elliptic Calogero system. So the solution involves the theta functions of n variables.

However, using the Weierstrass theorem [26], [51], [52], see also [1], we can obtain solution of equations of motion in terms of theta functions of $(n - 1)$ variables.

Theorem 7.1 ([19]) *The Krichever curve Γ_n is an n-sheeted covering of an elliptic curve $\Gamma_1 = \mathbb{C}/\{\mathbb{Z} + \tau \mathbb{Z}\}$. There exists a canonical homology basis and a* *normalized basis of holomorphic one-forms on Γ_n, such that the corresponding period matrix of Γ_n takes the form (I, B), where $I = \mathrm{diag}(1, 1, \ldots, 1)$, and*

$$B = \begin{pmatrix} \frac{\tau}{n} & \frac{1}{n} & 0 & \cdots & 0 \\ \frac{1}{n} & b_{22} & b_{23} & \cdots & b_{2n} \\ 0 & b_{32} & b_{33} & \cdots & b_{3n} \\ \vdots & \vdots & \vdots & \vdots & \vdots \\ 0 & b_{n2} & b_{n3} & \cdots & b_{nn} \end{pmatrix}. \tag{29}$$

In this basis the vectors U and V in (28) read as

$$U = (1, 0, \ldots, 0), \quad V = (0, V_2, \ldots, V_n).$$

Note that this Theorem can be proved also without using the Weierstrass theorem [19].

From now on we make the convention that $2\omega = 1$, so the period lattice of Γ_1 is $\mathbb{Z} + \tau \mathbb{Z}$, $\tau = 2\omega'/2\omega = 2\omega'$.

2. Case of $n = 2$. This case is reduced to the case of one degree of freedom, i.e. to the system with Hamiltonian

$$H = \frac{1}{2}p^2 + g^2 \, \mathrm{sn}^{-2}(q, k),$$

where $\mathrm{sn}(x, k)$ is the so-called elliptic sine:

$$\mathrm{sn}^{-2}((e_1 - e_3)^{1/2}q, k) = (e_1 - e_2)^{-1}(\wp(q) - e_2),$$
$$e_j = \wp(\omega_j), \quad j = 1, 2, \quad e_3 = \wp(\omega_1 + \omega_2).$$

For $E > E_0$, $E_0 = g^2$, the particle moves on the interval $a_1 < q < a_2$, $\mathrm{sn}(a_j, k) = \sqrt{g^2/E}$, $j = 1, 2$. Integrating equations of motions, we obtain

$$\mathrm{cn}(q, k) = a \, \mathrm{cn}(bt, \tilde{k}), \quad \text{or} \quad \mathrm{dn}(q, k) = c \, \mathrm{dn}(bt, \tilde{k}). \tag{30}$$

Here

$$a = \mathrm{cn}(q_0, k), \quad c = \mathrm{dn}(q_0, k), \quad b = \sqrt{2(E - g^2 k^2)}, \quad \tilde{k}^2 = k^2 \frac{E - g^2}{E - g^2 k^2}. \tag{31}$$

Period of oscillation is

$$T = \frac{4}{b} K(\tilde{k}), \quad K(k) = \int_0^1 \frac{dx}{\sqrt{((1 - x^2)(1 - k^2 x^2))}}, \quad k < 1.$$

Let us obtain this solution by the projection method [19]. The curve Γ_2 is 2-sheeted covering of the curve Γ_1 and the Riemann theta function associated with the curve has the form:

$$\theta(z_1, z_2) = \sum_{n_i, n_j} \exp\{i\pi \, [B_{ij} n_i n_j + 2n_j z_j]\}, \quad i, j = 1, 2,$$

where

$$B_{11} = \tau_1/2, \quad B_{22} = \tau_2/2, \quad B_{12} = B_{21} = 1/2.$$

A straightforward computation gives

$$\theta(z_1, z_2) = \sum_{n_1, n_2} \exp \left\{ i\pi \left[\tau_1 \frac{n_1^2}{2} + n_1 n_2 + \tau_2 \frac{n_2^2}{2} + 2n_1 z_1 + 2n_2 z_2 \right] \right\}$$

$$= \sum_{k_1, n_2 \in \mathbb{Z}} \exp\{i\pi \, [2\tau_1 k_1^2 + 4k_1 z_1]\} \exp \left\{ i\pi \left[\tau_2 \frac{n_2^2}{2} + 2n_2 z_2 \right] \right\}$$

$$+ \sum_{k_1, n_2 \in \mathbb{Z}} \exp \left\{ i\pi \left[2\tau_1 \left(k_1 + \frac{1}{2} \right)^2 + 4 \left(k_1 + \frac{1}{2} \right) z_1 \right] \right\}$$

$$\times \exp \left\{ i\pi \left[\tau_2 \frac{n_2^2}{2} + 2 \left(n_2 + \frac{1}{2} \right) z_2 \right] \right\}$$

$$= \theta_3(2z_1|2\tau_1) \, \theta_3 \left(z_2 | \frac{\tau_2}{2} \right) + \theta_2(2z_1|2\tau_1) \, \theta_4 \left(z_2 | \frac{\tau_2}{2} \right),$$

where $\theta_1, \theta_2, \theta_3$, and θ_4 are defined by formulae:

$$\theta_1(z|\tau) = 2q^{1/4} \sum_{n=1}^{\infty} (-1)^n q^{n(n+1)} \sin[(2n+1)\pi z],$$

$$\theta_2(z|\tau) = 2q^{1/4} \sum_{n=1}^{\infty} q^{n(n+1)} \cos[(2n+1)\pi z],$$

$$\theta_3(z|\tau) = 1 + 2 \sum_{n=1}^{\infty} q^{n^2} \cos(2\pi nz),$$

$$\theta_4(z|\tau) = 1 + 2 \sum_{n=1}^{\infty} (-1)^n q^{n^2} \cos(2\pi nz), \quad q = \exp(i\pi\tau).$$

So in this case, the equation $\theta(z_1, z_2) = 0$ is equivalent either to

$$A \, \text{dn} \, (2z_1|4\tau_1) \, \text{dn} \, (z_2|\tau_2) + \text{cn} \, (2z_1|4\tau_1) = 0,$$

or to

$$A \, \text{dn} \, (2z_2|4\tau_2) \, \text{dn} \, (z_1|\tau_1) + \text{cn} \, (2z_2|4\tau_2) = 0,$$

where

$$A = \frac{\theta_3(0|4\tau_1)\,\theta_3(0|\tau_2)}{\theta_2(0|4\tau_1)\,\theta_4(0|\tau_2)},$$

or

$$\mathrm{dn}\,(z_1|\tau_1) = B\,\mathrm{dn}\,(2iz_2 + K|\tilde{\tau}_2).$$

Let us give also a more symmetric form of the theta divisor for this case:

$$\mathrm{dn}(2z_1, k_1)\,\mathrm{dn}(2z_2, k_2) + \mathrm{dn}(2z_1, k_1)\,\mathrm{cn}(2z_2, k_2) + \mathrm{cn}(2z_1, k_1)\,\mathrm{dn}(2z_2, k_2)$$

$$- \,\mathrm{cn}(2z_1, k_1)\,\mathrm{cn}(2z_2, k_2) = 0.$$

Using the constraint $\theta(\mathbf{a}x + \mathbf{b}t + \mathbf{c}) = 0$ and taking $z_1 = q$, $z_2 = (1/2)K + ibt$, we get once again (30), (31).

3. Case of arbitrary n. Consider now the case of arbitrary n. Let $\theta(z_1, z_2, \ldots, z_n|B)$ be the Riemann theta function with period matrix as in (29). By a similar method we get

$$\theta(z_1, z_2, \ldots, z_n) = \sum_{j=0}^{n-1} \theta_j(z_1)\,\Theta_j(z_2, \ldots, z_n), \qquad (32)$$

where

$$\theta_j(z_1) = \theta\begin{bmatrix} j/n \\ 0 \end{bmatrix}(nz_1|n^2\tau_1),$$

$$\Theta_j(z_2, \ldots, z_N) = \Theta\begin{bmatrix} 0 & 0 & \cdots & 0 \\ j/n & 0 & \cdots & 0 \end{bmatrix}(z_2, \ldots, z_n|\hat{B}).$$

In this formula $\hat{B}$ is the right lower $(n-1) \times (n-1)$ minor of matrix B (29), and the theta functions with fractional characteristics are defined as in [27]. A reduction formula similar to (32), but containing n^2 terms, can be found in [4, Corollary 7.3].

Let us give the formulations of a few other propositions from [19].

Corollary 7.2 *The symmetric functions $f_k(t) = \sum_{i=1}^{n} \wp^{(k)}(q_i(t))$ are meromorphic in t and they are given by the formulae:*

$$f_0(t) = \frac{\partial^2}{\partial x^2} \log\theta(x, \mathbf{t})|_{x=0} - n\,\frac{\theta_1'''(0)}{3\theta_1'(0)},$$

$$f_k(t) = (-1)^k \frac{\partial^{k+2}}{\partial x^{k+2}} \log\theta(x, \mathbf{t})|_{x=0}, \quad k > 0,$$

where

$$\mathbf{t} = (V_2 t + W_2, V_3 t + W_3, \ldots, V_n t + W_n).$$

Our next construction is motivated by [38],[39] and [29]. Let us define the function

$$F(x,t) = \prod_{j=1}^{n} \frac{\sigma(x - q_j(t))}{\sigma(x)\sigma(q_j(t))} = [\theta_1'(0)]^{-n} \prod_{j=1}^{n} \frac{\theta_1(x - q_j(t))}{\theta_1(x)\theta_1(q_j(t))}, \quad \sum_{j=1}^{n} q_j(t) = 0,$$

where

$$q_j(t), \quad t \in \mathbb{C}, \quad j = 1, 2, \ldots, n,$$

is a solution of the elliptic Calogero system.

Lemma 7.3 $F(x,t)$ *is a meromorphic function in x on Γ_1 and meromorphic function in t on $\mathbb{C}$, explicitly given by*

$$F(x,t) = [-\theta_1'(0)]^{-n}] \frac{\theta(Ux + Vt + W)}{\theta_1(x)^n \theta(Vt + W)}.$$

The expansion of $F(x,t)$ on the basis of first order theta functions in x defines $(n-1)$ meromorphic functions in the variables $q_1, \ldots, q_n$ which are also meromorphic functions in t with simple poles only. Hence, we can take them as new "good" variables. The expansion of $F(x,t)$ can be obtained by making use of the addition formulae for elliptic functions. In the case $n = 2$ we have the following "addition formula" [5]

$$F(x,t) = \frac{\sigma(x - q)\,\sigma(x + q)}{\sigma^2(x)\sigma^2(q)} = \wp(q) - \wp(x),$$

which generalizes for arbitrary n.

Lemma 7.4 *For any $\mathbf{q} = (q_1, q_2, \ldots, q_n)$, x, such that $\sum q_j = 0$ define*

$$F(x, \mathbf{q}) = \prod_{j=1}^{n} \frac{\sigma(x - q_j)}{\sigma(x)\sigma(q_j)},$$

$$\Delta(\mathbf{q}) = (n-1)! \det \begin{vmatrix} 1 & \wp(q_1) & \wp'(q_1) & \cdots & \wp^{(n-3)}(q_1) \\ 1 & \wp(q_2) & \wp'(q_2) & \cdots & \wp^{(n-3)}(q_2) \\ \cdots & \cdots & \cdots & \cdots & \cdots \\ 1 & \wp(q_{n-1}) & \wp'(q_{n-1}) & \cdots & \wp^{(n-3)}(q_{n-1}) \end{vmatrix}.$$

The following identity holds

$$F(x, \mathbf{q})\Delta(\mathbf{q}) \equiv \det \begin{vmatrix} 1 & \wp(x) & \wp'(x) & \cdots & \wp^{(n-2)}(x) \\ 1 & \wp(q_1) & \wp'(q_1) & \cdots & \wp^{(n-2)}(q_1) \\ \cdots & \cdots & \cdots & \cdots & \cdots \\ 1 & \wp(q_{n-1}) & \wp'(q_{n-1}) & \cdots & \wp^{(n-2)}(q_{n-1}) \end{vmatrix}. \quad (33)$$

Remark. The substitution $x = q_n$ in (33) gives the following addition formula for the Weierstrass $\wp$-function

$$\det \begin{vmatrix} 1 & \wp(q_1) & \wp'(q_1) & \cdots & \wp^{(n-2)}(q_1) \\ 1 & \wp(q_2) & \wp'(q_2) & \cdots & \wp^{(n-2)}(q_2) \\ \cdots & \cdots & \cdots & \cdots & \cdots \\ 1 & \wp(q_n) & \wp'(q_n) & \cdots & \wp^{(n-2)}(q_n) \end{vmatrix} \equiv 0.$$

This implies the following

Corollary 7.5 *Let $f(x)$ be a meromorphic function on the elliptic curve Γ_1, and let S be a symmetric rational function in $n - 1$ variables. If $q_1(t)$, $q_2(t), \ldots, q_n(t), \sum q_i \equiv 0$ is a solution of the elliptic Calogero system, then $S(f(q_1(t)), f(q_2(t)), \ldots, f_{n-1}(q_{n-1}(t)))$ is a meromorphic function in t.*

8 Conclusion

In this article we have demonstrated that the projection method [38], [39], [19] gives explicit solution of equations of motion for systems of type I–VI. This solution is impossible to obtain by other known methods.

It would be interesting to extend the list of systems that can be integrated by this method.

I conclude this article with an open problem for non-hyperelliptic curves of genus 3. It is well-known that such a curve by birational transform can be reduced to a quartic curve and such a curve has 28 bitangents.

The case $g = 3$, $\tau_{12} = \frac{1}{2}$, $\tau_{13} = 0$ was investigated by S. Kovalevskaya [26]. She proved that the characteristic property for this case is crossing of four bitangents at one point.

It would be interesting to investigate the properties of bitangents for Krichever's curve $g = 3$, $\tau_{12} = \frac{1}{3}$, $\tau_{13} = 0$, (or maybe even for the more general case $g = 3$, $\tau_{12} = \frac{1}{k}$, $\tau_{13} = 0$, k is an arbitrary integer).

References

[1] Accola R.,Previato E., Lett. Math. Phys. **76**, 135–161 (2006)

[2] Arnold V.I., *Mathematical Methods of Classical Mechanics*, Berlin, Heidelberg, New York: Springer (1978)

[3] Adler M., van Moerbeke P., Vanhaecke P., *Algebraic Integrability, Painlevé Geometry and Lie Algebras*, Berlin, Heidelberg, New York: Springer (2004)

[4] Belokolos B.A., Bobenko A.I., Enolskii V.Z., Its A. R., Matveev V.B., *Algebro-Geometric Approach to Nonlinear Integrable Equations,* Berlin, Heidelberg, New York: Springer (1994)

[5] Bateman H., Erdelyi A., *Higher Transcendental Functions* V.3, New York: McGraw-Hill (1955)

[6] Bogoyavlensky O.I., Commun. Math. Phys. **51**, 201–209 (1976)

[7] Bour J., J. Math. Pures Appl. **20**, 185–200 (1855)

[8] Calogero F., J. Math. Phys. **10**, 2197–2200 (1969)

[9] Calogero F., J. Math. Phys. **12**, 419–436 (1971)

[10] Calogero F., Lett. Nuovo Cimento **13**, 411–416 (1975)

[11] Calogero F., *Classical Many-Body Problems Amenable to Exact Treatments*, Berlin, Heidelberg, New York: Springer (2001)

[12] Dubrovin B.A., Matveev V.B., Novikov S.P., Russ. Math. Surv. **31**, 59–146 (1976)

[13] Dubrovin B.A., Russ. Math. Surv. **36**, 11–92 (1981)

[14] Flaschka H., Phys. Rev. **B9**, 1924–1925 (1974)

[15] Flaschka H., Prog. Theor. Phys. **51**, 703–716 (1974)

[16] Gambardella P., J. Math. Phys. **16** 1172–1187 (1975)

[17] Gardner C., Greene J., Kruskal M., Miura R., Phys. Rev. Lett. **19** 1921–1923 (1967)

[18] Griffith P., Harris J., *Principles of Algebraic Geometry*, Wiley and Sons (1978)

[19] Gavrilov L., Perelomov A.M., J. Math. Phys. **40**, 6339–6352, arXiv: solv-int 9905011 (1999)

[20] Helgason S., *Differential Geometry, Lie Groups and Symmetric Spaces*, Acad. Press New York (1978)

[21] Jacobi C., *Vorlesungen über Dynamik* which Jacobi gave at 1842/43 at Königsberg University, in: Gesammelte Werke, SupplementBand, Berlin, Reimer (1884)

[22] Jacobi C., *Problema trium corporum mutuis attractionibus cubis distantiarum inverse proportionalibus recta linea se moventium* in: Gesammelte Werke, Bd. **4**, pp. 533–539 (1886)

[23] Kac M., van Moerbeke P., Proc. Nat. Acad. Sci. USA, **72**, 2879–2880 (1975)

[24] Kostant B., Adv. in Math. **39**, 195–338 (1979)

[25] Kamalin S.A., Perelomov A.M., Commun. Math. Phys. **97**, 553–568 (1985)

[26] Kowalevsky S.V., Acta Math. **4** , 392–414 (1884)

[27] Krazer A., *Lehrbuch der Thetafunctionen* , Lepzig: Teubner (1903)

[28] Krichever I.M., Russ. Math. Surv. **33**, 255–256 (1978)

[29] Krichever I.M., Funct. Anal. Appl. **14**, 282–290 (1980)

[30] Lax P., Commun. Pure Appl. Math. **21**, 467–490 (1968)

[31] Liouville J., J. Math. Pures Appl. **20**, 137–138 (1855)

[32] Manakov S.V., Sov. Phys. JETP **40**, 269–274 (1975)

[33] Moser J., Adv. Math. **16**, 197–220 (1975)

[34] Moser J., in: Lecture Notes in Physics **38**, 97–101 Berlin, Heidelberg, New York: Springer (1976)

[35] Moser J., *Various aspects of integrable Hamiltonian systems.*, in: *Progress in Mathematics* **8**, Dynamical Systems, 233–289, Birkhäuser (1980)

[36] Olshanetsky M.A., Perelomov A.M., Preprint ITEP No.103 (1975)

[37] Olshanetsky M.A., Perelomov A.M., Invent. Math. **37**, 93–108 (1976)

[38] Olshanetsky M.A., Perelomov A.M., Funct. Anal. Appl. **10** 237–239 (1976)

[39] Olshanetsky M.A., Perelomov A.M., Funct. Anal. Appl. **11** 66–68 (1976)

[40] Olshanetsky M.A., Perelomov A.M., Invent. Math. **54**, 261–269 (1979)

[41] Olshanetsky M.A., Perelomov A.M., Phys. Reps. **71**, No 5, 313–400 (1981)

[42] Olshanetsky M.A., Perelomov A.M., Theor. Math. Phys. **45**, 843–854 (1981)

[43] Olshanetsky M.A., Perelomov A.M., Phys. Reps. **94**, No 6, 313–404 (1983)

[44] Olshanetsky M.A., Rogov V.-B. K., Ann. Inst. H. Poincaré, **29**, 169–177 (1978)

[45] Perelomov A.M., Theor. Math. Phys. **6**, 263–283 (1971)

[46] Perelomov A.M., Preprint ITEP-27 (1976); arXiv: math-ph/0111018

[47] Perelomov A.M., Lett. Math. Phys. **1**, 531–534 (1977)

[48] Perelomov A.M., Commun. Math. Phys. **63**, 9–11 (1978)

[49] Perelomov A.M., Commun Math. Phys. **81** 239–241 (1981)

[50] Perelomov A.M., *Integrable Systems of Classical Mechanics and Lie Algebras.* **I**, Basel: Birkhäuser (1990)

[51] Poincaré H., Bull Soc. Math. France **12**, 124–143 (1884)

[52] Poincaré H., Amer. J. Math. **8**, 289–342 (1886)

[53] Sutherland B., Phys. Rev. **A4**, 2019–2021 (1971)

[54] Sutherland B., Phys. Rev. **A5**, 1372–1376 (1972)

[55] Toda M., J. Phys. Soc. Japan **29**, 431–436 (1967)

[56] Whittaker E.T., *A Treatise on the Analytical Dynamics of Particles and Rigid Bodies*, 4-th Ed., Cambridge Univ. Press: Cambridge (1937)

[57] Whittaker E.T., Watson G.N., A Course of Modern Analysis, 4th Ed., Cambridge Univ. Press: Cambridge (1927)

A. M. Perelomov

Institute of Theoretical and Experimental Physics, 117259 Moscow, Russia

7

Pencils of Quadrics, Billiard Double-Reflection and Confocal Incircular Nets

Vladimir Dragović, Milena Radnović and Roger Fidèle
Ranomenjanahary

Dedicated to Emma Previato on the occasion of her birthday.

Abstract. We present recent results about double reflection and incircular nets. The building blocks are pencils of quadrics, related billiards and quad graphs.

1 Introduction

Discrete differential geometry emerged quite recently (see [BS2008]), within the study of lattice geometry. Similarly, discrete integrable systems is a decades-old field that excites both mathematics and physics communities. So-called integrability conditions for the quad-graphs have a fundamental and connecting role between the two mentioned areas of modern mathematics. For detailed expositions on discrete integrable systems and discrete differential geometry, see for example [GKST2004, BS2008, Dui2010] and references therein.

The geodesics on an ellipsoid are one of the most important and exciting examples from the classical differential geometry. The billiard systems within quadrics are natural discretizations of the systems of the geodesics on ellipsoids. In this paper, we are going to employ the billiards within quadrics to develop some of the building blocks of the foundations of discrete differential geometry. In this work, we present and study classes of discrete systems which have recently arisen from billiards within confocal quadrics. Some of them, like *double reflection nets* and *incircular nets* are defined on cubic lattices, while some others are defined on a non-cubic lattice. The latter assign hyperplanes of a given projective space to the vertices of the lattice.

A silent hero in the background of the whole story is the celebrated Poncelet Theorem [Pon1822]. As it is well known, Emma Previato has been one of the leading devotees and fanciers of the Poncelet Theorem and integrable billiards, these evergreen topics, over the last twenty years, [Pre2018, Pre, Pre1999].

This paper is organised as follows. Section 2 contains an overview of confocal quadrics, related billiards, Darboux coordinates and quad-graphs. Double reflection nets are introduced and studied in Section 3. They are based on billiard algebra and quad-graphs. In Section 3.4, we introduce discrete systems that are naturally related to the double reflection nets, but form a non-cubic lattice of cuboctahedra and octahedra. Their basic properties and the cases when $m = 2$ and $m = 3$ are discussed. In Section 4 we present some classical and some very recent results on incircular nets and checkerboard incircular nets. In Section 4.2 we present a construction of so-called confocal incircular nets based on a generalization of the Darboux Theorem on Poncelet-Darboux grids, based on non-periodic billiard trajectories. In the last Section 5 we present yet another way to connect pencils of conics with quad-graphs, this time through so-called discriminantly separable polynomials.

2 Preliminaries and Basic Notions

2.1 Confocal Quadrics, Billiards, and Double Reflection Configurations

A general family of confocal quadrics in the d-dimensional Euclidean space is given by:

$$\frac{x_1^2}{b_1 - \lambda} + \cdots + \frac{x_d^2}{b_d - \lambda} = 1, \quad \lambda \in \mathbf{R} \tag{1}$$

with $b_1 > b_2 > \cdots > b_d > 0$, see Figure 7.1.

Such a family has the following properties:

- each point of the space $\mathbf{E}^d$ is the intersection of exactly d quadrics from (1); moreover, all these quadrics are of different geometrical types;
- the family (1) contains exactly d geometrical types of non-degenerate quadrics – each type corresponds to one of the disjoint intervals of the parameter λ: $(-\infty, b_d), (b_d, b_{d-1}), \ldots, (b_2, b_1)$;
- by the Chasles' theorem, each line in the space is touching exactly $d - 1$ quadrics from the family (1);
- two lines satisfying the billiard reflection law off any quadric from (1) are touching the same $d - 1$ confocal quadrics.

By the last property, any billiard trajectory within confocal quadrics have $d - 1$ caustics from the confocal family.

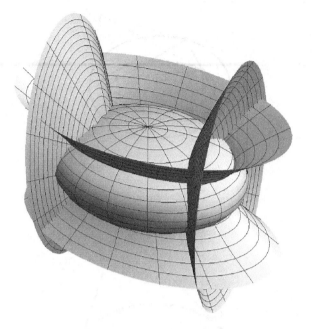

Figure 7.1 Confocal quadrics in the three-dimensional Euclidean space.

2.2 Double Reflection Configurations

Theorem 2.1 *Let a, b, c, d be lines and $\mathcal{Q}_\alpha$, $\mathcal{Q}_\beta$ confocal quadrics in the space such that (see Figure 7.2):*

* *the lines a, b and c, d meet on $\mathcal{Q}_\alpha$;*
* *the lines a, c and b, d meet on $\mathcal{Q}_\beta$;*
* *the tangent planes to the quadrics at the intersection points belong to a pencil.*

Then the pairs a, b and c, d satisfy the reflection law off $\mathcal{Q}_\alpha$ and the pairs a, c and b, d satisfy the reflection law off $\mathcal{Q}_\beta$.

Definition 2.2 *We will say that the quadruple of lines a, b, c, d satisfying the conditions of Theorem 2.1 constitutes* a double reflection configuration.

A double reflection configuration is shown on Figure 7.2.

Now, we list some of the basic facts about the double reflection configurations.

Proposition 2.3 *Let a, b, c be lines and $\mathcal{Q}_\alpha$, $\mathcal{Q}_\beta$ confocal quadrics. Suppose that a, b reflect to each other off $\mathcal{Q}_\alpha$, and a, c off $\mathcal{Q}_\beta$. Then there is a unique line d such that four lines a, b, c, d form a double reflection configuration.*

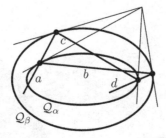

Figure 7.2 Double reflection configuration.

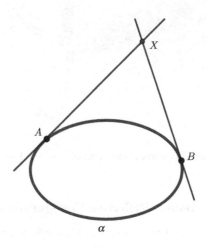

Figure 7.3 Draboux coordinates.

2.3 Darboux Coordinates Generated by a Conic

We are going to begin with the notion of Darboux coordinates generated by a conic. This notion can be found in [Dar1917] and, among recently published articles, in [Dra2010, LT2007, IT2017].

Let α be a given ellipse. This is a rational curve, so we fix one of its rational parameterizations and denote by x the parameter in this parameterization. Consider a point X outside of the ellipse α. The Darboux coordinates x_1 and x_2 of the point X are the values of the parameter x at the touching points A and B of the two tangent lines from the point X to the ellipse α, see Figure 7.3.

These Darboux coordinates [Dar1917], (see also [Dar1901, Dar1914]), should not be confused with well-known Darboux coordinates from symplectic geometry, [Arn1989]. We will employ the Darboux coordinates in Sections 4 and 5.

Following [LT2007] and [IT2017], the equations of confocal ellipses and confocal hyperbolas are given respectively by the following equations:

$$x_1 - x_2 = constant,$$

$$x_1 + x_2 = constant.$$

2.4 Graves-Chasles Theorem

The following theorem is the key tool of the proof of our main theorems in the section 4.2.

Theorem 2.4 (Graves-Chasles theorem, [AB2018]) *Consider a complete quadrilateral such that all its sides are lines tangent to a conic α and pairs of opposite vertices are a, c; b, d and e, f. Then the four following properties are equivalent:*

(1) quadrilateral (abcd) is circumscribed,
(2) points a and c lie on a conic confocal with α,
(3) points b and d lie on a conic confocal with α,
(4) points e and f lie on a conic confocal with α.

The big part of the proof of this theorem is based on two well-known geometrical theorems which are *the classical equal angle lemma* and *the optical properties of conics*. See [AB2018] and the references therein for the details.

2.5 Quad-Graphs

The main elements of the systems on the quad-graphs are the equations of the form $Q(x, x_1, x_2, x_{12}) = 0$ on quadrilaterals, where Q is a multiaffine

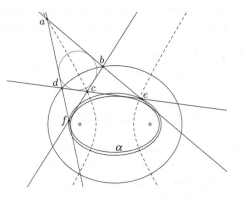

Figure 7.4 Graves-Chasles theorem.

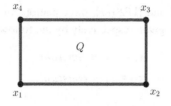

Figure 7.5 Quad-equation $Q(x_1, x_2, x_3, x_4) = 0$.

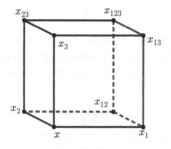

Figure 7.6 3D-consistency.

polynomial, that is a polynomial of degree one in each argument. Such equations are called *quad-equations*. The field variables x_i are assigned to the vertices of a quadrilateral as in Figure 7.5. The quad equation can be solved for each variable, and the solution is a rational function of the other three variables. Following [ABS2009a], we consider the idea of integrability as consistency, see Figure 7.6. We assign six quad-equations to the faces of the cube. The system is said to be *3D-consistent* if the three values for x_{123} obtained from the equations on right, back, and top faces coincide for arbitrary initial data $\{x, x_1, x_2, x_3\}$.

We will be interested in Section 3.1 in a geometric version of the integrable quad graphs, where lines play a role of the vertex fields, while in Section 5.2 we will deal with a more usual algebraic version of quad graphs.

Remark 2.5   Proposition 2.3 shows that the double reflection configuration is playing the role of the quad-equation for lines in the projective space.

3 Double Reflection Nets

Section 3.1 is devoted to a quad-graph interpretation of some results from [DR2008] obtained using the billiard algebra. In Section 3.2, we revise the notion of double reflection net from [DR2012b], making it broader, see Definition 3.3 and Remark 3.4. In Theorem 3.5 we show how nets from that broader class can be constructed starting from a few billiard trajectories within confocal quadrics and then present several new examples. In Section 3.3 we construct a Yang-Baxter map related to a family of confocal quadrics.

3.1 Billiard Algebra and Quad-Graphs

Let us start with a theorem about the confocal families of quadrics from [DR2008]:

Theorem 3.1 (Six-pointed star theorem) *Let $\mathcal{F}$ be a family of confocal quadrics in the three-dimensional space. There exist configurations consisting of twelve planes in the space with the following properties:*

- *The planes may be organized in eight triplets, such that each plane in a triplet is tangent to a different quadric from $\mathcal{F}$ and the three touching points are collinear. Every plane in the configuration is a member of two triplets.*
- *The planes may be organized in six quadruplets, such that the planes in each quadruplet belong to a pencil and are tangent to two different quadrics from $\mathcal{F}$. Every plane in the configuration is a member of two quadruplets.*

Moreover, such a configuration is determined by three planes tangent to three different quadrics from $\mathcal{F}$, with collinear touching points.

Such a configuration of planes in the dual space is shown in Figure 7.7: each plane corresponds to a vertex of the polygonal line.

To understand the notation used in Figure 7.7, let us recall the construction leading to the configurations from Theorem 3.1. Take $\mathcal{Q}_1$, $\mathcal{Q}_2$, $\mathcal{Q}_3$ to be quadrics from $\mathcal{F}$, and α, β, γ respectively their tangent planes such that the

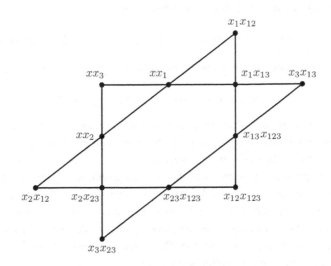

Figure 7.7 A configuration of planes from Theorem 3.1

touching points A, B, C are collinear. Denote by x the line containing these three points, and by x_1, x_2, x_3 the lines obtained from x by the reflections off Q_1, Q_2, Q_3 at A, B, C respectively.

Now, as in Proposition 2.3, determine lines x_{12}, x_{13}, x_{23}, x_{123} such that they respectively complete triplets $\{x, x_1, x_2\}$, $\{x, x_1, x_3\}$, $\{x, x_2, x_3\}$, $\{x_3, x_{13}, x_{23}\}$ to the double reflection configurations.

Notice the following objects in Figure 7.7:

twelve vertices: to each vertex, a plane tangent to one of the three quadrics Q_1, Q_2, Q_3 and a pair of lines are assigned — the lines of any pair are reflected to each other off the quadric at the touching point with the assigned plane;

eight triangles: in any triangle, the planes assigned to the vertices are touching the corresponding quadrics at three collinear points — thus to each triangle, the line containing these points is naturally assigned;

six edges: each edge contains four vertices — four planes assigned to these vertices are in the same pencil; thus a double reflection configuration corresponds to each edge.

Now, we can get the 3D-consistency of the quad-relation introduced via double reflection configurations. The meaning of the following theorem is that reflections off the three quadrics commute.

Theorem 3.2 *Let x, x_1, x_2, x_3 be lines in the projective space, such that x_1, x_2, x_3 are obtained from x by reflections off confocal quadrics Q_1, Q_2, Q_3 respectively. Introduce lines x_{12}, x_{13}, x_{23}, x_{123} such that the following quadruplets are double reflection configurations:*

$$\{x, x_1, x_{12}, x_2\}, \quad \{x, x_1, x_{13}, x_3\}, \quad \{x, x_2, x_{23}, x_3\}, \quad \{x_1, x_{12}, x_{123}, x_{13}\}.$$

Then the following quadruplets are also double reflection configurations:

$$\{x_2, x_{12}, x_{123}, x_{23}\}, \quad \{x_3, x_{13}, x_{123}, x_{23}\}.$$

Proof. Let us remark that the configuration described in Theorem 3.1 has obviously a combinatorial structure of the cube, with the planes corresponding to the edges of the cube. In this way, lines x, x_1, x_2, x_3, x_{12}, x_{13}, x_{23}, x_{123} will correspond to the vertices of the cube as shown in Figure 7.6. A pair of lines is represented by the endpoints of an edge if they reflect to each other off the plane joined to this edge. Faces of the cube represent double reflection configurations. Notice also that the planes joined to the parallel edges of the cube are tangent to a same quadric. The statement follows from Theorem 3.1 and the construction given after, see Figure 7.7.   □

3.2 Definition of Double Reflection Nets and Properties

Fix $d - 1$ quadrics from the family of confocal quadrics (1) and take $\mathcal{A} \subset \mathcal{L}^d$ to be the set of all lines touching these $d - 1$ quadrics.

Definition 3.3 A double reflection net *is a map*

$$\varphi : \mathbf{Z}^m \to \mathcal{A},$$

such that the lines $\varphi(\mathbf{n}_0)$, $\varphi(\mathbf{n}_0 + \mathbf{e}_i)$, $\varphi(\mathbf{n}_0 + \mathbf{e}_j)$, $\varphi(\mathbf{n}_0 + \mathbf{e}_i + \mathbf{e}_j)$ *form a double reflection configuration, for all* $i, j \in \{1, \ldots, m\}$, $i \neq j$ *and* $\mathbf{n}_0 \in \mathbf{Z}^m$.

If, in addition, for each $j \in \{1, \ldots, m\}$ *and* $\mathbf{n}_0 \in \mathbf{Z}^m$, *the sequence* $\{\varphi(\mathbf{n}_0 + i\mathbf{e}_j)\}_{i \in \mathbf{Z}}$ *represents a billiard trajectory within a quadric from the confocal family* (1) *then we will say that* φ *is a* billiard double reflection net.

Remark 3.4 Let us remark that billiard double reflection nets were introduced in [DR2012b], and called just double reflection nets there.

The following theorem shows how to construct a double reflection net if it is given on the coordinate axes.

Theorem 3.5 *Let* $(\ell_i^j)_{i \in \mathbf{Z}}$, $1 \leq j \leq m$, *be* m *sequences in* $\mathcal{A}$, *such that two consecutive lines in each sequence always coplanar and* $\ell_0^1 = \cdots = \ell_0^m$.

Then there is a unique double reflection net $\varphi : \mathbf{Z}^m \to \mathcal{A}$ *such that* $\varphi(n\mathbf{e}_j) = \ell_n^j$, *for each* $n \in \mathbf{Z}$.

Proof. The map φ can be expanded to whole $\mathbf{Z}$ using the request that lines assigned to the vertices of a unit square of the lattice form a double reflection configuration.

Consistency of the construction follows from Theorem 3.2. □

In the following Example 3.6 we use the generalised Full Poncelet Theorem and construct a finite double reflection net.

Example 3.6 *Let* $\mathcal{Q}_{\alpha_1}, \ldots, \mathcal{Q}_{\alpha_m}$ *be quadrics from (1) such that there is an* m-*periodic billiard trajectory with the consecutive reflections off* $\mathcal{Q}_{\alpha_1}, \ldots, \mathcal{Q}_{\alpha_m}$. *Take* $\mathcal{A}$ *to be the set of all lines in the space sharing the same* $d - 1$ *caustics as the given periodic trajectory The generalised Full Poncelet Theorem then states that there is such a periodic trajectory starting by any line in* $\mathcal{A}$; *moreover, the order of reflections can be taken arbitrarily* [CCS1993].

Fix a line $\ell_0 \in \mathcal{A}$. *Then, for each permutation* $(p_1, \ldots, p_m)$ *of the set* $\{1, \ldots, m\}$, *there is a periodic billiard trajectory starting with* ℓ_0 *with the consecutive reflections off* $\mathcal{Q}_{\alpha_{p_1}}, \ldots, \mathcal{Q}_{\alpha_{p_m}}$. *Starting from these trajectories, we construct by Theorem 3.5 a double reflection net* $\varphi_0 : \mathbf{Z}^{m!} \to \mathcal{A}$.

Two double reflection nets φ_0, φ_1 are said to be *related by an F-transformation* if the corresponding lines $\varphi_0(\mathbf{n})$ and $\varphi_1(\mathbf{n})$ intersect for each $\mathbf{n}$ [BS2008]. Next Example 3.7 shows that two double reflection nets constructed in Example 3.6 can be related to each other by a finite sequence of F-transformations.

Example 3.7 *With the assumptions of Example 3.6, take another line $\ell_1 \in \mathcal{A}$ and construct the corresponding double reflection net φ_1. Since ℓ_1 can be obtained from ℓ_0 by a sequence of at most $d-1$ reflections off quadrics from (1)* [DR2008, DR2011], *there is a mapping* $\Phi : \mathbf{Z}^{m!} \times \{0, \ldots, d-1\} \to \mathcal{A}$ *such that*

$$\Phi|_{\mathbf{Z}^{m!} \times \{0\}} = \varphi_0, \quad \Phi|_{\mathbf{Z}^{m!} \times \{d-1\}} = \varphi_1.$$

Such a mapping is unique and can be constructed as explained in Theorem 3.5. Two neighbouring double reflection nets $\Phi|_{\mathbf{Z}^{m!} \times \{i-1\}}$ and $\Phi|_{\mathbf{Z}^{m!} \times \{i\}}$ are related by an F-transformation, since their corresponding lines intersect.

In the last couple of years, the billiard dynamics within pencils of quadrics has also been developed in the pseudo-Euclidean setting, see [KT2009, DR2012a] and also [DR2013]. Some examples of double reflection nets in the pseudo-Euclidean spaces were constructed in [DR2012b, DR2014]. For more recent billiard constructions related to double-reflection configurations in the pseudo-Euclidean spaces see [JJ2015].

3.3 Yang-Baxter Map

A *Yang-Baxter map* is a map $R : \mathcal{X} \times \mathcal{X} \to \mathcal{X} \times \mathcal{X}$, satisfying the Yang-Baxter equation:

$$R_{23} \circ R_{13} \circ R_{12} = R_{12} \circ R_{13} \circ R_{23},$$

where $R_{ij} : \mathcal{X} \times \mathcal{X} \times \mathcal{X} \to \mathcal{X} \times \mathcal{X} \times \mathcal{X}$ acts as R on the i-th and j-th factor in the product, and as the identity on the remaining one, see [ABS2004] and references therein.

Here, we are going to construct an example of the Yang-Baxter map associated to the confocal families of quadrics. To begin, we fix a family of confocal quadrics in $\mathbf{CP}^n$:

$$\mathcal{Q}_\lambda : \quad \frac{z_1^2}{a_1 - \lambda} + \cdots + \frac{z_d^2}{a_d - \lambda} = z_{n+1}^2, \quad (2)$$

where $a_1, \ldots, a_d$ are constants in $\mathbf{C}$, and $[z_1 : z_2 : \cdots : z_{n+1}]$ are the homogeneous coordinates in $\mathbf{CP}^n$.

Take $\mathcal{X}$ to be the space $\mathbf{CP}^{n*}$ dual to the n-dimensional projective space, i.e. the variety of all hyper-planes in $\mathbf{CP}^n$. Note that a general hyper-plane

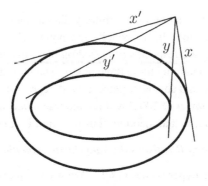

Figure 7.8 $R(x, y) = (x', y')$.

in the space is tangent to exactly one quadric from family (2). Besides, in a general pencil of hyper-planes, there are exactly two of them tangent to a fixed general quadric.

Now, consider a pair x, y of hyper-planes. They are touching respectively unique quadrics $\mathcal{Q}_\alpha$, $\mathcal{Q}_\beta$ from (2). Besides, these two hyper-planes determine a pencil of hyper-planes. This pencil contains unique hyper-planes x', y', other than x, y, that are tangent to $\mathcal{Q}_\alpha$, $\mathcal{Q}_\beta$ respectively, as shown in Figure 7.8.

We define $R : \mathbf{CP}^{n*} \times \mathbf{CP}^{n*} \to \mathbf{CP}^{n*} \times \mathbf{CP}^{n*}$, in such a way that $R(x, y) = (x', y')$ if (x', y') are obtained from (x, y) as just described, see Figure 7.8.

Maps

$$R_{12}, \ R_{13}, \ R_{23} \ : \ \mathbf{CP}^{n*} \times \mathbf{CP}^{n*} \times \mathbf{CP}^{n*} \to \mathbf{CP}^{n*} \times \mathbf{CP}^{n*} \times \mathbf{CP}^{n*}$$

are then defined as follows:

$$R_{12}(x, y, z) = (x', y', z) \quad \text{for} \quad (x', y') = R(x, y);$$
$$R_{13}(x, y, z) = (x', y, z') \quad \text{for} \quad (x', z') = R(x, z);$$
$$R_{23}(x, y, z) = (x, y', z') \quad \text{for} \quad (y', z') = R(y, z).$$

To prove the Yang-Baxter equation for the map R, we will need the following

Lemma 3.8 *Let $\mathcal{Q}_\alpha$, $\mathcal{Q}_\beta$, $\mathcal{Q}_\gamma$ be three non-degenerate quadrics from family (2) and x, y, z respectively their tangent hyper-planes. Take:*

$$(x_2, y_1) = R(x, y), \quad (x_3, z_1) = R(x, z), \quad (y_3, z_2) = R(y, z).$$

Let x_{23}, y_{13}, z_{12} be the joint hyper-planes of pencils determined by pairs (x_3, y_3) and (x_2, y_2), (x_3, y_3) and (y_1, z_1), (y_1, z_1) and (x_2, z_2) respectively.

Then x_{23}, y_{13}, z_{12} touch the quadrics $\mathcal{Q}_\alpha$, $\mathcal{Q}_\beta$, $\mathcal{Q}_\gamma$ respectively.

Proof. This statement, formulated for the dual space in dimension $n = 2$ is proved as [ABS2004, Theorem 5].

Consider the dual situation in an arbitrary dimension n. The dual quadrics $\mathcal{Q}^*_\alpha$, $\mathcal{Q}^*_\beta$, $\mathcal{Q}^*_\gamma$ belong to a linear pencil, and points x^*, y^*, z^*, dual to the hyper-planes x, y, z, are respectively placed on these quadrics. Take the two-dimensional plane containing these three points. The intersection of the pencil of quadrics with that, and any other plane as well, represents a pencil of conics. Thus, Theorem 5 from [ABS2004] will remain true in any dimension.

This lemma is dual to that theorem, thus the proof is complete. $\qquad\square$

Theorem 3.9 *The map R satisfies the Yang-Baxter equation.*

Proof. Let x, y, z be hyper-planes in $\mathbf{CP}^n$. We want to prove that

$$R_{23} \circ R_{13} \circ R_{12}(x, y, z) = R_{12} \circ R_{13} \circ R_{23}(x, y, z).$$

Denote by $\mathcal{Q}_\alpha$, $\mathcal{Q}_\beta$, $\mathcal{Q}_\gamma$ the quadrics from (2) touching x, y, z respectively. Let:

$$(x, y, z) \xrightarrow{R_{12}} (x_2, y_1, z) \xrightarrow{R_{13}} (x_{23}, y_1, z_1) \xrightarrow{R_{23}} (x_{23}, y_{13}, z_{12}),$$

$$(x, y, z) \xrightarrow{R_{23}} (x, y_3, z_2) \xrightarrow{R_{13}} (x_3, y_3, z'_{12}) \xrightarrow{R_{12}} (x'_{23}, y'_{13}, z'_{12}).$$

Now, apply Lemma 3.8 to the hyper-planes x, y, z_2. Since:

$$(x_2, y_1) = R(x, y), \quad (x_3, z'_{12}) = R(x, z_2), \quad (y_3, z) = R(y, z_2),$$

we have that the joint hyper-plane of the pencils (x_3, y_3) and (x_2, z) is touching $\mathcal{Q}_\alpha$ – therefore, this plane must coincide with x_{23} and x'_{23}, i.e. $x_{23} = x'_{23}$. Also, the joint hyper-plane of the pencils (y_1, z'_{12}) and (x_2, z) is touching $\mathcal{Q}_\gamma$ – therefore, this is z_1 and $z_{12} = z'_{12}$. Finally, the joint hyper-plane of pencils (x_3, y_3) and (y_1, z'_{12}) is tangent to $\mathcal{Q}_\beta$ – it follows this is $y_{13} = y'_{13}$, which completes the proof. $\qquad\square$

Remark 3.10 Instead of defining R to act on the whole space $\mathbf{CP}^{n*} \times \mathbf{CP}^{n*}$, we can restrict it to the product of two non-degenerate quadrics from (2), namely:

$$R(\alpha, \beta) \; : \; \mathcal{Q}^*_\alpha \times \mathcal{Q}^*_\beta \to \mathcal{Q}^*_\alpha \times \mathcal{Q}^*_\beta,$$

where a pair (x, y) of tangent hyper-planes is mapped into a pair (x_1, y_1) in such a way that x, y, x_1, y_1 belong to the same pencil.

The corresponding Yang-Baxter equation is:

$$R_{23}(\beta, \gamma) \circ R_{13}(\alpha, \gamma) \circ R_{12}(\alpha, \beta) = R_{12}(\alpha, \beta) \circ R_{13}(\alpha, \gamma) \circ R_{23}(\alpha, \beta),$$

where both sides of the equation represent maps from $\mathcal{Q}^*_\alpha \times \mathcal{Q}^*_\beta \times \mathcal{Q}^*_\gamma$ to itself.

In [ABS2004], for irreducible algebraic varieties $\mathcal{X}_1$ and $\mathcal{X}_2$, a quadrirational mapping $F \; : \; \mathcal{X}_1 \times \mathcal{X}_2$ is defined. For such a map F and any fixed pair $(x, y) \in \mathcal{X}_1 \times \mathcal{X}_2$, except from some closed subvarieties of codimension bigger or equal to 1, the graph $\Gamma_F \subset \mathcal{X}_1 \times \mathcal{X}_2 \times \mathcal{X}_1 \times \mathcal{X}_2$ intersects each of the sets $\{x\} \times \{y\} \times \mathcal{X}_1 \times \mathcal{X}_2$, $\mathcal{X}_1 \times \mathcal{X}_2 \times \{x\} \times \{y\}$, $\mathcal{X}_1 \times \{y\} \times \{x\} \times \mathcal{X}_2$, $\{x\} \times \mathcal{X}_2 \times \mathcal{X}_1 \times \{y\}$ exactly at one point (see [ABS2004, Definition 3]). In other words, Γ_F is the graph of four rational maps: F, F^{-1}, $\bar{F}$, $\bar{F}^{-1}$.

The following Proposition is a generalization of [ABS2004, Proposition 4].

Proposition 3.11 *Map* $R(\alpha, \beta) : \mathcal{Q}_\alpha^* \times \mathcal{Q}_\beta^* \to \mathcal{Q}_\alpha^* \times \mathcal{Q}_\beta^*$, *is quadrilateral. It is an involution and it coincides with its companion* $\bar{R}(\alpha, \beta)$.

3.4 Hyper-Plane Billiard Nets

The Six-pointed star theorem, i.e Theorem 3.1 of this manuscript, states that there exist configurations consisting of twelve planes which may be organised in eight triplets and six quadruplets, such that every plane in the configuration is a member of two such triplets and two such quadruplets. That is the same as the configuration of the edges of the cube. If we assign each plane to the mid-point of the corresponding edge of the cube, we will get that each plane corresponds to a vertex of the cuboctahedron, as shown in Figure 7.9. These cuboctahedra will represent basic building blocks for lattices of hyper-planes that we introduce in this section.

Consider a lattice $\mathbf{Z}^m$ in $\mathbf{R}^m$. That lattice generates a *honeycomb*, that is a filling of the space by polytopes [Cox1973]. In this case, it is a regular honeycomb consisting of m-cubes. The set of all midpoints of the edges of the cubes is

$$\mathcal{M}^m = \bigcup_{1 \le i \le m} \left(\mathbf{Z}^m + \frac{1}{2}\mathbf{e}_i \right).$$

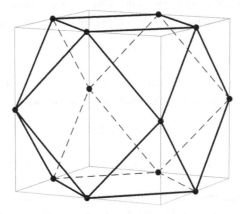

Figure 7.9 A configuration of planes from the Six-pointed star theorem.

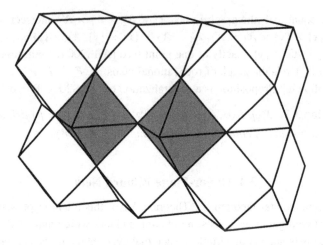

Figure 7.10 Honeycomb consisting of cuboctahedra and octahedra.

The lattice $\mathcal{M}^m$ determines a honeycomb containing two types of polytopes (see [Cox1973]):

- *rectified m-cubes* – the vertices of each polytope of this kind are midpoints of the edges of an m-cube in the lattice $\mathbf{Z}^m$;
- *cross polytopes* – the vertices of each cross polytope are midpoints of all edges with a common endpoint in $\mathbf{Z}^m$.

This is an example of a convex uniform honeycomb [Wel1977]. For $m = 3$, it is shown in Figure 7.10, and for $m = 2$ in Figure 7.11. Integrable systems on such lattices were studied in [KS2003, KS2006].

Assume that $\varphi : \mathbf{Z}^m \to \mathcal{L}$ is a given double reflection net. For each $\mathbf{n_0} \in \mathbf{Z}^m$ and $i \in \{1, \ldots, m\}$, the lines $\varphi(\mathbf{n_0})$ and $\varphi(\mathbf{n_0} + \mathbf{e}_i)$ are reflected to each other off the quadric $\mathcal{Q}_i^*$ from the confocal family. We define the map

$$\mathcal{H} : \mathcal{M}^m \to \mathbf{P}^{d*},$$

such that it assigns to the midpoint of the edge $(\mathbf{n_0}, \mathbf{n_0} + \mathbf{e}_i)$ the tangent plane to $\mathcal{Q}_i^*$ at the point of reflection. We introduce also the map

$$\mathcal{P} : \mathcal{M}^m \to \mathbf{P}^{d*},$$

which assigns the intersection point $\varphi(\mathbf{n_0}) \cap \varphi(\mathbf{n_0} + \mathbf{e}_i)$ to the midpoint of the edge $(\mathbf{n_0}, \mathbf{n_0} + \mathbf{e}_i)$.

Since each hyperplane of the space, except of a subset of measure 0, is touching exactly one quadric from the given confocal family. Thus, the touching point is uniquely determined and map $\mathcal{P}$ is uniquely determined by $\mathcal{H}$. The inverse, to determine $\mathcal{H}$ when $\mathcal{P}$ is given, is not straightforward, since each point of the d-dimensiona space belongs to d confocal quadrics.

Proposition 3.12 ([Rad2015]) *From the construction, $\mathcal{H}$ and $\mathcal{P}$ have the following properties:*

(1) *For each cross polytope of the honeycomb, $\mathcal{P}$ assigns to all its vertices collinear points. Moreover, the points joined to the opposite vertices are on the same quadric from the confocal family.*
(2) *For each square 2-face of any rectified m-cube, $\mathcal{H}$ assignes to its vertices hyperplanes that belong to one pencil and form a harmonic quadruple. The hyperplanes corresponding to the opposite vertices are tangent to the same quadric from the confocal family.*
(3) *The hyperplanes assigned to any two adjacent vertices of a square 2-face uniquely determine the hyperplanes assigned to the other two vertices.*

Remark 3.13 For $m = 2$, $\mathcal{M}^2$ determines a regular tessellation by squares, see Figure 7.11. For each square, if the values of $\mathcal{H}$ are given at two neighbouring vertices, it is possible to uniquely determine the values at the remaining two vertices. The hyperplanes joined to the opposite points of a square are always tangent to the same quadric from the confocal pencil. However, the discrete dynamics depends on the type of each square:

- For the squares with vertices of the form

$$\mathbf{n_0} + \frac{\mathbf{e_1}}{2}, \quad \mathbf{n_0} + \frac{\mathbf{e_2}}{2}, \quad \mathbf{n_0} + \mathbf{e_1} + \frac{\mathbf{e_2}}{2}, \quad \mathbf{n_0} + \frac{\mathbf{e_1}}{2} + \mathbf{e_2},$$

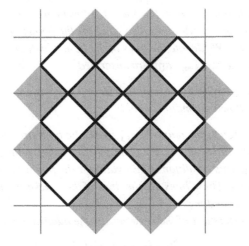

Figure 7.11 The lattice $\mathcal{M}^2$.

the corresponding hyperplanes are in a pencil and harmonically conjugated. Such squares are white in Figure 7.11.

- For the squares with vertices of the form

$$\mathbf{n_0} + \frac{\mathbf{e_1}}{2}, \quad \mathbf{n_0} + \frac{\mathbf{e_2}}{2}, \quad \mathbf{n_0} - \frac{\mathbf{e_1}}{2}, \quad \mathbf{n_0} - \frac{\mathbf{e_2}}{2},$$

the corresponding touching points are collinear, and in general not harmonically conjugated. Such squares are gray coloured in Figure 7.11.

We denoted by $\mathbf{n_0}$ a point of $\mathbf{Z}^2$ and by $\mathbf{e_1}$, $\mathbf{e_2}$ unit coordinate vectors.

It would be interesting to write explicitly recursive relations that maps $\mathcal{H}$, $\mathcal{P}$ and double reflection nets satisfy on the lattice $\mathcal{M}^m$ and to see how they fit in the classification from [ABS2003].

4 Incircular Nets

Consider a map of the square grid to the plane

$$f : \mathbb{Z}^2 \longrightarrow \mathbb{R}^2,$$

- the vertices of net are denoted by $f_{i,j} = f(i, j)$,
- the quadrilateral $(f_{i,j}, f_{i+c,j}, f_{i+c,j_c}, f_{i,j_c})$ is denoted by $\square_{i,j}^c$. We can also call this quadrilateral a *net-square*. If $c = 1$, then we will call it a unit net-square and denote it by $\square_{i,j}^1$ or just simply by $\square_{i,j}$.

Definition 4.1 ([Böh1970, AB2018]) *A map* $f : \mathbb{Z}^2 \longrightarrow \mathbb{R}^2$ *is called an* incircular net *or, shorter,* IC-net *if the following conditions are satisfied:*

(1) *For any integer i, the points* $\{f_{i,j} \mid j \in \mathbb{Z}\}$ *lie on a straight line and preserve the order, that is, the point $f_{i,j}$ lies between the points $f_{i,j-1}$ and $f_{i,j+1}$. This condition is also valid for the points* $\{f_{i,j} \mid i \in \mathbb{Z}\}$ *for any integer j. We denote the lines of the IC-net by l_i and m_j.*

(2) *All unit net-square $\square_{i,j}$ are circumscribed.*

Definition 4.2 ([AB2018]) *A map* $g : \mathbb{Z}^2 \longrightarrow \mathbb{R}^2$ *is called a* checkerboard IC-net *if the following conditions hold:*

(1) *For any integer i, the points* $\{g_{i,j} \mid j \in \mathbb{Z}\}$ *lie on a straight line and preserve the order, that is, the point $g_{i,j}$ lies between the points $g_{i,j-1}$ and $g_{i,j+1}$. This condition is also valid for the points* $\{g_{i,j} \mid i \in \mathbb{Z}\}$ *for any integer j. These lines are the lines of the checkerboard IC-net and denoted by l_i and m_j.*

(2) *For any integers i, j of the same parity, the quadrilateral*

$$(g_{i,j}, g_{i+1,j}, g_{i+1,j+1}, g_{i,j+1})$$

is circumscribed.

4.1 Geometric Properties

Main geometric properties of incircular nets are listed in the following theorems.

Theorem 4.3 (see [Böh1970], [AB2018]) *A given IC-net f has the following properties:*

(1) *Let k be a fixed integer, the points $f_{i,j}$, where $i + j = k$, belong to a confocal conic with α. The point $f_{i,j}$, where $i - j = k'$ (k' is fixed integer), also belong to a confocal conic with α.*

(2) *All net-squares of f are circumscribed.*

(3) *The cross ratio*

$$cr(f_{i,j_1}, f_{i,j_2}, f_{i,j_3}, f_{i,j_4}) = \frac{(f_{i,j_1} - f_{i,j_2})(f_{i,j_3} - f_{i,j_4})}{(f_{i,j_2} - f_{i,j_3})(f_{i,j_4} - f_{i,j_1})}$$

is independent of i. The cross ration $cr(f_{i_1,j}, f_{i_2,j}, f_{i_3,j}, f_{i_4,j})$ is also independent of j.

(4) *Let $f_{i,j}$, with $i + j = k$, be points of the IC-net such that these points belong to the conics C_k. Then, for any integer l there exists an affine transformation $A_{k,l} : C_k \longrightarrow C_{k+2l}$ such that $A_{k,l}(f_{i,j}) = f_{i+l,j_l}$. The same is still true for the conics passing through the points $f_{i,j}$, with $i - j = k'$, in such a case the affine transformation can be expressed as $A_{k',l'}(f_{i,j}) = f_{i-l',j+l'}$.*

Proof. We are providing a proof different that one from [AB2018] for the properties 1 and 4.

Property 1. We will show the first property of the IC-net for the vertices with $i + j = 5$. Regarding the other cases, the proof is similar. Here in particular, we need to show that the three points $f_{1,4}, f_{2,3}, f_{3,2}$ lie on the same conic confocal with α. Without loss of generality, we assume that the pairs of points $f_{1,2}$, $f_{2,3}; f_{2,1}, f_{3,2}$ lie on the ellipses confocal with α and the pairs of points $f_{1,2}$, $f_{2,1}; f_{2,3}, f_{3,2}$ lie on the hyperbolas confocal with α. Since the quadrilateral $(f_{1,3}f_{2,3}f_{2,4}f_{1,4})$ is circumscribed, then there exists a conic confocal with α passing through the vertices $f_{2,3}$ and $f_{1,4}$. Thus, this conic could be an ellipse or a hyperbola. Suppose that it is a hyperbola. Let denote by γ_1 and γ_2 the hyperbolas which contain the points $f_{3,2}, f_{2,3}$ and $f_{2,3}, f_{1,4}$ respectively. Clearly, γ_1 and γ_2 intersect at the point $f_{2,3}$. However, they are confocal, therefore they must be identical. Suppose now that the conic is an ellipse. Denote by β_1 and β_2 the respective ellipses that pass the points $f_{1,2}, f_{2,3}$ and $f_{2,3}, f_{1,4}$. Then the point $f_{2,3}$ is a common point of these ellipses. Therefore, β_1 and β_2 could be the same or different. It is impossible that these ellipses are different because they are confocal ellipses and the confocal ellipses cannot

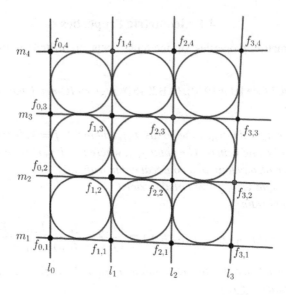

Figure 7.12 Three points belong to the same conic.

have a common point. Let x_0, x_1, x_2, y_0, y_1, y_2 be the canonical coordinates of the tangency points of the lines l_0, l_1, l_2, m_2, m_3, m_4 with α. Since the points $f_{1,2}$, $f_{2,3}$, $f_{1,4}$ belong to the same ellipse, then we obtain the following equations:

$$y_0 - x_1 = y_1 - x_2 \tag{3a}$$

$$y_2 - x_1 = y_1 - x_2 \tag{3b}$$

The equations (3a) and (3b) imply that $y_0 = y_2$, so the points $f_{1,4}$ and $f_{1,2}$ coincide. We obtain a contradiction.

Property 4. This part follows the exposition from [AB2018]. Let A, B, C, D and E be five fixed points on the conic α. Let l_{i_1} be a tangent line passing through the point E. Then the tangent lines at the points A, B, C, D intersect the tangent line l_{i_1} at the points f_{i_1,j_1}, f_{i_1,j_2}, f_{i_1,j_3}, f_{i_1,j_4} respectively. One has already demonstrated that

$$cr(f_{i_1,j_1}, f_{i_1,j_2}, f_{i_1,j_3}, f_{i_1,j_4}) = cr(A, B, C, D).$$

This equality means that there exists a projective transformation p_1 which maps the point A to f_{i_1,j_1}, B to f_{i_1,j_2}, C to f_{i_1,j_3}, D to f_{i_1,j_4}. Let us draw an other tangent line l_{i_2} to the conic α. Using again the argument above, we find another projective transformation p_2 which sends the point A to f_{i_2,j_1}, B to f_{i_2,j_2}, C to f_{i_2,j_3}, D to f_{i_2,j_4}. Since the set of projective transformations

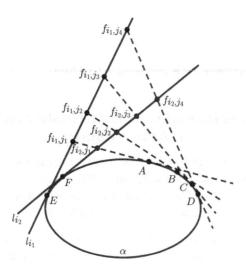

Figure 7.13 Projective map between two tangent lines to the conic α.

forms a group under the operation of composition of functions, then the map defined by $p := p_2 \circ p_1^{-1}$ is a projective transformation and it can be written as follows.

$$p : (f_{i_1,j_1}, f_{i_1,j_2}, f_{i_1,j_3}, f_{i_1,j_4}) \longmapsto (f_{i_2,j_1}, f_{i_2,j_2}, f_{i_2,j_3}, f_{i_2,j_4}).$$

Therefore, the cross ratio under the projective transformation p is preserved. This complete the proof of the property of IC-net. The second statement is proved in the same way. □

Theorem 4.4 (A. Akopyan, A. Bobenko) *A checkerboard IC-net g has the following geometric properties:*

(1) *All net-squares are circumscribed.*
(2) *The points $g_{i,j}$, where $i + j$ is an odd constant, lie on a conic. The points $g_{i,j}$, where $i - j$ is an even constant, also lie on a conic.*
(3) *For any $(i, j) \in \mathbb{Z}^2$, with $i + j$ is even, and any even integer c, then g satisfies*

$$d_C(\Box_{i-c,j}, \Box_{i+c,j}) = d_C(\Box_{i,j-c}, \Box_{i,j+c}),$$

where $d_C(\Box_{i-c,j}, \Box_{i+c,j})$ (resp. $d_C(\Box_{i,j-c}, \Box_{i,j+c})$) is the distance between the tangency points on a common exterior tangent line to the respective inscribed circles $\omega_{i-c,j}$, $\omega_{i+c,j}$ in $\Box_{i-c,j}$ and in $\Box_{i+c,j}$ (resp. $\omega_{i,j-c}$, $\omega_{i,j+c}$ in $\Box_{i,j-c}$ and in $\Box_{i,j+c}$).

(4) *The centers $o_{i,j}$ of the circles of a checkerboard IC-net g build a circle-conical net.*

The proof of theorem 4.4 is contained in [AB2018].

4.2 Poncelet-Darboux Grids and Incircular Nets Related to Billiard Trajectories

The following two theorems that we are going to present are a generalizations of Darboux theorem related to the periodic billiard trajectories within a conic, and have been proved in [DR2008, DR2015, Dar1914]. Other aspects of the Darboux theorem are presented in [Sch2007, LT2007].

Theorem 4.5 (V. Dragović, M. Radnović) *Let $(l_m)_{m \in \mathbb{Z}}$ be the sequence of segments of a billiard trajectory within the ellipse α. Define the sets:*

$$P_k = \bigcup_{i-j=k} l_i \cap l_j \quad \text{and} \quad Q_k = \bigcup_{i+j=k} l_i \cap l_j \quad \text{with} \quad k \in \mathbb{Z}.$$

Then, the following results hold:

(1) *Each of the sets P_k and Q_k belongs to a single conic confocal with α.*
(2) *If the caustic κ of the trajectory l_m is an ellipse, the sets P_k are situated on ellipses and Q_k on hyperbolas.*
(3) *If the caustic κ of the trajectory l_m is an hyperbola, then the sets P_k and Q_k are situated on ellipses for k even and on hyperbolas for k odd.*

Theorem 4.6 (V. Dragović and M. Radnović) *Let α be an ellipse in the Euclidean plane. Consider two sequences of segments of billiard trajectories $(l_m)_{m \in \mathbb{Z}}$ and $(t_m)_{m \in \mathbb{Z}}$ within α such that they share the same caustic κ. The following results hold:*

(1) *All the points $l_m \cap t_m$ belong to one conic β confocal with α.*
(2) *If the conic κ is an ellipse and both trajectories are winding in one direction about κ, then β is also an ellipse.*
(3) *If the conic κ is an ellipse and both trajectories are winding in opposite directions about κ, then β is an hyperbola.*
(4) *If the caustic κ is a hyperbola and the segments l_m and t_m intersect the major axis of α in the same direction then β is a hyperbola.*
(5) *If the caustic κ is a hyperbola and the segments l_m and t_m intersect the major axis of α in the opposite directions then β is a an ellipse.*

The main point of difference between the Darboux theorem and theorems 4.5 and 4.6 is that the latter do not assume periodicity of the billiard trajectories, while the former does.

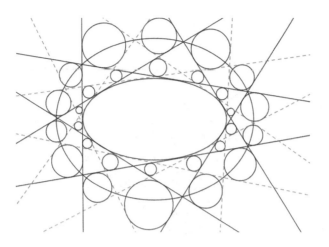

Figure 7.14 Confocal checkerboard IC-net.

Definition 4.7 ([Böh1970], [AB2018]) *An IC-net (or a checkerboard IC-net) is called confocal if all lines of it are tangent to a given conic.*

In the following statements, we are going to present a new method of constructing a confocal IC-net starting from two different billiard trajectories within the same conic and sharing the same caustic. We do not assume that these trajectories are periodic. They are also either winding in the same direction or in the opposite directions. And then, we will show how to obtain a confocal checkerboard IC-net from the confocal IC-net.

Theorem 4.8 *Let α be an ellipse in $\mathbb{E}^2$ and $(l_m)_{m\in\mathbb{Z}}$, $(t_m)_{m\in\mathbb{Z}}$ be two sequences of the segments of billiard trajectories within α, sharing the same caustic κ. Then all quadrilaterals made by the sequence of the segments $(l_m)_{m\in\mathbb{Z}}$ and $(t_m)_{m\in\mathbb{Z}}$, whose vertices are formed by the points $l_m \cap t_m$, are circumscribed.*

Proof. Since the boundary of both billiard trajectories $(l_m)_{m\in\mathbb{Z}}$, $(t_m)_{m\in\mathbb{Z}}$ is the ellipse α and they also share the same caustic κ, due to the theorem 4.6 all points $l_m \cap t_m$ lie on a conic β. It follows that two opposite vertices for each quadrilateral made by these billiard trajectories belong to β. Applying theorem 2.4, the quadrilateral is circumscribed. Thus, the global statement follows immediately. $\qquad\square$

Remark 4.9 The result of the theorem 4.8 does not change if we add more additional assumptions such as fixing the direction of both trajectories and specifying type of the caustic. For instance, if the caustic is an ellipse, then the conic β is either an ellipse or a hyperbola depending on the direction of the billiard trajectories. Likewise, if the caustic is a hyperbola and the both

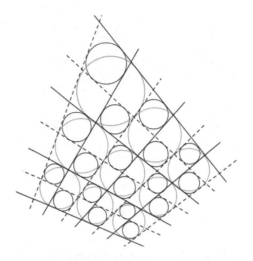

Figure 7.15 The quadrilaterals having vertices $l_m \cap t_m$ are circumscribed by the red circles.

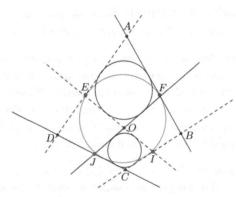

Figure 7.16 Circumscribed quadrilateral $(ABCD)$.

segments l_m and t_m intersect along the major axis of the boundary α in the same direction, then β is a hyperbola; otherwise it is an ellipse. Moreover, the type of the caustic and the direction of billiard trajectories determine the position of small circles in the figure 7.15.

Proposition 4.10 *Assume that the billiard trajectories satisfy all assumptions of theorem 4.8 and consider a quadrilateral $(ABCD)$ constructed by these billiard trajectories. This quadrilateral is divided into four small quadrilaterals (see figure 7.16). Then the quadrilateral $(ABCD)$ is circumscribed if and only if two small quadrilaterals at opposite corners are circumscribed.*

Proof. In figure 7.16, all lines are tangent to the same caustic κ since they are lines from the billiard trajectories. Now, let us denote by x_1, x_2, x_3, y_1, y_2, y_3 the respective coordinates of the tangency points of the tangent lines (BC), (FJ), (AD), (AB), (EI), (CD) on the caustic κ so that $A(x_3, y_1)$, $B(x_1, y_1)$, $C(x_1, y_3)$, $D(x_3, y_3)$, $E(x_3, y_2)$, $F(x_2, y_1)$, $I(x_1, y_2)$, $J(x_2, y_3)$ and $O(x_2, y_2)$. Suppose that the quadrilaterals $(AEOF)$ and $(OJCI)$ are circumscribed. We need to show that the quadrilateral $(ABCD)$ is circumscribed as well. Since the quadrilateral $(AEOF)$ is circumscribed, then by the Graves-Chasles theorem, the points E and F lie on the same conic confocal with κ. This conic is an ellipse because the points E and F belong to the corresponding set P_k and P_k' (as defined in theorem 4.5). Thus, the points A and O lie on the same hyperbola. Therefore, we obtain the following equations:

$$y_2 - x_3 = y_1 - x_2 \tag{4a}$$

$$y_1 + x_3 = y_2 + x_2. \tag{4b}$$

Notice that the equations (4a) and (4b) are identical. Using the argument above to the circumscribed quadrilateral $(OJCI)$, we also have:

$$y_2 - x_1 = y_3 - x_2 \tag{5a}$$

$$y_2 + x_2 = y_3 + x_1. \tag{5b}$$

Subtracting the equations (4a) and (5a), we have $y_1 - x_1 = y_3 - x_3$. This equation shows us that the points B and D lie on the same ellipse. Thus, the quadrilateral $(ABCD)$ is circumscribed according to the Graves-Chasles theorem.

Assume that the $(ABCD)$ in the figure 7.16 is circumscribed. Then, the points B, D lie on the same conic confocal with κ. Knowing the directions of the billiard trajectories and the type of conic through the points B and D, it is straightforward to identify the confocal conics passing through the pairs of point E, F and I, J. Then it follows again from the Graves-Chasles theorem that the quadrilaterals $(AEOF)$ and $(OJCI)$ are circumscribed. $\square$

Theorem 4.11 *We use the same assumptions and terminology of the theorem 4.8. The billiard trajectories $(l_m)_{m \in \mathbb{Z}}$ and $(t_m)_{m \in \mathbb{Z}}$ form a confocal IC-net having the points $l_m \cap t_m$ as vertices of its unit net-squares. The same trajectories also form a confocal checkerboard IC-net such that the vertices of each unit net-square are organized as follows: one vertex is the intersection point of the trajectory l_m to itself, one vertex also is the intersection point of t_m to itself, and two remaining vertices are of the form $l_n \cap t_m$.*

Proof. First of all, notice that all lines l_m and t_n are tangent to the caustic κ. The first condition in the definition of IC-net is clear because the orders of the points on the lines of confocal IC-net l_m and t_n are preserved (this result still holds for the confocal checkerboard IC-net) and the second condition follows from the theorem 4.8. Thus the billiard trajectories form a confocal IC-net. It remains to show that for any integer m and n with the same parity, the unit net-square $\square_{m,n}$ is circumscribed. It follows from the proposition 4.10. Therefore, the billiard trajectories $(l_m)_{m \in \mathbb{Z}}$ and $(t_m)_{m \in \mathbb{Z}}$ build also a confocal checkerboard IC-net. $\qquad\square$

5 Discriminantly Separable Polynomials

The concept of discriminantly separable polynomials has been introduced by one of the authors some years ago [Dra2010]. We review the basic notions and we emphasize here the relationships of discriminantly separable polynomials with pencils of conics and with quad-graphs. Several other relationships and applications to other areas of mathematics are left outside the scope of the present paper. For example among the basic examples of multivalued groups, there are n-valued additive group structures on $\mathbb{C}$. For $n = 2$, this is a two-valued group p_2 defined by the relation (see [Dra2010] and references therein)

$$m_2 : \mathbb{C} \times \mathbb{C} \to (\mathbb{C})^2$$
$$x *_2 y = [(\sqrt{x} + \sqrt{y})^2, (\sqrt{x} - \sqrt{y})^2] \tag{6}$$

The product $x *_2 y$ corresponds to the roots in z of the polynomial equation

$$p_2(z, x, y) = 0,$$

where

$$p_2(z, x, y) = (x + y + z)^2 - 4(xy + yz + zx).$$

One can easily check that p_2 is an example of a discriminantly separable polynomial, as defined below. One can calculate the discriminants of p_2 understood as polynomial of second degree in one of variables, say x with the other two variables assumed as parameters. Then the discriminant, as a polynomial in the other two variables, decomposes as a product of two polynomials of one variable only:

$$\mathcal{D}_x(p_2)(y, z) = P(y)P(y),$$

where $P(y) = 2y$.

Although purely algebraic in nature, the discriminantly separable polynomials originally emerged within an attempt to develop a novel approach to the classical, celebrated Kowalevski top and a geometrization of the Kowalevski integration procedure from [Kow1889]. Namely, it was observed in [Dra2010] that the so-called Kowalevski fundamental relation, see relation (15) in [Dra2010], is defined as $Q(w, x_1, x_2) = 0$, where $Q(w, x_1, x_2)$ is an example of a disciminantly separable polynomial as defined below. This property of discriminant separability is the key for the most of Kowalevski clever substitutions and algebraic miracles that accompanied her integration of the equations of motion of the Kowalevski top. After a change of variables in Q generated by Möbius transformations, which preserve discriminant separability, it can be transformed (see [Dra2010]) to a law which defines the two-valued group structure on an elliptic curve.

Suppose that a cubic W is given in the standard form

$$W : t^2 = J(s) = 4s^3 - g_2 s - g_3.$$

Consider the mapping $W \to S^2 = \hat{\mathbb{C}} : (s, t) \mapsto s$, where $\hat{\mathbb{C}}$ represents a complex line extended by ∞.

The curve W as a cubic curve has the group structure. Together with its canonical involution $\tau : (s, t) \mapsto (s, -t)$, it defines the standard two-valued group structure of coset type (see [BR1997], [Buc2006]) on S^2 with the unit at infinity in S^2. The product is defined by the formula:

$$[s_1] *_c [s_2] = \left[\left[-s_1 - s_2 + \left(\frac{t_1 - t_2}{2(s_1 - s_2)} \right)^2 \right], \left[-s_1 - s_2 + \left(\frac{t_1 + t_2}{2(s_1 - s_2)} \right)^2 \right] \right],$$

(7)

where $t_i = J(s_i), i = 1, 2$, and

$$[s_i] = \{(s_i, t_i), (s_i, -t_i)\}, \quad s_i = \wp(u_i), t_i = \wp'(u_i),$$

by using addition theorem for the Weierstrass function $\wp(u)$:

$$\wp(u_1 + u_2) = -\wp(u_1) - \wp(u_2) + \left(\frac{\wp'(u_1) - \wp'(u_2)}{2(\wp(u_1) - \wp(u_2))} \right)^2.$$

One can see this two-valued group structure as a deformation of the above two-valued group p_2, defined by a polynomial $\hat{Q}$ obtained from Q. Let us mention that the first development of two-valued group structures goes back to work of Buchstaber and Novikov in 1971, while the two-valued group structure on elliptic curves was introduced by Buchstaber in 1990, see [Buk1990].

It was just recently observed that discriminantly separable polynomials appear in the description of the caustics of elliptical billiards in the Euclidean plane, see [DR2019].

Beside continuous integrable systems, discrete integrable systems, namely integrable quad-graphs appear to be closely related to dicriminantly separable polynomials. Moreover, there is a full parallelism between a classification of discriminantly separable polynomials and a well-known ABS classification

[ABS2009b] of quad-graphs. The main results about discriminantly separable polynomials are obtained in [Dra2010], [DK2011], [DK2014b], [DK2014c], [DK2014a], [DK2014d], [Dra2013], [Dra2014], [DK2017].

5.1 Relationship with Pencils of Conics

Before giving a formal definition of the discriminantly separable polynomials, let us recall the equations of a pencil of conics. Denote such an equation as $\mathcal{F}(w, x_1, x_2) = 0$, where w is the pencil parameter; x_1 and x_2 are the Darboux coordinates (see Section 2.3 for the definition). The use of that classical, but mainly forgotten notion, instead of usual projective coordinates, appeared to have significant consequences to the development of the theory of discriminantly separable polynomials. We recall some of the details: given two conics C_1 and C_2 in a general position by their tangential equations

$$C_1 : a_0 w_1^2 + a_2 w_2^2 + a_4 w_3^2 + 2a_3 w_2 w_3 + 2a_5 w_1 w_3 + 2a_1 w_1 w_2 = 0;$$
$$C_2 : w_2^2 - 4w_1 w_3 = 0. \tag{8}$$

Then the conics of this general pencil $C(s) := C_1 + sC_2$ have four common tangent lines. Denote the matrix M:

$$M(s, z_1, z_2, z_3) = \begin{bmatrix} 0 & z_1 & z_2 & z_3 \\ z_1 & a_0 & a_1 & a_5 - 2s \\ z_2 & a_1 & a_2 + s & a_3 \\ z_3 & a_5 - 2s & a_3 & a_4 \end{bmatrix}. \tag{9}$$

The coordinate equations of the conics of the pencil are

$$F(s, z_1, z_2, z_3) := \det M(s, z_1, z_2, z_3) = 0,$$

which determines a quadratic polynomial in the pencil parameter s, namely $F := H + Ks + Ls^2$, with H, K, and L being quadratic expressions in (z_1, z_2, z_3).

Assume the standard projective coordinates $(z_1 : z_2 : z_3)$ in the plane, and choose, without loss of generality, a rational parametrization of the conic C_2 by $(1, \ell, \ell^2)$. The tangent line to the conic C_2 through a point of the conic with the parameter ℓ_0 is given by the equation

$$t_{C_2}(\ell_0) : z_1 \ell_0^2 - 2z_2 \ell_0 + z_3 = 0.$$

For a given point P outside the conic in the plane with the coordinates $P = (\hat{z}_1, \hat{z}_2, \hat{z}_3)$, there are two corresponding solutions x_1 and x_2 of the equation quadratic in ℓ

$$\hat{z}_1\ell^2 - 2\hat{z}_2\ell + \hat{z}_3 = 0. \tag{10}$$

The two solutions correspond to two tangent lines to the conic C_2 from the point P, and they define the pair (x_1, x_2) of the Darboux coordinates of the point P. One finds immediately the converse formulae $\hat{z}_1 = 1$, $\hat{z}_2 = (x_1 + x_2)/2$, $\hat{z}_3 = x_1 x_2$.

Changing the variables in the polynomial F from the projective coordinates $(z_1 : z_2 : z_3)$ to the Darboux coordinates, we rewrite its expression in the form

$$\mathcal{F}(s, x_1, x_2) = L(x_1, x_2)s^2 + K(x_1, x_2)s + H(x_1, x_2).$$

The key algebraic property of the pencil polynomial written in this form, as a quadratic polynomial in each of the three variables s, x_1, x_2 is: *all three of its discriminants are expressed as products of two polynomials in one variable each*:

$$\mathcal{D}_s(\mathcal{F})(x_1, x_2) = P(x_1)P(x_2), \ \mathcal{D}_{x_i}(\mathcal{F})(s, x_j) = J(s)P(x_j), i, j = 1, 2,$$

where J and P are polynomials of degree 3 and 4 respectively, and the elliptic curves

$$\Gamma_1 : y^2 = P(x), \quad \Gamma_2 : y^2 = J(s)$$

appear to be isomorphic (see Proposition 1 of [Dra2010]). Here, and below, we denote by $\mathcal{D}_{x_i}\mathcal{F}(x_j, x_k)$, the discriminant of $\mathcal{F}$ considered as a quadratic polynomial in x_i.

As a geometric interpretation of $F(s, x_1, x_2) = 0$ we may say that the point P in the plane, with the Darboux coordinates with respect to C_2 equal to (x_1, x_2) belongs to two conics of the pencil, with the pencil parameters equal to s_1 and s_2, such that

$$\mathcal{F}(s_i, x_1, x_2) = 0, \ i = 1, 2.$$

It is easy to see that the locus of points with a double conic passing through them coincides with the union of the four common tangent lines to the pencil. From there, one can directly conclude that

$$\mathcal{D}_s(\mathcal{F})(x_1, x_2) = P(x_1)P(x_2),$$

where $P(x) = (x - k_1)(x - k_2)(x - k_3)(x - k_4)$, and $k_i, i = 1, \ldots, 4$ are the parameters of the points of tangency of the four common tangents with the conic C_2.

Now we provide a general definition of the discriminantly separable polynomials. With $\mathcal{P}_m^n$ denote the polynomials of m variables of the degree n in each variable.

Definition 5.1 ([Dra2010]) *A polynomial* $F(x_1, \ldots, x_n)$ *is discriminantly separable if there exist polynomials* $f_i(x_i)$ *such that for every* $i = 1, \ldots, n$

$$\mathcal{D}_{x_i} F(x_1, \ldots, \hat{x}_i, \ldots, x_n) = \prod_{j \neq i} f_j(x_j).$$

It is symmetrically discriminantly separable if $f_2 = f_3 = \cdots = f_n$, *while it is strongly discriminantly separable if* $f_1 = f_2 = f_3 = \cdots = f_n$. *It is weakly discriminantly separable if there exist polynomials* $f_i^j(x_i)$ *such that for every* $i = 1, \ldots, n$

$$\mathcal{D}_{x_i} F(x_1, \ldots, \hat{x}_i, \ldots, x_n) = \prod_{j \neq i} f_j^i(x_j).$$

Theorem 5.2 [DK2014b] *Given a nonzero polynomial* $P(x)$. *The strongly discriminantly separable polynomials* $\mathcal{F}(x_1, x_2, x_3)$ *of degree two in each of the three variables, are exhausted modulo Möbius transformations, by the following list coded by the structure of the roots of the polynomial* $P(x)$:

(A): *If P has four simple zeros, it can be transformed to a canonical form* $P_A(x) = (k^2 x^2 - 1)(x^2 - 1)$, *and*

$$\mathcal{F}_A = \frac{1}{2}(-k^2 x_1^2 - k^2 x_2^2 + 1 + k^2 x_1^2 x_2^2)x_3^2 + (1 - k^2)x_1 x_2 x_3$$
$$+ \frac{1}{2}(x_1^2 + x_2^2 - k^2 x_1^2 x_2^2 - 1),$$

(B): *if P has two simple zeros and one double, it can be transformed to a canonical form* $P_B(x) = x^2 - e^2$, $e \neq 0$, *and*

$$\mathcal{F}_B = x_1 x_2 x_3 + \frac{e}{2}(x_1^2 + x_2^2 + x_3^2 - e^2),$$

(C): *If P has two double zeros, and the canonical form* $P_C(x) = x^2$, *then*

$$\mathcal{F}_{C_1} = \lambda x_1^2 x_2^2 + \mu x_1 x_2 x_3 + \nu x_3^2, \quad \mu^2 - 4\lambda\nu = 1,$$

$$\mathcal{F}_{C_2} = \lambda x_1^2 x_3^2 + \mu x_1 x_2 x_3 + \nu x_2^2, \quad \mu^2 - 4\lambda\nu = 1,$$

$$\mathcal{F}_{C_3} = \lambda x_2^2 x_3^2 + \mu x_1 x_2 x_3 + \nu x_1^2, \quad \mu^2 - 4\lambda\nu = 1,$$

$$\mathcal{F}_{C_4} = \lambda x_1^2 x_2^2 x_3^2 + \mu x_1 x_2 x_3 + \nu, \quad \mu^2 - 4\lambda\nu = 1,$$

(D): *if P has one simple and one triple zero, then the canonical form is* $P_D(x) = x$, *and,*

$$\mathcal{F}_D = -\frac{1}{2}(x_1 x_2 + x_2 x_3 + x_1 x_3) + \frac{1}{4}(x_1^2 + x_2^2 + x_3^2),$$

(E): *if P has one quadruple zero, then the canonical form is* $P_E(x) = 1$, *and*

$$\mathcal{F}_{E_1} = \lambda(x_1 + x_2 + x_3)^2 + \mu(x_1 + x_2 + x_3) + \nu, \quad \mu^2 - 4\lambda\nu = 1,$$

$$\mathcal{F}_{E_2} = \lambda(x_2 + x_3 - x_1)^2 + \mu(x_2 + x_3 - x_1) + \nu, \quad \mu^2 - 4\lambda\nu = 1,$$

$$\mathcal{F}_{E_3} = \lambda(x_1 + x_3 - x_2)^2 + \mu(x_1 + x_3 - x_2) + \nu, \quad \mu^2 - 4\lambda\nu = 1,$$

$$\mathcal{F}_{E_4} = \lambda(x_1 + x_2 - x_3)^2 + \mu(x_1 + x_2 - x_3) + \nu, \quad \mu^2 - 4\lambda\nu = 1.$$

The proof from [DK2014b] is performed by a straightforward calculation. Family (A) corresponds to a generic pencil of conics, see [Dra2010, DK2014b], while the relationship of degenerate pencils of conics to the families (B-E) has been discussed in [DK2014b].

5.2 Relationship with ABS Quad-Graphs

The discriminantly separable polynomials appear to be related to discrete integrable systems. We will show a relationship with integrable quad-equations, from [DK2014b]. The theory of quad-graphs and quad-equations emerged in works of Adler, Bobenko, Suris [ABS2009b], [ABS2009a], see also [BS2008], [BS2010].

As we have said before (see Section 2.5), the quad-equations are defined on quadrilaterals and they have the form

$$Q(x_1, x_2, x_3, x_4) = 0. \tag{11}$$

Here Q is a polynomial of degree one in each variable. Such a polynomial is said to be multiaffine. So-called field variables x_i are assigned to four vertices of a quadrilateral as in a Figure 6. The polynomial Q depends on the variables $x_1, \ldots, x_4 \in \mathbb{C}$, but also depends on two additional parameters $\alpha, \beta \in \mathbb{C}$ that are assigned to the edges of a quadrilateral. The opposite edges carry the same parameter.

The equation (11) solved for each variable, gives the solution as a rational function of the other three variables. A solution (x_1, x_2, x_3, x_4) of the equation (11) is said to be *singular* with respect to x_i if it also satisfies the equation $Q_{x_i}(x_1, x_2, x_3, x_4) = 0$.

Following [ABS2009b] we adopt the idea of integrability as *a consistency*, see Figure 7.7. We assign six quad-equations to the faces of the coordinate cube. The system is *3D-consistent* if the three values for x_{123} obtained from the equations on the right, back, and top faces coincide for arbitrary initial data x, x_1, x_2, x_3.

The discriminant-like operators are introduced in [ABS2009b]

$$\delta_{x,y}(Q) = Q_x Q_y - Q Q_{xy}, \quad \delta_x(h) = h_x^2 - 2h h_{xx}, \tag{12}$$

and one can make a descent from the faces to the edges and then to the vertices of the cube: in that way, from a multiaffine polynomial $Q(x_1, x_2, x_3, x_4)$ we pass to a biquadratic polynomial $h(x_i, x_j) := \delta_{x_k, x_l}(Q(x_i, x_j, x_k, x_l))$ and then, further, to a polynomial $P(x_i) = \delta_{x_j}(h(x_i, x_j))$ of degree up to four. Using the relative invariants of polynomials under fractional linear transformations, the formulae that express Q through the biquadratic polynomials of three edges, were obtained in [ABS2009b]:

$$\frac{2Q_{x_1}}{Q} = \frac{h_{x_1}^{12} h^{34} - h_{x_1}^{14} h^{23} + h^{23} h_{x_3}^{34} - h_{x_3}^{23} h^{34}}{h^{12} h^{34} - h^{14} h^{23}}. \tag{13}$$

A biquadratic polynomial $h(x, y)$ is said to be *nondegenerate* if no polynomial in its equivalence class with respect to the fractional linear transformations, is divisible by a factor of the form $x - c$ or $y - c$, with $c = const$. A multiaffine function $Q(x_1, x_2, x_3, x_4)$ is said to be of *type Q* if all four of its accompanying biquadratic polynomials h^{jk} are nondegenerate. Otherwise, it is of *type H*. Previous notions were introduced in [ABS2009b].

Take an arbitrary strongly discriminantly separable polynomial

$$\mathcal{F}(x_1, x_2, \alpha)$$

of degree two in each of the three variables. To relate that polynomial to the corresponding quad-equations, one needs to provide a biquadratic polynomial $h = h(x_1, x_2)$ and a multiaffine polynomial $Q = Q(x_1, x_2, x_3, x_4)$.

The requirement that the discriminants of $h(x_1, x_2)$ are independent on α, see [ABS2009a], [ABS2009b], is fulfilled if as a biquadratic polynomials $h(x_1, x_2)$ we select

$$\hat{h}(x_1, x_2) := \frac{\mathcal{F}(x_1, x_2, \alpha)}{\sqrt{P(\alpha)}}.$$

Proposition 5.3 [DK2014b] *The biquadratic polynomials*

$$\hat{h}_I(x_1, x_2) = \frac{\mathcal{F}_I(x_1, x_2, \alpha)}{\sqrt{P_I(\alpha)}} \tag{14}$$

satisfy

$$\delta_{x_1}(\hat{h}) = P_I(x_2), \quad \delta_{x_2}(\hat{h}) = P_I(x_1)$$

for $I = A, B, C, D, E$ and polynomials $P_I, \mathcal{F}_I$ from Theorem 5.2.

By using the formulae (13) and replacing the polynomials h^{ij} by $\hat{h}^{ij}$, one gets the quad-equations which correspond to representatives of discriminantly separable polynomials from Theorem 5.2. These equations are re-parameterizations of the quad-equations of type Q from the list obtained in [ABS2009b].

For the quad-equations obtained from the biquadratic polynomials $\hat{h}(x_1, x_2)$, that parameter α has a role symmetric to x_1 and x_2.

Acknowledgments

The research was partially supported by the Serbian Ministry of Education and Science, Project 174020 Geometry and Topology of Manifolds, Classical Mechanics and Integrable Dynamical Systems and by The University of Texas at Dallas.

References

[ABS2003] V. E. Adler, A. I. Bobenko, and Yu. B. Suris, *Classification of integrable equations on quad-graphs. The consistency approach*, Comm. Math. Phys. **233** (2003), no. 3, 513–543.

[ABS2004] ———, *Geometry of Yang-Baxter maps: pencils of conics and quadrirational mappings*, Comm. Anal. Geom. **12** (2004), no. 5, 967–1007.

[ABS2009a] V. E. Adler, A. I. Bobenko, and Y. B. Suris, *Integrable discrete nets in Grassmannians*, Lett. Math. Phys.**89** (2009), no. 2, 131–139.

[ABS2009b] V. È. Adler, A. I. Bobenko, and Yu. B. Suris, *Discrete nonlinear hyperbolic equations: classification of integrable cases*, Funktsional. Anal. i Prilozhen.**43** (2009), no. 1, 3–21 (Russian, with Russian summary); English transl., Funct. Anal. Appl. **43** (2009), no. 1, 3–17.

[AB2018] A. V. Akopyan and A. I. Bobenko, *Incircular nets and confocal conics*, Trans. Amer. Math. Soc. **370** (2018), no. 4, 2825–2854.

[Arn1989] V. I. Arnol'd, *Mathematical methods of classical mechanics*, 2nd ed., Graduate Texts in Mathematics, vol. 60, Springer-Verlag, New York, 1989. Translated from the Russian by K. Vogtmann and A. Weinstein.

[BS2008] A. I. Bobenko and Y. B. Suris, *Discrete differential geometry: Integrable structure*, Graduate Studies in Mathematics, vol. 98, American Mathematical Society, Providence, RI, 2008.

[BS2010] ———, *On the Lagrangian structure of integrable quad-equations*, Lett. Math. Phys. **92** (2010), no. 1, 17–31.

[Böh1970] W. Böhm, *Verwandte Sätze über Kreisvierseitnetze*, Arch. Math. (Basel) **21** (1970), 326–330 (German).

[Buc2006] V. M. Buchstaber, *n-valued groups: theory and applications*, Moscow Mathematical Journal **6** (2006), no. 1, 57–84.

[Buk1990] V. M. Bukhshtaber, *Functional equations that are associated with addition theorems for elliptic functions, and two valued group*, Uspekhi Math. Nauk **45** (1990), no. 3 (273), 185–186; English transl., Russian Math. Surveys **45**, no. 3, 213–215.

[BR1997] V. M. Buchstaber and E. G. Rees, *Multivalued groups, their representations and Hopf algebras*, Transform. Group **2** (1997), 225–249.

[CCS1993] S.-J. Chang, B. Crespi, and K.-J. Shi, *Elliptical billiard systems and the full Poncelet's theorem in n dimensions*, J. Math. Phys. **34** (1993), no. 6, 2242–2256.

[Cox1973] H. S. M. Coxeter, *Regular polytopes*, 3rd ed., Dover Publications, Inc., New York, 1973.

[Dar1901] G. Darboux, *Sur un Problème de Méchanique*, Archives Neerlandaises (2) **6** (1901), 371–376.

[Dar1914] ———, *Leçons sur la théorie générale des surfaces et les applications géométriques du calcul infinitesimal*, Vol. 2 and 3, Gauthier-Villars, Paris, 1914.

[Dar1917] ———, *Principes de géométrie analytique*, Gauthier-Villars, Paris, 1917.

[Dra2010] V. Dragović, *Geometrization and generalization of the Kowalevski top*, Communications in Mathematical Physics **298** (2010), no. 1, 37–64, available at arXiv: 0912.3027. DOI: 10.1007/s00220-010-1066-z.

[Dra2013] ———, *Pencils of conics as a classification code*, Geometric methods in physics, Trends Math., Birkhäuser/Springer, Basel, 2013, pp. 323–330.

[Dra2014] ———, *Pencils of conics and biquadratics, and integrability*, Topology, geometry, integrable systems, and mathematical physics, Amer. Math. Soc. Transl. Ser. 2, vol. 234, Amer. Math. Soc., Providence, RI, 2014, pp. 117–140.

[DK2011] V. Dragović and K. Kukić, *New examples of systems of the Kowalevski type*, Regul. Chaotic Dyn. **16** (2011), no. 5, 484–495.

[DK2014a] ———, *The Sokolov case, integrable Kirchhoff elasticae, and genus 2 theta functions via discriminantly separable polynomials*, Proc. Steklov Inst. Math. **286** (2014), no. 1, 224–239. Reprint of Tr. Mat. Inst. Steklova **286** (2014), 246–261.

[DK2014b] ———, *Discriminantly separable polynomials and quad-equations*, J. Geom. Mech. **6** (2014), no. 3, 319–333.

[DK2014c] ———, *Systems of Kowalevski type and discriminantly separable polynomials*, Regul. Chaotic Dyn. **19** (2014), no. 2, 162–184.

[DK2014d] ———, *Role of discriminantly separable polynomials in integrable dynamical systems*, AIP Conf. Proc. **1634** (2014), 3–8.

[DK2017] ———, *Discriminantly separable polynomials and the generalized Kowalevski top*, Theoretical and Applied Mechanics **44** (2017), no. 2, 229–236.

[DR2008] V. Dragović and M. Radnović, *Hyperelliptic Jacobians as Billiard Algebra of Pencils of Quadrics: Beyond Poncelet Porisms*, Adv. Math. **219** (2008), no. 5, 1577–1607.

[DR2011] ———, *Poncelet Porisms and Beyond*, Springer Birkhauser, Basel, 2011.

[DR2012a] ———, *Ellipsoidal billiards in pseudo-Euclidean spaces and relativistic quadrics*, Advances in Mathematics **231** (2012), 1173–1201.

[DR2012b] ———, *Billiard algebra, integrable line congruences, and double re ection nets*,Journal of Nonlinear Mathematical Physics **19** (2012), no. 3, 1250019.

[DR2013] V. Dragović and M. Radnović, *Minkowski plane, confocal conics, and billiards*, Publ. Inst. Math. (Beograd) (N.S.)**94(108)** (2013), 17–30.

[DR2014] ——, *Bicentennial of the great Poncelet theorem (1813–2013): current advances*, Bull. Amer. Math. Soc. (N.S.) **51** (2014), no. 3, 373–445.

[DR2015] ——, *Pseudo-integrable billiards and double reection nets*, (Russian) UspekhiMat. Nauk **70** (2015), no. 1(421), 3–34; English transl., translation in Russian Math. Surveys **70** (2015), no. 1, 1–31.

[DR2019] ——, *Caustics of Poncelet polygons and classical extremal polynomials*, (Regular and Chaotic Dynamics **24** (2019), no. 1, 1–35.

[Dui2010] J. J. Duistermaat, *Discrete integrable systems: QRT maps and elliptic surfaces*, Springer Monographs in Mathematics, Springer, New York, 2010.

[GKST2004] B. Grammaticos, Y. Kosmann-Schwarzback, and K. M Tamizhmani, *Discrete Integrable Systems*, Springer-Verlag, 2004.

[IT2017] I. Izmestiev and S. Tabachnikov, *Ivory's theorem revisited*, J. Integrable Syst. **2** (2017), no. 1, xyx006, 36.

[JJ2015] B. Jovanović and V. Jovanović, *Geodesic and billiard ows on quadrics in pseudo-Euclidean spaces: L–A pairs and Chasles theorem*, Int. Math. Res. Not. IMRN **15** (2015), 6618–6638.

[KT2009] B. Khesin and S. Tabachnikov, *Pseudo-Riemannian geodesics and billiards*, Advances in Mathematics **221** (2009), 1364–1396.

[KS2003] A. D. King and W. K. Schief, *Tetrahedra, octahedra and cubo-octahedra: integrable geometry of multi-ratios*, J. Phys. A **36** (2003), no. 3, 785–802.

[KS2006] ——, *Application of an incidence theorem for conics: Cauchy problem and integrability of the dCKP equation*, J. Phys. A **39** (2006), no. 8, 1899–1913.

[Kow1889] S. Kowalevski, *Sur le problème de la rotation d'un corp solide autoar d'un point fixe*, Acta Math. **12** (1889), 177–232.

[LT2007] M. Levi and S. Tabachnikov, *The Poncelet grid and billiards in ellipses*, Amer. Math. Monthly **114** (2007), no. 10, 895–908.

[Pon1822] J. V. Poncelet, *Traité des propriétés projectives des figures*, Mett, Paris, 1822.

[Pre1999] E. Previato, *The Poncelet's theorem and generalizations*, Proc. Amer. Math. Soc. **127** (1999), 2547–2556.

[Pre2018] E. Previato, *Poncelet's porism and projective fibrations. Higher genus curves in mathematical physics and arithmetic geometry*, Contemp. Math. **703** (2018), 157–169.

[Pre] E. Previato, *Some integrable billiards*, SPT2002: Symmetry and Perturbation Theory (S. Abenda, G. Gaeta, and S. Walcher, eds.), World Scientific, Singapore, 2002,pp. 181–195.

[Rad2015] M. Radnović, *Integrable lattices of hyperplanes related to billiards within confocal quadrics*, Publ. Inst. Math. (Beograd) (N.S.) **98(112)** (2015).

[Sch2007] R. Schwartz, *The Poncelet grid*, Advances in Geometry **7** (2007), 157–175.

[Wel1977] A. F. Wells, *Three-dimensional nets and polyhedra*, Wiley-Interscience [John Wiley & Sons], New York-London-Sydney, 1977. Wiley Monographs in Crystallography.

Vladimir Dragović
The University of Texas at Dallas, Department of Mathematical Sciences, USA.
Mathematical Institute, Serbian Academy of Sciences and Arts (SANU), Belgrade.
Email address: vladimir.dragovic@utdallas.edu

Milena Radnović
The University of Sydney, School of Mathematics and Statistics, Australia.
Mathematical Institute, Serbian Academy of Sciences and Arts (SANU), Belgrade.
Email address: milena.radnovic@sydney.edu.au

Roger Fidèle Ranomenjanahary
The University of Texas at Dallas, Department of Mathematical Sciences, USA.
Email address: roger@utdallas.edu

8

Bi-Flat F-Manifolds: A Survey

Alessandro Arsie and Paolo Lorenzoni

*To Emma Previato, on the occasion of her 65th birthday, with admiration
and with our best wishes*

Abstract. We present a survey of the work done by the authors in the last
few years developing the theory of bi-flat F-manifolds and exploring their
relationships with integrable hierarchies (dispersionless and dispersive), with
Painlevé transcendents, and with complex reflection groups.

1 Introduction

The modern theory of integrable systems has developed an amazing array
of interconnections with a wide variety of areas in mathematics, including
algebraic and differential geometry, special functions and Painlevé equations, solitons and moduli spaces, twistor and knot theory, tropical geometry,
enumerative geometry and combinatorics, Lie algebras, cluster algebras and
representation theory, quantum groups and non-commutative algebras, spectral
theory and Riemann-Hilbert problems, probability theory and random matrix
theory.

At the beginning of the 1990s, Witten's conjecture [43] on the relation
between the KdV hierarchy and intersection numbers on the moduli space of
stable curves (later proved by Kontsevich in [29]) opened a new era in the
theory of integrable systems. Motivated by this discovery, Dubrovin developed
the multi-faceted theory of Frobenius manifolds, with important ramifications
in singularity theory, Coxeter groups, Painlevé transcendents, bi-Hamiltonian
integrable hierarchies, enumerative geometry and Gromov-Witten invariants,
just to name a few.

In the last few years we have pursued a research program with the specific
goal to introduce and study a suitable generalization of Frobenius manifolds.
Our motivations came from the theory of dispersionless integrable hierarchies

and, in particular from the observation that the recursive procedure that allows to define such hierarchies starting from a Frobenius manifold can be easily adapted to describe non-Hamiltonian integrable hierarchies starting from a special class of F-manifolds, (these are a generalization of Frobenius manifolds introduced in [23]). We called *twisted Lenard-Magri chains* the non-Hamiltonian version of the classical bi-Hamiltonian recursive relations and bi-flat F-manifolds the geometric objects that substitute Frobenius manifolds in this more general setting. We realized later that other important results in the theory of Frobenius manifolds such as relationships with reflection groups and Painlevé transcendents (see for instance [15] and [16]) have also a natural generalization in this framework.

This work is a survey of our main results in this area. In Section 2 we introduce bi-flat and multi-flat F structures and compare them to Frobenius and F-manifolds. In Section 3 we show how germs of bi-flat F-structures are related to Painlevé transcendents. In Section 4 we show how to construct bi-flat F-structures on orbit spaces of suitable complex reflection groups. In Section 5 we show how to construct an integrable dispersionless hierarchy starting from the geometric data of a bi-flat F-manifold and we also briefly consider the problem of deforming the obtained integrable dispersionless hierarchy to an integrable dispersive hierarchy.

2 Bi-Flat and Multi-Flat F-Manifolds

In this Section we are going to introduce our main characters, bi-flat F-manifolds. There are different ways to introduce these structures. Here we use an approach that highlights their relations with Manin's flat F-manifolds on one hand, and Dubrovin's Frobenius manifolds on the other.

The manifold we consider will be either smooth or analytic. If M is a manifold, we denote with $\mathcal{X}(M)$ the $C^\infty(M)$-module of vector fields on M.

2.1 Flat F-Manifold

Definition 2.1 ([35]) *A flat F-manifold (or F-manifold with compatible flat structure) is a quadruple $(M, \circ, \nabla, e)$ where M is a manifold, $\circ : \mathcal{X}(M) \times \mathcal{X}(M) \to \mathcal{X}(M)$ is a product, ∇ is a connection on the tangent bundle TM and e is a distinguished vector field, satisfying the following axioms:*

1 the one parameter family of connections $\nabla_{(\lambda)} := \nabla + \lambda\circ$ is flat and torsionless for any $\lambda \in \mathbb{R}$.

2 e is the unit of the product $\circ$, i.e. for any $X \in \mathcal{X}(M)$, one has
$e \circ X = X \circ e = X$.

3 e is flat: $\nabla e = 0$.

Manifolds equipped with a product $\circ$, a connection ∇ and a vector field e satisfying conditions 1 and 2 above will be called *almost flat F-manifolds*.

To elucidate the condition 1 above, let us fix a coordinate system $\{u^1, \ldots u^n\}$ and let us denote with c^i_{jk} the structure constants of $\circ$, i.e. $\partial_j \circ \partial_k = c^i_{jk}\partial_i$, where Einstein's summation convention is enforced. In the same coordinate system, let Γ^i_{jk} be the Christoffel symbols of ∇. Then it is easy to see that the torsion tensor and the curvature tensor of the deformed connection $\nabla_{(\lambda)}$ are given respectively by

$$T^{(\lambda)k}_{ij} = \Gamma^k_{ij} - \Gamma^k_{ji} + \lambda(c^k_{ij} - c^k_{ji}), \tag{1}$$

$$R^{(\lambda)k}_{ijl} = R^k_{ijl} + \lambda(\nabla_i c^k_{jl} - \nabla_j c^k_{il}) + \lambda^2(c^k_{im}c^m_{jl} - c^k_{jm}c^m_{il}), \tag{2}$$

where R^k_{ijl} is the curvature tensor of ∇. We get immediately the following properties from equations (1), (2), the condition 1 above, and the identity principle of polynomials:

(a) the connection ∇ is torsionless,
(b) the product $\circ$ is commutative,
(c) the connection ∇ is flat,
(d) the tensor field $\nabla_l c^k_{ij}$ is symmetric in the lower indices,
(e) the product $\circ$ is associative.

It turns out that in a system of coordinates for which the Christoffel symbols of ∇ vanish identically (flat coordinates) the structure constants c^k_{ij} have a special form. Indeed from conditions (a), (b), (c) and (d) above, in flat coordinates we have the following description for the structure constants:

$$c^i_{jk} = \partial_j \partial_k A^i. \tag{3}$$

The vector field whose components are A^i, $i = 1, \ldots, n$ is called *vector potential* of $\circ$. Furthermore, the condition (e) above forces the vector potential to satisfy a non-trivial system of PDEs called *generalized WDVV equations* or *oriented associativity equations*:

$$\partial_j \partial_l A^i \partial_k \partial_m A^l = \partial_k \partial_l A^i \partial_k \partial_m A^l. \tag{4}$$

2.2 Frobenius Manifolds and Almost Duality

Adding a suitable pseudo-Riemannian metric to a flat F-manifold leads to the notion of Frobenius manifold. We say that a pseudo-Riemannian metric η is an *invariant metric* for a flat F-manifold $(M, \circ, \nabla, e)$ if

$$\nabla \eta = 0, \tag{5}$$

$$\eta(X \circ Y, Z) = \eta(X, Y \circ Z), \quad \forall X, Y, Z \in \mathcal{X}(M). \tag{6}$$

In coordinates, equation (6) reads $\eta_{il}c^l_{jk} = \eta_{jl}c^l_{ik}$. An important consequence of the existence of an invariant metric η for a flat F-manifold is that the vector potential A itself can be expressed as a derivative of a scalar function (in flat coordinates from ∇). Indeed, if there is an invariant metric, it turns out that $A^i = \eta^{il}\partial_l F$ for a scalar function F and the associativity equations (4) become the WDVV associativity equations:

$$\partial_j\partial_h\partial_i F\eta^{il}\partial_l\partial_k\partial_m F = \partial_j\partial_k\partial_i F\eta^{il}\partial_l\partial_h\partial_m F. \tag{7}$$

Before introducing Frobenius manifolds, we need one more notion, that of Euler vector field.

Definition 2.2 *A vector field E satisfying the conditions*

$$[e, E] = e, \quad \mathrm{Lie}_E\circ = \circ$$

is called an Euler vector field.

In the literature flat F-manifolds have been introduced by Manin as generalization of Frobenius manifolds. Following this non-chronological path we can define Dubrovin's Frobenius manifolds as flat F-manifolds equipped with some additional structures.

Definition 2.3 *A Frobenius manifold M is a flat F-manifold $(M, \circ, \nabla, e)$ equipped with an invariant metric η and a* linear *Euler vector field E (i.e. $\nabla\nabla E = 0$), which acts as a conformal Killing vector for the metric η:*

$$\mathrm{Lie}_E\eta = (2 - d)\eta,$$

where the constant d is called the charge of the Frobenius manifold.

Given a Frobenius manifold, it is possible to construct out of it an almost flat F-manifold, called the almost dual structure of the Frobenius manifold. This construction is due to Dubrovin.

Indeed we have the following

Theorem 2.4 *[17] Given a Frobenius manifold $(M, \circ, e, E, \eta, \nabla)$, consider the open set U where the endomorphism of the tangent bundle $E\circ$ is invertible. Consider also the contravariant metric $g := (E\circ)\,\eta^{-1}$, called intersection form. Then on U, the data given by*

1 the Levi-Civita connection $\tilde{\nabla}$ of g,
2 the Euler vector field E, and
3 a dual product defined as $X * Y = (E\circ)^{-1} X \circ Y, \quad \forall X, Y \in \mathcal{X}(U)$,

define an almost flat F-manifold with unit E and invariant metric g^{-1}.

In general, this is only an almost flat F-manifold since $\tilde{\nabla} E \neq 0$. However, replacing $\tilde{\nabla}$ with $\nabla^* := \tilde{\nabla} + \bar{\lambda}*$ (for a suitable value of $\bar{\lambda}$) one obtains a flat connection ∇^* satisfying $\nabla^* E = 0$. (Of course, gauging $\tilde{\nabla}$ with any multiple of $*$ gives rise always to a flat torsionless connection, since this is one of the conditions that define an almost flat F-structure.)

2.3 Bi and Multi-Flat F-Manifolds

From our discussion, for any given Frobenius manifold $(M, \eta, \circ, e, E)$, the open set $U \subset M$ where $(E \circ)$ is invertible is equipped with two flat F-manifold structures:

- the flat structure $(\nabla, \circ, e)$,
- the flat structure $(\nabla^*, *, E)$.

These two structures are intertwined in a peculiar way. It turns out that the two connections ∇ and ∇^* satisfy the condition:

$$(d_\nabla - d_{\nabla^*})(X \circ) = 0, \ \forall X \in \mathcal{X}(U), \tag{8}$$

where d_∇ is the exterior covariant derivative. Two connections ∇ and ∇^* satisfying (8) are called *almost hydrodynamically equivalent* (see Definition 6.1 in [2]).

This motivates to introduction of the following

Definition 2.5 *[3] A bi-flat F-manifold M is a manifold equipped with two different flat F-structures $(\nabla, \circ, e)$ and $(\nabla^*, *, E)$ related by the following conditions*

1 E is an Euler vector field.
2 $*$ is the dual product defined by E.
3 ∇ and ∇^* are almost hydrodynamically equivalent.

The connection ∇ is called *the natural connection* while the connection ∇^* is called *the dual connection*. It is possible to prove that the dual connection is uniquely determined by the above conditions in term of the natural connection, the dual product and the Euler vector field. Moreover the compatibility of the dual connection with the dual product is a consequence of the other axioms (see [5] for details).

The above definition can be easily extended considering additional compatible flat structures:

Vector field	Associated product	Associated connection
e	$\circ$	∇
E	$\circ_{(1)}$	$\nabla^{(1)}$
$E \circ E$	$\circ_{(2)}$	$\nabla^{(2)}$
$E \circ E \circ E$	$\circ_{(3)}$	$\nabla^{(3)}$
$\ldots$	$\ldots$	$\ldots$

Definition 2.6 *[5] An N-multi-flat F-manifold M is a manifold equipped with N different flat F-structures* $(\nabla^{(0)}, \circ_{(0)} = \circ, E_{(0)} = e)$, $(\nabla^{(1)}, \circ_{(1)} = *, E_{(1)} = E)$, $(\nabla^{(2)}, \circ_{(2)}, E_{(2)}), \ldots, (\nabla^{(N-1)}, \circ_{(N-1)}, E_{(N-1)})$ *satisfying the following conditions:*

1 E is an Euler vector field, i.e. $[e, E] = e$, $\text{Lie}_E \circ = \circ$.
2 $E_{(l)} = E^{\circ l} := E \circ E \circ \ldots \circ E$ *l-times,* $l = 0, \ldots, N-1$, *(by definition* $E_{(0)} = e$, $E_{(1)} =)$, *and* $\circ_{(l)}$ *is defined as* $X \circ_{(l)} Y = (E_{(l)})^{-1} \circ X \circ Y$, *for all* $X, Y \in \mathcal{X}(M)$.
3 $(d_{\nabla^l} - d_{\nabla^{l'}})(X \circ) = 0$, $\quad \forall X \in \mathcal{X}(M)$, $\forall l, l' \in \{0, \ldots, N-1\}$.   (9)

Let us remark that it has been observed in [5] that not all compatibility conditions required in the above definition need to be satisfied, since some of them are automatically fulfilled.

It turns out that an N-multi-flat F-manifold can be defined as *a flat F-manifold equipped with an Euler vector field E and N − 1 additional almost hydrodynamically equivalent flat torsionless connections* $\nabla_{(1)}, \ldots, \nabla_{(N-1)}$ *satisfying the conditions* $\nabla_{(l)} E_{(l)} = 0$ *where* $E_{(l)} = E^{\circ l} := E \circ E \circ \cdots \circ E$ *l-times.*

Let us recall that a flat F-manifold M is called semisimple on an open connected U if there exists a local system of coordinates $\{u^1, \ldots, u^n\}$ on U (called *canonical coordinates*) such that the tensor field c representing $\circ$ has the special form $c_{ij}^k = \delta_i^k \delta_j^k$. The same definition applies for a multi-flat or bi-flat F-manifold M, where we require that $c_{ij}^k = \delta_i^k \delta_j^k$, c being the tensor field that represents $\circ = \circ_{(0)}$.

We have the following:

Theorem 2.7 *[5] Semisimple (non-trivial) multi-flat F-manifolds with more than three flat structures do not exist.*

The previous theorem is proved reducing the problem of finding multiple compatible flat F-structures to a problem in the theory of distributions (in the sense of Differential Geometry).

However, if one removes the semisimplicity assumptions, then it is possible to have more than three compatible structure. In [5] there is an explicit construction of a flat F-manifold in dimension three equipped with infinitely many compatible flat structures.

3 Bi-Flat F-Manifolds and Painlevé Transcendents

We discuss now some relations between bi-flat F-manifolds and Painlevé transcendents. They were originally discovered using an extended Egorov-Darboux system, essentially relaxing the symmetry condition for the Ricci rotations coefficients (see [3] and also [33]). Later, in [5] we considered a different approach more suitable for the study of the non-semisimple case.

3.1 The Semisimple Case

In the semisimple case, due the compatibility condition (8), the natural connection and the dual connection share part of their Christoffel symbols. In canonical coordinates we have

$$\Gamma_{ij}^{(1)i} = \Gamma_{ij}^{(2)i} = \Gamma_{ij}^{i} \qquad i \neq j.$$

It turns out that once the functions Γ_{ij}^{i} are given all the remaining data entering the definition of bi-flat F-manifolds are uniquely determined.

For this reason, in order to parameterize semisimple bi-flat F-manifolds it is enough to study the set of conditions satisfied by the functions Γ_{ij}^{i}.

To obtain this set of conditions one can follows two approaches:

- The first approach is based on a generalization of the classical Darboux-Egorov system [3, 33], as we mentioned before. Indeed, writing the Christoffel symbols as $\Gamma_{ij}^{i} = \frac{H_j}{H_i}\beta_{ij}\, i \neq j$, it turns out that the functions β_{ij} satisfy the system

$$\partial_k \beta_{ij} = \beta_{ik}\beta_{kj}, \qquad k \neq i \neq j \neq k \tag{10}$$

$$e(\beta_{ij}) = 0, \tag{11}$$

$$E(\beta_{ij}) = (d_i - d_j - 1)\beta_{ij}. \tag{12}$$

Given a solution β_{ij} of the above system the functions H_i are obtained from the system

$$\partial_j H_i = \beta_{ij} H_j, \qquad i \neq j \tag{13}$$

$$e(H_i) = 0, \tag{14}$$

$$E(H_i) = d_i H_i. \tag{15}$$

If the rotation coefficients are asymmetric $\beta_{ij} = \beta_{ji}$ the functions β_{ij} are the classical rotations coefficients of a diagonal metric with Lamé coefficients H_i.

- The second approach is based on the study of the bi-flatness conditions written as a system of PDEs for the $n(n-1)$ unknown functions $\Gamma_{ij}^{i}(\mathbf{u})$:

$$\partial_k \Gamma_{ij}^{i} = -\Gamma_{ij}^{i}\Gamma_{ik}^{i} + \Gamma_{ij}^{i}\Gamma_{jk}^{j} + \Gamma_{ik}^{i}\Gamma_{kj}^{k}, \qquad i \neq k \neq j \neq i, \tag{16}$$

$$\sum_{i=1}^{n} \partial_i(\Gamma_{ij}^{i}) = 0, \qquad i \neq j \tag{17}$$

$$\sum_{i=1}^{n} u^i \partial_i(\Gamma_{ij}^{i}) = -\Gamma_{ij}^{i}, \qquad i \neq j \tag{18}$$

The above system is compatible and thus its general solution depends on $n(n-1)$ arbitrary constants [5].

For $n = 3$ the above systems of PDEs can be reduced to a system of six first order ODEs that turns out to be equivalent to the sigma form of the generic Painlevé VI equation [33, 5].

3.2 The General Case

The second approach is more flexible and can be used to perform a similar analysis in the regular non-semisimple case. Regular simply means that at each point p the endomorphism $L_p := E_p \circ$ has exactly one Jordan block for each distinct eigenvalue. In the regular case, the role of canonical coordinates is played by a distinguished set of coordinates found by David and Hertling [14].

Theorem 3.1 *[5] Three dimensional regular bi-flat F-manifolds are locally parameterized by solutions of the full Painlevé IV, V, and VI equations according to the Jordan canonical form J of $L = E \circ$. More precisely,*

- *PVI in the case*

$$J = \begin{pmatrix} \lambda_1 & 0 & 0 \\ 0 & \lambda_2 & 0 \\ 0 & 0 & \lambda_3 \end{pmatrix},$$

- *PV in the case*

$$J = \begin{pmatrix} \lambda_1 & 1 & 0 \\ 0 & \lambda_1 & 0 \\ 0 & 0 & \lambda_3 \end{pmatrix},$$

- *PIV in the case*

$$J = \begin{pmatrix} \lambda_1 & 1 & 0 \\ 0 & \lambda_1 & 1 \\ 0 & 0 & \lambda_1 \end{pmatrix}.$$

Due to the above theorem, the counterpart of the confluences of the Painlevé equations is the collision of eigenvalues and the creation of non-trivial Jordan blocks in the operator of multiplication by the Euler vector field. Using a different approach based on Okubo type systems, recently Kawakami and Mano proved that PII–PVI appear, as special cases, in dimension 4 [27].

Remark 3.2 In the case of Frobenius manifolds, Dubrovin proved that germs of semisimple 3-dimensional Frobenius structures are parameterized by a one-parameter subfamily of PVI (see [15] and [16] for instance). In this sense, the previous results extend Dubrovin's correspondence to bi-flat F-manifolds. In particular, semisimple 3-dimensional bi-flat F-structures are parameterized by the full PVI family.

4 Bi-Flat F-Manifolds and Complex Reflection Groups

In this Section we are going to introduce what are the most interesting examples, at least up to now, of bi-flat F-structures. We will see that the orbit spaces of some classes of complex reflection groups are equipped with a natural bi-flat F-manifold structure and that moreover, in some cases, this structure appears in a family, namely it depends on parameters. We will also present a Conjecture the relates the number of parameters to the properties of the relevant complex reflection group. In the case of Coxeter groups this conjectured has been formulated in [8].

First we recall the definition of the main characters of our construction, i.e. finite complex reflection groups.

Definition 4.1 *A complex (pseudo)-reflection is a unitary transformation of $\mathbb{C}^n$ of finite period that leaves invariant a hyperplane.*

Therefore, a complex (pseudo)-reflection is characterized by the property that all the eigenvalues of the associated matrix representation are equal to 1, except for one. The remaining eigenvalue is a k-th primitive root of unity, where k is the period of the transformation. Irreducible finite complex reflection groups were classified by Shephard and Todd and consist of an infinite family depending on 3 positive integers and 34 exceptional cases. All finite complex reflection groups considered here will act on their reflection representation.

It is known that the ring of invariant polynomials of an irreducible finite complex reflection group is generated by n algebraically independent invariant polynomials, where n is the dimension of the complex vector space on which the group acts. We are interested mostly in well-generated complex reflection groups:

Definition 4.2 *Well-generated irreducible complex reflection groups are irreducible complex reflection groups of rank n, whose minimal generating set consists of n reflections.*

4.1 Flat Structures Associated with Coxeter Groups

Recall in particular that any finite *real* reflection group is automatically well-generated (so finite Coxeter groups are well-generated). Before introducing a bi-flat F-manifold structure on the orbit spaces of some classes of irreducible well-generated complex reflection groups, let us review some relevant constructions associated to Coxeter groups. First we recall the following result:

Theorem 4.3 (Dubrovin, [18]) *The orbit space of a finite Coxeter group is equipped with a Frobenius manifold structure* $(\eta, \circ, e, E)$ *where*

1 *The flat coordinates for* η *are basic invariants* $(u^1, \ldots, u^n)$ *of the group called Saito flat coordinates.*
2 *In the Saito flat coordinates*

$$e = \frac{\partial}{\partial u^n}, \quad E = \sum_{i=1}^n \left(\frac{d_i}{d_n}\right) u^i \frac{\partial}{\partial u^i}.$$

where d_i *are the degrees of the invariant polynomials* u_i *and* $2 = d_1 < d_2 \le d_3 \le \ldots \le d_{n-1} < d_n$ *(d_n is the Coxeter number).*

Dubrovin's construction relies on the existence of a flat pencil of metrics associated with any Coxeter group. One is the euclidean metric of the euclidean space where the reflections act and one is the Saito flat metric [38, 37].

To the orbits spaces of finite Coxeter groups it is possible to associate another structure, constructed directly from the reflecting hyperplanes.

Theorem 4.4 *Let G be a finite Coxeter group acting on a euclidean space $\mathbb{E}^n$ with euclidean coordinates* $(p^1, \ldots, p^n)$. *Let g be the euclidean metric and ∇^* the associated Levi-Civita connection. Then the data*

$$\left(\nabla^*, \quad * = \sum_{H \in \mathcal{H}} \frac{d\alpha_H}{\alpha_H} \otimes \sigma_H \pi_H, \quad E = \sum p^k \frac{\partial}{\partial p^k}\right)$$

where

- $\mathcal{H}$ *is the collection of the reflecting hyperplanes H,*
- α_H *is a linear form defining a reflecting hyperplane H,*
- π_H *is the orthogonal projection onto the orthogonal complement of H,*
- *the collection of weights σ_H is G-invariant and satisfies*

$$\sum_{H \in \mathcal{H}} \sigma_H \pi_H = Id.$$

Define an almost flat structure with invariant metric g.

Proof. This an equivalent reformulation of a result of Veselov [41, 42] and it is essentially an example of $\vee$-system. Equivalence between flatness coditions and the definition of $\vee$-system is discussed in [4, 22].   □

The above almost flat structure for a special choice of the weights is related to the Frobenius manifold structure of Theorem 4.3. Indeed Dubrovin proved that choosing all the weights σ_H equal to each other, the almost flat structure of Theorem 4.4 coincides with the almost dual structure associated to the Frobenius structure of Theorem 4.3, (see [17]).

4.2 Flat Structures Associated with Complex Reflection Groups

Like in the case of the orbit spaces of finite Coxeter groups, it is possible to define two flat structures on the orbit spaces of some classes of finite complex reflection groups. The first structure generalizes the Dubrovin-Saito construction to well-generated finite irreducible complex reflection groups. The second one is obtained starting from a Dunkl-Kohno-type connection associated with complex reflection groups, which can be thought as generalization of Veselov's ∨-systems. In general, these two flat F structures do not possess corresponding invariant metrics and as such that they do *not* originate from a Frobenius manifold (and its almost dual).

For the first F-flat structure we have:

Theorem 4.5 *[25] The orbit space of a well-generated complex reflection group is equipped with a flat F-structure $(\nabla, \circ, e, E)$ with linear Euler vector field where*

1 *The flat coordinates for ∇ are basic invariants $(u^1, \ldots, u^n)$ of the group (generalized Saito coordinates).*
2 *In the Saito flat coordinates*

$$e = \frac{\partial}{\partial u^n}, \quad E = \sum_{i=1}^{n} \left(\frac{d_i}{d_n}\right) u^i \frac{\partial}{\partial u^i}.$$

Remark 4.6 Almost flat F-structures with linear Euler vector field are also called *Saito structures without metrics* [36].

The second flat F-structure (more precisely family of flat structures) is described by the following:

Theorem 4.7 *Let G be an irreducible complex reflection group acting on $\mathbb{C}^n$. Then the data*

$$\left(\nabla^* = \nabla^0 - \sum_{H \in \mathcal{H}} \frac{d\alpha_H}{\alpha_H} \otimes \tau_H \pi_H, \ * = \sum_{H \in \mathcal{H}} \frac{d\alpha_H}{\alpha_H} \otimes \sigma_H \pi_H, \ E = \sum p^k \frac{\partial}{\partial p^k}\right)$$

where

- *$\mathcal{H}$ is the collection of the reflecting hyperplanes H,*
- *α_H is a linear form defining a reflecting hyperplane H,*
- *π_H is the unitary projection onto the unitary complement of H,*

- *the collections of weights σ_H and τ_H are G-invariant and satisfy*

$$\sum_{H \in \mathcal{H}} \sigma_H \pi_H = \sum_{H \in \mathcal{H}} \tau_H \pi_H = Id. \qquad (19)$$

- ∇^0 *is the standard flat connection on* $\mathbb{C}^n$,

define a flat F-structure on the orbit space of the action of G on $\mathbb{C}^n$.

Proof. Flatness of the one parameter family of connections $\nabla^* - \lambda *$ follows from the G-invariance of the collections of the weights [32]. Torsionless is obvious. The conditions $c^i_{jk} E^k = \delta^i_j$ and $\nabla E = 0$ follow immediately from (19). □

Remark 4.8   In the construction of π_H one uses the unique (up to a scalar multiple) G-invariant Hermitian metric g on $\mathbb{C}^n$.

Remark 4.9   The fact that the Euler vector field is linear turns out to be equivalent to the existence of a second compatible flat structure. This was proved in the semisimple case in [8] and later in the general regular case in [28].

Due to the Theorem 4.7, the flat structures above might depend on parameters depending on the number of the orbits of the group on the collection $\mathcal{H}$ of all reflecting hyperplanes.

In order to elucidate this last point, let us consider first the case of a single orbit for the action of G on $\mathcal{H}$. In this case G-invariance implies $\sigma_H = \sigma_{H'}$ and $\tau_H = \tau_{H'}$ for any $H, H' \in \mathcal{H}$. If $\sigma_H > 0$ for all $H \in \mathcal{H}$, then

$$\sum_{H \in \mathcal{H}} \sigma_H \alpha_H \otimes \bar{\alpha}_H = \lambda g$$

where g is an invariant Hermitian metric and $\lambda > 0$. Indeed the above sum is by construction G-invariant and positive definite. As a consequence

$$\sum_{H \in \mathcal{H}} \sigma_H \pi_H = \lambda I,$$

where I is the identity endomorphism. If there is only one orbit in $\mathcal{H}$, then all the weights have to be equal and therefore it it is clear that there is only one choice of the weights $\sigma_H = \tau_H$ such that

$$\sum_{H \in \mathcal{H}} \sigma_H \pi_H = I. \qquad (20)$$

Let us consider the case of two orbits. In this case we can assign different weight to hyperplanes belonging to different orbits. Let us denote by $(\mathcal{H}_1, \sigma_1)$ the set of the hyperplanes in the first orbit and the corresponding weight and by $(\mathcal{H}_2, \sigma_2)$ the set of the hyperplanes in the second orbit and the corresponding weight. Let (σ_1, σ_2) and (σ'_1, σ'_2) two linearly independent choices of the weights. If $\sigma_i > 0, \sigma'_i > 0, i = 1, 2$ then

$$\sum_{H \in \mathcal{H}} \sigma_H \alpha_H \otimes \bar{\alpha}_H = \sigma_1 \sum_{H \in \mathcal{H}_1} \alpha_H \otimes \bar{\alpha}_H + \sigma_2 \sum_{H \in \mathcal{H}_2} \alpha_H \otimes \bar{\alpha}_H = \lambda g,$$

$$\sum_{H \in \mathcal{H}} \sigma'_H \alpha_H \otimes \bar{\alpha}_H = \sigma'_1 \sum_{H \in \mathcal{H}_1} \alpha_H \otimes \bar{\alpha}_H + \sigma'_2 \sum_{H \in \mathcal{H}_2} \alpha_H \otimes \bar{\alpha}_H = \mu g,$$

with $\lambda, \mu > 0$. As a consequence choosing $w_H = \mu \sigma_1 - \lambda \sigma'_1 \; \forall H \in \mathcal{H}_1$ and $w_H = \mu \sigma_2 - \lambda \sigma'_2 \; \forall H \in \mathcal{H}_2$ we have

$$\sum_{H \in \mathcal{H}} w_H \alpha_H \otimes \bar{\alpha}_H = 0, \quad \sum_{H \in \mathcal{H}} w_H \pi_H = 0.$$

Notice that by assumption, we cannot have $w_H = 0, \; \forall H \in \mathcal{H}$. Moreover if $w_H = w'_1 \; \forall H \in \mathcal{H}_1$ and $w_H = w'_2 \; \forall H \in \mathcal{H}_2$ and $\sum_{H \in \mathcal{H}} w'_H \pi_H = 0$ then the vectors (w_1, w_2) and (w'_1, w'_2) are linearly dependent otherwise we would obtain

$$w'_1 \sum_{H \in \mathcal{H}_1} \pi_H + w'_2 \sum_{H \in \mathcal{H}_2} \pi_H = 0$$

for any choice of the weights (w'_1, w'_2).

In order to define the weights τ_H we observe that from (19) it follows that

$$\sum_{H \in \mathcal{H}} (\sigma_H - \tau_H) \pi_H = 0.$$

In other words once the weights (σ_1, σ_2) are chosen the choice of the weights (τ_1, τ_2) is given by:

$$\tau_i = \sigma_i + c w_i, \quad i = 1, 2,$$

where c is an arbitrary constant, and (w_1, w_2) is defined as the only non-zero vector, (up to a proportionality constant), such that

$$w_1 \sum_{H \in \mathcal{H}_1} \pi_H + w_2 \sum_{H \in \mathcal{H}_2} \pi_H = 0.$$

With similar arguments, it is easy to prove that if we have d orbits we can construct $d - 1$ linearly independent vectors

$$(w_1^{(1)}, \ldots, w_d^{(1)}), \ldots, (w_1^{(d-1)}, \ldots, w_d^{(d-1)})$$

such that

$$w_1^{(k)} \sum_{H \in \mathcal{H}_1} \pi_H + w_2^{(k)} \sum_{H \in \mathcal{H}_2} \pi_H + \cdots + w_d^{(k)} \sum_{H \in \mathcal{H}_d} \pi_H = 0, \quad k = 1, \ldots, d - 1.$$

In this case once the weights $(\sigma_1, \ldots, \sigma_d)$ are chosen, the choice of the weights $(\tau_1, \ldots, \tau_d)$ depends on $d - 1$ parameters $c_1, \ldots, c_{d-1}$:

$$\tau_i = \sigma_i + \sum_{k=1}^{d-1} c_k w_i^{(k)}, \quad i = 1, \ldots, d. \tag{21}$$

Having in mind the Dubrovin-Saito construction and Veselov's form of dual structure in the Coxeter case some natural questions arise: is it possible to construct a bi-flat F-manifold structure on the orbit space of complex reflection groups having a dual flat structure of the form described in Theorem 4.7? If this is so, then:

1 Which values of the weights are allowed?
2 What it the relation of the natural flat structure with the flat structure described in Theorem 4.5? The remark 4.9 suggests that such a relation should exist.

In all the cases of well-generated complex reflection groups we analyzed in [8], it is indeed possible to construct a bi-flat F-manifold structure on the relevant orbit spaces of the above form, see Theorem 4.10 below. Furthermore, in all the examples of well-generated complex reflection groups we studied each of the weights σ_H coincides with the order of the corresponding reflection (up to a normalizing factor) while the weights τ_H are of the form (21). In the case $\tau_H = \sigma_H$ the natural flat structure coincides with the flat structure described in Theorem 4.5. A remarkable fact is that even in the case of Coxeter groups the associated bi-flat structure might not be unique.

In order to illustrate the above theory we consider a simple example.

4.3 A Simple Example: B_2

Step 1. The dual product $*$

We start from the product

$$* = \sum_{H \in \mathcal{H}} \frac{d\alpha_H}{\alpha_H} \otimes \sigma_H \pi_H$$

where

$$\alpha_1 = [1, 0], \qquad \alpha_2 = [0, 1], \qquad \alpha_3 = [1, -1] \qquad \alpha_4 = [1, 1]$$

and the collection of the weights σ_H defines a G-invariant function. In practice this means to assign the same weight $\frac{x}{x+y}$ to the first two mirrors and the same weight $\frac{y}{x+y}$ to the last two mirrors (the factor $x+y$ is just a normalizing factor). We get

$$c_{11}^{*1} = \frac{(x+y)p_1^2 - xp_2^2}{(x+y)p_1(p_1^2 - p_2^2)}, \quad c_{11}^{*2} = \frac{-yp_2}{(x+y)(p_1^2 - p_2^2)} = c_{21}^{*1} = c_{21}^{*1}$$

$$c_{12}^{*2} = \frac{yp_1}{(x+y)(p_1^2 - p_2^2)} = c_{21}^{*2} = c_{22}^{*1}, \quad c_{22}^{*2} = \frac{xp_1^2 - (x+y)p_2^2}{(x+y)p_2(p_1^2 - p_2^2)}.$$

Step 2. The connection ∇

We assume that the basic invariants are flat coordinates of ∇. For B_2 up to a constant factor they depend on a single parameter c:

$$u_1 = p_1^2 + p_2^2, \qquad u_2 = p_1^4 + p_4^2 + cu_1^2.$$

Writing the connection ∇ in the coordinates p_1, p_2 we get

$$\Gamma_{11}^1 = \frac{2cp_1^2 + 3p_1^2 - p_2^2}{p_1(p_1^2 - p_2^2)}, \quad \Gamma_{11}^2 = \frac{-2p_1^2(c+1)}{p_2(p_1^2 - p_2^2)}, \quad \Gamma_{21}^1 = \frac{2cp_2}{(p_1^2 - p_2^2)} = \Gamma_{21}^1$$

$$\Gamma_{12}^2 = \frac{-2cp_1}{(p_1^2 - p_2^2)} = \Gamma_{21}^2, \quad \Gamma_{22}^1 = \frac{2p_2^2(c+1)}{p_1(p_1^2 - p_2^2)}, \quad \Gamma_{22}^2 = -\frac{2cp_2^2 + 3p_2^2 - p_1^2}{p_2(p_1^2 - p_2^2)}.$$

Step 3. The unit vector field e

We assume that in the basic invariant $e = \frac{\partial}{\partial u_2}$.

Step 4. The product $\circ$

From $*$ and e we can define $\circ$ in the usual way as

$$X \circ Y = (e*)^{-1} X * Y, \quad \forall X, Y.$$

We get

$$c_{11}^1 = \frac{4p_1(xp_1^2 - (x+y)p_2^2)}{x+y}, \quad c_{11}^2 = \frac{-4yp_1^2 p_2}{x+y} = c_{21}^1 = c_{21}^1$$

$$c_{12}^2 = \frac{-4yp_2^2 p_1}{x+y} = c_{21}^2 = c_{22}^1, \quad c_{22}^2 = \frac{4p_2((x+y)p_2^2) - xp_2^2}{x+y}.$$

Step 5. The constraint on the weights

Imposing the compatibility between ∇ and $\circ$:

$$\nabla_k c_{jl}^i = \nabla_j c_{lk}^i$$

we get the constraint $x = y$, that is $\sigma_1 = \sigma_2 = \sigma_3 = \sigma_4 = \frac{1}{2}$.

Step 6. The dual connection ∇^*

This is uniquely defined by the condition $\nabla^* E = 0$ and by the condition

$$(d_\nabla - d_{\nabla^*})(X *) = 0, \qquad \forall X \text{ vector field} \tag{22}$$

In local coordinates the above condition reads

$$\Gamma^{*k}_{lj} c^{*l}_{im} - \Gamma^{*k}_{li} c^{*l}_{jm} = \Gamma^{k}_{lj} c^{*l}_{im} - \Gamma^{k}_{li} c^{*l}_{jm}$$

This immediately implies

$$\Gamma^{*k}_{ij} = \Gamma^{k}_{ij} - c^{*l}_{ji} \nabla_l E^k \tag{23}$$

In our case we get

$$\Gamma^{*1}_{11} = -\frac{(2c+1)p_2^2 + p_1^2}{p_1(p_1^2 - p_2^2)}, \quad \Gamma^{*2}_{11} = \frac{2(c+1)p_2}{p_1^2 - p_2^2} = \Gamma^{*1}_{21} = \Gamma^{*1}_{21}$$

$$\Gamma^{*2}_{12} = \frac{-2(c+1)p_1}{p_1^2 - p_2^2} = \Gamma^{*2}_{21} = \Gamma^{*1}_{22}, \quad \Gamma^{*2}_{22} = \frac{(2c+1)p_1^2 + p_2^2}{p_2(p_1^2 - p_2^2)}.$$

The above connection can be also written in the form

$$\nabla^* = \nabla - \sum_{H \in \mathcal{H}} \frac{d\alpha_H}{\alpha_H} \otimes \tau_H \pi_H,$$

where ∇ is the trivial connection and $\tau_1 = \tau_2 = 2c + 1$, $\tau_3 = \tau_4 = -2 - 2c$.

Step 7. The vector potential

The above data and the Euler vector field $E = \sum_n \frac{\partial}{\partial p_n}$ define a bi-flat structure $(\nabla, \circ, e, \nabla^*, *, E)$ for any choice of the parameter c. Fixed c, writing the structure constants of the product $\circ$ in the basic invariants associated with this choice of c we get

$$c^i_{jk} = \partial_j \partial_k A^i_{B_2},$$

where

$$A^1_{B_2} = -\frac{2}{3}\left(c + \frac{3}{4}\right) u_1^3 + u_1 u_2, \quad A^2_{B_2} = -\frac{1}{6}(c+1)(2c+1)u_1^4 + \frac{1}{2}u_2^2.$$

For $c = -\frac{3}{4}$ (and only in this case) the vector potential comes from a Frobenius potential and the flat basic invariants coincide with the standard Saito flat coordinates.

4.4 A Conjecture

The above construction is based on the assumption that the dual strucuture has the form described Theorem 4.7 and the natural flat structure has the form described in Theorem 4.5 (even if it does not necessarily coincide with the flat structure found in [25]).

The main result of [8] is the following:

Theorem 4.10 *[8] Let G be a Weyl group of rank 2, 3, or 4, or the dihedral groups $I_2(m)$, or any of the exceptional well-generated complex reflection groups of rank 2 and 3 or any of the groups $G(m, 1, 2)$ and $G(m, 1, 3)$. Suppose G acts on V via its reflection representation. Then there exists a family of bi-flat F-structures on the orbit spaces V/G depending on μ parameters, where $\mu = |\mathcal{H}/G| - 1$, and $|\mathcal{H}/G|$ is the number of orbits for the action of G on the collection of reflecting hyperplanes $\mathcal{H}$.*

Theorem 4.10 and the study of several examples lead us to formulate the following conjecture

Conjecture 4.11 *Let G any finite well-generated irreducible complex reflection group acting on a vector space V via its reflection representation. Then the orbit space V/G is equipped with a family of bi-flat structures (of the form described above) depending on μ parameters, where $\mu = |\mathcal{H}/G| - 1$, and $|\mathcal{H}/G|$ is the number of orbits for the action of G on the collection of reflecting hyperplanes $\mathcal{H}$.*

This conjecture has been formulated in the case of Coxeter groups in [8]. In the case of Coxeter groups with two orbits (for their action on $\mathcal{H}$) the only relevant open case is the one of B_n for general n.

Remark 4.12 In the case of dihedral groups $I_2(m)$ it is possible to construct a one-parameter family of bi-flat structures also when m is odd. However, in such cases, only for one value of the parameter the bi-flat structure has the form described above.

Remark 4.13 We have seen that bi-flat F-manifolds are related to Painlevé transcendents and to complex reflection groups. It is known that Painlevé transcendents are related to complex reflection groups. In the case of Coxeter groups this problem was addressed in [19] (see also [24]) and for general complex reflection groups in [10, 11, 12].

5 Bi-Flat *F*-Manifolds and Integrable Hierarchies

In this Section we show how to construct integrable hierarchies of dispersionless quasilinear PDEs from the geometric structures introduced in the previous Section.

5.1 The Principal Hierarchy

Given a flat F-manifold M, choose local coordinates $v := \{v^1, \ldots, v^n\}$. To it, we associate the following collection of quasilinear systems of evolutionary PDEs:

$$v^i_{t_{(p,l)}} = c^i_{jk} X^k_{(p,l)} v^k_x, \qquad p = 1, \ldots, n \qquad l = 0, 1, 2, 3, \ldots \qquad (24)$$

For each l (level of the hierarchy), there are n systems of n quasilinear PDEs. The vector fields $X_{(p,l)}$ used to define this collection of PDEs are the coefficients of the formal expansion in λ of flat sections of the deformed flat connection $\nabla^\lambda := \nabla - \lambda \circ$ (the minus sign here is just due to avoid a minus sign in the recurrence relations). Concretely,

$$(\nabla - \lambda \circ) \left(X_{(p,0)} + X_{(p,1)}\lambda + X_{(p,2)}\lambda^2 + \ldots \right) = 0. \qquad (25)$$

Isolating terms containing equal powers of λ, from (25) one obtains immediately that the vector fields $X_{(p,0)}$ (the fields used to define the primary flows) are covariantly constant with respect to ∇, while the others are obtained through the recurrence relation:

$$\nabla X_{(p,l+1)} = X_{(p,l)} \circ . \qquad (26)$$

The integrable hierarchy (24) is a generalization of Dubrovin's principal hierarchy associated to a Frobenius manifold. The construction of (24) in the case of a flat F-manifold and the proof of its integrability (i.e. the proof of the consistency of the recurrence relations (26)) is in [34]. We call (24) the Principal Hierarchy associated to a flat F-manifold.

Writing (24) in a general coordinates system $\{v^1, \ldots, v^n\}$ one obtains:

$$v^i_{t_{(p,l)}} = c^i_{jk} X^k_{(p,l)} v^k_x, \qquad p = 1, \ldots, n \qquad l = 0, 1, 2, 3, \ldots$$

where for the primary flows (those that correspond to $t_{(p,0)}$) one has $\nabla_j X^i_{(p,0)} = 0$, while for the higher flows (corresponding to times $t_{(p,l)}, l > 0$) $\nabla_j X^i_{(p,l+1)} = c^i_{jk} X^k_{(p,l)}$. Furthermore, if one chooses a coordinate system $\{u^1, \ldots, u^n\}$ in which the Christoffel symbols vanish identically, the flows (24) are systems of conservation laws:

$$u^i_{t_{(p,l)}} = c^i_{jk} X^j_{(p,l)} u^k_x = \partial_x X^i_{(p,l+1)}. \qquad (27)$$

On the other hand, if the product $\circ$ is semisimple, then there exists a distinguished coordinates system $\{r^1, \ldots, r^n\}$ such that $c^k_{ij} = \delta^k_i \delta^k_j$. Therefore in canonical coordinates we have that (24) assumes the form

$$r^i_{t_{(p,l+1)}} = V^i_{(p,l)}(\mathbf{r}) r^i_x, \qquad i = 1, \ldots n \qquad (28)$$

which means that the canonical coordinates are the *Riemann invariants* for the conservation laws (27). In the case of flat F-manifolds with invariant metric the principal hierarchy becomes Hamiltonian w.r.t. the Dubrovin-Novikov bracket associated with η^{-1} [1]. In the case of Frobenius manifolds the principal hierarchy becomes bi-Hamiltonian: the second Hamiltonian structure the Dubrovin-Novikov bracket associated with the intersection form.

5.2 A Simple Example: B_2

In flat coordinates the flows of the hierarchy have the form

$$u^i_{t_{(p,l)}} = \left(\partial_j \partial_k A^i_{B_2}\right) X^j_{(p,l)} u^k_x, \qquad i = 1, 2.$$

The recursion relations read

$$\partial_j X^i_{(p,0)} = 0, \qquad \partial_j X^i_{(p,l+1)} = \left(\partial_j \partial_k A^i_{B_2}\right) X^k_{(p,l)}.$$

By straightforward computations we get:

$$X_{(1,0)} = \begin{pmatrix} 1 \\ 0 \end{pmatrix}, \; X_{(1,1)} = \begin{pmatrix} u_2 \\ \frac{1}{12} u_1^3 \end{pmatrix}, \; X_{(1,2)} = \begin{pmatrix} \frac{1}{2} u_2^2 + \frac{1}{48} u_1^4 \\ \frac{1}{12} u_1^3 u_2 \end{pmatrix}, \dots$$

$$X_{(2,0)} = \begin{pmatrix} 0 \\ 1 \end{pmatrix}, \; X_{(2,1)} = \begin{pmatrix} u_1 \\ u_2 \end{pmatrix}, \; X_{(2,2)} = \begin{pmatrix} u_1 u_2 \\ \frac{1}{2} u_2^2 + \frac{1}{16} u_1^4 \end{pmatrix}, \dots$$

For generic values of c the above hierarchy is not Hamiltonian (w.r.t. a local Hamiltonian structure). More precisely:

- for $c \neq -1, -\frac{1}{2}, -\frac{3}{4}$ (generic case): the principal hierarchy is not Hamiltonian.
- for $c = -\frac{3}{4}$: the principal hierarchy is (bi)-Hamiltonian. For this value of the parameter we have a Frobenius manifold.
- for $c = -1, c = -\frac{1}{2}$: one of the chains of the principal hierarchy is degenerate.

5.3 Twisted Lenard-Magri Chains

Any bi-flat F-manifold, being in particular a flat F-manifold possesses an associated principal hierarchy, defined via (24). However, in a bi-flat F-manifold the presence of a second flat F-structure $(\nabla^*, *, E)$ intertwined with the first $(\nabla, \circ, e)$ equips the principal hierarchy (24) with an additional system of recurrence relations, called *twisted Lenard-Magri chains* which generalize the standard bi-Hamiltonian recursive relations in this non-Hamiltonian framework:

Theorem 5.1 *[2] Let M be a semisimple bi-flat F-manifold with flat F-structures $(\nabla, \circ, e)$ and $(\nabla^*, *, E)$. Then the following twisted version of the Lenard-Magri chain holds:*

$$d_\nabla X_{(p+1,\alpha)} = d_{\nabla^*}\left(E \circ X_{(p,\alpha)}\right)$$

where $(X_{(0,1)}, \dots, X_{(0,n)})$ is a frame of covariantly constant vector fields for ∇. The corresponding equations of the associated hierarchy are

$$u^i_{t_{(p,\alpha)}} = [d_{\nabla^*}\left(E \circ X_{(p-1,\alpha)}\right)]^i_j\, u^j_x = (d_\nabla X_{(p,\alpha)})^i_j\, u^j_x,$$

$$i = 1, \ldots, n, \quad p = 1, 2, 3, \ldots. \tag{29}$$

It turns out that the above recurrence relations are well defined due to the condition (8). This fact is the at the root of the definition of bi-flat F-manifolds.

5.4 Integrable Deformations: an Open Conjecture

For integrable systems of evolutionary PDEs admitting the dispersionless limit there are several definition of integrability. The most general definition requires the existence of infinitely many formal symmetries.

Definition 5.2 *[9] A system of evolutionary PDEs*

$$u^i_t = A^i_j(\mathbf{u})u^j_x + \epsilon(B^i_j(\mathbf{u})u^j_{xx} + B^i_{jk}(\mathbf{u})u^j_x u^k_x) + \ldots, \quad i = 1, \ldots, n, \tag{30}$$

is said to be integrable (or formally integrable*) if*

- *The dispersionless limit is integrable in the sense of Tsarev [40].*
- *Every symmetry (commuting flow) of the dispersionless limit can be extended to a symmetry of the complete system* (30).

The above definition of integrability in the scalar case coincides with the definition of *formal integrability* given in [31].

The first condition is equivalent (see [39]) to the existence of two sets of distinguished coordinates:

1. a set of coordinates $(v^1, \ldots, v^n)$ reducing the system $u^i_t = A^i_j(\mathbf{u})u^j_x$ to a system of conservation laws:

$$v^i_t = \partial_x X^i(\mathbf{v}), \quad i = 1, \ldots, n.$$

2. the Riemann invariants $(r^1, \ldots, r^n)$. By definition this is a system of coordinates reducing the system $u^i_t = A^i_j(\mathbf{u})u^j_x$ to diagonal form

$$r^i_t = V^i(\mathbf{r})r^i_x, \quad i = 1, \ldots, n.$$

Due to the results of the Subsection 5.1 all the flows of the principal hierarchy associated with a semisimple F-manifold satisfy the above conditions.

The second condition is not too restrictive. Even fixing the dispersionless limit there is a huge group of transformations, called Miura transformations preserving the form of the equations. They have the form

$$u^i \to \tilde{u}^i = u^i + \sum_k \epsilon^k F^i_k(\mathbf{u}, \mathbf{u}_x, \ldots), \tag{31}$$

where $\det\left(\partial_j F_0^i\right) \neq 0$ and $\deg F_k^i = k$. The last condition means

$$F_1^i - F_{11j}^i(\mathbf{u})u_x^j$$
$$F_2^i = F_{21j}^i(\mathbf{u})u_{xx}^j + F_{22jk}^i(\mathbf{u})u_x^j u_x^k$$
$$F_3^i = F_{31j}^i(\mathbf{u})u_{xxx}^j + F_{32jk}^i(\mathbf{u})u_{xx}^j u_x^k + F_{33jkl}^i(\mathbf{u})u_x^j u_x^k u_x^l$$

and so on. All the coefficients F_{11j}^i, F_{21j}^i, F_{22jk}^i, ... are arbitrary functions. For this reason the aim of the classification problem is to compute integrable systems of PDEs up to the action of the Miura group (in the bi-Hamiltonian setting the classification problem has been formulated in this way by Dubrovin and Zhang [21]).

In order to solve this problem one needs first to select a special set of invariants which parametrize equivalence classes of integrable systems (admitting the same dispersionless limit). It turns out that the good candidates in the general case are the *Miura invariants*, introduced in [9]. They are defined in the following way: let

$$X_{quasilinear}^i = A_j^i(\mathbf{u})u_x^j + \epsilon B_j^i(\mathbf{u})u_{xx}^j + \epsilon^2 C_j^i(\mathbf{u})u_{xxx}^j + \cdots$$
$$= \left(A_j^i(\mathbf{u}) + \epsilon B_j^i(\mathbf{u})\partial_x + \epsilon^2 C_j^i(\mathbf{u})\partial_x^2 + \cdots\right)u_x^j$$

be the quasilinear part of the system (30). Introduce the *Miura matrix*

$$M_j^i(\mathbf{u}, p) := A_j^i(\mathbf{u}) + B_j^i(\mathbf{u})p + C_j^i(\mathbf{u})p^2 + \cdots$$

Definition 5.3 *We call* Miura invariants *of the system* (30) *the eigenvalues* $\lambda_i(\mathbf{u}, p)$ *of the Miura matrix.*

Due to the invariance of the functions $\lambda_i(\mathbf{u}, p)$ with respect to Miura transformations of the form (31) if two systems admitting the same dispersionless limit are related by a Miura transformation then their Miura invariants coincide. In the scalar case [7, 6] we have conjectured that also the converse statement is true. The case of systems is much more complicated due to highly non trivial computations involved. However the first results of [9] seem suggest a similar conjecture also in the general case.

Conjecture 5.4 *Two integrable systems of the form* (30) *admitting the same dispersionless limit are related by a Miura transformation* (31) *if and only if they have the same Miura invariants.*

Remark 5.5 In the bi-Hamiltonian setting the relevant set of invariants are called *central invariants* and have been introduced in [20]. In the same paper the authors proved that two deformations of the same bi-Hamiltonian structure of hydrodynamic type are related by a Miura transformation (31) if and only if they have the same central invariants.

5.5 A Simple Example: B_2

The validity of the above conjecture does not tell us the number of functional parameters we really need to parameterize integrable deformations. In the table below we summarize the result of the computations (up to the second order in ϵ) in the example of B_2. In the second and third order columns we have the number of non (Miura)-trivial functional parameters appearing at the first and second order respectively. We skip details of computations for which we refer to [9].

Values of c	First order deformations	Second order deformations
$c \neq -\frac{3}{4}, -1, -\frac{1}{2}$	0	2
$c = -\frac{3}{4}$	0	2
$c = -1$	1	2
$c = -\frac{1}{2}$	1	2

For special values of the parameter c and for special choices of the Miura invariants one gets (up to the second order and up to Miura transformation) known examples:

- for $c = -\frac{3}{4}$ the Drinfeld-Sokolov hierarchy associated with B_2 and the hierarchy of topological type associated with B_2 [30].
- for $c = -1$: the double ramification hierarchy for the extended 2-spin theory [13].

In all these examples the functional parameters appearing in the Miura invariants turn out to be constant.

It is a completely open problem to determine the existence at all order in ϵ of non-trivial Miura integrable deformations of the Principal Hierarchy associated with a (bi)-flat F-manifold. Unfortunately, in the case of hierarchies coming from (bi)-flat F-manifolds the powerful algebraic tools related to the existence of a bi-Hamiltonian structure (that are available in the case of Frobenius manifolds) can not be directly put to fruition.

Ackowledgements

We would like to thank Oleg Chalykh and Misha Feigin for useful discussions and Gwyn Bellamy and Ulrich Thiel for useful information about the number of orbits for the action of a complex reflection group on the set of reflecting hyperplanes.

References

[1] A. Arsie and P. Lorenzoni, *Poisson bracket on 1-forms and evolutionary PDEs*, Journal of Physics A, Mathematical and Theoretical, Vol 45, no. 47, (2012).

[2] A. Arsie and P. Lorenzoni, *F-Manifolds with Eventual Identities, Bidifferential Calculus and Twisted Lenard-Magri Chains*, International Mathematics Research Notices, Volume 2013, Issue **17**.

[3] A. Arsie, P. Lorenzoni, *From the Darboux– Egorov system to bi-flat F-manifolds*, J. of Geom. and Physics, Vol **70**, (2013), Pages 98–116.

[4] A. Arsie and P. Lorenzoni, *Purely non-local Hamiltonian formalism, Kohno connections and ∨-systems*, J. Math. Phys. 55 (2014).

[5] A. Arsie, P. Lorenzoni, *F-manifolds, multi-flat structures and Painlevé transcendents*, arXiv:submit/1685937 (2015), accepted for publication in Asian Journal of Mathematics.

[6] A. Arsie, P. Lorenzoni, A. Moro, *Integrable viscous conservation laws*, Nonlinearity, Volume 28, Number 6, (2015).

[7] A. Arsie, P. Lorenzoni, A. Moro, *On integrable conservation laws*, Proc. Royal Society A. 471 (2015).

[8] A. Arsie and P. Lorenzoni, *Complex reflection groups, logarithmic connections and bi-flat F-manifolds*, Lett. Math Phys (2017).

[9] A. Arsie and P. Lorenzoni, *Flat F-manifolds, Miura invariants and integrable systems of conservation laws*, Journal of Integrable Systems, Vol **3**, Issue 1, (2018).

[10] P. Boalch, *Painlevé equations and complex reflections*, Ann. Inst. Fourier 53 (2003) no.4, 1009–1022.

[11] P. Boalch, *From Klein to Painlevé via Fourier, Laplace and Jimbo*, Proc. London Math. Soc. (3) 90 (2005) 167–208.

[12] P. Boalch, *The fifty-two icosahedral solutions to Painlevé VI*, J. Reine Angew. Math. 596 (2006) 183–214.

[13] A. Buriak and P. Rossi, *DR hierarchy for the extended 2-spin theory*, arXiv:1806.09825.

[14] L. David and C. Hertling, *Regular F-manifolds: initial conditions and Frobenius metrics*, Annali della Scuola Normale di Pisa, Classe di Scienze Vol. XVII, issue 3 (2017) pp 1121–1152.

[15] B. Dubrovin, *Painlevé transcendents in two-dimensional topological field theory*, in The Painlevé Property- One Century Later, Editor R. Conte, Springer-Verlag, 1999.

[16] B. Dubrovin, *Integrable Systems and Classification of 2-Dimensional Topological Field Theories* in Integrable Systems - The Verdier Memorial Conference, Actes du Colloque International de Luminy, Editors O. Babelon, Y. Kosmann-Schwarzbach, P. Cartier, Birkhäuser-Verlag, 1993.

[17] B. Dubrovin, *On almost duality for Frobenius manifolds* in Geometry, Topology, and Mathematical Physics, Editors. V. M. Buchstaber and I. M. Krichever, American Mathematical Society Translations: Series 2 (2004), Volume 212.

[18] B. Dubrovin, *Differential Geometry of the Space of Orbits of a Coxeter Group*, Surveys in Differential Geometry , Vol. IV (1999), p. 181–212.

[19] B. Dubrovin and M. Mazzocco, *Monodromy of certain Painlevé -VI transcendents and reflection groups*, Invent. Math. 141 (2000), no. 1, 55–147.

[20] B. Dubrovin, S.Q. Liu, Y. Zhang, *Frobenius manifolds and central invariants for the Drinfeld-Sokolov bihamiltonian structures*, Adv. Math. **219** (2008), 780–837.

[21] B. Dubrovin, Y. Zhang, *Normal forms of integrable PDEs, Frobenius manifolds and Gromov-Witten invariants*, math.DG/0108160.

[22] M. V. Feigin and A. P. Veselov, $\vee$-*systems, holonomy Lie algebras and logarithmic vector fields*, IMRN, Volume 2018, Issue 7, 11 April 2018, p. 2070–2098

[23] C. Hertling and Yu. Manin, *Weak Frobenius Manifolds*, Int Math Res Notices (1999) (6) pp 277–286.

[24] N.J. Hitchin, Poncelet Polygons and the Painlevé Transcendents, Geometry and Analysis, edited by Ramanan, Oxford University Press (1995) 151–185.

[25] M. Kato, T. Mano and J. Sekiguchi, *Flat Structures without Potentials*, Rev. Roumaine Math. Pures Appl. **60** (2015), no. 4, 481–505

[26] M. Kato, T. Mano and J. Sekiguchi *Flat structure on the space of isomonodromic deformations*, arXiv:1511.01608v1

[27] H. Kawakami and T. Mano, *Regular flat structure and generalized Okubo system*, arXiv:1702.03074.

[28] Y. Konishi, S. Minabe and Y. Shiraishi, *Almost duality for Saito structure and complex reflection groups*, Journal of Integrable Systems, Volume 3, Issue 1, 1 January 2018.

[29] M. Kontsevich, *Intersection theory on the moduli space of curves and the matrix Airy function*, Comm. Math. Phys. 147:1 (1992), 1–23.

[30] S.-Q. Liu, Y. Ruan and Y. Zhang, *BCFG Drinfeld-Sokolov Hierarchies and FJRW-Theory* Inven. Math. 201 (2015), 711–772

[31] S-Q. Liu, Y. Zhang *On Quasitriviality and Integrability of a Class of Scalar Evolutionary PDEs* J. Geom. Phys. **57**, 101–119, (2006).

[32] E. Looijenga, *Arrangements, KZ systems and Lie algebra homology*, in: Singularity Theory, B. Bruce and D. Mond eds., London Math. Soc. Lecture Note Series 263, CUP (1999), 109–130.

[33] P. Lorenzoni, *Darboux-Egorov System, Bi-flat F-Manifolds and Painlevé VI*, IMRN, (2014), **12**, 3279–3302.

[34] P. Lorenzoni, M. Pedroni and A. Raimondo, *F-manifolds and integrable systems of hydrodynamic type*, Archivum Mathematicum **47** (2011), 163–180.

[35] Y.I. Manin, *F-manifolds with flat structure and Dubrovin's duality*, Adv. Math. **198** (2005), no. 1, 5–26.

[36] C. Sabbah, *Isomonodromic deformationsa and Frobenius manifolds, An Introduction*, Universitext, Springer Verlag, (2008).

[37] K. Saito K., T. Yano and J. Sekiguchi, *On a certain generator system of the ring of invariants of a finite reflection group*, Comm. in Algebra 8(4) (1980) 373–408.

[38] K. Saito, *On a linear structure of a quotient variety by a finite reflexion group*, RIMS Kyoto preprint 288, 1979.

[39] B. Sevennec, *Géométrie des systèmes hyperboliques de lois de conservation*, Mémoires de la Société Mathématique de France, **56**, pag 1–125 (1994).

[40] S.P. Tsarev, *Geometry of Hamiltonian systems of hydrodynamic type. Generalized hodograph method*, Izvestija AN USSR Math. **54**, no. 5 (1990) 1048–1068.

[41] A. P. Veselov, *Deformations of the root systems and new solutions to generalised WDVV equations*, Physics Letters A, **261**, (1999), 297–302.

[42] A. P. Veselov, *On geometry of a special class of solutions to generalized WDVV equations*, Integrability: the Seiberg-Witten and Whitham equations (Edinburgh, 1998), 125–135, Gordon and Breach.

[43] E. Witten, *Two-dimensional gravity and intersection theory on moduli space*, Surveys in differential geometry (Cambridge, MA 1990), Lehigh Univ., Bethlehem, PA 1991, pp. 243–310.

Alessandro Arsie
Department of Mathematics and Statistics
The University of Toledo, 2801 W. Bancroft St., 43606 Toledo, OH, USA
alessandro.arsie@utoledo.edu

Paolo Lorenzoni
Dipartimento di Matematica e Applicazioni
Università di Milano-Bicocca, Via Roberto Cozzi 53, I-20125 Milano, Italy
paolo.lorenzoni@unimib.it

9

The Periodic 6-Particle Kac-van Moerbeke System

Pol Vanhaecke[†]

Abstract. We study some algebraic-geometrical aspects of the periodic 6-particle Kac-van Moerbeke system. This system is known to be algebraically integrable, having the affine part of a hyperelliptic Jacobian of a genus two curve as the generic fiber of its momentum map. Particular attention goes to the divisor needed to complete this fiber into an Abelian variety: it consists of six copies of the curve, intersecting according to a pattern which we will determine. We will also compare this divisor to the divisor which appears in some natural *singular* compactification of the fiber.

1 Introduction

The periodic n-particle Kac-van Moerbeke system (KM system) is given by the following quadratic vector field on $\mathbb{C}^n$:

$$\dot{x}_i = x_i(x_{i-1} - x_{i+1}), \qquad i = 1, \ldots, n, \tag{1}$$

where $x_0 := x_n$ and $x_{n+1} := x_1$. It is a Hamiltonian system with respect to the quadratic Poisson structure defined by

$$\{x_i, x_j\} := x_i x_j(\delta_{i,j+1} - \delta_{i+1,j}), \qquad i, j = 1, \ldots, n.$$

Indeed, taking $H := x_1 + x_2 + \cdots + x_n$ as Hamiltonian, the Hamiltonian vector field $\mathcal{X}_H := \{\cdot, H\}$ is precisely (1). The Poisson structure has rank $n - 1$ when n is odd and $n - 2$ otherwise. The system was introduced by Kac and van Moerbeke in [6] who constructed this system as a discretization of the Korteweg-de Vries equation and who also showed its Liouville integrability. Several first integrals are produced from the Lax operator $L(\mathfrak{h})$, which is obtained from the Lax operator of the classical n-particle Toda lattice by replacing all diagonal elements by zero; see [5] for a precise account of this in terms of Poisson geometry. They yield $s := [(n + 3)/2]$ independent

2000 *Mathematics Subject Classification.* 53D17, 37J35

first integrals, in involution, which is the exact number required to assure the Liouville integrability of the periodic n-particle KM system.

It was shown in [5] that all periodic KM systems are algebraically integrable (a.c.i.). It means that for generic $\mathbf{c} := (c_1, \ldots, c_s) \in \mathbb{C}^s$ the fiber $\mathbf{F_c} := \mathbf{F}^{-1}(\mathbf{c})$ of the momentum map (the map $\mathbf{F} : \mathbb{C}^n \to \mathbb{C}^s$, defined by the first integrals) is an affine part[1] of an Abelian variety $\mathbf{T_c}$ and that the integrable vector fields are translation invariant (with respect to the group structure on the Abelian varieties) on these fibers. In the present case of the periodic n-particle KM system, the Abelian variety $\mathbf{T_c}$ has two (equivalent) descriptions:

- As the Prym variety of the spectral curve $|\mathfrak{z}\mathrm{Id}_n - L_\mathbf{c}(\mathfrak{h})| = 0$, equipped with the involution $(\mathfrak{z}, \mathfrak{h}) \mapsto (-\mathfrak{z}, \mathfrak{h})$; here, $L_\mathbf{c}(\mathfrak{h})$ denotes the Lax operator $L(\mathfrak{h})$, restricted to $\mathbf{F_c}$;
- As hyperelliptic Jacobians, associated to the quotient of the above spectral curve by the involution $(\mathfrak{z}, \mathfrak{h}) \mapsto (-\mathfrak{z}, -\mathfrak{h})$.

Moreover, it is shown that the divisor which needs to be adjoined to the generic fiber $\mathbf{F_c}$ in order to complete it into $\mathbf{T_c}$ consists of n translates of the theta divisor.

In the case of $n = 5$ and of $n = 6$, the Abelian varieties are surfaces. For these cases an alternative proof of algebraic integrability can be given using the systematic method which was developed by Adler and van Moerbeke and presented in its final form in [2]. In fact, it is precisely the periodic 5-particle KM system which is used as a running example in [2] to present the method. Accessorily it provides an alternative proof of the algebraic integrability in the case $n = 5$. Similarly, such an alternative proof can be given in the case of $n = 6$. In this paper we will *not* present such an alternative proof, but study two compactifications of the fibers of the momentum map, using the known fact that the system is algebraically integrable. Notice however that our presentation will contain most elements, in particular all essential formulas, needed for providing the alternative proof which we just mentioned.

The two compactifications which we will consider of the generic fiber $\mathbf{F_c}$ are of a quite different character. The first one is the one which compactifies $\mathbf{F_c}$ into the torus $\mathbf{T_c}$. The six translates of the theta divisor (which is in this case a curve, which can be identified with a quotient of the spectral curve) intersect each other in a quite particular, symmetric pattern: if we order the translates cyclically, every curve intersects its two neighboring curves in two different points, is tangent to its two second nearest neighbors and has two different intersection points with the remaining curve (its furthest neighbor). The other

---

[1] When n is even, the fiber contains one or two isomorphic components, depending on the precise choice of momentum map $\mathbf{F}$; see Section 2 and in particular diagram (7) for details in the case of $n = 6$.

compactification is the standard homogeneous compactification of the fiber, obtained by first compactifying $\mathbb{C}^6$ in the standard way to $\mathbb{P}^6$ and then taking the closure $\bar{\mathbf{F}}_\mathbf{c}$ of $\mathbf{F}_\mathbf{c}$ in $\mathbb{P}^6$. The resulting surface is singular: the divisor which has been added consists of 6 non-singular conics and three singular conics which are double lines and it is precisely along these lines that the surface $\bar{\mathbf{F}}_\mathbf{c}$ is singular. It will be clear that we rely heavily on the KM vector field for obtaining the first compactification, but not for the second one. Since the two compactifications are birationally isomorphic, it would be interesting to obtain the first compactification in a purely algebraic-geometrical way, i.e., without using the periodic 6-particle KM vector field. Since in this case the singularities are clearly identified and not too complicated, doing this may be feasible.

The plan of the paper is as follows. In Section 2 we recall the main results on the integrability and algebraic integrability of the periodic n-particle KM system and add some extra observations. An essential ingredient in the study of algebraic integrable systems is the family of Laurent solutions to the vector field(s) of the system. We give explicit formulas for (the first terms of) all Laurent solutions in Section 3. The principal balances (Laurent solution depending on 5 free parameters) are used in Section 4 to construct an embedding of the generic Abelian surfaces $\mathbf{T}_\mathbf{c}$ which compactify the generic fibers $\mathbf{F}_\mathbf{c}$ of the momentum map. The embedding allows us to compute an equation for the 6 (isomorphic) curves which make up the Painlevé divisor. This is done in Section 5, where we also relate the Painlevé curves to the spectral curve. In the final Section 6 we present the two compactifications of the generic surfaces $\mathbf{F}_\mathbf{c}$, with special attention to the geometry of the divisors which are adjoined in both cases.

2 The Periodic 6-Particle KM System

In this section we recall from [6] and [5] the main results on the Liouville, respectively algebraic integrability of the periodic n-particle Kac-van Moerbeke (KM) system, which we specialize to the case of $n = 6$. We also add a few extra observations which are specific to this case. The notions and notations which are introduced here will be used throughout the paper.

The periodic 6-particle KM system is given by the following quadratic vector field on $\mathbb{C}^6$:

$$\dot{x}_i = x_i(x_{i-1} - x_{i+1}), \qquad i = 1, \ldots, 6. \tag{2}$$

Here, $x_1, \ldots, x_6$ are the standard linear coordinates on $\mathbb{C}^6$; also, $x_7 = x_1$ and $x_0 = x_6$, i.e., all indices are taken modulo 6. The latter accounts for the adjective *periodic*. It is a Hamiltonian system with linear Hamiltonian

$$H := x_1 + x_2 + \cdots + x_6, \tag{3}$$

and with a quadratic Poisson structure, defined by the following formulas:

$$\{x_i, x_j\} := x_i x_j (\delta_{i,j+1} - \delta_{i+1,j}), \qquad i, j = 1, \ldots, 6.$$

Some basic constants of motion of (2) are found by using a Lax operator. Indeed, (2) can be written as the following Lax equation with a spectral parameter (which we denote by $\mathfrak{h}$),

$$\dot{L}(\mathfrak{h}) = [L(\mathfrak{h}), M(\mathfrak{h})],$$

where $L(\mathfrak{h})$ and $M(\mathfrak{h})$ are given by

$$L(\mathfrak{h}) = \begin{pmatrix} 0 & x_1 & 0 & 0 & 0 & \mathfrak{h}^{-1} \\ 1 & 0 & x_2 & 0 & 0 & 0 \\ 0 & 1 & 0 & x_3 & 0 & 0 \\ 0 & 0 & 1 & 0 & x_4 & 0 \\ 0 & 0 & 0 & 1 & 0 & x_5 \\ \mathfrak{h}x_6 & 0 & 0 & 0 & 1 & 0 \end{pmatrix} \tag{4}$$

and

$$M(\mathfrak{h}) = \begin{pmatrix} 0 & 0 & x_1 x_2 & 0 & 0 & 0 \\ 0 & 0 & 0 & x_2 x_3 & 0 & 0 \\ 0 & 0 & 0 & 0 & x_3 x_4 & 0 \\ 0 & 0 & 0 & 0 & 0 & x_4 x_5 \\ \mathfrak{h}x_5 x_6 & 0 & 0 & 0 & 0 & 0 \\ 0 & \mathfrak{h}x_6 x_1 & 0 & 0 & 0 & 0 \end{pmatrix}.$$

The characteristic polynomial of $L(\mathfrak{h})$ is given by

$$|{}_{3}\mathrm{Id}_6 - L(\mathfrak{h})| = {}_{3}^6 - F_1 {}_{3}^4 + F_2 {}_{3}^2 - \frac{1}{\mathfrak{h}}(1 + F_3 \mathfrak{h})(1 + F_4 \mathfrak{h}), \tag{5}$$

where the coefficients F_i are polynomial functions on $\mathbb{C}^6$; they are explicitly given by the following formulas:

$$F_1 = x_1 + x_2 + x_3 + x_4 + x_5 + x_6,$$
$$F_2 = x_1 x_4 + x_2 x_5 + x_3 x_6 + x_1 x_3 + x_2 x_4 + x_3 x_5 + x_4 x_6 + x_1 x_5 + x_2 x_6,$$
$$F_3 = x_1 x_3 x_5,$$
$$F_4 = x_2 x_4 x_6. \tag{6}$$

Notice that F_1 is just H, the Hamiltonian of the system. By a basic property of Lax equations, every coefficient (in $_3$ and $\mathfrak{h}$) of (5) is a constant of motion of (2), hence the functions $F_1, \ldots, F_4$ are constants of motion of (5). Both F_3 and F_4 are Casimir functions of the Poisson structure, whose rank is 4 at a generic point of $\mathbb{C}^6$ (to be precise: the rank is 4 at all points, except at those satisfying $x_i = x_{i+2} = 0$ for some $i \in \{1, \ldots, 6\}$). Also, F_1 and F_2 are in involution since F_2 is a constant of motion of (2) and since $H = F_1$; it follows that the functions $F_1, \ldots, F_4$ are pairwise in involution. Finally, they are also independent, so $(\mathbb{C}^6, \{\cdot\,,\cdot\}, (F_1, \ldots, F_4))$ defines a Liouville integrable system. We view $\mathbf{F} := (F_1, F_2, F_3, F_4)$ as a polynomial map $\mathbf{F} : \mathbb{C}^6 \to \mathbb{C}^4$, the *momentum map* of the integrable system. By what precedes, for a generic point $\mathbf{c} \in \mathbb{C}^4$ the fiber $\mathbf{F_c} := \mathbf{F}^{-1}(\mathbf{c})$ of $\mathbf{F}$ is a smooth complex surface to which the commuting Hamiltonian vector fields $\mathcal{X}_{F_1}$ and $\mathcal{X}_{F_2}$ are tangent; moreover, these vector fields generate the tangent space to the fiber $\mathbf{F_c}$ at each point. Finally, notice that the system is *homogeneous*, i.e., the Poisson structure $\{\cdot\,,\cdot\}$, the constants of motion $F_1, \ldots, F_4$ and the vector field $\mathcal{X}_H$ are weight homogeneous when all variables x_i are given weight 1 and time t is given weight -1.

We now turn to the algebraic integrability of the system, which yields a more precise description of the generic fiber $\mathbf{F_c}$ of $\mathbf{F}$. By construction, the characteristic polynomial of $L(\mathfrak{h})$ is constant on the fibers of $\mathbf{F}$ and we have the following commutative triangle:

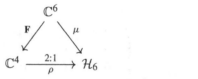

$$(7)$$

In this diagram, $\mathcal{H}_6$ stands for the following space of Laurent polynomials:

$$\left\{ f_{\mathbf{c}}(\mathfrak{z}, \mathfrak{h}) := \mathfrak{z}^6 - c_1 \mathfrak{z}^4 + c_2 \mathfrak{z}^2 - \frac{1}{\mathfrak{h}}(1 + c_3 \mathfrak{h})(1 + c_4 \mathfrak{h}) \mid \mathbf{c} = (c_1, \ldots, c_4) \in \mathbb{C}^4 \right\}$$

and ρ is defined for $\mathbf{c} \in \mathbb{C}^4$ by $\rho(\mathbf{c}) := f_{\mathbf{c}}(\mathfrak{z}, \mathfrak{h})$; the definition of μ follows from it: $\mu := \rho \circ \mathbf{F}$. For $\mathbf{c} \in \mathbb{C}^4$ the Laurent polynomial $f_{\mathbf{c}}$ defines an algebraic curve, to wit the curve $f_{\mathbf{c}}(\mathfrak{z}, \mathfrak{h}) = 0$, which is called the *spectral curve*. Setting

$$v := 2c_3 c_4 \mathfrak{h} - \mathfrak{z}^6 + c_1 \mathfrak{z}^4 - c_2 \mathfrak{z}^2 + c_3 + c_4,$$

one easily computes that the spectral curve is birationally isomorphic to the affine algebraic curve $\Gamma_{\mathbf{c}}$, defined by equation $v^2 = g_{\mathbf{c}}(\mathfrak{z}^2)$, where

$$g_{\mathbf{c}}(\tau) := \tau(\tau^2 - c_1\tau + c_2)(\tau^3 - c_1\tau^2 + c_2\tau - 2(c_3 + c_4)) + (c_3 - c_4)^2.$$

From this equation it is clear that $\Gamma_{\mathbf{c}}$ is for generic $\mathbf{c}$ a smooth hyperelliptic curve of genus $\left[\frac{2 \deg g_{\mathbf{c}} - 1}{2}\right] = \deg g_{\mathbf{c}} - 1 = 5$. Denoting the smooth compactification of $\Gamma_{\mathbf{c}}$ by $\bar{\Gamma}_{\mathbf{c}}$ we have a ramified double cover $\pi : \bar{\Gamma}_{\mathbf{c}} \to \mathbb{P}^1$. It is also clear from the equation of $\Gamma_{\mathbf{c}}$ that $\bar{\Gamma}_{\mathbf{c}}$ has three different involutions: first there is the hyperelliptic involution, defined on $\Gamma_{\mathbf{c}}$ by $\iota(\mathfrak{z}, v) := (\mathfrak{z}, -v)$; a second involution is defined by $\sigma(\mathfrak{z}, v) := (-\mathfrak{z}, v)$; since ι and σ commute, a third involution is defined by their composition, $\tau(\mathfrak{z}, v) := (-\mathfrak{z}, -v)$. Setting

$\bar{\Gamma}_{\mathbf{c}}^{\sigma} := \bar{\Gamma}_{\mathbf{c}}/\sigma$ and $\bar{\Gamma}_{\mathbf{c}}^{\tau} := \bar{\Gamma}_{\mathbf{c}}/\tau$ the different curves can be represented by the following diagram:

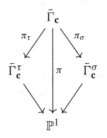

All maps in this diagram are double covers, with 12, 4, 0 ramification points for π, π_{τ} and π_{σ} respectively. It follows that $\bar{\Gamma}_{\mathbf{c}}^{\tau}$ has genus 2 and that $\bar{\Gamma}_{\mathbf{c}}^{\sigma}$ has genus 3 (and that $\bar{\Gamma}_{\mathbf{c}}$ has genus 5, but we know that already). An explicit equation for an affine part of the quotient curves $\bar{\Gamma}_{\mathbf{c}}^{\tau}$ and $\bar{\Gamma}_{\mathbf{c}}^{\sigma}$ is respectively given by

$$\Gamma_{\mathbf{c}}^{\tau} : v^2 = g_{\mathbf{c}}(u), \qquad \Gamma_{\mathbf{c}}^{\tau} : v^2 = u g_{\mathbf{c}}(u).$$

Of course these quotient curves are also hyperelliptic, with their hyperelliptic involution $(u, v) \mapsto (u, -v)$ induced by ι. The three involutions and the corresponding quotient curves play an important role in the description of the fibers of the momentum map $\mathbf{F}$ of the periodic 6-particle KM system:

Proposition 2.1 ([5]) *For generic* $\mathbf{c} \in \mathbb{C}^4$, *the fiber* $\mathbf{F}_{\mathbf{c}}$ *of the momentum map* $\mathbf{F} = (F_1, \dots, F_4) : \mathbb{C}^6 \to \mathbb{C}^4$, *with the* F_i *given by* (6), *is an affine part of*

$$\mathrm{Prym}(\bar{\Gamma}_{\mathbf{c}}/\Gamma_{\mathbf{c}}^{\sigma}) \simeq \mathrm{Jac}(\bar{\Gamma}_{\mathbf{c}}^{\tau}),$$

obtained by removing 6 translates of the theta divisor. Moreover, the vector fields $\mathcal{X}_{F_1}$ *and* $\mathcal{X}_{F_2}$ *are translation invariant on these tori.*

We denote the divisor consisting of the 6 translates of the theta divisor by $\mathcal{D}_{\mathbf{c}}$ and we denote the complete Abelian surface, which we may think of as a Prym variety or as a hyperelliptic Jacobian, by $\mathbf{T}_{\mathbf{c}}$. Thus, $\mathbf{F}_{\mathbf{c}} = \mathbf{T}_{\mathbf{c}} \setminus \mathcal{D}_{\mathbf{c}}$.

The genericity condition on $\mathbf{c}$ in Proposition 2.1 can be made precise: the statement of the proposition holds precisely for those $\mathbf{c} \in \mathbb{C}^4$ for which the affine curve $\Gamma_{\mathbf{c}}$ is smooth. In what follows we will not need this precise description: we will only use that for generic $\mathbf{c} \in \mathbb{C}^4$ the curves $\Gamma_{\mathbf{c}}$, $\Gamma_{\mathbf{c}}^{\tau}$ and $\Gamma_{\mathbf{c}}^{\sigma}$ are smooth and that Proposition 2.1 holds for such $\mathbf{c}$. Also, in view of Diagram (7) and Proposition 2.1, the fibers of the momentum map μ over a generic Laurent polynomial $f_{\mathbf{c}} \in \mathcal{H}_6$ consist of the disjoint union of the isomorphic fibers $\mathbf{F}_{\mathbf{c}}$ and $\mathbf{F}_{\mathbf{c}'}$, where $\mathbf{c}'$ is obtained from $\mathbf{c}$ by permuting c_3 and c_4; thus, it is sufficient to study the fibers of $\mathbf{F}$ and we will not consider the fibers of μ in what follows.

3 Laurent Solutions

In this section we determine all Laurent solutions of the periodic 6-particle KM vector field

$$\dot{x}_i = x_i(x_{i-1} - x_{i+1}), \qquad i = 1, \ldots, 6, \tag{8}$$

where we recall that the indices are taken modulo 6 (so that for example $x_7 = x_1$ and $x_0 = x_6$). A *Laurent solution* to (8) is a 6-tuple of convergent (for small $t \neq 0$) Laurent series

$$x_i(t) = \frac{1}{t^{r_i}} \sum_{j=0}^{\infty} a_i^{(j)} t^j, \quad r_i \in \mathbb{Z}, \quad i = 1, \ldots, 6,$$

which yield a formal solution to (8). As we will see, the coefficients of these series depend polynomially on several parameters, where the parameter space is an affine variety which is not irreducible. The Laurent solutions to (8) are therefore naturally organized in *irreducible* families, i.e., families of Laurent solutions, parametrized by an irreducible affine variety. An irreducible family parametrized by an affine variety of dimension $n - 1$ is called a *principal balance*; the other balances are called *lower balances*. According to the following theorem, known as the *Kowalevski-Painlevé Criterion*, every irreducible[2] a.c.i. system, such as the periodic 6-particle KM system, admits one or several principal balances.

Theorem 3.1 ([2, Theorem 6.13]) *Let* $(\mathbb{C}^n, \{\cdot, \cdot\}, \mathbf{F})$ *be an irreducible, polynomial a.c.i. system, where* $\mathbf{F} = (F_1, \ldots, F_s)$ *and let* $(x_1, \ldots, x_n)$ *be a system of linear coordinates on* $\mathbb{C}^n$. *Let* $\mathcal{X}$ *be any one of the integrable vector fields* $\mathcal{X}_{F_1}, \ldots, \mathcal{X}_{F_s}$. *For every* $1 \leqslant i \leqslant n$ *such that* x_i *is not constant along the integral curves of* $\mathcal{X}$ *there exists a principal balance* $x(t) = (x_1(t), \ldots, x_n(t))$ *for which* $x_i(t)$ *has a pole.*

Since the KM system is homogeneous, it is natural to look for *weight homogeneous* Laurent solutions of (8), i.e., Laurent solutions for which the pole order of $x_i(t)$ is at most the weight of the variable x_i, which is 1 for all variables x_i of the KM system (see Section 2; also, see [2, Section 7] for more information on (weight) homogeneous systems and Laurent solutions). We show in the following proposition by a simple argument that *all* Laurent solutions to the KM system are weight homogeneous.

---

[2] An a.c.i. system is said to be *irreducible* if for generic $\mathbf{c}$ the Abelian variety compactifying the fiber $\mathbf{F_c}$ of its momentum map $\mathbf{F}$ is a *simple* Abelian variety, i.e., it contains no proper Abelian subvarieties.

Proposition 3.2 *Let*

$$x_i(t) = \frac{1}{t^r} \sum_{j=0}^{\infty} a_i^{(j)} t^j, \qquad i = 1, \ldots, 6, \tag{9}$$

be a strict Laurent solution to the periodic n-particle KM system, where $a_i^{(0)} \neq 0$ for at least one index i. Then $r = 1$.

Proof. Suppose that a (9) is a *strict* Laurent solution, i.e., with $r \geqslant 1$ (otherwise it would be a Taylor solution). Notice that not all $a_i^{(0)}$ with i odd can be different from zero because the product $x_1 x_3 x_5$ is a constant of motion of (9), and so $x_1(t) x_3(t) x_5(t)$ is independent of t. Similarly, not all $a_i^{(0)}$ with i even can be different from zero. It follows that there exists an index i such that exactly one of $a_{i-1}^{(0)}$ and $a_{i+1}^{(0)}$ has a pole order r (and so the pole order of the other one is smaller). Since $x_i(t)$ is not identically zero (x_i is not a constant of motion), we may consider $\dot{x}_i(t)/x_i(t)$ which has at most a simple pole, but is in view of (8) equal to $x_{i-1}(t) - x_{i+1}(t)$, which has pole of order r. Hence $r = 1$ as was to be shown. $\qquad\square$

It follows that all Laurent solutions to (8) are weight homogeneous, can be algorithmically computed (see [2, Proposition 7.6]) and are convergent (see [2, Theorem 7.25]). Setting

$$x_i(t) = \frac{1}{t} \sum_{j=0}^{\infty} a_i^{(j)} t^j, \qquad i = 1, \ldots, 6, \tag{10}$$

one first solves the *indicial equations* which are the non-linear equations obtained by substituting the Laurent solutions (10) in (8) and equating the lowest order terms, i.e., the terms in t^{-2}. The result is the following system of quadratic equations:

$$-a_i^{(0)} = a_i^{(0)} \left(a_{i-1}^{(0)} - a_{i+1}^{(0)} \right), \qquad i = 1, \ldots, 6. \tag{11}$$

All non-zero solutions of (11) are easily found as follows. First recall from the proof of Proposition 3.2 that $a_i^{(0)} = 0$ for at least one odd and for at least one even value of i. We may therefore assume (by a cyclic permutation of the indices, if needed) that $a_1^{(0)} \neq 0$ and that $a_6^{(0)} = 0$. Then $a_3^{(0)} = 0$ or $a_5^{(0)} = 0$. When $a_3^{(0)} = 0$, the remaining equations in (11) lead to $a_1^{(0)} = -a_2^{(0)} = -1$, and

$$(a_5^{(0)} = 1 \text{ or } a_4^{(0)} = 0) \quad \text{and} \quad (a_4^{(0)} = -1 \text{ or } a_5^{(0)} = 0),$$

which leads to two solutions with $a_1^{(0)} \neq 0$ and $a_6^{(0)} = 0$, to wit $a^{(0)} = (-1, 1, 0, -1, 1, 0)$ and $a^{(0)} = (-1, 1, 0, 0, 0, 0)$. Similarly, when $a_5^{(0)} = 0$

we find a single new solution $a^{(0)} = (-2, 1, -1, 2, 0, 0)$. The upshot is that the indicial equations have 15 non-trivial solutions, to wit $(-1, 1, 0, 0, 0, 0)$, $(-1, 1, 0, -1, 1, 0)$, $(-2, 1, -1, 2, 0, 0)$ and their cyclic permutations (notice that the second solution has only three *different* cyclic permutations).

Having determined all possibilities for the leading coefficients of the Laurent series $x_i(t)$, we need to investigate the existence of the subsequent terms in the series as well as their dependence on free parameters. This has to be done seperately for each solution $a^{(0)}$ to the indicial equations. Thanks to the order 6 symmetry, we only need to consider the above three particular solutions. In each case, the subsequent terms $a_i^{(k)}$ are for $k = 1, 2, 3, \ldots$ determined by the following linear problem:

$$\left(k\mathrm{Id}_6 - \mathcal{K}(a^{(0)})\right) a^{(k)} = R^{(k)}, \tag{12}$$

where $\mathcal{K}(a^{(0)})$ is the Kowalevski matrix, evaluated at $a^{(0)}$ and $R^{(k)}$ is a column vector of polynomials which depends on the coefficients $a_1^{(j)}, \ldots, a_6^{(j)}$ with $0 \leqslant j < k$ only. Thanks to homogeneity, the Kowalevski matrix is just the sum of the Jacobian matrix of the right hand side of (8), evaluated at $a^{(0)}$, plus the identity matrix (see [2, Proposition 7.6] for the formula for the Kowalevski matrix in the general weight homogeneous case), namely $\mathcal{K}(a^{(0)})$ is given by

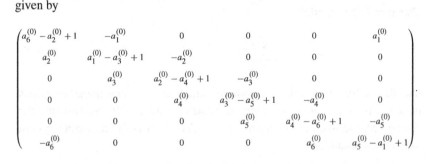

For most values of k, namely for those which do not belong to the spectrum of $\mathcal{K}(a^{(0)})$ the linear equation (12) has a unique solution. When k is an eigenvalue of $\mathcal{K}(a^{(0)})$ of multiplicity μ_k, we can get at most μ_k free parameters at step k; in fact, it may happen that for the solvability of (12) one needs to impose conditions on $R^{(k)}$, i.e. on the free parameters which have been introduced in the previous steps, or it may even happen that (12) has no solution at all, independently of the values of those free parameters, which means that there is no Laurent solution with $a^{(0)}$ as leading coefficients. We will see that in the present case, the number of free parameters at each step k is equal to the multiplicity of k as an eigenvalue of $\mathcal{K}(a^{(0)})$, for any of the values of $a^{(0)}$ that we have found.

In order to do this, we first compute the characteristic polynomial of $\mathcal{K}(a^{(0)})$ for the above three particular values of $a^{(0)}$. To start with, consider

$$\mathcal{K}(-1,1,0,0,0,0) = \begin{pmatrix} 0 & 1 & 0 & 0 & 0 & -1 \\ 1 & 0 & -1 & 0 & 0 & 0 \\ 0 & 0 & 2 & 0 & 0 & 0 \\ 0 & 0 & 0 & 1 & 0 & 0 \\ 0 & 0 & 0 & 0 & 1 & 0 \\ 0 & 0 & 0 & 0 & 0 & 2 \end{pmatrix}.$$

Thanks to the almost upper triangular form of this matrix, we obtain at once the following formula for its characteristic polynomial:

$$\chi(\mathcal{K}(-1,1,0,0,0,0),\lambda) = (\lambda+1)(\lambda-1)^3(\lambda-2)^2.$$

Similarly, one obtains

$$\chi(\mathcal{K}(-2,1,-1,2,0,0),\lambda) = (\lambda+2)(\lambda+1)(\lambda-1)(\lambda-2)(\lambda-3)^2,$$
$$\chi(\mathcal{K}(-1,1,0,-1,1,0),\lambda) = (\lambda+1)^2(\lambda-1)^2(\lambda-3)^2.$$

In the first case we have 5 positive eigenvalues, while there are only 4 positive eigenvalues in the two other cases. So only the first case can lead to principal balances; however we know that the system is a.c.i., hence it must have principal balances (Theorem 3.1), and so $a^{(0)} = (-1,1,0,0,0,0)$ leads — just like its cyclic permutations — to a principal balance. We exhibit the first few terms of it, for future use:

$$x_1(t) = -\frac{1}{t} + a - \frac{1}{3}(a^2 - 2d + e)t - \frac{1}{8}(8ad - be - 3cd)t^2 + \mathcal{O}(t^3),$$

$$x_2(t) = \frac{1}{t} + a + \frac{1}{3}(a^2 + d - 2e)t - \frac{1}{8}(8ae - 3be - cd)t^2 + \mathcal{O}(t^3),$$

$$x_3(t) = et + e(a-b)t^2 + \mathcal{O}(t^3),$$

$$x_4(t) = b - bct - \frac{b}{2}(bc - c^2 - e)t^2 + \mathcal{O}(t^3),$$

$$x_5(t) = c + bct + \frac{c}{2}(b^2 - bc + d)t^2 + \mathcal{O}(t^3),$$

$$x_6(t) = -dt + d(a-c)t^2 + \mathcal{O}(t^3). \tag{13}$$

In these formulas, $a, b, \ldots, e$ are the five free parameters; a, b and c appear at the first step, while c and d appear at the second step, in agreement with the eigenvalues of the Kowalevski matrix. The subsequent terms are completely determined by the displayed terms because 2 is the largest eigenvalue.

For $a^{(0)} = (-2,1,-1,2,0,0)$ we get a lower balance, depending on 4 free parameters a, b, c, d, appearing at steps 1, 2 and 3. For future use, we also give its first few terms:

$$x_1(t) = -\frac{2}{t} + 2a - 2bt - 2ct^2 + \mathcal{O}(t^3),$$

$$x_2(t) = \frac{1}{t} + a + (a^2 - 2b)t + (a^3 - 3ab - 3c + d)t^2 + \mathcal{O}(t^3),$$

$$x_3(t) = -\frac{1}{t} + a - (a^2 - 2b)t + (3ab - a^3 + 7c - 3d)t^2 + \mathcal{O}(t^3),$$

$$x_4(t) = \frac{2}{t} + 2a + 2bt + (12ab - 4a^3 + 18c - 8d)t^2 + \mathcal{O}(t^3),$$

$$x_5(t) = (4a^3 - 12ab - 20c + 9d)t^2 + \mathcal{O}(t^3),$$

$$x_6(t) = dt^2 + \mathcal{O}(t^3). \tag{14}$$

Finally, for $a^{(0)} = (-1, 1, 0, -1, 1, 0)$ we also get a lower balance, depending on 4 free parameters a, b, c, d, which appear at steps 1 and 3. Its first few terms are given by

$$x_1(t) = -\frac{1}{t} - b - \frac{b^2}{3}t + (c - 3d)t^2 + \mathcal{O}(t^3),$$

$$x_2(t) = \frac{1}{t} - b + \frac{b^2}{3}t - dt^2 + \mathcal{O}(t^3),$$

$$x_3(t) = ct^2 + \mathcal{O}(t^3),$$

$$x_4(t) = -\frac{1}{t} + a - \frac{a^2}{3}t - dt^2 + \mathcal{O}(t^3),$$

$$x_5(t) = \frac{1}{t} + a + \frac{a^2}{3}t + (c - 3d)t^2 + \mathcal{O}(t^3),$$

$$x_6(t) = (8d - 3c)t^2 + \mathcal{O}(t^3). \tag{15}$$

4 Embedding the Abelian Surfaces

We now construct a projective embedding of the Abelian surfaces $\mathbf{T_c}$ which compactify the generic fiber $\mathbf{F_c}$ of the momentum map of the periodic 6-particle KM system. To do this, we use the methods developed in [2], which we recall here and which we adapt to this system; there are some simplifications due to the fact that we know already that the latter system is an irreducible a.c.i. system.

First, recall that by definition every complex Abelian variety embeds in projective space and that such an embedding can be constructed by using the sections of a very ample line bundle on it; in the case of Abelian variety the third power of any ample line bundle suffices. In the present case, the generic fiber $\mathbf{F_c}$ of the momentum map is an affine part of a hyperelliptic Jacobian and the divisor to be adjoined to $\mathbf{F_c}$ to complete it into the torus $\mathbf{T_c}$ consists of 6 translates of the theta divisor, so it is very ample. We will therefore look for a basis of the sections of the line bundle defined by the divisor at infinity. Said differently, we look for meromorphic functions on the fiber, having at most a simple pole along the divisor at infinity. According to [2, Proposition 6.14] this can be done using the Laurent solutions: if we denote in the present case

Table 9.1. *The divisor of zeros and poles of the coordinate functions* $x_1, \ldots, x_6$, *restricted to the generic Abelian surface* $\mathbf{T_c}$.

	$\mathcal{D}_c^1$	$\mathcal{D}_c^2$	$\mathcal{D}_c^3$	$\mathcal{D}_c^4$	$\mathcal{D}_c^5$	$\mathcal{D}_c^6$
x_1	-1	1	0	0	1	-1
x_2	-1	-1	1	0	0	1
x_3	1	-1	-1	1	0	0
x_4	0	1	-1	-1	1	0
x_5	0	0	1	-1	-1	1
x_6	1	0	0	1	-1	-1

by $x(t; \mathcal{D}^i)$ the family of Laurent solutions corresponding to a Painlevé wall[3] $\mathcal{D}^i$ and f is a rational function of $x_1, \ldots, x_6$ then the pole order (in t) of $f(x(t; \mathcal{D}^i))$ equals, for generic $\mathbf{c} \in \mathbb{C}^4$, the pole order of f, viewed as a meromorphic function on $\mathbf{T_c}$, along $\mathcal{D}_c^i$. For example, it suffices to look at the pole orders of the Laurent series (13) to determine the divisor of zeros and poles of the coordinate functions $x_1, \ldots, x_6$, restricted to the generic Abelian surface $\mathbf{T_c}$. The result is displayed in Table 9.1.

In it, the labelings of the Painlevé walls $\mathcal{D}^i$ are chosen such that $x(t; \mathcal{D}^1)$ is the principal balance (13) and the other labelings are obtained by a cyclic permutation of the variables, i.e., $x_i(t, \mathcal{D}^2) = x_{i-1}(t, \mathcal{D}^1)$, and so on. In a single formula, the table can be summarized by

$$(x_i)|_{\mathbf{T_c}} = \mathcal{D}_c^{i+1} - \mathcal{D}_c^i - \mathcal{D}_c^{i-1} + \mathcal{D}_c^{i-2}.$$

From this formula we can for example conclude that the divisors $\mathcal{D}_c^i + \mathcal{D}_c^{i-1}$ and $\mathcal{D}_c^{i+1} + \mathcal{D}_c^{i-2}$ are linearly equivalent for all i.

Of course, the coordinate functions $x_1, \ldots, x_6$ and the constant function 1 are the first elements of the polynomial functions we are looking for. In order to find the other such polynomials f, one proceeds by the degree d of f, looking for the most general polynomial of degree d such that $f(x(t; \mathcal{D}^i))$ has at most a simple pole (in t) for all i. Notice that we can take f to be homogeneous, because when f has the desired property, then by homogeneity every homogeneous component of f will also have this property. A more delicate issue is that we will find some polynomials which are dependent on the previously found polynomials, when restricted to the tori; in fact, if one multiplies any polynomial with the desired property with a constant of motion, the product will still have the desired property, without leading to

---

[3] Roughly speaking, the Painlevé wall $\mathcal{D}^i$ is the collection of Painlevé divisors $\mathcal{D}_c^i$, with $\mathbf{c} \in \mathbb{C}^4$ generic; see [2, Chapter 6] for a precise description of $\mathcal{D}^i$ as a divisor on a partial compactification of phase space.

Table 9.2. *Computing a basis for the polynomials of degree at most 6 which have a simple pole at most when any principal balance $x(t)$ is substituted in them.*

k	dim $\mathcal{F}^{(k)}$	dim $\mathcal{H}^{(k)}$	dim $\mathcal{Z}^{(k)}$	# dep	ζ_k	indep. functions
0	1	1	1	0	1	z_0
1	6	1	6	1	5	$z_1, \ldots, z_5$
2	21	2	15	7	8	$z_6, \ldots, z_{13}$
3	56	4	32	22	10	$z_{14}, \ldots, z_{23}$
4	126	5	57	51	6	$z_{24}, \ldots, z_{29}$
5	252	7	96	90	6	$z_{30}, \ldots, z_{35}$
6	462	11	144	144	0	—

a new meromorphic function, when restricted to the tori. The results of the process are summarized in Table 9.2: In the table are displayed, for small k, the following data, corresponding to the different columns (in that order).

(1) dim $\mathcal{F}^{(k)}$, the number of linearly independent monomials of degree k; it is computed from the formula dim $\mathcal{F}^{(k)} = \binom{k+5}{5}$;

(2) dim $\mathcal{H}^{(k)}$, the number of linearly independent constants of motion of degree k; it is the coefficient in t^k of $((1-t)(1-t^2)(1-t^3)^2)^{-1}$;

(3) dim $\mathcal{Z}^{(k)}$, the number of linearly independent polynomials having a simple pole at most when the principal balances are substituted in them; for computing this, a computer program is very useful;

(4) The number of linearly independent elements in $\mathcal{Z}^{(k)}$ that are dependent of the previous ones over $\mathcal{H}$. This is number is computed from the previous data by the formula $\sum_{j=0}^{i-1} \zeta_j$ dim $\mathcal{H}^{(i-j)}$;

(5) ζ_k, the number of linearly independent elements in $\mathcal{Z}^{(k)}$ that are independent of the previous ones over $\mathcal{H}$; it is computed as the difference of the two previous columns;

(6) The last column gives a choice of these new functions; their explicit expressions are given below.

We now list the functions z_i and explain why we need not look at polynomials of higher degree. In degree zero we have the constant function $z_0 := 1$; in degree one all coordinate functions $x_1, \ldots, x_6$ have a simple pole at most when the principal balances are substituted in them, but they are not independent over the Hamiltonians, since their sum is the Hamiltonian H_1. So we set $z_i := x_i$ for $i = 1, \ldots, 5$. The 8 quadratic polynomials are given by

$$z_6 := x_1 x_4, \quad z_7 := x_2 x_5, \quad z_8 := x_3 x_6, \quad z_9 := x_1 x_3,$$
$$z_{10} := x_2 x_4, \quad z_{11} := x_3 x_5, \quad z_{12} := x_4 x_6, \quad z_{13} := x_1 x_5,$$

while the 10 cubic polynomials are given by

$$z_{14} := x_1 x_2 x_3, \qquad z_{15} := x_2 x_3 x_4, \qquad z_{16} := x_3 x_4 x_5,$$
$$z_{17} := x_4 x_5 x_6, \qquad z_{18} := x_1 x_5 x_6, \qquad z_{19} := x_1 x_2 x_6,$$
$$z_{20} := x_1 x_3 (x_1 + x_6), \qquad z_{21} := x_2 x_4 (x_1 + x_2),$$
$$z_{22} := x_3 x_5 (x_3 + x_2), \qquad z_{23} := x_4 x_6 (x_4 + x_3).$$

Next follow the 6 quartic polynomials,

$$z_{24} := x_1 x_2 x_3 x_4, \quad z_{25} := x_2 x_3 x_4 x_5, \quad z_{26} := x_3 x_4 x_5 x_6,$$
$$z_{27} := x_1 x_4 x_5 x_6, \quad z_{28} := x_1 x_2 x_5 x_6, \quad z_{29} := x_1 x_2 x_3 x_6,$$

and the 6 quintic polynomials

$$z_{30} := x_1^2 x_2 x_3^2, \quad z_{31} := x_2^2 x_3 x_4^2, \quad z_{32} := x_3^2 x_4 x_5^2,$$
$$z_{33} := x_4^2 x_5 x_6^2, \quad z_{34} := x_5^2 x_6 x_1^2, \quad z_{35} := x_1 x_2 x_3 x_4 (x_1 + x_2).$$

Notice that these polynomials are all either monomials or binomials, which makes it very easy to determine their leading behaviour; in fact, for the 30 monomials it suffices to look at Table 9.1 to verify that along any of the six curves $\mathcal{D}_{\mathbf{c}}^i$ their pole order is at most one!

On the generic torus $\mathbf{T_c}$, which is a Jacobian surface, we have 36 independent functions with a simple pole at most along the divisor at infinity. Since this divisors consists of 6 translates of the theta divisor, it defines a polarization of type $(6, 6)$ on the Abelian surface, and so $\dim H^0(\mathbf{T_c}, \mathcal{D}_{\mathbf{c}}) = 6^2 = 36$. Therefore, we have constructed a basis of this space and the given function provide an embedding of the generic torus $\mathbf{T_c}$ in $\mathbb{P}^{35}$. We will use this embedding to determine the intersection pattern of these 6 theta translates.

5　The Spectral and Painlevé Curves

Recall that for generic $\mathbf{c} \in \mathbb{C}^4$ we denote by $\mathbf{F_c}$ the fiber $\mathbf{F}^{-1}(c)$ of the momentum map $\mathbf{F} : \mathbb{C}^6 \to \mathbb{C}^4$, by $\mathbf{T_c}$ its completion into an Abelian surface and by $\mathcal{D}_{\mathbf{c}}$ the divisor of $\mathbf{T_c}$ which has been added to do this completion. As we have seen, the divisor $\mathcal{D}_{\mathbf{c}}$ has six irreducible components, which are the Painlevé curves $\mathcal{D}_{\mathbf{c}}^i$, $i = 1, \ldots, 6$. These 6 curves are isomorphic, so in order to compute an affine equation for the Painlevé curves, it suffices to compute the equation for one of them; yet, as we will see in the next section, some care has to be taken when considering the 6 different projective embeddings of these curves, rather than the isomorphism class which they define.

In order to compute an affine equation for one of the $\mathcal{D}_{\mathbf{c}}^i$, one fixes $\mathbf{c} = (c_1, \ldots, c_4)$ and substitutes the principal balance (13) of (8) in the equations $H_j = c_j$, for $j = 1, \ldots, 4$. Since H_j is a constant of motion, $H_j(x(t))$ is independent of t, hence depends on the free parameters $a, \ldots, e$

only. We therefore get 4 polynomial equations in these parameters, and they give equations for an affine part of any one of the Painlevé curves. Notice that due to the simple form of the Hamiltonians, we only need to substitute the first two terms of the Laurent series (13) in $H_1 = c_1$, the first three terms in $H_2 = c_2$ and the leading terms in $H_3 = c_3$ and in $H_4 = c_4$. One obtains the following equations:

$$2a + b + c = c_1,$$
$$2a(b + c) - d - e = c_2,$$
$$-ce = c_3,$$
$$-bd = c_4. \qquad (16)$$

This curve is called the *(abstract) Painlevé curve*; we denote it by $\Delta_{\mathbf{c}}$. Solving the first and last two equations linearly for a, b and c and substituting the results in the second equation yields, after clearing the denominator, the following equation for a curve, birationally equivalent to $\Delta_{\mathbf{c}}$:

$$d^2 e^2 (e + d + c_2) + (c_3 d + c_4 e)(c_1 d e + c_3 d + c_4 e) = 0. \qquad (17)$$

We show that this curve is also birationally equivalent to the curve $\Gamma_{\mathbf{c}}^\tau$, which we constructed as a quotient of the spectral curve $\Gamma_{\mathbf{c}}$. Recall from Section 2 that an equation for $\Gamma_{\mathbf{c}}^\tau$ is given by

$$v^2 = u(u^2 - c_1 u + c_2)(u^3 - c_1 u^2 + c_2 u - 2(c_3 + c_4)) + (c_3 - c_4)^2. \qquad (18)$$

The birational map is given by

$$u = -\frac{c_3}{e} - \frac{c_4}{d},$$
$$v = -\frac{d^2 e(d + e + c_2) + c_1 d(c_3 d + c_4 e) + c_4(2c_3 d + (c_3 + c_4)e)}{dc_3},$$

with inverse map

$$d, e = -\frac{1}{2}\left(u^2 - c_1 u + c_2 \pm \frac{v - c_3 + c_4}{u}\right),$$

where the plus sign corresponds to d and the minus sign to e. These formulas are easily checked by direct computation; to find the above map, one can for example write (17) in Weierstraß form and then rescale the variables so as to make the equation match with the Weierstraß form (18) of $\Gamma_{\mathbf{c}}^\tau$. The fact that the compactified Painlevé curve corresponding to $\mathbf{c}$ is isomorphic to $\bar{\Gamma}_{\mathbf{c}}^\tau$ is not surprizing since on the one hand the Painlevé curve is a divisor of the torus $\mathbf{T}_{\mathbf{c}}$ and on the other hand $\mathbf{T}_{\mathbf{c}}$ is isomorphic to the Jacobian of $\bar{\Gamma}_{\mathbf{c}}^\tau$.

We will need in the next section the points at infinity of (16), i.e., the points needed to complete the affine curve defined by (16) into a compact Riemann surface. We will in fact need a local parametrization around each of these points. It is important that we do this with the representation of the curve in terms of the parameters which appear in the Laurent series, rather than using some (possibly simpler) birational model, such as (17), because the embedding of $\mathbf{T_c}$ was constructed by using the Laurent solutions, and so the corresponding embeddings of the curve which we will construct will also be expressed in terms of these parameters. In order to find these parametrizations, it suffices to first observe that $bcde \neq 0$ for any affine point (recall that $\mathbf{c}$ is generic), so that for the points at infinity at least one of the parameters b, c, d, e must be zero; also notice that b and d cannot vanish at the same time, and similarly for c and e. In fact, out of c and e exactly one has to vanish, and similarly for b and d. For each of the four possibilities we find a single parametrization, except when d and e vanish, in which case we find two parametrizations. Thus we have five points at infinity. Local parametrizations around these points are given by the following list (we only give the parametrization for two of the variables; one easily derives from them parametrizations for the other variables by using (16)):

$$\infty_1 : \quad e = -c_3\tau, \quad d = c_4\tau(1 + \beta\tau) + \mathcal{O}(\tau^3),$$

$$\infty_2 : \quad e = -c_3\tau, \quad d = c_4\tau(1 + (c_1 - \beta)\tau) + \mathcal{O}(\tau^3),$$

$$\infty_3 : \quad b = c_4\tau, \quad c = -c_3\tau(1 + c_2\tau + (c_1c_3 - c_1c_4 + c_2^2)\tau^2) + \mathcal{O}(\tau^4),$$

$$\infty_4 : \quad d = c_4\tau, \quad c = c_3\tau^2(1 - c_1\tau) + \mathcal{O}(\tau^4),$$

$$\infty_5 : \quad e = c_3\tau, \quad b = c_4\tau^2(1 - c_1\tau) + \mathcal{O}(\tau^4).$$

In the first two formulas, β stands for the same root of the quadratic polynomial $\beta^2 - c_1\beta + c_2$; picking the other root just amounts to permuting the two points ∞_1 and ∞_2.

It is also useful to restrict the lower balances to the generic fibers $\mathbf{F_c}$. When doing so, one gets explicit formulas for the four free parameters in terms of the 4 constants $c_1, \ldots, c_4$, so that the Laurent solutions can be entirely expressed in terms of the latter constants. For the lower balances (14), the resulting Laurent solutions are given by

$$x_1(t) = -\frac{2}{t} + \frac{c_1}{3} - \frac{1}{18}(c_1^2 - 3c_2)t$$

$$+ \frac{1}{540}(2c_1^3 - 9c_1c_2 + 27c_3 - 243c_4)t^2 + \mathcal{O}(t^3),$$

$$x_2(t) = \frac{1}{t} + \frac{c_1}{6} - \frac{1}{36}(c_1^2 - 6c_2)t$$

$$- \frac{1}{1080}(4c_1^3 - 18c_1c_2 - 81c_3 + 189c_4)t^2 + \mathcal{O}(t^3),$$

$$x_3(t) = -\frac{1}{t} + \frac{c_1}{6} + \frac{1}{36}(c_1^2 - 6c_2)t$$
$$- \frac{1}{1080}(4c_1^3 - 18c_1c_2 + 189c_3 - 81c_4)t^2 + \mathcal{O}(t^3),$$

$$x_4(t) = \frac{2}{t} + \frac{c_1}{3} + \frac{1}{18}(c_1^2 - 3c_2)t$$
$$+ \frac{1}{540}(2c_1^3 - 9c_1c_2 - 243c_3 + 27c_4)t^2 + \mathcal{O}(t^3),$$

$$x_5(t) = \frac{c_3}{2}t^2 + \mathcal{O}(t^3),$$

$$x_6(t) = \frac{c_4}{2}t^2 + \mathcal{O}(t^3),$$

while for the lower balances (15) they are given by

$$x_1(t) = -\frac{1}{t} + \frac{\beta}{2} - \frac{\beta^2}{12}t + \frac{1}{8}(3c_3 + c_4)t^2 + \mathcal{O}(t^3),$$

$$x_2(t) = \frac{1}{t} + \frac{\beta}{2} + \frac{\beta^2}{12}t + \frac{1}{8}(c_3 + 3c_4)t^2 + \mathcal{O}(t^3),$$

$$x_3(t) = -c_4t^2 + \mathcal{O}(t^3),$$

$$x_4(t) = -\frac{1}{t} - \frac{1}{2}(\beta - c_1) - \frac{1}{12}(\beta - c_1)^2t + \frac{1}{8}(c_3 + 3c_4)t^2 + \mathcal{O}(t^3),$$

$$x_5(t) = \frac{1}{t} - \frac{1}{2}(\beta - c_1) + \frac{1}{12}(\beta - c_1)^2t + \frac{1}{8}(3c_3 + c_4)t^2 + \mathcal{O}(t^3),$$

$$x_6(t) = -c_3t^2 + \mathcal{O}(t^3),$$

where β is any root of he quadratic polynomial $\beta^2 - c_1\beta + c_2$. Notice that this means that, restricted to the generic fiber $\mathbf{F_c}$ we do not have just three but six of the latter lower balances. This will be reflected in the geometry of the divisor at infinity.

In order to obtain from these formulas the formulas for all lower balances one uses the order six automorphism, but one should not forget that it permutes also the constants c_3 and c_4.

6 The Configuration of Painlevé Curves

In this section we use the embedding of the Abelian surfaces $\mathbf{T_c}$ in $\mathbb{P}^{35}$ to construct six projective embeddings of the smooth Painlevé curve $\Delta_{\mathbf{c}}$, which we recall is birationally isomorphic to the smooth genus 2 curve $\Gamma_{\mathbf{c}}^\tau$. We will then be able to determine the intersection pattern of the 6 completed image curves which make up the Painlevé divisor $\mathcal{D}_{\mathbf{c}}$ of $\mathbf{T_c}$. To do this, we first substitute the principal balance (13) in the embedding functions $z_0, \ldots, z_{35}$, constructed in Section 4, which gives an embedding of a neighborhood in $\mathbf{T_c}$ of an affine part of the embedded curve, times a neighborhood of 0, corresponding to time t (the parameter of the integral curves of the vector field $\mathcal{X}_H$). Setting $t = 0$ in this embedding yields an embedding of an affine part of the curve $\mathcal{D}_{\mathbf{c}}^1$. Notice that the components of this embedding are just the residues of the Laurent series $z_0(t), \ldots, z_{35}(t)$ since all these series have a simple pole at worst for $t = 0$. Writing P as a shorthand for (a, b, c, d, e), the resulting map, which we denote by γ_1, is given by

$$\gamma_1(P) = (0 : -1 : 1 : 0_3 : -b : c : 0_2 : b : 0_2 : -c : -e : 0_4 : d : e :$$
$$2ab : 0_2 : -be : 0_3 : cd : 0 : e^2 : b^2e : 0_2 : -c^2d : -2abe).$$

We have used the convenient notation 0_i to denote that i successive coordinates are zero. Notice that γ_1 is clearly injective, and so is indeed an embedding of the affine curve. Similarly, the embedding γ_i of $\mathcal{D}_{\mathbf{c}}^i$ is found by substituting the corresponding principal balance in the embedding functions $z_0, \ldots, z_{35}$; to determine this principal balance, it suffices to do a cyclic permutation in (13) of the indices of the variables x_j, just replacing x_1 by x_i and so on. The five other embeddings that one obtains are given by

$$\gamma_2(P) = (0_2 : -1 : 1 : 0_3 : -b : c : 0_2 : b : 0_2 : d : -e : 0_5 : e :$$
$$2ab : 0_2 : -be : 0_3 : cd : -d^2 : e^2 : b^2e : 0_2 : -ed),$$

$$\gamma_3(P) = (0_3 : -1 : 1 : 0 : c : 0 : -b : -c : 0_2 : b : 0_2 : d : -e : 0_3 : -c(b+c) :$$
$$0 : e : 2ab : cd : 0 : -be : 0_3 : -c^2d : -d^2 : e^2 : b^2e : 0 : c^2d),$$

$$\gamma_4(P) = (0_4 : -1 : 1 : -b : c : 0_2 : -c : 0_2 : b : 0_2 : d : -e : 0_3 : -c(c+b) :$$
$$0 : e : 0 : cd : 0 : -be : 0_3 : -c^2d : -d^2 : e^2 : b^2e : 0),$$

$$\gamma_5(P) = (0_5 : -1 : 0 : -b : c : 0_2 : -c : 0_5 : d : -e : 0_3 : -c(b+c) :$$
$$0_3 : cd : 0 : -be : 0_3 : -c^2d : -d^2 : e^2 : 0),$$

$$\gamma_6(P) = (0 : 1 : 0_4 : c : 0 : -b : b : 0_2 : -c : 0_5 : d : -e : 2ab : 0_2 :$$
$$-c(c+b) : 0_3 : cd : 0 : -be : b^2e : 0_2 : -c^2d : -d^2 : bce).$$

It is easy to see that the different images of these embeddings do not intersect. For example, $\text{Im}(\gamma_1)$ and $\text{Im}(\gamma_2)$ cannot intersect because all points in $\text{Im}(\gamma_1)$ have their second coordinate different from zero, while that coordinate vanishes for all points of $\text{Im}(\gamma_2)$.

Since the polynomials $z_0, \ldots, z_{35}$ provide (upon restriction) an embedding of $\mathbf{T_c}$, the embeddings $\gamma_1, \ldots, \gamma_6$ of the affine curve can be holomorphically extended to its smooth compactification; as we will see, the extension is an embedding of the complete curve, so that the six image curves are non-singular, but these image curves will intersect in several points according to a pattern which we will determine.

To do this, recall from Section 5 that we have determined parametrizations of a neighborhood of each one of points at infinity $\infty_1, \dots, \infty_5$ of the Painlevé curves $\Delta_{\mathbf{c}}$. If we substitute the parametrization of one of these points ∞_i in either one of the embeddings γ_j we get an embedding of a punctured neighborhood of ∞_i in $\mathbb{P}^{35}$ and it suffices to let the parameter τ of the parametrization tend to 0 to find the image point in $\mathbb{P}^{35}$. Doing this for the embedding γ_1 we find 5 *different* points, which confirms that the six irreducible components of $\mathcal{D}_{\mathbf{c}}$ are non-singular curves (of genus 2), isomorphic to $\Gamma_{\mathbf{c}}$. Namely, we find the following images:

$$\infty_1 \mapsto (0_6 : 1 : 1 : 0_2 : -1 : 0_2 : -1 : 0_7 : \beta - c_1 : 0_9 : -c_3 : 0_2 : -c_4 : 0)$$

$$\infty_2 \mapsto (0_6 : 1 : 1 : 0_2 : -1 : 0_2 : -1 : 0_7 : -\beta : 0_9 : -c_3 : 0_2 : -c_4 : 0)$$

$$\infty_3 \mapsto (0_{30} : 1 : 0_5)$$

$$\infty_4 \mapsto (0_{30} : 1 : -1 : 0_3 : -1)$$

$$\infty_5 \mapsto (0_{34} : 1 : 0)$$

When the other embeddings γ_j are used we find in total 30 image points, but they are not all different, as some appear twice and the others three times. In total we find 12 different image points, as indicated in Table 9.3. The coordinates of the 6 points of the form P_{ij} with $j - i \in \{2, 4\}$ are given by

$$P_{13} = (0_{30} : 1 : -1 : 0_3 : -1), \qquad P_{24} = (0_{31} : 1 : -1 : 0_3),$$
$$P_{15} = (0_{34} : 1 : 0), \qquad P_{26} = (0_{30} : 1 : 0_5),$$
$$P_{35} = (0_{32} : 1 : -1 : 0_2), \qquad P_{46} = (0_{33} : 1 : -1 : 0),$$

Table 9.3. *The irreducible components $\mathcal{D}_{\mathbf{c}}^i$ of the Painlevé divisor intersect in two points which may coincide, in which case the two components are tangent.*

	$\mathcal{D}_{\mathbf{c}}^1$	$\mathcal{D}_{\mathbf{c}}^2$	$\mathcal{D}_{\mathbf{c}}^3$	$\mathcal{D}_{\mathbf{c}}^4$	$\mathcal{D}_{\mathbf{c}}^5$	$\mathcal{D}_{\mathbf{c}}^6$
$\mathcal{D}_{\mathbf{c}}^1$	$=$	P_{26}, P_{13}	P_{13}	P_{14}, P_{14}'	P_{15}	P_{26}, P_{15}
$\mathcal{D}_{\mathbf{c}}^2$	P_{26}, P_{13}	$=$	P_{13}, P_{24}	P_{24}	P_{25}, P_{25}'	P_{26}
$\mathcal{D}_{\mathbf{c}}^3$	P_{13}	P_{13}, P_{24}	$=$	P_{24}, P_{35}	P_{35}	P_{36}, P_{36}'
$\mathcal{D}_{\mathbf{c}}^4$	P_{14}, P_{14}'	P_{24}	P_{24}, P_{35}	$=$	P_{35}, P_{46}	P_{46}
$\mathcal{D}_{\mathbf{c}}^5$	P_{15}	P_{25}, P_{25}'	P_{35}	P_{35}, P_{46}	$=$	P_{46}, P_{15}
$\mathcal{D}_{\mathbf{c}}^6$	P_{26}, P_{15}	P_{26}	P_{36}, P_{36}'	P_{46}	P_{46}, P_{15}	$=$

while the points $P_{i,i+3}$ have as coordinates

$$P_{14} = (0_6 : 1 : 1 : 0_2 : -1 : 0_2 : -1 : 0_7 : -\beta : 0_9 : -c_3 : 0_2 : -c_4 : 0),$$

$$P_{25} = (0_7 : 1 : 1 : 0_2 : -1 : 0_{10} : -\beta : 0_9 : -c_4 : 0_3),$$

$$P_{36} = (0_6 : 1 : 0 : 1 : -1 : 0_2 : -1 : 0_7 : -\beta : 0_2 : \beta - c_1 :$$
$$0_6 : -c_4 : 0_2 : -c_3 : 0 : c_4).$$

For $i = 1, \ldots, 3$, the coordinates of the point $P'_{i,i+3}$ are obtained by replacing β by $c_1 - \beta$ in the coordinates of the point $P_{i,i+3}$ (i.e., replace β, which is a root of the quadratic polynomial $\beta^2 - c_1\beta + c_2$, by the other root). These 12 points are also obtained when substituting the 12 lower balances in the embedding. Namely, the points P_{ij} with $j - i \neq 3$ are obtained by substituting the 6 cyclic permutations of (14) in the embedding, the points $P_{i,i+3}$ are obtained similarly by using the 3 cyclic permutations of (15) while the points $P'_{i,i+3}$ are obtained as the points $P_{i,i+3}$, but with β replaced by the other root $c_1 - \beta$ of the polynomial $\beta^2 - c_1\beta + c_2$.

It is clear that we have chosen the notations as follows: when $j - i \in \{2, 4\}$ then the point P_{ij} is the unique intersection point of $\mathcal{D}_c^i$ and $\mathcal{D}_c^j$, so the curves $\mathcal{D}_c^i$ and $\mathcal{D}_c^j$ are tangent at P_{ij}; another component passes transversally through this point, namely $\mathcal{D}_c^{i+1}$ in case $j = i + 2$ and $\mathcal{D}_c^{j+1}$ when $j = i + 4$. Also, the points $P_{i,i+3}$ and $P'_{i,i+3}$ are the unique intersection points of $\mathcal{D}_c^i$ and $\mathcal{D}_c^{i+3}$ and no other curve of the divisor passes through them. With this notation, $\mathcal{D}_c^i$ contains the points $P_{i,i+2}, P_{i-2,i}, P_{i-1,i+1}, P_{i,i+3}$ and $P'_{i,i+3}$.

Though the table contains all information on the intersection pattern of the curves $\mathcal{D}_c^i$ a few pictures may help to visualize this pattern. First, Figure 9.1 shows the intersection pattern of $\mathcal{D}_c^1$ with a neighbor, a second nearest neighbor and its furthest neighbor.

Secondly, we display the intersection pattern of $\mathcal{D}_c^1$ with two other curves (see Figure 9.2). There are three essentially different possibilities, according to whether the configuration contains three, two or no consecutive curves.

It is also instructive to picture one single $\mathcal{D}_c$, say $\mathcal{D}_c^1$, with its 5 points at infinity, as well as arcs of the other $\mathcal{D}_c^i$ passing through them (see Figure 9.3).

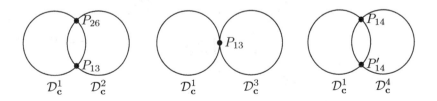

Figure 9.1 Each Painlevé curve $\mathcal{D}^i$ is tangent to its second nearest neighbors and intersects the other Painlevé curves in two points.

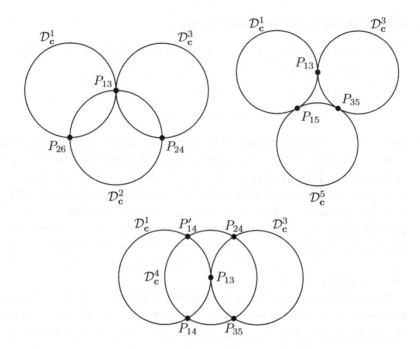

Figure 9.2 For three Painlevé curves there are three possible intersection patterns.

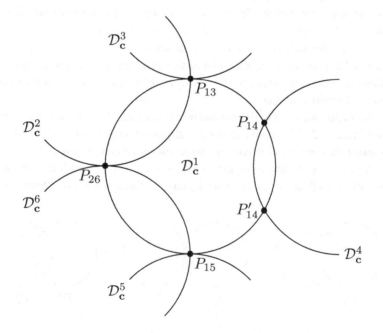

Figure 9.3 The intersection pattern of one of the Painlevé curves with all the other Painlevé curves.

We compare this configuration to a similar configuration of the divisor needed for another natural — but singular — compactification of the generic fiber $\mathbf{F_c}$ of the momentum map $\mathbf{F} : \mathbb{C}^6 \to \mathbb{C}^4$. It is obtained by introducing an extra variable x_0 and making the affine equations for $\mathbf{F_c}$ homogeneous, i.e., to consider the compact surface $\bar{\mathbf{F}}_{\mathbf{c}} \subset \mathbb{P}^6$, given by

$$x_1 + x_2 + x_3 + x_4 + x_5 + x_6 = c_1 x_0,$$
$$x_1 x_4 + x_2 x_5 + x_3 x_6 + x_1 x_3 + x_2 x_4 + x_3 x_5 + x_4 x_6 + x_1 x_5 + x_2 x_6 = c_2 x_0^2,$$
$$x_1 x_3 x_5 = c_3 x_0^3,$$
$$x_2 x_4 x_6 = c_4 x_0^3. \tag{19}$$

Since $\mathbf{c}$ is generic in the sense that the affine part $\mathbf{F_c}$ of $\bar{\mathbf{F}}_{\mathbf{c}}$ is non-singular, all singularities of $\mathbf{F_c}$ are at infinity, i.e., are contained in the divisor $\mathcal{C} := \bar{\mathbf{F}}_{\mathbf{c}} \setminus \mathbf{F_c}$. Equations for $\mathcal{C}$ are obtained by intersecting $\bar{\mathbf{F}}_{\mathbf{c}}$ with the hyperplane $x_0 = 0$, giving the following equations (they are independent of $\mathbf{c}$, which is the reason why we do not add an index $\mathbf{c}$ to $\mathcal{C}$):

$$x_1 + x_2 + x_3 + x_4 + x_5 + x_6 = 0,$$
$$x_1 x_4 + x_2 x_5 + x_3 x_6 + x_1 x_3 + x_2 x_4 + x_3 x_5 + x_4 x_6 + x_1 x_5 + x_2 x_6 = 0,$$
$$x_1 x_3 x_5 = 0,$$
$$x_2 x_4 x_6 = 0. \tag{20}$$

Starting from the last two equations, it is clear how to determine the irreducible components of the divisor $\mathcal{C}$: we need to pick an odd index i and an even index j and set $x_i = x_j = 0$ in the other two equations which are easily rewritten as a single quadratic equation in five variables, so they are conics. Since each choice of i and j leads to a different conic, we get 9 conics in total. For example, setting $x_1 = x_2 = 0$ we get the following non-singular conic:

$$\mathcal{C}^1 : \begin{cases} x_3 + x_4 + x_5 + x_6 = 0, \\ x_3 x_6 + x_3 x_5 + x_4 x_6 = 0. \end{cases} \tag{21}$$

In view of the order 6 automorphism there are six such conics, which we denote by $\mathcal{C}^i$, with $i = 1, \ldots, 6$, where $\mathcal{C}^i$ is the conic contained in the subspace $x_i = x_{i+1} = 0$. The three remaining conics are obtained by setting $x_i = x_{i+3} = 0$. For example, setting $x_3 = x_6 = 0$ we get the following conic:

$$\mathcal{L}^1 : \begin{cases} x_1 + x_2 + x_4 + x_5 = 0, \\ (x_1 + x_2)(x_4 + x_5) = 0. \end{cases}$$

The conic is degenerate, consisting of the double line $x_1 + x_2 = x_4 + x_5 = 0$. Two other such lines are obtained by using the order 6 automorphism. They are denoted by $\mathcal{L}^i$ where $\mathcal{L}^i$ is the (double) line contained in the subspace $x_{i+2} = x_{i-1} = 0$.

The singularities of $\mathbf{F_c}$ which are contained in $\mathcal{C}$ are the points $(0 : x_1 : x_2 : \cdots : x_6)$ where the rank of the following Jacobian matrix is at most 3:

$$\begin{pmatrix} -c_1 & 1 & 1 & 1 & 1 & 1 & 1 \\ 0 & x_3+x_4+x_5 & x_4+x_5+x_6 & x_1+x_5+x_6 & x_1+x_2+x_6 & x_1+x_2+x_3 & x_2+x_3+x_4 \\ 0 & x_3x_5 & 0 & x_1x_5 & 0 & x_1x_3 & 0 \\ 0 & 0 & x_4x_6 & 0 & x_2x_6 & 0 & x_2x_4 \end{pmatrix}.$$

Consider the following parametrization of $\mathcal{L}^1$:

$$(u : v) \mapsto (0 : u : -u : 0 : v : -v : 0) \tag{22}$$

and substitute it in the Jacobian matrix, to see that except for two columns, all columns are a multiple of the first column, and so the rank is at most 3 (it is in fact 3 at all points where $uv \neq 0$; the rank drops to 2 at the points where $u = 0$ or $v = 0$). By symmetry, the same holds true for the lines $\mathcal{L}^2$ and $\mathcal{L}^3$. The suface $\mathbf{F_c}$ is therefore singular at all points of the three lines $\mathcal{L}^1, \ldots, \mathcal{L}^3$. With some extra work it can be shown that $\mathbf{F_c}$ has for generic $\mathbf{c}$ no other singularities.

Because of the simple equations for the conics and lines it is easy to determine how they intersect. It is clear that the lines $\mathcal{L}^i$ do not intersect. In order to find out how $\mathcal{C}^1$ and $\mathcal{L}^1$ intersect, we set $x_1 = x_2 = 0$ in the parametrization (22) of $\mathcal{L}^1$ to find that $u = 0$, yielding $Q_1 := (0_4 : 1 : -1 : 0)$ as the unique intersection point. Since the tangent line to the conic $\mathcal{C}^1$ at this intersection point has the parametrization $(u : v) \mapsto (0_3 : u : -v : v - 2u : u)$ it is different from $\mathcal{L}^1$, so that the point $(0_4 : 1 : -1 : 0)$ is a simple intersection point; the latter fact also follows from the fact that the line $\mathcal{L}^1$ is not contained in the subspace $x_1 = x_2 = 0$ containing the conic $\mathcal{C}^1$.

The intersection points between the conics and lines are given in Table 9.4. There are 6 intersection points which we denote by $Q_1, \ldots, Q_6$, where Q_i is the unique intersection point of $\mathcal{C}^i$ and $\mathcal{L}^i$ (or $\mathcal{L}^{i-3}$ when $i > 3$). The coordinates of Q_i are all zero, except for the $(i + 3)$-th and $(i + 4)$-th which are equal with opposite sign. Notice that each line $\mathcal{L}^i$ contains two of these special points, to wit Q_i and Q_{i+3} and that each conic contains three of them, to wit Q_{i-1}, Q_i and Q_{i+1}.

One determines in a similar way how the conics intersect. For example, to determine how $\mathcal{C}^1$ and $\mathcal{C}^2$ intersect one sets $x_3 = 0$ in (21) to find two intersection points, namely $Q_1 = (0 : 0 : 0 : 0 : 1 : -1 : 0)$ and $Q_2 = (0 : 0 : 0 : 0 : 0 : 1 : -1)$. The intersection points of the curves

Table 9.4. *Each conic $\mathcal{C}^i$ intersects each line $\mathcal{L}^j$ transversally in a single point.*

	$\mathcal{C}^1$	$\mathcal{C}^2$	$\mathcal{C}^3$	$\mathcal{C}^4$	$\mathcal{C}^5$	$\mathcal{C}^6$
$\mathcal{L}^1$	Q_1	Q_1	Q_4	Q_4	Q_4	Q_1
$\mathcal{L}^2$	Q_2	Q_2	Q_2	Q_5	Q_5	Q_5
$\mathcal{L}^3$	Q_6	Q_3	Q_3	Q_3	Q_6	Q_6

Table 9.5. *The non-singular conics C^i are either disjoint, they intersect transversally in one point or they intersect in two points.*

	C^1	C^2	C^3	C^4	C^5	C^6
C^1	=	Q_1, Q_2	Q_2	—	Q_6	Q_1, Q_6
C^2	Q_1, Q_2	=	Q_2, Q_3	Q_3	—	Q_1
C^3	Q_2	Q_2, Q_3	=	Q_3, Q_4	Q_4	—
C^4	—	Q_3	Q_3, Q_4	=	Q_4, Q_5	Q_5
C^5	Q_6	—	Q_4	Q_4, Q_5	=	Q_5, Q_6
C^6	Q_1, Q_6	Q_1	—	Q_5	Q_5, Q_6	=

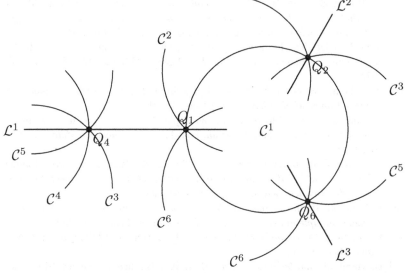

Figure 9.4 The intersection pattern of one of the lines and one of the conics with all other lines and conics.

C^i are indicated in Table 9.5. Notice that the curves C^i and C^{i+3} are disjoint, and that the conics intersect (only) at the intersection points $Q_1, \ldots, Q_6$ of the conics and the lines. Since every conic C^i contains three of the points Q_j and through every point Q_j pass three of the conics C^i, these 6 conics and 6 points form a 6_3 configuration.

In order to better visualize the configuration of conics and lines, we represent them in Figure 9.4, where we draw for the line $\mathcal{L}^1$ and for the conic $\mathcal{L}^1$ all lines and conics passing through them. This diagram is to be compared with Figure 9.3, which represents the Painlevé divisor, which appears in the compactification of $\mathbf{F_c}$ into an Abelian variety. It would be interesting to obtain the latter compactification — including a complete description of the divisor added in the process of compactification — by purely algebraic geometric means, i.e., without using the periodic 6-particle KM vector field which we have used in a very essential way.

References

[1] M. Adler and P. van Moerbeke. The complex geometry of the Kowalewski-Painlevé analysis. *Invent. Math.*, 97(1):3–51, 1989.

[2] M. Adler, P. van Moerbeke, and P. Vanhaecke. *Algebraic integrability, Painlevé geometry and Lie algebras*, volume 47 of *Ergebnisse der Mathematik und ihrer Grenzgebiete. 3. Folge. A Series of Modern Surveys in Mathematics [Results in Mathematics and Related Areas. 3rd Series. A Series of Modern Surveys in Mathematics]*. Springer-Verlag, Berlin, 2004.

[3] C. Birkenhake and H. Lange. *Complex abelian varieties*, volume 302 of *Grundlehren der Mathematischen Wissenschaften [Fundamental Principles of Mathematical Sciences]*. Springer-Verlag, Berlin, second edition, 2004.

[4] S. G. Dalaljan. The Prym variety of a two-sheeted covering of a hyperelliptic curve with two branch points. *Mat. Sb. (N.S.)*, 98(140)(2 (10)):255–267, 334, 1975.

[5] R. L. Fernandes and P. Vanhaecke. Hyperelliptic Prym varieties and integrable systems. *Comm. Math. Phys.*, 221(1):169–196, 2001.

[6] M. Kac and P. van Moerbeke. On an explicitly soluble system of nonlinear differential equations related to certain Toda lattices. *Advances in Math.*, 16:160–169, 1975.

[7] C. Laurent-Gengoux, A. Pichereau, and P. Vanhaecke. *Poisson structures*, volume 347 of *Grundlehren der Mathematischen Wissenschaften [Fundamental Principles of Mathematical Sciences]*. Springer, Heidelberg, 2013.

[8] D. Mumford. Prym varieties I. In *Contributions to analysis: a collection of papers dedicated to Lipman Bers,* pages 325–350. Academic Press, New York, 1974.

[9] D. Mumford. *Curves and their Jacobians.* The University of Michigan Press, Ann Arbor, Mich., 1975.

[10] D. Mumford. *Tata lectures on theta. II.* Modern Birkhäuser Classics. Birkhäuser Boston, Inc., Boston, MA, 2007. Jacobian theta functions and differential equations, With the collaboration of C. Musili, M. Nori, E. Previato, M. Stillman and H. Umemura, Reprint of the 1984 original.

[11] E. Previato. Flows on r-gonal Jacobians. In *The legacy of Sonya Kovalevskaya (Cambridge, Mass., and Amherst, Mass., 1985)*, volume 64 of *Contemp. Math.*, pages 153–180. Amer. Math. Soc., Providence, RI, 1987.

[12] E. Previato. Seventy years of spectral curves: 1923–1993. In *Integrable systems and quantum groups (Montecatini Terme, 1993)*, volume 1620 of *Lecture Notes in Math.*, pages 419–481. Springer, Berlin, 1996.

[13] P. Vanhaecke. *Integrable systems in the realm of algebraic geometry*, volume 1638 of *Lecture Notes in Mathematics*. Springer-Verlag, Berlin, second edition, 2001.

Pol Vanhaecke

Université de Poitiers, Laboratoire de Mathématiques et Applications, UMR 7348 du CNRS, Bât. H3, Boulevard Marie et Pierre Curie, Site du Futuroscope, TSA 61125, 86073 POITIERS Cedex 9
E-mail address: pol.vanhaecke@math.univ-poitiers.fr

10

Integrable Mappings from a Unified Perspective

Tova Brown and Nicholas M. Ercolani

Abstract. Two discrete dynamical systems are discussed and analyzed whose trajectories encode significant explicit information about a number of problems in combinatorial probability, including graphical enumeration on Riemann surfaces and random walks in random environments. The two models are integrable and our analysis uncovers the geometric sources of this integrability and uses this to conceptually explain the rigorous existence and structure of elegant closed form expressions for the associated probability distributions. Connections to asymptotic results are also described. The work here brings together ideas from a variety of fields including dynamical systems theory, probability theory, classical analogues of quantum spin systems, addition laws on elliptic curves, and links between randomness and symmetry.

1 Introduction

The modern subject of dynamical systems can often be described in terms of a number of fundamental dichotomies, such as discrete versus continuous or integrable versus chaotic. The diversity that these dichotomies represent often invites comparisons across the divides that actually provide new insights into the respective extremes that can, in turn, lead to new mathematical developments. For example, in the case of the former dichotomy, methods stemming from combinatorial analysis or arithmetic can be brought to bear on discrete systems that have no immediate analogue in the continuous setting but may still suggest mathematical themes that transcend the dichotomy. Such was the case, in the middle of the last century, when dynamic mappings on the interval and associated period doubling cascades led to notions of universality and

Supported by NSF grant DMS-1615921

The authors thank Joceline Lega for orignally encouraging us to study the problems considered here from a dynamical systems perspective, both theoretically and numerically, and for many helpful conversations in this regard along the way.

phase transitions that were reminiscent of themes from statistical mechanics. Similarly, studies of chaotic systems or systems of more general ergodic type were spurred on by studying perturbations of *nonlinear* integrable dynamical systems, providing many concrete examples of so-called KAM phenomena with suggestions of connections to stochastic analysis.

More recently, new bridges across these divides have emerged stimulated by developments in random matrix theory, analytical combinatorics and related areas of probability theory and mathematical physics. In broad terms the realization that underlies these developments is that random settings in the presence of symmetry lead to new universality classes in the probabilistic or statistical mechanical sense; moreover, the symmetries present here frequently lead to the asymptotic distributions of these universality classes being describable in terms of integrable systems theory. A perhaps by now classical example of this is the asymptotic analysis of the longest increasing sequence in a random permutation [AD99], a long-standing problem in the area of asymptotic combinatorial probability, first posed by Erdös. It was solved in terms of powerful analytical methods of Riemann-Hilbert analysis arising in the theory of integrable PDE and whose universal distribution, due to Tracy and Widom, is expressible in terms of a Painlevé transcendent. This celebrated result has had ramifications in the study of random walks on Lie algebras[BJ02], quantum integrable systems [O'C14] and non-equilibrium statistical mechanics [PS00, Joh98].

What we will discuss in this paper is certainly less extensive in scope than the results just mentioned; however, it is illustrative of these trends and it does bring together in a novel and unified way themes from combinatorial probability theory, dynamical systems theory, symmetry groups, algebraic geometry and asymptotic analysis.

1.1 Outline

We will begin in Section 2 with a careful description of the two dynamical systems we want to study, together with an indication of the source of their integrability and a numerical study of their trajectories, particularly the ones of interest to us. Then, in Section 3, we will reveal the combinatorial and probabilistic significance of the special orbits we have identified, as well as the hierarchy of combinatorial generating functions whose elegant closed form expressions will be the object of our analysis in the remainder of the paper. That analysis, in terms of elliptic function theory, its solitonic degenerations, and associated algebro-geometric features of the dynamical phase space, will be detailed in Sections 4 and 5. Finally in Section 6 we will amplify upon the combinatorial and probabilistic relevance of our work and indicate some directions for related future research.

2 Background

The explicit focus of this paper is on two two-dimensional autonomous (that is, whose coefficients are n-independent) discrete dynamical systems of the form

$$x_{n+1} = \frac{x_n - 1}{g x_n} - c x_n - y_n \tag{1a}$$

$$y_{n+1} = x_n, \tag{1b}$$

equivalently given by the second-order recurrences

$$x_n = 1 + g x_n (x_{n+1} + c x_n + x_{n-1}), \tag{2}$$

for c equal to either 0 or 1, and with system parameter g. Both of these systems are integrable in a sense that will be described shortly; but despite that and the fact that superficially they look quite similar, the structure of this integrability looks quite different. However, in the remainder of this subsection we will focus on the similarities.

An orbit, or trajectory, in either of these systems refers to the infinite sequence of points $\{(x_n, y_n)\}$ generated from an initial condition (x_0, y_0) according to equations (1). Everything is considered as a function of the system parameter g.

Both systems have the same fixed point structure: two fixed points, one hyperbolic and the other a center. A fixed point is of the form (x^*, x^*) where x^* satisfies

$$x^* = 1 + (2 + c) g x^{*2}. \tag{3}$$

Therefore, the two fixed points (x_+^*, x_+^*) and (x_-^*, x_-^*) are located at the solutions to this quadratic equation:

$$x_\pm^* = \frac{1 \pm \sqrt{1 - 4(2 + c)g}}{2(2 + c)g}. \tag{4}$$

Classification of the fixed points is determined from the Jacobian evaluated at these points:

$$J(x^*, x^*) = \begin{pmatrix} \frac{1}{g x^{*2}} - c & -1 \\ 1 & 0 \end{pmatrix} = \begin{pmatrix} \frac{1 - 2(c+1)g x^*}{g x^*} & -1 \\ 1 & 0 \end{pmatrix}.$$

With $\tau = \mathrm{Tr}(J)$ and $\Delta = \det(J)$, the eigenvalues of the Jacobian satisfy $|\lambda_+||\lambda_-| = \Delta = 1$ and are given by

$$\lambda_\pm = \frac{\tau \pm \sqrt{\tau^2 - 4\Delta}}{2}.$$

At (x_+^*, x_+^*) the determinant $\tau^2 - 4\Delta < 0$ for all appropriate values of g (that is, $0 < g < \frac{1}{4(2+c)}$, in order that the fixed points are real), so both eigenvalues are complex. Therefore they are complex conjugates of each other, forcing that $|\lambda_+| = |\lambda_-| = 1$, therefore the fixed point (x_+^*, x_+^*) is a center of the system. At (x_-^*, x_-^*), the determinant is positive for all g, so the eigenvalues are real and distinct. Their magnitudes are thus not constrained to be equal but are inversely proportional, and so there are attracting and repelling directions and the fixed point (x_-^*, x_-^*) is hyperbolic, a saddle.

Besides their fixed point structure, both systems also share the key property of being integrable, in the sense that each one has an integral of motion, or invariant, $I(x, y)$ satisfying $I(x_n, y_n) = I(x_{n+1}, y_{n+1})$ for all n. The invariants have different degrees:

$$I_{c=1}(x, y) = xy\left(x + y - \frac{1}{g}\right) + \frac{1}{g}(x + y) - \frac{1}{g^2} \qquad (5)$$

and

$$I_{c=0}(x, y) = xy(1 - gx)(1 - gy) + gxy - x - y + \frac{1}{g}. \qquad (6)$$

Level sets $I(x, y) = e$ for constants e we call "energies" foliate the plane with curves; an orbit of the system is restricted to lie on one such curve. The level sets are in general the real loci of elliptic curves (degree three and four curves respectively) with no singularities in the affine (finite) plane; there are also special energy values for which the curves become singular in the affine plane. Among these latter we will be primarily interested in those special values for which the singular degeneration of the curve corresponds to an irreducible curve containing a single separatrix of the dynamical system. In Table 10.1 these correspond to the energy value e_1.

One should also note that there are various compactifications of the phase plane which yield an extension of the level sets to infinity where the associated curve can become singular. If one completes to the projective plane $\mathbb{P}^2$ then, in the $c = 1$ case, the curve extends to be smooth at infinity. However, in the $c = 0$ case there are two singularities at infinity for all values of the energy. However, if one extends the affine plane to $\mathbb{P}^1 \times \mathbb{P}^1$ then, for $c = 0$, there

Table 10.1. *Energy values giving singular level sets of the invariant.*

	$c = 1$ system	$c = 0$ system
e_0	0	0
e_1	$\dfrac{1 + 36g + (12g - 1)\sqrt{1 - 12g}}{-54g^3}$	$\dfrac{1 + 20g - 8g^2 + (8g - 1)\sqrt{1 - 8g}}{32g^2}$
e_2	$\dfrac{1 + 36g - (12g - 1)\sqrt{1 - 12g}}{-54g^3}$	$\dfrac{1 + 20g - 8g^2 - (8g - 1)\sqrt{1 - 8g}}{32g^2}$

are no singularities on this completion of the level curve to infinity. This will play a key role in our analysis of the $c = 0$ case with more details provided in Section 5.

2.1 The QRT Mapping and the Sakai Classification

There is a systematic construction that one may look to in trying to understand the integrable structure which underlies the types of dynamical systems we consider here. This construction grew out of earlier studies of classical analogues of quantum spin systems of Heisenberg type which, in its current form, is usually attributed to Quispel, Roberts and Thompson from which it derives its name, *the QRT mapping*. The mapping provides a construction of integral invariants of the type (5) for a large class of systems of discrete Painlevé type. Our case of $c = 1$ is in fact the discrete Painlevé I equation (dPI). For full details about the QRT mapping we refer the reader to [QRT89], but for our purposes here we just state the general QRT form for systems of dPI type:

$$x_{n+1} + x_{n-1} = -\frac{\beta x_n^2 + \epsilon x_n + \zeta}{\alpha x_n^2 + \beta x_n + \gamma}. \tag{7}$$

The case of $c = 0$ does not fit directly within this scheme, but it does correspond to a degeneration in which the numerator and denominator on the RHS of (7) have a common factor [RG96]. It is not immediately clear how to find an invariant when the QRT mapping degenerates, but in our case it was possible to do so.

QRT turns out to be the first step to an even deeper insight into the integrability of the discrete Painlevé systems due to Sakai [Dui10] who established a correspondence between the classification of rational and elliptically fibered algebro-geometric surfaces on the one hand and Painlevé equations, both discrete and continuous, on the other. This correspondence encodes relations between the dynamics, Bäcklund transformations, and generalized Lie algebras. The $c = 0$ case of our study does not fit immediately into this corresponence either, but just recently a more elaborate treatement of Sakai's point configuration method was put forward [KNY17] which can be applied to this case. It will be of future interest to relate this extension to the results we present in this paper.

2.2 Numerical Study

Numerical investigations of the systems (1) were done in Python, with simulations taking as input a value for the system parameter g lying in the region $0 < g < \frac{1}{4(2+c)}$, an initial condition (x_0, y_0) lying in the finite plane, and the maximum value of n to which the calculation should be run. A sequence of points $\{(x_n, y_n)\}_{n=0}^{n_{max}}$ is calculated using (1), and the numbers are output as scatter plots of the orbit. Invariant level sets are viewed via contour plots.

We note that orbits near the hyperbolic fixed point experience a strong pull in the unstable directions, rapidly accumulating numerical error and producing artifacts in the image. In order to combat this source of error, a multiprecision library in Python [J+13] was used so that computations could all be done to any level of precision desired (100 decimal places of accuracy was sufficient to balance survival of an accurate simulation with computation time, in contrast to the standard approximately 13 digits of accuracy).

The two fixed points of the system are plotted in all simulation images. Orbits are plotted discretely, and the initial conditions of each simulation appear in a lighter shade of grey. Only one orbit is shown per image. Level sets of the integral of motion are plotted as curves.

Due to the integrable nature of the systems, the shape of each orbit is constrained to a curve given by the level set $I(x, y) = I(x_0, y_0)$. Initial conditions which are close to one another yield orbit shapes which are close to one another, except where a separatrix lies between them due to a singularity occurring on an intermediate curve.

2.3 Features of the $c = 1$ System, dPI

Four generic orbits are displayed in Figure 10.1. We observe the presence of closed orbits encircling the center, and of three-branched (cubic elliptic curve) orbits in this system.

The three singular level sets of this system are shown in Figure 10.2, separately and superimposed on one set of axes.

For sufficiently small positive values of the system parameter g, the qualitative shape of the entire system is preserved by continuous deformation. As g approaches a critical value $g_c = \frac{1}{12} = 0.08\bar{3}$, the nearby elliptic and hyperbolic fixed points approach each other. Figure 10.3 illustrates this movement and the changing shape of orbits via plotting of the e_1 separatrix.

2.4 Features of the $c = 0$ System

In comparing Figures 10.4–10.6 with Figures 10.1–10.3, we point out that, in terms of what immediately meets the eye, the respective trajectories show many similarities with the exception of the four-branched nature of this system, in contrast to the three-branched nature of the $c = 1$ system.

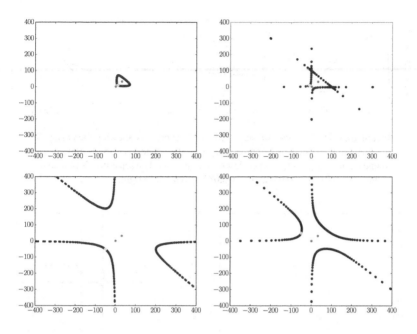

Figure 10.1 Four orbits of the system, with their two fixed points, and initial conditions at $(10, 10)$, $(-10, 10)$, $(-50, -50)$, and $(-50, 50)$, listed left to right and top to bottom.

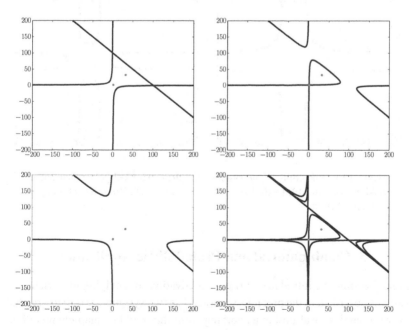

Figure 10.2 The three level sets $I(x, y) = 0, \approx -9897.9$, and ≈ -40472.5 at the degenerate energies when $g = 0.01$. The fourth image is the superposition of the others.

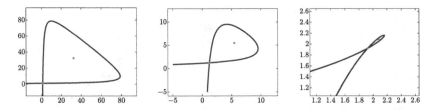

Figure 10.3 Three images of the $I(x, y) = e_1$ separatrix showing the change in the shape of the system as g approaches $g_c = 0.08\bar{3}$. The fixed points move closer to one another as $g \to g_c$ and coalesce when $g = g_c$. The separatrix is plotted for values $g = 0.01$, $g = 0.05$, and $g = 0.0833$ from left to right.

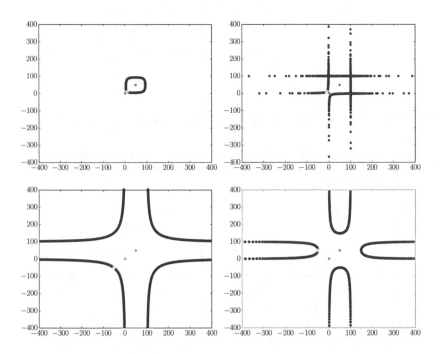

Figure 10.4 Four orbits of the system, with their two fixed points, and initial conditions at $(10, 10)$, $(-10, 10)$, $(-50, -50)$, and $(-50, 50)$, listed left to right and top to bottom.

3 Combinatorial and Probabilistic Applications

In this section, we introduce a primary motivation for studying the dynamical systems described in the previous sections. The two systems are directly related to two combinatorial problems coming from the world of map enumeration. A *map* is a graph which is embedded into a surface so that (i) the vertices are distinct points in the surface, (ii) the edges are curves on the surface

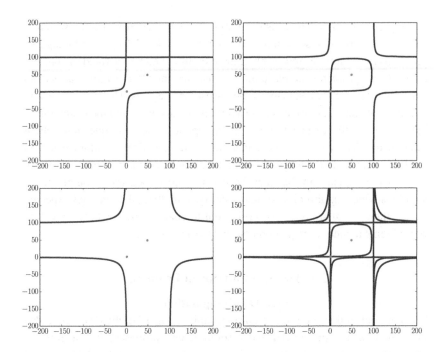

Figure 10.5 The three level sets $I(x, y) = 0, \approx 98.9897,$ and ≈ 650.5103 at the degenerate energies when $g = 0.01$. The fourth image is the superposition of the others.

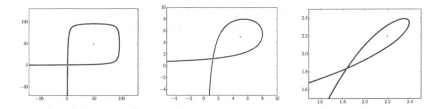

Figure 10.6 Three images of a separatrix showing the change in the shape of the system as g approaches $g_c = 0.125$. The fixed points move closer to one another as $g \to g_c$ and coalesce when $g = g_c$. The separatrix is plotted for values $g = 0.01, g = 0.08,$ and $g = 0.124$ from left to right.

intersecting only at vertices, and (iii) if the surface is cut along the graph, what remains is a disjoint union of connected components, called *faces*, each homeomorphic to an open disk. By *graph* we mean a collection of vertices, edges, and incidence relations, with loops and multi-edges allowed, taken to be connected unless stated otherwise. By *surface* we mean a compact connected complex-analytic manifold of complex dimension one, up to orientation-preserving homeomorphism.

Map enumeration has a history dating back to the early 1960s, which is briefly summarized in section 3.2. We begin in section 3.1 by introducing two problems and solutions appearing in the literature which motivated the present study. These problems are basic in the sense that they treat *triangulations* and *quadrangulations*, that is, maps of the lowest face degrees. They are interesting because both families of maps have a notion of a distance, and the enumeration knows about this distance, opening up opportunities for questions about random metrics, statistical mechanics on random lattices, random walks in random environments and random combinatorial structures in general. These problems are special in that they have very nice closed form combinatorial solutions which exhibit hints of integrability, something that was not expected during their initial study but is turning out to characterize a whole collection of problems related to map enumeration (see section 3.2).

3.1 Two Combinatorial Problems

3.1.1 Quadrangulations

The material in this section is from the seminal paper [BDFG03], where this problem was first introduced and a solution was proposed. It concerns the enumeration of 4-valent 2-legged (that is, there are also two 1-valent vertices) planar maps, sorted according to a notion of geodesic distance that will be defined presently. It is equivalent to counting rooted planar quadrangulations (that is, the face degree, rather than the vertex degree, is 4) with an origin vertex, again counting according to an appropriate distance [BFG03a].

Let $R = R(g)$ denote the generating function for the family of 2-legged 4-valent planar maps:

$$R(g) = \sum_{k \geq 0} r_k g^k$$

where r_k denotes the number of 2-legged 4-valent planar maps on k vertices. By Schaeffer's bijection [Sch97], this is also the generating function for 4-valent blossom trees. *Blossom trees* are rooted trees whose external vertices (called leaves) are either black or white, and in which a certain number of leaves at every internal vertex are black, depending on the valence of the trees (1 in a 4-valent blossom tree). The subtree structure, shown in Figure 10.7, is that each internal vertex of a 4-valent k-vertex blossom tree has exactly one descendent black leaf and two descendent subtrees on $k - 1$ vertices, where a single white leaf is a subtree with zero vertices. The number of ways to choose two such subtrees and arrange them in the ordered tree is

$$r_k = 3 \sum_{i=0}^{k-1} r_i r_{k-1-i}.$$

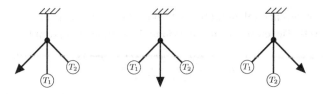

Figure 10.7 Decomposition of a 4-valent blossom tree at its root: the single black leaf has three possible positions, and the other two descendants are blossom sub-trees.

Therefore, the generating function $R(g)$ satisfies

$$R = 1 + 3gR^2 \tag{8}$$

with analytic solution

$$R(g) = \frac{1 - \sqrt{1 - 12g}}{6g}. \tag{9}$$

The *geodesic distance* of a 2-legged map is the minimal number of edges crossed by a path on the surface connecting the two legs. Schaeffer's bijection was extended in [BDFG03] to keep track of the geodesic distance, preserved through the bijection and called *root depth* on the tree side, of even-valent blossom trees. The subtree structure decomposition in Figure 10.7 was again applied to recursively build a blossom tree from its subtrees, and the root depth of the pieces and the whole kept track of using contour walks; the resulting generating function is for maps with an upper bound on the geodesic distance. Furthermore, defining

$$R_n(g) := \sum_{k=0}^{\infty} r_{n,k} g^k$$

for $r_{n,k} = \#\{$2-leg 4-valent k-vertex planar maps with goedesic distance $\leq n\}$ the derivation yields a second-order three-term recurrence relation for the generating functions instead of the quadratic functional relation that determined $R(g)$:

$$R_n = 1 + gR_n \left(R_{n-1} + R_n + R_{n+1}\right). \tag{10}$$

The recurrence relation has two boundary conditions:

$$R_{-1} = 0 \tag{11a}$$

$$\lim_{n \to \infty} R_n = R, \tag{11b}$$

the initial condition stating that there are no maps with negative geodesic distance, and the limit condition stating that the large n limit amounts to removing the upper bound on geodesic distance.

The recurrence relation (10) with boundary conditions (11) was solved in [BDFG03]. The closed-form solution to the recursion is given by

$$R_n = R\frac{(1 - x^{n+1})(1 - x^{n+4})}{(1 - x^{n+2})(1 - x^{n+3})}.$$ (12)

where x is the solution with modulus less than 1 to the characteristic equation

$$x + \frac{1}{x} = \frac{1 - 4gR}{gR}.$$ (13)

The original paper did not contain a proof of this result; it can be directly checked that the closed-form solves the recursion, and the authors provided the formal means by which they discovered this amazing formula. The first proof of this closed-form appeared in [BG12], by continued fraction expansions and exploiting a remarkable connection to the problem of enumerating planar maps with a boundary. This leads to an expression for the generating functions in terms of symplectic Schur functions. Our approach provides an alternative proof that emphasizes the connection to integrable dynamical systems and natural extensions to Painlevé equations and Riemann-Hilbert analysis (see section 3.2); however, it will be very interesting to study relations between these two approaches.

3.1.2 Triangulations

The second problem comes from the paper [BM06] and is an enumeration problem of labeled plane trees which are bijectively equivalent to certain triangulations of the sphere (see section 6.2). A *plane tree* is a planar map with only one face, and the enumeration is of the family of labeled plane trees, rooted with the root labeled 0 and with the labels of two adjacent vertices differing by ±1. The valence of these trees is unrestricted. Figure 10.8 shows a tree in this family, with its root labeled in boldface. Such trees are also called *embedded*, because trees labeled in this way can be naturally embedded into a one-dimensional integer lattice, with the labeling of each vertex dictating its location in the lattice. The maximal label, or equivalently the largest point in

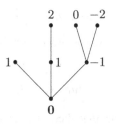

Figure 10.8 A labeled plane tree rooted at 0; reproduction of an example in [BM06].

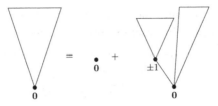

Figure 10.9 Recursive breakdown at the root of a labeled plane tree: a tree is either simply a root or it can be decomposed as discussed above.

the lattice having positive mass, is called the *label height*. The label height is the appropriate notion of distance for this enumeration problem, corresponding to the geodesic distance of the previous problem.

Define the generating functions

$$T(g) = \sum_{k=0}^{\infty} t_k g^k \quad \text{and} \quad T_j(g) = \sum_{k=0}^{\infty} t_{j,k} g^k$$

where $t_k = \#\{\text{trees in the family with } k \text{ edges}\}$ and $t_{j,k} = \#\{\text{trees in the family having } k \text{ edges and label height} \le j\}$. A rooted tree either has zero edges (and is simply the root), or it has at least one edge; in the latter case, it can be decomposed (see Figure 10.9) into two subtrees connected by an edge incident to the root, and the sum of the edges in the two subtrees will be one less than the number of edges in the overall tree. The subtree which doesn't contain the root is still in the family, by shifting all its labels by $+1$ or -1.

The recurrence relation for t_k,

$$t_k = 2 \sum_{j=0}^{k-1} t_j t_{k-1-j},$$

is summed to see that the generating function for this family of trees is also given by a quadratic equation:

$$T = 1 + 2g T^2. \tag{14}$$

Keeping track of label height through this enumeration, the recursion relation obeyed by T_j is

$$T_j = 1 + g T_j \left(T_{j-1} + T_{j+1} \right) \tag{15}$$

with boundary conditions

$$T_{-1} = 0 \tag{16a}$$

$$\lim_{j \to \infty} T_j = T. \tag{16b}$$

Then [BM06] contains the following closed-form solution (formally derived, and verifiable by insertion into the recursion) for this family of generating functions:

$$T_j = T\frac{(1 - Z^{j+1})(1 - Z^{j+5})}{(1 - Z^{j+2})(1 - Z^{j+4})},\tag{17}$$

where $Z \equiv Z(g)$ is implicitly given by

$$x_* = \frac{(1 + Z)^2}{1 + Z^2},\tag{18}$$

where $x_* = x_-^*$ as defined by (4) which also specifies its g-dependence.

A discrete integral of motion

$$f(x, y) = xy(1 - gx)(1 - gy) + gxy - x - y$$

satisfying $f(T_{n-1}, T_n) = f(T_n, T_{n+1})$ with convergence of T_n to its limit T assured by $f(T_n, T_{n+1}) = f(T, T)$ was given in [BFG03b].

3.2 Combinatorics Context and Applications

Systematic enumeration of families of maps began with Tutte in [Tut62b, Tut62c, Tut62a, Tut63]. He studied rooted maps (maps having a distinguished directed edge) and, by decomposing maps at their root, derived functional equations for generating functions which are often multi-variate due to the combinatorial complexity of the maps. Tutte's Quadratic Method along with Lagrange Inversion gives the asymptotic behavior of many families of maps (see [BR86] for a summary). This foundation has made the study of map enumeration a signature component of the modern theory of random combinatorial structures. Further research has followed two methodological streams, one algebraic and the other analytic. In the latter category, one has the subject of analytic combinatorics amply illustrated in [FS09] on the one hand and, on the other hand, methods of Riemann-Hilbert Analysis coming from integrable systems theory that solved the Erdös problem on longest increasing sequences in permutations mentioned earlier. However we believe the most promising developments will come from a combination of both methods. The problems discussed in this paper are a perfect example of that. On the algebraic side was the breakthrough result of Schaeffer [Sch97], already mentioned, which built upon pioneering work by Cori and Vacquelin [CV81] to establish a bijection between the families of planar quadrangulations/triangulations and appropriate families of trees. On the analytic side was the asymptotic analysis of families of orthogonal polynomials which motivated the postulation in [BDFG03] of formulas like (12). The rigorous asymptotic analysis of the Riemann-Hilbert problem for orthogonal polynomials with exponential weights that has been carried out in [EM03, EMP08, EP12, Erc11] provides a rigorous background for the analysis and extension of (12). Indeed our original motivation was to understand *non-autonomous* extensions of (1) that arose from the analysis in the above cited papers. More will be said about that in section 6.1. Here we just close by noting that besides providing a natural derivation of the formulas (12) and (17), the results in this paper provide for a rigorous global analysis of these generating functions as well as the determination of their maximal domains of holomorphy.

4 Elliptic Parametrization, c = 1

4.1 The Generic Orbit

In this section we show how the combinatorial problem, corresponding to $c = 1$, embeds into a more general integrable dynamical system and provide an elliptic parametrization of the general orbits of this system. This is a discrete analogue of action-angle variables of continuous integrable systems. The approach used here is primarily that of classical function theory using canonical coordinates based on the Weierstrass $\wp$ function. In this perspective the separation of variables provided by (21) may be viewed as a discrete analogue of the Hamilton-Jacobi equation. In the next theorem we will make use of such an elliptic parametrization for generic orbits of the dPI system that has already been developed by [BTR01]. Our main results concern the degeneration of this parametrization and are presented in the next subsection.

Theorem 4.1 *The discrete dynamical system*

$$x_{n+1} = \frac{1}{g} - \frac{1}{gx_n} - x_n - y_n \tag{19a}$$

$$y_{n+1} = x_n \tag{19b}$$

with invariant

$$I(x, y) = xy\left(x + y - \frac{1}{g}\right) + \frac{1}{g}(x + y) - \frac{1}{g^2}$$

satisfying $I(x_{n+1}, x_n) = I(x_n, x_{n-1})$ *has an elliptic parametrization through the Weierstrass $\wp$-function as*

$$x(\xi) = \frac{1}{2}\left(\frac{1}{g} + \frac{\wp'(v) - \wp'(\xi)}{\wp(v) - \wp(\xi)}\right) \tag{20a}$$

$$y(\xi) = \frac{1}{2}\left(\frac{1}{g} + \frac{\wp'(v) + \wp'(\xi)}{\wp(v) - \wp(\xi)}\right) \tag{20b}$$

where v is such that $\wp(v) = \frac{4g+1}{12g^2}$. The discrete dynamics are recovered in this continuous parametrization via addition of v.

Remark: $x_n = x(\xi + nv)$.

Proof. On a level set $I(x, y) = e$ of the invariant, we have the factorization

$$
I(x, y) = xy\left(x + y - \frac{1}{g}\right) + \frac{1}{g}(x + y) - \frac{1}{g^2}
$$
$$
= \left(x + y - \frac{1}{g}\right)\left(xy + \frac{1}{g}\right)
$$
$$
= e
$$

into which can be introduced a new parameter, t, so that

$$
t = xy + \frac{1}{g}
$$
$$
\frac{e}{t} = x + y - \frac{1}{g}.
$$

Then

$$
xy = t - \frac{1}{g}
$$
$$
x + y = \frac{e}{t} + \frac{1}{g},
$$

so x and y can be recovered as the roots of the quadratic with coefficients the elementary symmetric polynomials xy and $x + y$:

$$
X^2 - (x + y)X + xy = 0. \tag{21}
$$

Thus

$$
x, y = \frac{x + y \mp \sqrt{(x + y)^2 - 4xy}}{2}
$$
$$
= \frac{\frac{e}{t} + \frac{1}{g} \mp \sqrt{\left(\frac{e}{t} + \frac{1}{g}\right)^2 - 4\left(t - \frac{1}{g}\right)}}{2}
$$
$$
= \frac{e + \frac{t}{g} \mp \sqrt{\left(e + \frac{t}{g}\right)^2 - 4t^2\left(t - \frac{1}{g}\right)}}{2t}.
$$

Set

$$
w^2 = \left(e + \frac{t}{g}\right)^2 - 4t^2\left(t - \frac{1}{g}\right)
$$
$$
= -4t^3 + \frac{1}{g^2}(4g + 1)t^2 + \frac{2e}{g}t + e^2.
$$

This curve is put into Weierstrass normal form,

$$
w^2 = -4z^3 + g_2 z - g_3,
$$

by the translation $t = z + \alpha$ for $\alpha = \frac{1}{12g^2}(4g + 1)$, giving the following for Weierstrass invariants:

$$g_2 = 12\alpha^2 + \frac{2e}{g} \tag{22a}$$

$$g_3 = -8\alpha^3 - \frac{2e}{g}\alpha - e^2. \tag{22b}$$

The Weierstrass normal form of an elliptic curve is equivalent to the differential equation

$$(\wp')^2 = 4\wp^3 - g_2\wp - g_3$$

defining the Weierstrass $\wp$-function and its derivative, under the identification $z = -\wp(\xi)$ and $w = \wp'(\xi)$. Make the change of variables

$$t = z + \alpha = -\wp(\xi) + \alpha \tag{23a}$$

$$w = \wp'(\xi) \tag{23b}$$

$$\alpha = \wp(v) \tag{23c}$$

$$e = \wp'(v) \tag{23d}$$

where (23c) is a definition of v and (23d) follows from the calculation

$$\wp'(v) = \sqrt{4\wp^3(v) - g_2\wp(v) - g_3}$$

$$= \sqrt{4\alpha^3 - \left(12\alpha^2 + \frac{2e}{g}\right)\alpha - \left(-8\alpha^3 - \frac{2e}{g}\alpha - e^2\right)}$$

$$= e.$$

Under this change of variables, points $(x, y) = (x(\xi), y(\xi))$ on the curve $I(x, y) = e$ are given by the parametrization

$$x, y = \frac{\wp'(v) + \frac{\wp(v) - \wp(\xi)}{g} \mp \wp'(\xi)}{2(\wp(v) - \wp(\xi))}, \tag{24}$$

equivalent to equation (20). We separate out the proof of discrete dynamics recovery into Lemma 4.3. □

The following corollary is for computational use later on, and is simply a reformulation of the elliptic parametrization in terms of a different function in the Weierstrass function family.

Corollary 4.2 *The function $x(\xi)$ of Theorem 4.1 can be expressed entirely in terms of the Weierstrass σ-function:*

$$x(\xi) = \frac{1}{2}\left[\frac{1}{g} - \frac{\sigma(2v)\sigma^4(\xi) - \sigma(2\xi)\sigma^4(v)}{\sigma^2(v)\sigma^2(\xi)\sigma(\xi - v)\sigma(\xi + v)}\right]. \tag{25}$$

Proof. We make use of the following two identities ([Mar77], equation III.5.52 and exercise III.5.16) which relate the Weierstrass σ-function to the $\wp$-function:

$$\wp(z) - \wp(Z) = -\frac{\sigma(z-Z)\sigma(z+Z)}{\sigma^2(z)\sigma^2(Z)} \tag{26}$$

if Z is not a period of $\wp(z)$ and

$$\wp'(z) = -\frac{\sigma(2z)}{\sigma^4(z)}. \tag{27}$$

Then

$$
\begin{aligned}
x(\xi) &= \frac{1}{2}\left(\frac{1}{g} + \frac{\wp'(v) - \wp'(\xi)}{\wp(v) - \wp(\xi)}\right) \\
&= \frac{1}{2}\left[\frac{1}{g} - \left(\frac{\sigma(2v)}{\sigma^4(v)} - \frac{\sigma(2\xi)}{\sigma^4(\xi)}\right)\left(-\frac{\sigma^2(v)\sigma^2(\xi)}{\sigma(v-\xi)\sigma(v+\xi)}\right)\right] \\
&= \frac{1}{2}\left[\frac{1}{g} + \left(\frac{\sigma(2v)\sigma^4(\xi) - \sigma(2\xi)\sigma^4(v)}{\sigma^4(v)\sigma^4(\xi)}\right)\left(\frac{\sigma^2(v)\sigma^2(\xi)}{\sigma(v-\xi)\sigma(v+\xi)}\right)\right] \\
&= \frac{1}{2}\left[\frac{1}{g} + \left(\frac{1}{\sigma^2(v)\sigma^2(\xi)}\right)\left(\frac{\sigma(2v)\sigma^4(\xi) - \sigma(2\xi)\sigma^4(v)}{\sigma(v-\xi)\sigma(v+\xi)}\right)\right] \\
&= \frac{1}{2}\left[\frac{1}{g} - \frac{\sigma(2v)\sigma^4(\xi) - \sigma(2\xi)\sigma^4(v)}{\sigma^2(v)\sigma^2(\xi)\sigma(\xi-v)\sigma(\xi+v)}\right],
\end{aligned}
$$

the last line since σ is an odd function. $\qquad\square$

Lemma 4.3 *Under the parametrization of Theorem 4.1, the discrete dynamics of the system are recovered via addition of v. That is,*

$$x(\xi + v) = \frac{1}{g} - \frac{1}{gx(\xi)} - x(\xi) - y(\xi) \tag{28a}$$

$$y(\xi + v) = x(\xi) \tag{28b}$$

where v is such that $\wp(v) = \frac{4g+1}{12g^2}$.

Proof. The lemma rests solely on the use of certain properties of the Weierstrass functions, including the Weierstrass ζ-function given by $\zeta'(u) = -\wp(u)$: the fact that ζ is an odd function, the identity ([Mar77] equation (5.59))

$$\frac{\wp'(u)}{\wp(u) - \wp(v)} = \zeta(u+v) + \zeta(u-v) - 2\zeta(u), \tag{29}$$

and the addition law for $\wp$ ([Mar77] Theorem 5.15)

$$\wp(u+v) + \wp(u) + \wp(v) = \frac{1}{4}\left(\frac{\wp'(u) - \wp'(v)}{\wp(u) - \wp(v)}\right)^2. \tag{30}$$

We first calculate that

$$x(\xi), y(\xi) = \frac{1}{2g} + \zeta(\pm\xi + v) \mp \zeta(\xi) - \zeta(v), \tag{31}$$

starting from (20), expanding, and using (29):

$$x(\xi), y(\xi) = \frac{1}{2g} + \frac{1}{2}(\zeta(v+\xi) + \zeta(v-\xi) - 2\zeta(v))$$

$$\pm \frac{1}{2}(\zeta(\xi+v) + \zeta(\xi-v) - 2\zeta(\xi))$$

$$= \frac{1}{2g} + \frac{1}{2}(\zeta(\xi+v) \pm \zeta(\xi+v))$$

$$+ \frac{1}{2}(\zeta(-\xi+v) \mp \zeta(-\xi+v)) - (\zeta(v) \pm \zeta(\xi))$$

$$= \frac{1}{2g} + \zeta(\pm\xi+v) \mp \zeta(\xi) - \zeta(v).$$

Demonstration of the result (28b) is now direct via evaluation at $\xi + v$:

$$y(\xi + v) = \frac{1}{2g} + \zeta(-(\xi+v)+v) + \zeta(\xi+v) - \zeta(v)$$

$$= \frac{1}{2g} + \zeta(-\xi) + \zeta(\xi+v) - \zeta(v)$$

$$= \frac{1}{2g} - \zeta(\xi) + \zeta(\xi+v) - \zeta(v)$$

$$= x(\xi).$$

Demonstration of the other part of the result proceeds in a few more steps. First compute the product $x(\xi)y(\xi)$ using the forms in (20), expanding along the way $\wp'(v)^2 - \wp'(\xi)^2 = -4(2\wp(v)+\wp(\xi))(\wp(v)-\wp(\xi))^2 - \frac{2\wp'(v)}{g}(\wp(v) - \wp(\xi))$ using the differential equation; thus

$$x(\xi)y(\xi) = \frac{1}{4g^2} - (2\wp(v) + \wp(\xi)). \tag{32}$$

Starting with the left-hand side of the following equality, expanding $\wp(\xi + v)$ using the addition formula (30), and then making a substitution based on (20a) give that

$$\wp(v) - \wp(\xi + v) = 2\wp(v) + \wp(\xi) - \left(x(\xi) - \frac{1}{2g}\right)^2, \tag{33}$$

which is inserted into

$$x(\xi + v) + x(\xi) = \frac{1}{g} + \frac{\wp'(v)}{\wp(v) - \wp(\xi + v)}, \tag{34}$$

the result of summing x and y from (20) evaluated at $\xi + v$, along with (28b). Inserting (32) into (33) yields $\wp(v) - \wp(\xi + v) = -x(\xi)\left(x(\xi) + y(\xi) - \frac{1}{g}\right)$, and recall that

$$x + y - \frac{1}{g} = \frac{\wp'(v)}{\wp(v) - \wp(\xi)} \quad \text{and} \quad xy + \frac{1}{g} = \wp(v) - \wp(\xi).$$

Putting all of these together into (34) the remaining result (28a) is concluded:

$$
\begin{aligned}
x(\xi + v) &= -x(\xi) + \frac{1}{g} + \frac{\wp'(v)}{-x(\xi)\left(x(\xi) + y(\xi) - \frac{1}{g}\right)} \\
&= -x(\xi) + \frac{1}{g} + \frac{\wp'(v)}{-x(\xi)\frac{\wp'(v)}{\wp(v) - \wp(\xi)}} \\
&= -x(\xi) + \frac{1}{g} - \frac{x(\xi)y(\xi) + \frac{1}{g}}{x(\xi)} \\
&= \frac{1}{g} - \frac{1}{gx(\xi)} - x(\xi) - y(\xi).
\end{aligned}
$$

$\square$

4.2 Separatrix Degeneration to the Combinatorial Orbit

In the previous section, elliptic parametrizations were given for the $c = 1$ system, relying on the invariant I. Orbits of the system lie on level sets $I(x, y) = e$, and we have so far treated the generic case in which $I(x, y) = e$ is a nonsingular curve. It is in this generic setting that results of the previous section hold. In this section, we study certain cases in which $I(x, y) = e$ is singular, and we extend the parametrizations to certain such curves, in particular, the ones of combinatorial interest. Recall that Table 10.1 contains all the energy values resulting in singular level sets for the two invariants. We will show at the appropriate time that the energy e_1 in the $c = 1$ case is the one corresponding to geodesic distance for quadrangulations. However, this can be observed in Figure 10.2 as it is the e_1 separatrix which contains the hyperbolic fixed point, with coordinates given by (x_-^*, x_-^*) and in the combinatorial problems the limit $x^* = \lim_{n \to \infty} x_n$ of the generating functions is the analytic function given by the root x_-^*.

Theorem 4.4 *The discrete integrable recurrence*

$$x_n = 1 + gx_n(x_{n-1} + x_n + x_{n+1}) \tag{35}$$

with invariant

$$I(x_n, x_{n-1}) = x_n x_{n-1}\left(x_n + x_{n-1} - \frac{1}{g}\right) + \frac{1}{g}(x_n + x_{n-1}) - \frac{1}{g^2} \qquad (36)$$

and limit $\lim_{n \to \infty} x_n = x_-^*$ *satisfying* $x_-^* = 1 + 3gx_-^{*2}$ *possesses a solution*

$$x_n = x(\xi + nv) = x_-^* \cdot \frac{\left(1 - e^{2\xi \sqrt{-3r_2}} \chi^{n-1}\right)\left(1 - e^{2\xi \sqrt{-3r_2}} \chi^{n+2}\right)}{\left(1 - e^{2\xi \sqrt{-3r_2}} \chi^n\right)\left(1 - e^{2\xi \sqrt{-3r_2}} \chi^{n+1}\right)}, \qquad (37)$$

where v *is such that* $\sinh(v\sqrt{-3r_2}) = \sqrt{\frac{-3r_2}{\alpha + r_2}}$ *and* $\chi = \exp\{-2v\sqrt{-3r_2}\}$.

In particular, the combinatorial generating functions $R_n(g) = x_n$ and $R(g) = x_-^*$ specialized by the additional boundary condition $R_{-1} = 0$ have the closed form

$$R_n = R\frac{(1 - \chi^{n+1})(1 - \chi^{n+4})}{(1 - \chi^{n+2})(1 - \chi^{n+3})} \qquad (38)$$

for χ *satisfying* $\chi + \frac{1}{\chi} = \frac{1 - 4gR}{gR}$, *by specializing* $\xi = -2v$.

An orbit $\{(x_n, y_n)\}$ satisfying $\lim_{n \to \infty} x_n = x_-^*$ limits to the hyperbolic fixed point (x_-^*, x_-^*), and thus the energy of the invariant level set on which this orbit lies can be calculated by $e = I(x_-^*, x_-^*) = I(R, R)$.

$$I(R, R) = \left(2R - \frac{1}{g}\right)\left(R^2 + \frac{1}{g}\right)$$

$$= \left[2\left(\frac{1 - \sqrt{1 - 12g}}{6g}\right) - \frac{1}{g}\right]\left[\left(\frac{1 - \sqrt{1 - 12g}}{6g}\right)^2 + \frac{1}{g}\right]$$

$$= \left(\frac{-2 - \sqrt{1 - 12g}}{3g}\right)\left(\frac{1 + 12g - \sqrt{1 - 12g}}{18g^2}\right)$$

$$= -\frac{1 + 36g + (12g - 1)\sqrt{1 - 12g}}{54g^3}$$

$$= e_1 \qquad (39)$$

In the following Proposition, the elliptic parametrization is extended to orbits on the level set $I(x, y) = e_1$, and the proof of the Theorem is deferred until after this degeneration of the parametrization is established.

Proposition 4.5 *The elliptic parametrization of Theorem 4.1 and Corollary 4.2 extends to a parametrization of the singular curve*

$$I(x, y) = e_1 = -\frac{1 + 36g + (12g - 1)\sqrt{1 - 12g}}{54g^3}$$

as

$$x(\xi + n\nu) = \frac{1}{2}\left[\frac{1}{g} - \frac{\sigma_d(2\nu)\sigma_d^4(\xi + n\nu) - \sigma_d(2(\xi + n\nu))\sigma_d^4(\nu)}{\sigma_d^2(\nu)\sigma_d^2(\xi + n\nu)\sigma_d(\xi + (n-1)\nu)\sigma_d(\xi + (n+1)\nu)}\right],$$
(40)

where the function σ_d is a degeneration of the Weierstrass σ-function, given explicitly by the formula

$$\sigma_d(\xi) = \exp\left\{\frac{r_2}{2}\xi^2\right\}\frac{\sinh(\xi\sqrt{-3r_2})}{\sqrt{-3r_2}},$$
(41)

where

$$r_2 = \frac{-4(1 - 12g + 2\sqrt{1 - 12g})}{12^2 g^2}.$$

The related functions in this degeneration of the family of Weierstrass functions are given explicitly by

$$\wp_d(\xi) = -r_2 - 3r_2\text{csch}^2(\xi\sqrt{-3r_2})$$
$$\wp_d'(\xi) = -2(-3r_2)^{3/2}\text{csch}^2(\xi\sqrt{-3r_2})\coth(\xi\sqrt{-3r_2})$$
$$\zeta_d(\xi) = r_2\xi + \sqrt{-3r_2}\coth(\xi\sqrt{-3r_2}).$$

Proof. With the energy e specialized to $e_1(g)$, the two variables g and e in the system are tied together, and we pursue further calculations after expressing everything in sight in terms of a new fundamental variable, γ, which suppresses the square root appearing in e_1:

$$\gamma = \sqrt{1 - 12g}$$
$$g = \frac{1 - \gamma^2}{12}$$
$$e_1 = \frac{-32(\gamma - 1)(\gamma + 2)^2}{(\gamma + 1)^3(\gamma - 1)^3}$$
$$g_2 = 12\frac{4^2\gamma^2(\gamma + 2)^2}{(\gamma + 1)^4(\gamma - 1)^4}$$
$$g_3 = -8\frac{4^3\gamma^3(\gamma + 2)^3}{(\gamma + 1)^6(\gamma - 1)^6}$$
$$\alpha = \frac{-4(\gamma + 2)(\gamma - 2)}{(\gamma + 1)^2(\gamma - 1)^2}$$
$$R = \frac{2}{\gamma + 1}.$$

On the singular curve $I(x, y) = e_1$, at least two roots of $-4z^3 + g_2z - g_3$ coalesce. We therefore seek a factorization

$$w^2 = -4z^3 + g_2z - g_3 = -4(z - r_1)(z - r_2)^2,$$

where r_1 and r_2 are not assumed to be distinct. Equating coefficients in the above equation gives that $r_1 + 2r_2 = 0$, $g_2 = -4(2r_1r_2 + r_2^2)$, and $g_3 = -4r_1r_2^2$. Therefore, $r_1 = -2r_2$, $g_2 = 12r_2^2$, and $g_3 = 8r_2^3$. Thus,

$$r_2 = \frac{-4\gamma(\gamma + 2)}{(\gamma + 1)^2(\gamma - 1)^2} = \frac{-4(1 - 12g + 2\sqrt{1 - 12g})}{12^2 g^2}$$

$$r_1 = \frac{8\gamma(\gamma + 2)}{(\gamma + 1)^2(\gamma - 1)^2} = \frac{8(1 - 12g + 2\sqrt{1 - 12g})}{12^2 g^2}.$$

The identification $z = -\wp_d(\xi)$ and $w = \wp_d'(\xi)$ is made just as before, to a function $\wp_d(\xi)$ and its derivative satisfying the same differential equation as the Weierstrass $\wp$-function,

$$(\wp_d')^2 = 4\wp_d^3 - g_2\wp_d - g_3 = -4(-\wp_d(\xi) - r_1)(-\wp_d(\xi) - r_2)^2.$$

The $\wp$-function is given as the inverse of an elliptic integral ([Mar77], Eq. (5.28))

$$\xi = \int_\infty^{\wp(\xi)} \frac{dt}{\sqrt{4t^3 - g_2 t - g_3}},$$

and the related, degenerate, function $\wp_d(\xi)$ is calculated directly by the same integral:

$$\xi = \int_\infty^{\wp_d(\xi)} \frac{dt}{-2(t + r_2)\sqrt{t + r_1}}.$$

The result of the integration is

$$\xi = -\frac{1}{\sqrt{r_1 - r_2}} \tanh^{-1}\left(\sqrt{\frac{t + r_1}{r_1 - r_2}}\right)\Bigg|_\infty^{\wp_d(\xi)},$$

and inverting the equation gives an explicit formula for this degeneration of the Weierstrass elliptic function:

$$\wp_d(\xi) = -r_2 - 3r_2\operatorname{csch}^2(\xi\sqrt{-3r_2}). \tag{42}$$

The derivative (with respect to ξ) is given by

$$\wp_d'(\xi) = -2(-3r_2)^{3/2}\operatorname{csch}^2(\xi\sqrt{-3r_2})\coth(\xi\sqrt{-3r_2}). \tag{43}$$

The Weierstrass ζ-function, an anti-derivative of $\wp$, is defined by ([Mar77] 5.40)

$$\frac{d}{d\xi}\zeta(\xi) = -\wp(\xi) \quad \text{and} \quad \lim_{\xi \to 0}\left[\zeta(\xi) - \frac{1}{\xi}\right] = 0.$$

Integrating $\wp_d(\xi)$ and pinning down the integration constant by the limit condition using the first few terms of a Taylor expansion for $\coth(y)$ about $y = 0$, we have the degeneration

$$\zeta_d(\xi) = r_2\xi + \sqrt{-3r_2}\coth(\xi\sqrt{-3r_2}). \qquad (44)$$

The Weierstrass σ-function is a logarithmic antiderivative of ζ ([Mar77] 5.44) given by

$$\frac{d}{d\xi}\log\sigma(\xi) = \frac{\sigma'(\xi)}{\sigma(\xi)} = \zeta(\xi) \quad \text{and} \quad \lim_{\xi\to 0}\frac{\sigma(\xi)}{\xi} = 1,$$

which can be calculated by the integral ([Mar77] 5.45)

$$\sigma(\xi) = \xi\exp\left\{\int_0^\xi\left[\zeta(t) - \frac{1}{t}\right]dt\right\}.$$

Thus by integrating the degenerate ζ_d,

$$\sigma_d(\xi) = \xi\exp\left\{C + \frac{r_2}{2}\xi^2 + \log\left(\frac{\sinh(\xi\sqrt{-3r_2})}{\xi}\right)\right\}$$

with integration constant $C = \log\left(\frac{1}{\sqrt{-3r_2}}\right)$, and the degenerate $\sigma_d(\xi)$ is given by

$$\sigma_d(\xi) = \exp\left\{\frac{r_2}{2}\xi^2\right\}\frac{\sinh(\xi\sqrt{-3r_2})}{\sqrt{-3r_2}}. \qquad (45)$$

Theorem 4.1 now goes through with all of the Weierstrass functions replaced by their appropriate degenerations. The proofs of Corollary 4.2 and Lemma 4.3 rely on three identities of the Weierstrass functions: equations (26), (29), (30), and the oddness of ζ, which can all be verified to hold in the degenerate case as well. The proofs are just algebra along with hyperbolic trigonometric identities; one could, for example, use the Pythagorean Theorem to write all hyperbolic trigonometric functions in terms of the hyperbolic cotangent, and then use simply the addition laws for the hyperbolic cotangent. Once these identities are established, all of the general parametrization and dynamics machinery extend directly to the $I(x, y) = e_1$ separatrix and Equation (40) is immediate.   □

Proof of Theorem 4.4. The results of Proposition 4.5 give the degeneration of the elliptic parametrization appropriate for the Theorem, as established by the calculation of the energy $I(x_-^*, x_-^*) = e_1$ in equation (39). Inserting the explicit formula for σ_d given in equation (45) into the parametrization of equation (40), observe that all of the exponentials cancel out and an overall factor of $\sqrt{-3r_2}$ rests in the numerator of the large fraction. Then, scaling the variables temporarily to suppress factors of $\sqrt{-3r_2}$ appearing in every argument, via

$$\phi = \xi\sqrt{-3r_2}$$
$$\eta = v\sqrt{-3r_2},$$

and simplifying using the double-angle formula for the hyperbolic sine, we have

$x(\xi + nv)$

$$= \frac{1}{2g} - \frac{\sqrt{-3r_2}}{2}\left(\frac{\sinh(2\eta)\sinh^3(\phi + n\eta)}{\sinh^2(\eta)\sinh(\phi + n\eta)\sinh(\phi + (n-1)\eta)\sinh(\phi + (n+1)\eta)}\right.$$

$$\left. + \frac{\sinh(2(\phi + n\eta))\sinh^3(\eta)}{\sinh(\eta)\sinh^2(\phi + n\eta)\sinh(\phi + (n-1)\eta)\sinh(\phi + (n+1)\eta)}\right)$$

$$= \frac{1}{2g} - \sqrt{-3r_2}\left(\frac{\cosh(\eta)\sinh^3(\phi + n\eta)}{\sinh(\eta)\sinh(\phi + n\eta)\sinh(\phi + (n-1)\eta)\sinh(\phi + (n+1)\eta)}\right.$$

$$\left. + \frac{\cosh(\phi + n\eta)\sinh^3(\eta)}{\sinh(\eta)\sinh(\phi + n\eta)\sinh(\phi + (n-1)\eta)\sinh(\phi + (n+1)\eta)}\right)$$

$$= \frac{1}{2g} - \sqrt{-3r_2}\left(\coth(\eta)\frac{\sinh^3(\phi + n\eta)}{\sinh(\phi + n\eta)\sinh(\phi + (n-1)\eta)\sinh(\phi + (n+1)\eta)}\right.$$

$$\left. + \coth(\phi + n\eta)\frac{\sinh^3(\eta)}{\sinh(\eta)\sinh(\phi + (n-1)\eta)\sinh(\phi + (n+1)\eta)}\right).$$

By definition of each hyperbolic trigonometric function in terms of exponentials, and defining

$$\chi = e^{-2\eta} = e^{-2v\sqrt{-3r_2}},$$

then

$$x(\xi + nv) = \frac{1}{2g} - \sqrt{-3r_2}\left[\left(\frac{\chi + 1}{\chi - 1}\right)\cdot\frac{(1 - e^{2\phi}\chi^n)^2}{(1 - e^{2\phi}\chi^{n-1})(1 - e^{2\phi}\chi^{n+1})}\right.$$

$$\left. + \left(\frac{1 + e^{2\phi}\chi^n}{1 - e^{2\phi}\chi^n}\right)\cdot\frac{e^{2\phi}\chi^n(1 - \chi)^2}{\chi(1 - e^{2\phi}\chi^{n-1})(1 - e^{2\phi}\chi^{n+1})}\right].$$

We Taylor expand for large n (note that χ^n is small). First, the various pieces:

$$\frac{(1 - e^{2\phi}\chi^n)^2}{(1 - e^{2\phi}\chi^{n-1})(1 - e^{2\phi}\chi^{n+1})} = (1 - e^{2\phi}\chi^n)^2 \sum_{j=0}^{\infty} \left(e^{2\phi}\chi^{n-1}\right)^j \sum_{k=0}^{\infty} \left(e^{2\phi}\chi^{n+1}\right)^k$$

$$= 1 + \sum_{k=1}^{\infty} e^{2\phi k}\chi^{nk}$$

$$\times \left(\sum_{j=0}^{k-1} \left(\chi^{k-2j} - 2\chi^{k-1-2j} + \chi^{k-2-2j} \right) \right)$$

$$\left(\frac{1 + e^{2\phi}\chi^n}{1 - e^{2\phi}\chi^n} \right) = 1 + \sum_{k=1}^{\infty} 2e^{2\phi k}\chi^{nk}$$

$$\frac{e^{2\phi}\chi^n(1 - x)^2}{\chi(1 - e^{2\phi}\chi^{n-1})(1 - e^{2\phi}\chi^{n+1})} = e^{2\phi}\chi^n \left(x - 2 + \frac{1}{x} \right) \sum_{k=0}^{\infty} e^{2\phi k}\chi^{nk} \left(\sum_{j=0}^{k} \chi^{k-2j} \right)$$

$$= \left(x - 2 + \frac{1}{x} \right) \sum_{k=1}^{\infty} e^{2\phi k}\chi^{nk} \left(\sum_{j=0}^{k-1} \chi^{k-1-2j} \right)$$

$$= \sum_{k=1}^{\infty} e^{2\phi k}\chi^{nk}$$

$$\times \left(\sum_{j=0}^{k-1} \chi^{k-2j} - 2\chi^{k-1-2j} + \chi^{k-2-2j} \right).$$

The product of the last two simplifies as

$$\left(\frac{1 + e^{2\phi}\chi^n}{1 - e^{2\phi}\chi^n} \right) \cdot \frac{e^{2\phi}\chi^n(1 - x)^2}{\chi(1 - e^{2\phi}\chi^{n-1})(1 - e^{2\phi}\chi^{n+1})}$$

$$= \sum_{k=1}^{\infty} e^{2\phi k}\chi^{nk} \sum_{l=0}^{k-1} \left(\chi^{k-2l} - 2\chi^{k-1-2l} + \chi^{k-2-2l} \right)$$

$$+ \sum_{k=2}^{\infty} \sum_{m=1}^{k-1} e^{2\phi m}\chi^{nm}$$

$$\times \sum_{j=0}^{m-1} \left(\chi^{m-2j} - 2\chi^{m-1-2j} + \chi^{m-2-2j} \right) 2e^{2\phi(k-m)}\chi^{n(k-m)}$$

$$= \sum_{k=1}^{\infty} e^{2\phi k}\chi^{nk} \sum_{l=0}^{k-1} \left(\chi^{k-2l} - 2\chi^{k-1-2l} + \chi^{k-2-2l} \right)$$

$$+ \sum_{k=2}^{\infty} e^{2\phi k} \chi^{nk} \sum_{m=1}^{k-1} \sum_{j=0}^{m-1} 2 \left(\chi^{m-2j} - 2\chi^{m-1-2j} + \chi^{m-2-2j} \right)$$

$$= \sum_{k=1}^{\infty} e^{2\phi k} \chi^{n} \left(\chi^{k} - 2 + \chi^{-k} \right)$$

since for $k \geq 2$ the following sum holds by telescoping the series:

$$\sum_{l=0}^{k-1} \left(\chi^{k-2l} - 2\chi^{k-1-2l} + \chi^{k-2-2l} \right)$$

$$+ \sum_{m=1}^{k-1} \sum_{j=0}^{m-1} 2 \left(\chi^{m-2j} - 2\chi^{m-1-2j} + \chi^{m-2-2j} \right) = \chi^{k} - 2 + \chi^{-k}.$$

Now putting all these pieces together, we have

$$x(\xi + nv)$$

$$= \frac{1}{2g} - \sqrt{-3r_2} \left[\frac{\chi + 1}{\chi - 1} + (\chi + 1) \sum_{k=1}^{\infty} e^{2\phi k} \chi^{nk} \right.$$

$$\times \sum_{j=0}^{k-1} \frac{\chi^{k-2j} - 2\chi^{k-1-2j} + \chi^{k-2-2j}}{\chi - 1}$$

$$\left. + \sum_{k=1}^{\infty} e^{2\phi k} \chi^{n} \left(\chi^{k} - 2 + \chi^{-k} \right) \right]$$

$$= \frac{1}{2g} - \sqrt{-3r_2} \left[\frac{\chi + 1}{\chi - 1} + (\chi + 1) \sum_{k=1}^{\infty} e^{2\phi k} \chi^{nk} \sum_{j=0}^{k-1} \left(\chi^{k-2j-1} - \chi^{k-2-2j} \right) \right.$$

$$\left. + \sum_{k=1}^{\infty} e^{2\phi k} \chi^{n} \left(\chi^{k} - 2 + \chi^{-k} \right) \right]$$

$$= \frac{1}{2g} - \sqrt{-3r_2} \left[\frac{\chi + 1}{\chi - 1} + \sum_{k=1}^{\infty} e^{2\phi k} \chi^{nk} \sum_{j=0}^{k-1} \left(\chi^{k-2j} - \chi^{k-2-2j} \right) \right.$$

$$\left. + \sum_{k=1}^{\infty} e^{2\phi k} \chi^{n} \left(\chi^{k} - 2 + \chi^{-k} \right) \right]$$

$$= \frac{1}{2g} - \sqrt{-3r_2} \left[\frac{\chi + 1}{\chi - 1} + \sum_{k=1}^{\infty} e^{2\phi k} \chi^{nk} \left(\chi^{k} - \chi^{-k} \right) \right.$$

$$\left. + \sum_{k=1}^{\infty} e^{2\phi k} \chi^{n} \left(\chi^{k} - 2 + \chi^{-k} \right) \right]$$

$$= \frac{1}{2g} - \sqrt{-3r_2} \left[\frac{\chi + 1}{\chi - 1} + \sum_{k=1}^{\infty} e^{2\phi k} \chi^{nk} 2 \left(\chi^{k} - 1 \right) \right].$$

The constants out front are simplified by first observing that

$$\sqrt{-3r_2} = R \cdot \frac{\chi^2 - 1}{2\chi},$$

by writing the fraction in χ as a product of the hyperbolic sine and cosine and squaring both sides of the equation, so that everything can be reduced to its expression in terms of γ. Next, we claim that

$$\chi + \frac{1}{\chi} = \frac{1 - 4gR}{gR},$$

which is easily proved by expressing $\sinh\left(v\sqrt{-3r_2}\right)$ in terms of χ and then reducing everything in sight to its expression as a function of γ. Using this second identity, it is also easily proved that

$$\frac{1}{2gR} - \frac{(\chi+1)^2}{2\chi} = 1,$$

from which the expression for $x(\xi + nv)$ is finally re-summable:

$x(\xi + nv)$

$$= R\left[1 - \frac{\chi^2 - 1}{\chi} \sum_{k=1}^{\infty} e^{2\phi k}\left(\chi^{(n+1)k} - \chi^{nk}\right)\right]$$

$$= R\left[1 - \frac{\chi^2 - 1}{\chi}\left(\frac{1}{1 - e^{2\phi}\chi^{n+1}} - \frac{1}{1 - e^{2\phi}\chi^n}\right)\right]$$

$$= R \cdot \frac{\chi(1 - e^{2\phi}\chi^{n+1})(1 - e^{2\phi}\chi^n) - (\chi^2 - 1)(1 - e^{2\phi}\chi^n) + (\chi^2 - 1)(1 - e^{2\phi}\chi^{n+1})}{\chi(1 - e^{2\phi}\chi^n)(1 - e^{2\phi}\chi^{n+1})}$$

$$= R \cdot \frac{1 - e^{2\phi}\chi^n - e^{2\phi}\chi^{n+1} + e^{4\phi}\chi^{2n+1} + e^{2\phi}\chi^{n+1} - e^{2\phi}\chi^{n-1} - e^{2\phi}\chi^{n+2} + e^{2\phi}\chi^n}{(1 - e^{2\phi}\chi^n)(1 - e^{2\phi}\chi^{n+1})}$$

$$= R \cdot \frac{1 + e^{4\phi}\chi^{2n+1} - e^{2\phi}\chi^{n-1} - e^{2\phi}\chi^{n+2}}{(1 - e^{2\phi}\chi^n)(1 - e^{2\phi}\chi^{n+1})}$$

$$= R \cdot \frac{(1 - e^{2\phi}\chi^{n-1})(1 - e^{2\phi}\chi^{n+2})}{(1 - e^{2\phi}\chi^n)(1 - e^{2\phi}\chi^{n+1})}.$$

$\square$

5 Elliptic Parametrization, $c = 0$

5.1 The Generic Orbit

The discrete system (1) with $c = 0$ has the invariant, (6),

$$e = (1 + g)xy - y(1 + gx^2) - x(1 + gy^2) + g^2x^2y^2 + \frac{1}{g}, \qquad (46)$$

which is quartic, or in fact bi-quadratic. This does not lend itself naturally to a canonical Weierstrass parametrization as in the case of $c = 1$ where the invariant was cubic. Therefore we will be taking a more geometric approach to the analysis of the $c = 0$ case.

For notational convenience we will sometimes pass to the translated energy variable

$$E = e - \frac{1}{g}.$$

To understand the elliptic parametrization of a generic level set of (6) we pass to a projective completion of the $x - y$ plane as $\mathbb{P}^1 \times \mathbb{P}^1$ in which the dynamical system becomes

$$[\bar{x}_0 : \bar{x}_1] = [gx_1y_0 : x_1y_0 - x_0y_0 - gx_1y_1] \tag{47}$$
$$[\bar{y}_0 : \bar{y}_1] = [x_0 : x_1].$$

This stems from its more usual form in affine coordinates:

$$\bar{x} = \frac{x - 1}{gx} - y \tag{48}$$

$$\bar{y} = x. \tag{49}$$

This mapping is reversible with inverse given by

$$x = \bar{y}$$
$$y = \frac{\bar{y} - 1}{g\bar{y}} - \bar{x}.$$

The invariant (46) for this system can also be expressed in homogeneous coordinates:

$$I([\bar{x}_0 : \bar{x}_1], [\bar{y}_0 : \bar{y}_1]) = (1 + g)x_0x_1y_0y_1 - y_0y_1(x_0^2 + gx_1^2)$$
$$- x_0x_1(y_0^2 + gy_1^2) + g^2x_1^2y_1^2 - Ex_0^2y_0^2 = 0. \tag{50}$$

One may check directly that the level sets of this invariant are generically smooth and, in particular, they are all smooth along the lines at infinity: $x_0 = 0$ and $y_0 = 0$.

Now we may consider the Segre embedding of $\mathbb{P}^1 \times \mathbb{P}^1$ into $\mathbb{P}^3$ given by

$$([\bar{x}_0 : \bar{x}_1], [\bar{y}_0 : \bar{y}_1]) \rightarrow [Z_0 : Z_1 : Z_2 : Z_3] \tag{51}$$
$$= [x_0y_0 : x_0y_1 : x_1y_0 : x_1y_1]. \tag{52}$$

It is immediate from this representation that the Segre map embeds $\mathbb{P}^1 \times \mathbb{P}^1$ as a quadratic surface in $\mathbb{P}^3$ whose equation is given by

$$F([Z_0 : Z_1 : Z_2 : Z_3]) = Z_0Z_3 - Z_1Z_2 = 0.$$

It is also straightforward to check that, for each value of E, the invariant curve (50) corresponds to the intersection of the surface $\{F = 0\}$ with another quadric surface in $\mathbb{P}^3$ explicitly given by

$$G_E([Z_0 : Z_1 : Z_2 : Z_3]) = (1 + g)Z_0 Z_3 - (Z_0 Z_1 + g Z_2 Z_3)$$
$$- (Z_0 Z_2 + g Z_1 Z_3) + g^2 Z_3^2 - E Z_0^2 = 0.$$

It is well known that the intersection of two smooth quadrics in $\mathbb{P}^3$ is, generically, a smooth elliptic curve: a smooth space curve of degree 4, which we will sometimes refer to as a space quartic.

Returning to the energy parameter e, the invariant is seen to factor as

$$G_e([Z_0 : Z_1 : Z_2 : Z_3]) = (g Z_3 + Z_0)\left(g Z_3 + \frac{1}{g} Z_0 - Z_1 - Z_2\right) - e Z_0^2.$$

We work with two additional models of the level set associated to this fundamental model of the elliptic space curve. These are based on two projections:

$$\pi_1 : \mathbb{P}^3 \to \mathbb{P}^2$$
$$[Z_0 : Z_1 : Z_2 : Z_3] \to [Z_0 : Z_1 : Z_2];$$

$$\pi_2 : \mathbb{P}^3 \to \mathbb{P}^1$$
$$[Z_0 : Z_1 : Z_2 : Z_3] \to [Z_0 : Z_3].$$

To understand the geometric meaning of these projections, set $t = (g Z_3 + Z_0)$, and also note that from (51–52) one has

$$x = \frac{Z_2}{Z_0} \tag{53}$$

$$y = \frac{Z_1}{Z_0} \tag{54}$$

$$\frac{t - 1}{g} = \frac{Z_3}{Z_0} = xy \tag{55}$$

$$\frac{g t^2 + (1 - g)t - eg}{gt} = \frac{Z_1 + Z_2}{Z_0} = x + y \tag{56}$$

$$\Delta \doteq \frac{Z_2 - Z_1}{Z_0} = x - y. \tag{57}$$

(53–54) show that the image of the elliptic space curve under π_1 is nothing but the invariant level set in the original coordinates, (46), completed in the projective plane $\mathbb{P}^2$. This image is, for generic values of E, a smooth quartic curve in the affine plane but with two double points on the line at infinity at $[0 : 1 : 0]$ and $[0 : 0 : 1]$ respectively. Since the space quartic is, generically, a smooth space curve, it follows that the pull-back π_1^{-1} effectively desingularizes the plane quartic; in fact, it amounts to a blow-up of $\mathbb{P}^2$ at $[0 : 1 : 0]$ and $[0 : 0 : 1]$.

On the other hand, (55) shows that π_2 is projection onto the t-line where $t = 1 + gxy$, which presents the space curve as a double cover of $\mathbb{P}^1$. To see this we note from (55) that fixing t determines $\frac{Z_3}{Z_0} = xy$; then, (56) determines $\frac{Z_1+Z_2}{Z_0} = x + y$. So then $Z_2/Z_0 = x$ and $Z_1/Z_0 = y$ are determined up to two choices by the sign of Δ in (57). This shows that the *elliptic involution* of the underlying abstract elliptic curve is realized on the space quartic by the rational involution which interchanges Z_1 and Z_2 in $\mathbb{P}^3$. (Consistent with this, we note that this involution preserves both F and G_E and hence must induce an involution of the space quartic.) We can make use of this involution to explicitly define the double cover of the t-line.

Recall that the Weierstrass points of an elliptic curve are the fixed points of its elliptic involution. This is realized in different ways in the various models of the curve that we have been describing. In the Segre (space) model, we have just seen that the involution is realized by exchanging Z_1 and Z_2. Hence the fixed points in this model are the four points of intersection of the space quartic with the plane $Z_1 - Z_2 = 0$. Making the consequent substitutions $Z_2 = Z_1, Z_0 = 1, Z_3 = Z_1^2$ in the factored form G_e, one reduces to

$$(gZ_1^2 + 1)(gZ_1^2 - 2Z_1 + 1/g) = e.$$

Setting the first factor equal to t and the second factor to e/t and solving for t in terms of e and g one derives the equation for the Weierstrass points in terms of the zeroes of

$$\Delta(t) = g^2 t^4 - 2g(g+1)t^3 + ((1+g)^2 - 2eg^2)t^2 + 2eg(g-1)t + e^2 g^2. \tag{58}$$

More commonly these are referred to as the *branch points* of the projection π_2 from the space quartic onto the t-line. We have denoted the function in (58) by $\Delta(t)$ because it is straightforward to check that this quartic is proportional to Δ defined in (57), consistent with the fact that the elliptic involution is given by interchanging Z_1 and Z_2. Indeed, the elliptic irrationality corresponds to $\sqrt{\Delta}$.

The elliptic involution in the plane curve model is realized by exchanging x and y: $(x, y) \to (y, x)$ on the curve. The four Weierstrass points of the invariant curve in the plane are the fixed points determined by setting $x = y$ in (46). So these fixed points are of the form (x, x) where x solves

$$g^2 x^4 - 2gx^3 + (1+g)x^2 - 2x - E = 0. \tag{59}$$

We can relate these points to the zeroes of $\Delta(t)$ which we denote by $t_i, i = 0, \ldots, 3$. Using (56) these respective coordinates of the Weierstrass points can be related by setting $x = y$:

$$x_i = y_i = \frac{1}{2}\left(t_i + \frac{1}{g} - 1 - \frac{e}{t_i}\right),$$

which then must also be the solutions of (59).

We may further make use of these observations about the elliptic involution to directly parametrize x and y in terms of elliptic functions. The function Z_i restricted to the quartic curve which is the intersection $\{F = 0\} \cap \{G_E = 0\}$ is necessarily a fourth order theta function on the curve [FK01]. It has four zeroes on the curve which we can determine by algebra. We illustrate this in the case of Z_0. If $Z_0 = 0$, it follows from the form of F that one must also have that either $Z_1 = 0$ or $Z_2 = 0$. It then also follows from the form of G_e that either $Z_3 = 0$ or $gZ_3 - Z_1 - Z_2 = 0$. Hence the points of intersection in $\{Z_0 = 0\} \cap \{F = 0\} \cap \{G_e = 0\}$ are $[0 : 0 : 1 : 0] + [0 : 1 : 0 : 0] + [0 : 0 : g : 1] + [0 : g : 0 : 1]$, representing them as a formal linear combination of the intersection points. We call this the divisor of Z_0 and denote it by (Z_0). One can similarly work out the divisors of the other Z_i. Here are the results:

$(Z_0) = [0 : 0 : 1 : 0] + [0 : 1 : 0 : 0] + [0 : 0 : g : 1] + [0 : g : 0 : 1].$

$(Z_1) = 2[0 : 0 : 1 : 0] + [0 : 0 : g : 1] + [1 : 0 : -E : 0].$

$(Z_2) = 2[0 : 1 : 0 : 0] + [1 : -E : 0 : 0] + [0 : g : 0 : 1].$

$(Z_3) = [0 : 0 : 1 : 0] + [0 : 1 : 0 : 0] + [1 : -E : 0 : 0] + [1 : 0 : -E : 0].$

From (53–54) we can then also determine the divisors of x and y for which there is significant cancellation:

$(x) = [0 : 1 : 0 : 0] + [1 : -E : 0 : 0] - [0 : 0 : 1 : 0] - [0 : 0 : g : 1]$ (60)

$(y) = [0 : 0 : 1 : 0] + [1 : 0 : -E : 0] - [0 : 1 : 0 : 0] - [0 : g : 0 : 1].$ (61)

Now let us pass to the projection π_2 and consider the uniformizing variable ξ given by the abelian integral of the first kind

$$\xi = \int^t \frac{gd\tau}{\sqrt{\Delta(\tau)}}. \tag{62}$$

This sets up a mapping

$$(x, y) \rightarrow (gxy + 1, sgn(x - y)) = \left(t, sgn\sqrt{\Delta(t)}\right) \rightarrow \xi$$

from a point on the space quartic to the parameter ξ on the universal cover of the elliptic curve. Modulo the periods of ξ, this map is a globally analytic homeomorphism.

By standard results on the Weierstrass sigma function, σ [Mar77], the uniformization (62) and the divisor information (60–61), it follows that the functions x and y can be parameterized as

$$x = c_1 \frac{\sigma(\xi - a_1)\sigma(\xi - a_2)}{\sigma(\xi - b_1)\sigma(\xi - b_2)}$$

$$y = c_2 \frac{\sigma(\xi - b_1)\sigma(\xi - a_3)}{\sigma(\xi - a_1)\sigma(\xi - b_3)}$$

where we have the correspondence

$$a_1 \leftrightarrow [0:1:0:0]$$

$$b_1 \leftrightarrow [0:0:1:0]$$

$$a_2 \leftrightarrow [1:-E:0:0]$$

$$b_2 \leftrightarrow [0:0:g:1]$$

$$a_3 \leftrightarrow [1:0:-E:0]$$

$$b_3 \leftrightarrow [0:g:0:1]$$

with ξ-coordinates on the left and points on the space quartic on the right. The c_i are some constants to be determined. We mentioned earlier that the π_1^{-1} desingularizes the plane quartic by pulling back to the Segre embedding. We can now make this more precise by observing from the above that it is the two points a_1 and b_3 that lie over $[0:1:0]$, and b_1 and b_2 that lie over $[0:0:1]$. (We will observe in the next paragraph that $b_1 = -a_1$ and $b_3 = -b_2$ so that these pairs are elliptic involutes of one another just as their base points are.)

Using the fact that the elliptic involution on the space quartic is given by the interchange of Z_1 and Z_2, it follows that the following pairs are in involution on the ξ-plane

$$a_1 \quad b_1$$

$$a_2 \quad a_3$$

$$b_2 \quad b_3.$$

But in ξ the elliptic involution is given by sign change, meaning, that

$$b_1 = -a_1$$

$$a_3 = -a_2$$

$$b_3 = -b_2$$

so that the sigma function representations reduce to the following fundamental parametrization

$$x(\xi) = c_1 \frac{\sigma(\xi - a_1)\sigma(\xi - a_2)}{\sigma(\xi + a_1)\sigma(\xi - b_2)} \tag{63}$$

$$y(\xi) = c_2 \frac{\sigma(\xi + a_1)\sigma(\xi + a_2)}{\sigma(\xi - a_1)\sigma(\xi + b_2)}. \tag{64}$$

We also record here, for later use, an application of the fundamental divisor relation for elliptic functions [Mar77]:

$$2a_1 + a_2 - b_2 \equiv 0 \mod \{2\omega_1, 2\omega_2\} \tag{65}$$

where $2\omega_1, 2\omega_2$ are fundamental periods for the elliptic space curve.

To help pin down the coefficients c_i, let us observe from the $\xi - (x, y)$ correspondence above that

$$\xi = a_2 \quad \leftrightarrow \quad (x, y) = (0, -E)$$
$$\xi = -a_2 \quad \leftrightarrow \quad (x, y) = (-E, 0).$$

Using the fact that σ is an odd function in (63–64) one has

$$\begin{aligned}
x(-\xi) &= c_1 \frac{\sigma(-\xi - a_1)\sigma(-\xi - a_2)}{\sigma(-\xi + a_1)\sigma(-\xi - b_2)} \\
&= c_1 \frac{\sigma(\xi + a_1)\sigma(\xi + a_2)}{\sigma(\xi - a_1)\sigma(\xi + b_2)} \\
&= \frac{c_1}{c_2} y(\xi).
\end{aligned}$$

Then setting $\xi = a_2$ in this relation and using the previous identifications yields

$$\begin{aligned}
-E &= x(-a_2) \\
&= \frac{c_1}{c_2} y(a_2) = -\frac{c_1}{c_2} E.
\end{aligned}$$

Hence, $c_1 = c_2 \doteq c$. It follows that $x(\xi)$ and $y(\xi)$ are in fact elliptic involutes of each other. As a corollary we see that elliptic involution of the level set in the (x, y)-plane is induced by the coordinate exchange $(x, y) \to (y, x)$ and that the second component (49) of the dynamic map is itself induced by elliptic involution. This calculation shows that

$$c = -\frac{cE}{x(-a_2)} = E \frac{\sigma(a_1 - a_2)\sigma(a_2 + b_2)}{\sigma(a_1 + a_2)\sigma(2a_2)}. \tag{66}$$

Let us now relate the above observations to the parametrization of the map (48–49). The fact is that this dynamical system expresses an elegant geometric construction.

Consider an initial condition (x_0, y_0) that corresponds to a point ξ_0 on the abstract elliptic curve and its first iterate (x_1, y_1) corresponding to another point ξ_1 on the abstract elliptic curve. In other words

$$(x_0, y_0) = (x(\xi_0), y(\xi_0))$$
$$(x_1, y_1) = (x(\xi_1), y(\xi_1))$$

where $x(\xi)$ is defined by (63).

Starting with the initial point one fixes the vertical line $x = x_0$. This line meets the plane quartic (46) in four points, two of which are the (multiplicity 2) double point $[0 : 1 : 0]$ at infinity. A third point is, of course, the initial point (x_0, y_0), which we know lies on the plane quartic. To find the remaining point is a matter of algebra; its y-coordinate is the other root of

$$gx_0(gx_0 - 1)y^2 - (gx_0^2 - (1 + g)x_0 + 1)y$$
$$- \left(gx_0(gx_0 - 1)y_0^2 - (gx_0^2 - (1 + g)x_0 + 1)y_0\right).$$

After cancellations we find that this root is simply

$$\frac{x_0 - 1}{gx_0} - y_0;$$

So the residual point of intersection of the line $x = x_0$ with the invariant curve is

$$\left(x_0, \frac{x_0 - 1}{gx_0} - y_0\right).$$

Finally, applying the elliptic involution to this point yields our map

$$(x_1, y_1) = \left(\frac{x_0 - 1}{gx_0} - y_0, x_0\right).$$

So, in summary, our dynamic map can be described in completely geometric terms: Starting with an initial point (x_0, y_0) on the planar invariant curve one finds the residual point of intersection of the line $x = x_0$ with the curve and then flips it across the diagonal $x = y$ to produce (x_1, y_1). Clearly this process can be arbitrarily iterated. It can also be reversed to yield the inverse map by starting with the vertical line $y = y_0$ and proceeding analogously.

Hence our map determines an algebro-geometric addition law on each of the affinely smooth level sets of the invariant. (We shall see in the next section that this extends to singular affine level sets as well.) Indeed, this type of addition law has been studied before [Gri76] and shown to linearize on the abstract elliptic curve. Hence, there is a phase v such that $\xi_1 = \xi_0 + v$ and the higher iterates $x_n = x(\xi_n)$ are determined by $\xi_n = \xi_0 + nv$. To pin down this phase we apply Abel's theorem, as described in [Gri76], which in our setting implies that if A, B, C, D are four points of intersection of a line in the plane with the planar quartic (46) then

$$\xi(A) + \xi(B) + \xi(C) + \xi(D) = K$$

where K is a constant *independent* of the planar line one considers. So for our case, in choosing the line $x = x_0$ one has

$$\xi_0 + 2\xi([0:1:0]) - \xi_1 = K.$$

(The factor of 2 here comes from the fact that $[0:1:0]$ is a double point on the curve.) But in fact, as we observed, this linear relation holds for any iterate and so one has in general

$$\xi_n + 2\xi([0:1:0]) - \xi_{n+1} = K. \tag{67}$$

To pin down the constant K one can consider the line $x = 0$ in $\mathbb{P}^2$. Starting with the standard projectivization in coordinates $[x:y:z]$ of (46),

$$(1+g)xyz^2 - yz(z^2+gx^2) - xz(z^2+gy^2) + g^2x^2y^2 - Ez^4 = 0. \tag{68}$$

Considering this in the vicinity of the point $[0:1:0]$ and setting $y = 1$, the invariant has the form

$$(1+g)xz^2 - z(z^2+gx^2) - xz(z^2+g) + g^2x^2 - Ez^4 = 0. \tag{69}$$

To leading order near $[0:1:0]$ this has the form $0 = gx(gx - z) +$ higher order terms, and so this point is a double point of the curve (as we had already observed) and the two branches of the curve there are $x = 0$ and $z = gx$. Hence the line $x = 0$ is tangent to the first branch and the line $x = 0$ must meet this double point with multiplicity 3. Indeed, this is confirmed by setting $x = 0$ in the LHS of (69) which evaluates to $-z^3$. There is therefore just one further point of intersection of the line with the affine part of the curve (68). By setting $z = 1$ and $x = 0$ to get

$$y = -E,$$

one sees that the third point of intersection is $[0:-E:1]$ which corresponds to a_2 in the Segre model. The points at infinity in the intersection of $x = 0$ with the curve correspond to $2a_1$, because of the tangency, and b_3. It then follows from Abel's theorem that

$$2a_1 + a_2 + b_3 = K$$
$$2a_1 + a_2 - b_2 = K$$
$$0 \equiv K$$

where the second line follows from the involution pairings and the third line follows from (65). Therefore the RHS of (67) vanishes modulo periods.

Putting all this together one has

$$\begin{aligned}
v = \xi_{n+1} - \xi_n &\equiv 2\xi([0:1:0]) \\
&= a_1 + b_3 \\
&= a_1 - b_2 \\
&\equiv -(a_1 + a_2),
\end{aligned} \tag{70}$$

which is clearly independent of n. The evaluations in the second and third equalities are made using the definitions of the a_j, b_j and the relations among them are determined by involution and (65).

Finally, we are in a position to present an elliptic parametrization of the map (48–49):

$$x(\xi + v) = \frac{1}{g} - \frac{1}{g}\frac{1}{c}\frac{\sigma(\xi + a_1)\sigma(\xi - b_2)}{\sigma(\xi - a_1)\sigma(\xi - a_2)} - c\frac{\sigma(\xi + a_1)\sigma(\xi + a_2)}{\sigma(\xi - a_1)\sigma(\xi + b_2)} \tag{71}$$

$$y(\xi + v) = c\frac{\sigma(\xi - a_1)\sigma(\xi - a_2)}{\sigma(\xi + a_1)\sigma(\xi - b_2)}. \tag{72}$$

Making use of (63) in (71) we see that the poles on the LHS of (71) are located at $\xi = -v - a_1 = a_2$ and $\xi = -v + b_2 = 3a_1 + 2a_2$ while those on the RHS must occur among the values $\xi = a_1, \xi = a_2$ and $\xi = -b_2$. (One of these latter values cannot occur as a pole in order to maintain the required balance between the number of poles on the two sides of equation (71); i.e. upon passing to a common denominator on the LHS, one of the zeroes of the numerator must cancel one of these possible polar values.) So upon comparison we see that a_2 is a pole. On the other hand, a_1 must be the potential pole on the RHS that gets cancelled, since otherwise we would deduce that $a_2 = -a_1$ which we know from our earlier analysis is not the case. Hence we may deduce that

$$\begin{aligned}
3a_1 + 2a_2 &= -b_2 \\
3a_1 + 2a_2 &= -2a_1 - a_2 \\
5a_1 + 3a_2 &= 0.
\end{aligned} \tag{73}$$

5.2 Separatrix Degeneration to the Combinatorial Orbit

We now specialize our considerations to the case of the invariant level curve which passes through the unique hyperbolic fixed point of the map (47) which has energy level e_1. This energy is given as a function of g in Table 10.1:

$$e = \frac{1 + 20g - 8g^2 + (8g - 1)\sqrt{1 - 8g}}{32g^2}$$

which we will continue to denote by e in this section. Setting $g = \dfrac{1 - \gamma^2}{8}$ we may express this and other parameters relevant to the degenerate level set purely in terms of γ:

$$g = \frac{1 - \gamma^2}{8} \tag{74}$$

$$\gamma = \sqrt{1 - 8g} \tag{75}$$

$$e = \frac{(\gamma + 3)^3}{4(1 + \gamma)(1 - \gamma^2)} \tag{76}$$

$$x_* = \frac{1 - \sqrt{1 - 8g}}{4g} \tag{77}$$

$$= \frac{2}{1 + \gamma} \tag{78}$$

where x_* denotes the fixed point, (4), of the separatrix. Given these definitions we also have a concise expression for the eigenvalues of the linearized combinatorial orbit at x_*

$$\chi^{\pm 1} = \frac{1 \pm \sqrt{1 - 4g^2 x_*^4}}{2g x_*^2} \tag{79}$$

$$= \left(\frac{1 + \sqrt{\gamma}}{1 - \sqrt{\gamma}} \right)^{\pm 1}. \tag{80}$$

In addition, the t-equation degenerates to

$$0 = [2(1 - \gamma)t - (3 - \gamma)]^2 [4(\gamma^2 - 1)^2 t^2$$

$$- 4(1 - \gamma^2)(\gamma - 3)(\gamma + 5)t + (\gamma - 3)^4]$$

whose roots are

$$t_0, t_1 = \frac{\gamma - 3}{2(\gamma^2 - 1)} \left(5 + \gamma \pm 4\sqrt{(1 + \gamma)} \right)$$

$$t_2 = t_3 = \frac{3 - \gamma}{2(1 - \gamma)} = \frac{1}{2} \left(1 + \frac{2}{1 - \gamma} \right).$$

By completing the square in the quadratic factor, the equation for t may be rewritten as

$$0 = (z + r_1)^2 \left(z^2 - r_2^2 \right) \tag{81}$$

$$r_1 = \frac{6}{1 - \gamma^2} \tag{82}$$

$$r_2^2 = 4 \frac{(\gamma - 3)^2}{(\gamma - 1)^2 (1 + \gamma)} \tag{83}$$

$$\sqrt{r_1^2 - r_2^2} = \frac{2\sqrt{-\gamma(\gamma^2 - 5\gamma + 3)}}{1 - \gamma^2} \tag{84}$$

$$z = t - \frac{1}{2} \frac{(3 - \gamma)(5 + \gamma)}{1 - \gamma^2}. \tag{85}$$

Here σ_d denotes the generalized sigma function that σ degenerates to on the singular curve that (46) limits to. We need to determine the form of σ_d. We observe that the equation (59), determining the Weierstrass points, degenerates in this limit to

$$((1 + \gamma)x - 2)^2((1 - \gamma)^2 x^2 - 4(1 - \gamma^2)(\gamma + 3)x + 4(\gamma^2 + 10\gamma + 5)) = 0.$$

The sigma function may be explicitly expressed in terms of the odd Jacobi theta function ϑ_1 [Mar77] as

$$\sigma(\xi) = 2\omega_1 \frac{\vartheta_1(v)}{\vartheta_1'(0)} e^{2\omega_1 \eta_1 v^2}$$

where $v = \frac{\xi}{2\omega_1}$ and ϑ_1 depends on the modular parameter $\tau = \omega_2/\omega_1$ where ω_1 is real. η_1 is an integration constant of the Weierstrass ζ-function. The separatrix limit here in which t_2 and t_3 coalesce (or equivalently $x(\omega_2)$ and $x(\omega_3)$ coalesce) corresponds to the infinite period limit in which $\tau \to \infty$. In this limit, ϑ_1 must converge to a linear combination of exponentials of the form $e^{s\xi}$ and $e^{-s\xi}$ which is odd since ϑ_1 remains odd throughout the limit. Hence in fact it must be a multiple of $\sinh(s\xi)$. Then, setting $s = \lim \frac{1}{2\omega_1}$ and $\rho = \lim \frac{\eta_1}{2\omega_1}$, the limiting form is

$$\sigma_d(\xi) = e^{\rho \xi^2} \frac{\sinh(s\xi)}{s}. \tag{86}$$

One has the implicit integral inversion:

$$\xi = \int \frac{dz}{(z + r_1)\sqrt{z^2 - r_2^2}}. \tag{87}$$

Setting $z = r_2 \cosh(x)$ this transforms to

$$\xi = \int \frac{dx}{r_2 \cosh(x) + r_1}$$

$$= \frac{2}{r_2} \int \frac{du}{u^2 + \frac{2r_1}{r_2}u + 1}$$

$$= \frac{2}{\sqrt{r_2^2 - r_1^2}} \tan^{-1} \frac{r_2 u + r_1}{\sqrt{r_2^2 - r_1^2}} + \alpha$$

where $u = e^x$. Unravelling the substitutions and setting $z = r_2 w$ yields

$$\sqrt{1 - (r_1/r_2)^2} \, \tan\left(\frac{\sqrt{r_2^2 - r_1^2}}{2} (\xi - \alpha) \right) = (w + r_1/r_2) \pm \sqrt{w^2 - 1}$$

$$\sqrt{(r_1/r_2)^2 - 1} \, \tanh\left(\frac{\sqrt{r_1^2 - r_2^2}}{2} (\xi - \alpha) \right) - r_1/r_2 = w \pm \sqrt{w^2 - 1}$$

$$\frac{\sqrt{-\gamma(\gamma^2 - 5\gamma + 3)}}{(3 - \gamma)\sqrt{1 + \gamma}} \, \tanh\left(\frac{\sqrt{-\gamma(\gamma^2 - 5\gamma + 3)}}{1 - \gamma^2} (\xi - \alpha) \right) - \frac{3}{(\gamma - 3)\sqrt{1 + \gamma}} = w \pm \sqrt{w^2 - 1}$$

$$\frac{1}{(3 - \gamma)\sqrt{1 + \gamma}} \left(\sqrt{-\gamma(\gamma^2 - 5\gamma + 3)} \, \tanh\left(\frac{\sqrt{-\gamma(\gamma^2 - 5\gamma + 3)}}{1 - \gamma^2} (\xi - \alpha) \right) + 3 \right)$$

$$= w \pm \sqrt{w^2 - 1}.$$

This last expression presents ξ as a *closed form* branched 2:1 cover of the w plane or, by scaling and translation, the t-plane. We can also pin down the values at branch points:

$$z = -r_1 \iff \xi - \alpha = \pm\infty$$

$$z = \pm r_2 \iff \xi - \alpha = \frac{2}{\sqrt{r_1^2 - r_2^2}} \tanh^{-1} \left(\frac{r_1 + r_2}{r_1 - r_2} \right)^{\pm 1/2}.$$

We can apply the above inversion analysis to the near-separatix linearization. Setting $z = -r_1 + q$ in (87) gives

$$\xi = \int_\epsilon \frac{dq}{q \sqrt{(r_1 - q)^2 - r_2^2}}$$

$$-\sqrt{r_1^2 - r_2^2} \, \xi = \log \epsilon (1 + \mathcal{O}(\epsilon / \log \epsilon))$$

$$\epsilon = e^{-\sqrt{r_1^2 - r_2^2} \, \xi} (1 + o(1)).$$

Comparing to the inversion of (87) detailed above, one may infer that

$$s = \frac{1}{2}\sqrt{r_1^2 - r_2^2} \tag{88}$$

and relate this to the eigenvalues at the linearization at the separatrix: the linearization realized in (70) presents the dynamics as a phase increment on the covering space of the planar quartic. This fact extends to the degenerate limit of the separatrix. This flow is the tangent space flow near the fixed point and so must coincide with the linearized flow in the directions of the stable or the unstable manifolds. But these flows, we know, are just the power flows of the respective eigenvalues, which are the exponentials of these phase translations. It follows that

$$\chi^{\pm} = e^{\mp 2sv}. \tag{89}$$

One can now say more about the dynamical system (48–49) on the separatrix level set. The parametrization on this level set becomes

$$x(\xi) = c \; e^{\rho(a_2^2 - b_2^2)} e^{-2\rho(2a_1 - b_2 + a_2)\xi} \frac{\sinh(s(\xi - a_1)) \sinh(s(\xi - a_2))}{\sinh(s(\xi + a_1)) \sinh(s(\xi - b_2))}$$

$$= c \; e^{\rho(a_2^2 - b_2^2)} e^{-2\rho(2a_1 - b_2 + a_2)\xi} e^{-s(2a_1 + a_2 - b_2)}$$

$$\times \frac{\left(1 - e^{-2s(\xi - a_1)}\right) \left(1 - e^{-2s(\xi - a_2)}\right)}{\left(1 - e^{-2s(\xi + a_1)}\right) \left(1 - e^{-2s(\xi - b_2)}\right)}$$

$$y(\xi) = x(-\xi)$$

We note that $2a_1 - b_2 + a_2$ is the limit of divisors of an elliptic function as noted in (65) and therefore must equal the limit of a linear combination of periods. By re-centering the origin of ξ we may assume this sum is zero. Independently, this is also verified from the dynamical systems perspective since, otherwise, x would blow up or vanish as n goes to infinity which our phase plane analysis has shown is not the case. (Indeed, this argument shows that the real part of $2a_1 - b_2 + a_2$ must already be zero, before any centering.) Hence, the expression for the first component reduces (using $b_2 = 2a_1 + a_2$) to

$$x(\xi) = c \; e^{-4\rho a_1(a_1 + a_2)} \frac{\left(1 - (e^{-2sv})^{\frac{\xi - a_1}{v}}\right) \left(1 - (e^{-2sv})^{\frac{\xi - a_2}{v}}\right)}{\left(1 - (e^{-2sv})^{\frac{\xi + a_1}{v}}\right) \left(1 - (e^{-2sv})^{\frac{\xi - 2a_1 - a_2}{v}}\right)}.$$

In the separatrix limit, as $\xi \to \infty$, we must have

$$ce^{-4\rho a_1(a_1 + a_2)} = x_*. \tag{90}$$

Hence, one finally has

$$x(\xi) = x_* \frac{\left(1 - (e^{-2sv})^{\frac{\xi - a_1}{v}}\right)\left(1 - (e^{-2sv})^{\frac{\xi - a_2}{v}}\right)}{\left(1 - (e^{-2sv})^{\frac{\xi + a_1}{v}}\right)\left(1 - (e^{-2sv})^{\frac{\xi - 2a_1 - a_2}{v}}\right)} \tag{91}$$

$$y(\xi) = x(-\xi). \tag{92}$$

To establish (17) it remains to impose the initial condition (16). In our current framework this means that we need to find ξ_0 such that $x(\xi_0 - v) = 0$. This is achieved in terms of (91) by setting

$$\xi_0 - v - a_1 = 0$$

$$\xi_0 + a_1 + a_2 - a_1 = 0$$

$$\xi_0 = -a_2.$$

With this in place, the combinatorial orbit is given by

$$x(\xi_0 + nv) = x_* \frac{\left(1 - (e^{-2sv})^{n+1}\right)\left(1 - (e^{-2sv})^{n + \frac{2a_2}{a_1 + a_2}}\right)}{\left(1 - (e^{-2sv})^{n + \frac{a_2 - a_1}{a_1 + a_2}}\right)\left(1 - (e^{-2sv})^{n+2}\right)}$$

$$= x_* \frac{\left(1 - (e^{-2sv})^{n+1}\right)\left(1 - (e^{-2sv})^{n + \frac{a_2 - a_1}{a_1 + a_2} + 1}\right)}{\left(1 - (e^{-2sv})^{n + \frac{a_2 - a_1}{a_1 + a_2}}\right)\left(1 - (e^{-2sv})^{n+2}\right)}$$

$$= x_* \frac{\left(1 - (e^{-2sv})^{n+1}\right)\left(1 - (e^{-2sv})^{n+5}\right)}{\left(1 - (e^{-2sv})^{n+4}\right)\left(1 - (e^{-2sv})^{n+2}\right)}$$

where the last line is justified by (73) which persists under degeneration because Abel's theorem does. In particular one sees that $\frac{a_2 - a_1}{a_1 + a_2} = 4$ is equivalent to $5a_1 + 3a_2 = 0$.

Finally we need to relate this to Bousquet-Melou's Z in (17). Comparing to (89) one has $\chi = \chi^+ = e^{-2sv}$ and so one may rewrite the combinatorial flow as

$$x(\xi_0 + nv) = x_* \frac{(1 - \chi^{n+1})(1 - \chi^{n+5})}{(1 - \chi^{n+4})(1 - \chi^{n+2})}.$$

It then suffices to show that the eigenvalue χ satisfies the equation (18) for Z. Starting with (79), and using the identity $2gx_*^2 = x_* - 1$ for the fixed point one has

$$\chi^{\pm} = \frac{1 \pm \sqrt{1 - (x_* - 1)^2}}{x_* - 1}$$

which is easily inverted by appropriate squaring to yield

$$x_* = \frac{(\chi + 1)^2}{\chi^2 + 1}$$

which finally establishes (17).

We close with a summary of the key results in this section:

Theorem 5.1 *The discrete integrable recurrence*

$$x_n = 1 + g x_n (x_{n-1} + x_{n+1}) \tag{93}$$

with invariant

$$I(x_n, x_{n-1}) = x_n x_{n-1} (1 - g x_n)(1 - g x_{n-1}) + g x_n x_{n-1} - x_n - x_{n-1} + \frac{1}{g} \tag{94}$$

and limit $\lim_{n \to \infty} x_n = x_-^*$ *satisfying* $x_-^* = 1 + 2g x_-^{*2}$ *possesses a solution*

$$x_n = x(\xi_0 + n\nu) = x_* \frac{\left(1 - \chi^{n+1}\right)\left(1 - \chi^{n+5}\right)}{\left(1 - \chi^{n+4}\right)\left(1 - \chi^{n+2}\right)}, \tag{95}$$

where $\nu = 2\xi([0 : 1 : 0])$ *under the composite Abel map determined by* (87), π_1 *and* π_2. *Here* $\chi = \left(\frac{1 + \sqrt{\gamma}}{1 - \sqrt{\gamma}}\right) = e^{-2sv}$ *where* $\gamma = \sqrt{1 - 8g}$ *and* $s = \frac{1}{2}\sqrt{r_1^2 - r_2^2}$. ξ_0 *is determined by the initial condition* $x(\xi_0 - \nu) = 0$.

Finally, making the combinatorial identification $T_n(g) = x_n$ *and* $T(g) = x_-^*$ *one has*

$$T_n = T \frac{(1 - \chi^{n+1})(1 - \chi^{n+5})}{(1 - \chi^{n+4})(1 - \chi^{n+2})} \tag{96}$$

for χ *satisfying* $T = \frac{(1+\chi)^2}{(1+\chi^2)}$.

6 Concluding Remarks

In this last section we mention a few directions that are currently under investigation for building on what was derived in this paper.

6.1 Combinatorial Problems in Non-Autonomous Extensions

The continuous Painlevé equations that give rise to Painlevé's famous six transcendents are non-autonomous differential equations. In the *autonomous limits* of these equations the transcendent limits to a classical function such as the elliptic functions discussed in this paper.

The discrete Painlevé equations also have natural non-autonomous extensions. In the case of dPI, we consider such an extension of (1), with $c = 1$ having the general form

$$x_{n+1} + x_{n-1} = \frac{n}{N} \frac{1}{g x_n} - \frac{\zeta}{g} - x_n, \tag{97}$$

g, N and ζ are parameters. Setting $\zeta = 1 = \frac{n}{N}$ and changing the sign of g, one recovers our autonomous dPI equation. One may now ask, are there orbits of this non-autonomous system having combinatorial significance related to what we saw in the autonomous case? The answer is yes and it comes from the analysis of the asymptotics of orthogonal polynomials and their relation to the enumeration of quadrangulations of surfaces of general genus. (The combinatorial problem studied in the autonomous case was for just planar maps; i.e., maps of genus 0.) This connection is mediated by the analysis of the Riemann-Hilbert problem for orthogonal polynomials as it relates to hermitian random matrix models, carried out in [EM03, EMP08, Erc11]. We briefly outline the essentials of this as it relates to (97).

Define the orthogonal polynomials $p_n(\lambda) = \gamma_n \lambda^n + \cdots$ for positive γ_n with respect to the exponential weight

$$w_\zeta(\lambda) = \exp\left(-N\left(\frac{\zeta}{2}\lambda^2 + \frac{g}{4}\lambda^4\right)\right),$$

for $g > 0$, so that

$$\int p_n(\lambda)p_m(\lambda)w_\zeta(\lambda)d\lambda = \delta_{nm}$$

for $n, m \geq 0$. These polynomials satisfy the three-term recurrence relation

$$\lambda p_{n,N}(\lambda) = b_{n+1,N} p_{n+1}(\lambda) + a_{n,N} p_n(\lambda) + b_{n,N} p_{n-1}(\lambda)$$

for $n \geq 0$ and $p_{-1} = 0$. Since the weight is even, the recurrence coefficients $a_{n,N}$ are all zero, and the polynomials are entirely defined by the recurrence coefficients $b_{n,N}$. These coefficients satisfy a *nonlinear* recurrence of their own:

$$g b_{n,N}^2 \left(b_{n+1,N}^2 + b_{n,N}^2 + b_{n-1,N}^2\right) + \zeta b_{n,N}^2 = \frac{n}{N}. \tag{98}$$

In approximation theory, (98) is referred to as Freud's equation. It is straightforward to see that this coincides with (97) if one sets $x_n = b_{n,N}^2$. The initial conditions that then relate these orthogonal polynomials recurrence coefficients to an orbit within our non-autonomous dPI system, (97), are $x_0 = 0$ and $x_1 = \frac{c_2}{c_0}$ where

$$c_j = \int \lambda^j w_\zeta(\lambda) d\lambda$$

is the jth moment of the measure. The latter is derived from the first equation in the recurrence relations,

$$\lambda p_0(\lambda) = b_1 p_1(\lambda) + b_0 p_{-1}(\lambda)$$
$$\lambda \gamma_0 = b_1 p_1(\lambda)$$

by the normalization requirement that $p_1(\lambda) = \frac{\lambda \gamma_0}{b_1}$ has norm 1:

$$\int p_1(\lambda) p_1(\lambda) w_\zeta(\lambda) d\lambda = \int \left(\frac{\lambda \gamma_0}{b_1} \right)^2 w_\zeta(\lambda) d\lambda$$
$$= \frac{\gamma_0^2}{b_1^2} \int \lambda^2 w_\zeta(\lambda) d\lambda$$
$$= 1,$$

so that

$$x_1 = b_1^2 = \gamma_0^2 \int \lambda^2 w_\zeta(\lambda) d\lambda.$$

Similarly, it is required that $p_0(\lambda) \gamma_0$ have norm 1:

$$\int p_0(\lambda) p_0(\lambda) w_\zeta(\lambda) d\lambda = \gamma_0^2 \int w_\zeta(\lambda) d\lambda$$
$$= 1.$$

So finally one has the second initial condition:

$$x_1 = \frac{\int \lambda^2 w_\zeta(\lambda) d\lambda}{\int w_\zeta(\lambda) d\lambda}.$$

The connection to the combinatorial problem of enumerating quadrangulations now stems from a result in [EMP08] where it is shown that for N large and with $X \doteq \frac{n}{N} \sim 1$, one has a full asymptotic expansion of $x_n = b_{n,N}^2$:

$$x_n \sim X \left(z_0(s) + \frac{1}{n^2} z_1(s) + \frac{1}{n^4} z_2(s) + \cdots \right)$$

where $s = -X \frac{g}{4}$. This expansion is uniformly valid on compact subsets of complex s with $\Re s < 0$. The coefficients $z_g(s)$ are the generating functions for 4-valent maps (whose dual maps are the quadrangulations in question) of genus g.

It will be of interest to determine how the combinatorial features just described are related to the dynamic properties of the orbit, corresponding to the orthogonal polynomial recurrence relations, of non-autonomous dPI. It will also be interesting to see if this can be related to the analysis of the autonomous case discussed in this paper. There is already one indication that such relations do hold: the generating function, $z_0(s)$, which enumerates planar 4-valent maps solves the same functional equation, (8), as does the limit $R(g)$ of the distance generating funcitons $R_n(g)$. Thus the non-autonomous x_n and the autonomous R_n agree at leading order in large n. These questions are currently under further investigation and will be reported on elsewhere.

As indicated at the start of section 3.1.2 there is an analogous relation between the combinatorics of (Eulerian) triangulations and the T_j generating functions [BFG03a]. The connection to recurrence relations for (generalized) orthogonal polynomials is more complicated on several levels in this case. For one thing, the coefficients a_n will now no longer automatically be zero. However, the analysis in [EP12] has shown how to handle this additional complication. Based on this we are currently exploring non-autonomous extensions of the $c = 0$ system.

6.2 Elliptic Combinatorics

The combinatorial focus of this paper has been on the separatrix orbit of (1). Is there a related combinatorial significance for the other, generic, elliptically parametrized orbits? One affirmative answer to this question has been provided in [BFG03a]. There the study of random embedded trees, corresponding to the sytem (1) with $c = 0$ but with the added structure of *walls* is discussed. A wall introduces a conditioning on the random system that strictly bounds the size of the labels/positions that the random process can attain. A "one-wall" case, which is that the half-line is bounded below at 0 or -1, reproduces the model we have been studying in this paper (introduced in section 3.1.2) that corresponds to the separatrix orbit. By contrast, a "two-wall" conditioning would require labels to take their values in a finite subinterval of $\mathbb{Z}$; i.e., it would replace the asymptotic boundary condition (16 b) by another finite boundary condition. From the dynamical point of view this corresponds to a two-point *finite* boundary value problem. The one-wall case we studied in this paper corresponded to a two-point *semi-infinite* boundary value problem. One may derive recursive formulas for the generating functions of such trees in the general two point boundary value problem of a two-wall conditioning. The effect of this is to select one of the generic orbits we described in section 5 whose closed form solutions for the generating functions will now be in terms of elliptic functions. These generating functions may be used to study the continuum scaling limit in terms of the periods of the associated elliptic curve, which may be applied to the probabilistic analysis of population spreading. We are studying how the geometric analysis developed in section 5 might be used in this application.

References

[AD99] D. Aldous and P. Diaconis. Longest increasing sequences: From patience sorting to the baik-deift-johansson theorem. *Bulletin (New Series) of the AMS*, 36:413–432, 1999.

[BDFG03] J. Bouttier, P. Di Francesco, and E. Guitter. Geodesic distance in planar graphs. *Nuclear Phys. B*, 663(3):535–567, 2003.

[BFG03a] J Bouttier, P Di Francesco, and E Guitter. Random trees between two walls: exact partition function. *Journal of Physics A: Mathematical and General*, 36(50):12349, 2003.

[BFG03b] J. Bouttier, P. Di Francesco, and E. Guitter. Statistics of planar graphs viewed from a vertex: a study via labeled trees. *Nuclear Physics B*, 675(3):631 – 660, 2003.

[BG12] J. Bouttier and E. Guitter. Planar maps and continued fractions. *Comm. Math. Phys.*, 309(3):623–662, 2012.

[BJ02] P. Bougerol and T. Jeulin. Paths in weyl chambers and random matrices. *Probab.Theory Related Fields*, 124:517–543, 2002.

[BM06] Mireille Bousquet-Mélou. Limit laws for embedded trees: applications to the integrated superBrownian excursion. *Random Structures Algorithms*, 29(4):475–523, 2006.

[BR86] E. Bender and L. B. Richmond. A survey of the asymptotic behaviour of maps. *Journal of Combinatorial Theory, Series B*, 40:297–329, 1986.

[BTR01] M Bernardo, TT Truong, and G Rollet. The discrete painlevé i equations: transcendental integrability and asymptotic solutions. *Journal of Physics A: Mathematical and General*, 34(15):3215, 2001.

[CV81] Robert Cori and Bernard Vauquelin. Planar maps are well labeled trees. *Canad. J. Math.*, 33(5):1023–1042, 1981.

[Dui10] J. J. 1942-(Johannes Jisse) Duistermaat. *Discrete integrable systems: QRT Maps and Elliptic Surfaces*. Springer, New York, 2010.

[EM03] N. M. Ercolani and K. D. T.-R. McLaughlin. Asymptotics of the partition function for random matrices via Riemann-Hilbert techniques and applications to graphical enumeration. *Int. Math. Res. Not.*, (14):755–820, 2003.

[EMP08] N. M. Ercolani, K. D. T-R McLaughlin, and V. U. Pierce. Random matrices, graphical enumeration and the continuum limit of Toda lattices. *Comm. Math. Phys.*, 278(1):31–81, 2008.

[EP12] Nicholas M. Ercolani and Virgil U. Pierce. The continuum limit of Toda lattices for random matrices with odd weights. *Commun. Math. Sci.*, 10(1):267–305, 2012.

[Erc11] N. M. Ercolani. Caustics, counting maps and semi-classical asymptotics. *Nonlinearity*, 24(2):481–526, 2011.

[FK01] Hershel Farkas and Irwin Kra. *Theta Constants, Riemann Surfaces and the Modular Group*. American Mathematical Society, Graduate Studies in Mathematics, 2001.

[FS09]   Philippe Flajolet and Robert Sedgewick. *Analytic combinatorics.* Cambridge University Press, Cambridge, 2009.

[Gri76] Phillip Griffiths. Variations on a theorem of abel. *Inventiones Math.,* 35:321–390, 1976.

[J$^+$13] Fredrik Johansson et al. *mpmath: a Python library for arbitrary-precision floating-point arithmetic (version 0.18),* December 2013. http://mpmath.org/.

[Joh98] Kurt Johansson. On fluctuations of eigenvalues of random Hermitian matrices. *Duke Math. J.,* 91(1):151–204, 1998.

[KNY17] K. Kajiwara, M. Noumi, and Y. Yamada. Geometric aspects of painlevé equations. *J. Phys. A: Math. Theor.,* 50:1–163, 2017.

[Mar77] A. I. Markushevich. *Theory of functions of a complex variable. Vol. I, II, III.* Chelsea Publishing Co., New York, english edition, 1977. Translated and edited by Richard A. Silverman.

[O'C14] Neil O'Connell. Whittaker functions and related stochastic processes. *MSRI Publications (Random Matrices),* 65:385–409, 2014.

[PS00]   P. Prähofer and H. Spohn. Universal distributions for growth processes in 1+1 dimensions and random matrices. *Phys. Rev. Lett.,* 84:4882–4885, 2000.

[QRT89] G. R. W. Quispel, J. A. G. Roberts, and C. J. Thompson. Integrable mappings and soliton-equations .2. *PHYSICA D,* 34(1-2):183–192, 1989.

[RG96] A. Ramani and B. Grammaticos. Discrete painlevé equations: coalescences, limits and degeneracies. *Physica A: Statistical Mechanics and its Applications,* 228:160–171, 1996.

[Sch97] Gilles Schaeffer. Bijective census and random generation of eulerian planar maps with prescribed vertex degrees. *Electron. J. Combin,* 4(1):20, 1997.

[Tut62a] William T Tutte. A census of slicings. *Canad. J. Math,* 14(4):708–722, 1962.

[Tut62b] William Thomas Tutte. A census of planar triangulations. *Canad. J. Math,* 14(1):21–38, 1962.

[Tut62c] WT Tutte. A census of hamiltonian polygons. *Canad. J. Math,* 14:402–417, 1962.

[Tut63] William Thomas Tutte. A census of planar maps. *Canad. J. Math,* 15(2):249–271, 1963.

Tova Brown
Mathematics Department, Wisconsin Lutheran College, Milwaukee, WI
E-mail address: tova.brown@wlc.edu

Nicholas M. Ercolani
Department of Mathematics, University of Arizona, Tucson, AZ
E-mail address: ercolani@math.arizona.edu

11

On an Arnold-Liouville Type Theorem for the Focusing NLS and the Focusing mKdV Equations

T. Kappeler and P. Topalov

Abstract. For the focusing NLS and the focusing mKdV equation on the circle we present an infinite dimensional version of the Arnold-Liouville theorem.

1 Introduction

Let us first review the classical Arnold-Liouville theorem ([25, 28, 1]) in the most simple setup: assume that the phase space M is an open subset of $\mathbb{R}^d \times \mathbb{R}^d$, $d \geq 1$, endowed with the standard Poisson bracket $((x, y) \in M)$

$$\{G, H\} = \begin{pmatrix} \nabla_x G \\ \nabla_y G \end{pmatrix} \begin{pmatrix} 0 & \mathrm{Id} \\ -\mathrm{Id} & 0 \end{pmatrix} \begin{pmatrix} \nabla_x H \\ \nabla_y H \end{pmatrix},$$

where $G, H : M \to \mathbb{R}$ are C^∞-smooth functions and $(\nabla_x G, \nabla_y G)(x, y)$ denotes the standard gradient of G at $(x, y) \in M$. Furthermore, assume that $F : M \to \mathbb{R}^d$ is a C^∞-smooth map whose components F_j, $1 \leq j \leq d$, satisfy

$$\{F_j, F_k\}(x, y) = 0, \qquad \forall 1 \leq j, k \leq d, \forall (x, y) \in M.$$

The map F is referred to as moment map. In this setup the Arnold-Liouville theorem reads as follows (cf. e.g. [1] or [29]):

Theorem 1.1 (Arnold-Liouville) *Assume that $c \in \mathbb{R}^d$ is a regular value of the momentum map F and N_c is a connected component of the preimage $F^{-1}(c)$ of c. If N_c is compact, then N_c is an d-dimensional torus and a neighborhood*

T.K. is partially supported by the Swiss NSF

P.T. is partially supported by the Simons Foundation, Award #526907

Keywords: Normal form, focusing NLS equation, focusing mKdV equation, Arnold-Liouville theorem, Birkhoff coordinates.

MSC 2010: 37K10, 37K20, 35B10, 35B15

of N_c in M is foliated by such tori. In more detail, there exists a neighborhood W of N_c in M, an open subset $D \subseteq \mathbb{R}^d$, and a C^∞-smooth diffeomorphism

$$\Phi : W \to (\mathbb{R}/2\pi\mathbb{Z})^d \times D, \quad (x, y) \mapsto \big(\theta(x, y), I(x, y)\big)$$

so that Φ is canonical, meaning that $\{\theta_j, I_k\}(x, y) = \delta_{jk}$ on W for any $1 \leq j, k \leq d$, whereas the Poisson brackets between all other coordinate functions vanish. The coordinates $I = (I_j)_{1\leq j\leq d}$ and $\theta = (\theta_j)_{1\leq j\leq d}$ are referred to as action angle coordinates. The level sets of the actions are the level sets of the moment map F, contained in W.

Remark 1.2   By translating D, if needed, we can assume without loss of generality that $D \subset \mathbb{R}^d_{>0}$. The complex coordinates

$$z_j := \sqrt{I_j}e^{-i\theta_j}, \quad w_j := \sqrt{I_j}e^{i\theta_j}, \qquad 1 \leq j \leq d,$$

are then well defined on W. They satisfy $I_j = z_j w_j$ and $\{z_j, w_j\} = -i$ for any $1 \leq j \leq d$, whereas the Poisson bracket between any other complex coordinates vanish. We remark that such coordinates play an important role in the study of (Hamiltonian) perturbations of integrable systems.

Applications of Theorem 1.1: Theorem 1.1 has the following two immediate applications. The first one concerns the Hamiltonian equations with Hamiltonian in the *local Poisson algebra*,

$$\mathcal{A}_W := \big\{G : W \to \mathbb{R} \,|\, G \in C^\infty(W, \mathbb{R}), \{G, F_j\} = 0 \,\forall\, 1 \leq j \leq d\big\}.$$

Note that any Hamiltonian $G \in \mathcal{A}_W$ gives rise to an integrable system on W, given by the Hamiltonian vector field $\big(\nabla_y G, -\nabla_x G\big)$ of G,

$$(\dot{x}, \dot{y}) = \big(\nabla_y G, -\nabla_x G\big) \tag{1}$$

where here and in the sequel, if not stated otherwise, ˙ stands for the partial derivative ∂_t with respect to the time variable t. The functions $F_1, \ldots, F_d$ are first integrals of (1) and for any $I \in D$, solutions of this equation leave the torus $\Phi^{-1}\big((\mathbb{R}/2\pi\mathbb{Z})^d \times \{I\}\big)$ invariant. In particular, for any $1 \leq j \leq d$, the action variable I_j is an element in the Poisson algebra $\mathcal{A}_W$.

Corollary 1.3 *For any $G \in \mathcal{A}_W$, $G \circ \Phi^{-1}$ is in Birkhoff normal form, meaning that $G \circ \Phi^{-1}$ is a $C^\infty-$smooth function of the actions alone. When expressed in the coordinates z_j, w_j, $1 \leq j \leq d$, the Hamiltonian equations corresponding to G read*

$$\dot{z}_j = -i\omega_j z_j, \quad \dot{w}_j = i\omega_j w_j, \qquad 1 \leq j \leq d \tag{2}$$

where $\omega_j := \partial_{I_j}(G \circ \Phi^{-1})$ denotes the jth frequency of the Hamiltonian G. Since the actions are integrals of G, $\dot{I}_j = \{I_j, G\} = 0$, $1 \le j \le d$, and the frequencies only depend on the actions, (2) can be solved by quadrature,

$$z_j(t) = z_j(0)e^{-i\omega_j t}, \quad w_j(t) = w_j(0)e^{i\omega_j t}, \quad 1 \le j \le d.$$

In particular, all solutions of (1) are quasi-periodic.

The second application concerns the moment map F and follows from Sard's theorem. Denote by $\mathcal{R}_F$ the subset of regular values of F.

Corollary 1.4 *If in addition to the assumptions made in Theorem 1.1, the moment map F is proper and*

$$\left\{(x, y) \in M \mid \nabla_{x,y} F_1, \ldots, \nabla_{x,y} F_d \text{ are linearly independent}\right\}$$

is dense in M, then $M_0 := \cup_{c \in \mathcal{R}_F} F^{-1}(c)$ is open and dense in M.

Comments about Theorem 1.1: First note that the level sets of F in the neighborhood W of Theorem 1.1 are tori of *maximal* dimension d and their tangent space at any given point $(x, y) \in W$ is spanned by the Hamiltonian vector fields $X_{F_1}(x, y), \ldots, X_{F_n}(x, y)$, evaluated at x, y. Informally, one can say that the Arnold-Liouville theorem asserts that generically there is no room for other types of dynamics besides quasi-periodic motion (cf Corollary 1.4 for a more precise statement). On the other hand, typically, action-angle coordinates cannot be extended globally. One reason for this is the existence of a singular value of the momentum map $F : M \to \mathbb{R}^d$. Such a situation appears e.g. in the case of an elliptic fixed point $\xi \in M$. In the case where smooth action angle coordinates $I = (I_j)_{1 \le j \le d}$, $\theta = (\theta_j)_{1 \le j \le d}$ can be constructed on a dense open subset of a neighborhood U of ξ in M, it might be possible to smoothly extend the coordinates $z_j := \sqrt{2I_j}e^{-i\theta_j}$, $w_j := \sqrt{2I_j}e^{i\theta_j}$, $1 \le j \le d$, to all of U. If z_j, w_j, $1 \le j \le d$, parametrize U, we refer to such coordinates as (complex) Birkhoff coordinates and the Hamiltonian H, when expressed in these coordinates, is said to be in Birkhoff normal form. For results in this direction we refer to [31, 9, 33] and references therein. In special cases, such as systems of coupled oscillators, Birkhoff coordinates can be defined on the entire phase space. In such a case we say they are *global* Birkhoff coordinates. In what follows we will keep this terminology. In particular, we will call the coordinates $(z_j)_{1 \le j \le d}$, $(w_j)_{1 \le j \le d}$, defined in terms of the action angle coordinates as above, again Birkhoff coordinates even if they are non necessarily related to an elliptic fixed point. Other possible obstructions to extend action angle coordinates globally are the presence of hyperbolic or focus-focus fixed points, which might cause a non-trivial monodromy of the action variables (cf [8, 9, 34]).

The aim of this paper is to report on a version of the Arnold-Liouville theorem in infinite dimension allowing to conclude that on an open dense subset of the phase space, the Hamiltonian of the focusing nonlinear Schrödinger (NLS) equation

$$i\partial_t u = -\partial_x^2 u - 2|u|^2 u, \qquad x \in \mathbb{T} = \mathbb{R}/\mathbb{Z} \qquad (3)$$

and the one of the focusing modified Korteweg-deVries (mKdV) equation

$$\partial_t v = -\partial_x^3 v - 6v^2 \partial_x v, \qquad x \in \mathbb{T} = \mathbb{R}/\mathbb{Z} \qquad (4)$$

can locally be brought into Birkhoff normal form in a sense which will be explained in detail in the subsequent section. We remark that in contrast to the defocusing NLS and the defocusing mKdV equations, the focusing versions of these equations are known to have hyperbolic features and hence cannot admit global Birkhoff coordinates. See [20] for more details (cf also [24]).

2 Results

To state our results we first need to introduce some notation and review some facts about the focusing NLS equation. Consider the NLS system

$$\partial_t \varphi_1 = -i\left(-\partial_x^2 \varphi_1 + 2\varphi_1^2 \varphi_2\right), \quad \partial_t \varphi_2 = i\left(-\partial_x^2 \varphi_2 + 2\varphi_1 \varphi_2^2\right) \qquad (5)$$

where $t \in \mathbb{R}$ (time variable) and where $\varphi_j : \mathbb{R} \to \mathbb{C}$, $j = 1, 2$, are one periodic,

$$\varphi_j(x + 1) = \varphi_j(x), \qquad \forall x \in \mathbb{R}, \ j = 1, 2.$$

The system (5) is a Hamiltonian PDE with phase space $L_c^2 := L^2 \times L^2$ where $L^2 := L^2(\mathbb{T}, \mathbb{C})$ is the Hilbert space of square-integrable complex valued functions on the unit torus $\mathbb{T} := \mathbb{R}/\mathbb{Z}$ and

$$X_{nls}(\varphi) = -i\left(\partial_{\varphi_2} \mathcal{H}_{nls}, -\partial_{\varphi_1} \mathcal{H}_{nls}\right), \qquad \varphi = (\varphi_1, \varphi_2)$$

is the Hamiltonian vector field with Hamiltonian $\mathcal{H}_{nls}$, given by

$$\mathcal{H}_{nls}(\varphi) := \int_0^1 \left(\partial_x \varphi_1 \partial_x \varphi_2 + \varphi_1^2 \varphi_2^2\right) dx$$

and Poisson bracket

$$\{F, G\}(\varphi) := -i \int_0^1 \left(\partial_{\varphi_1} F \cdot \partial_{\varphi_2} G - \partial_{\varphi_2} F \cdot \partial_{\varphi_1} G\right) dx. \qquad (6)$$

The NLS Hamiltonian $\mathcal{H}_{nls}$ is defined on the Sobolev space $H^1(\mathbb{T}, \mathbb{C}) \times H^1(\mathbb{T}, \mathbb{C})$ and $\partial_{\varphi_j} F$ denotes the L^2-gradient of F with respect to φ_j, $j = 1, 2$. The functionals F and G are defined on L_c^2 or on a dense subset of it and their

L^2-gradients are supposed to be sufficiently regular so that the integral in (6), when viewed as a dual pairing, is well defined on L_c^2 or on a dense subset of it. When (5) is restricted to the real subspace

$$L_r^2 := \{(\varphi_1, \varphi_2) \in L_c^2 \mid \varphi_2 = \overline{\varphi}_1\}$$

one obtains the *defocusing* NLS equation $i\partial_t u = -\partial_x^2 u + 2|u|^2 u$ (where $\varphi = (u, \overline{u})$) whereas the restriction to iL_r^2 yields the *focusing* NLS equation $i\partial_t u = -\partial_x^2 u - 2|u|^2 u$ (where $\varphi = (u, -\overline{u})$).

Actually, we will consider the scales of Sobolev spaces (N integer with $N \geq 0$)

$$H_c^N := H^N(\mathbb{T}, \mathbb{C}) \times H^N(\mathbb{T}, \mathbb{C}), \quad H_r^N := H_c^N \cap L_r^2, \quad iH_r^N := H_c^N \cap iL_r^2,$$

and the corresponding weighted ℓ^2-sequence spaces $h_c^N = h^N \times h^N, h_r^N$, and ih_r^N, obtained from the version $\mathcal{F}_{nls}$ of the Fourier transform, adapted to the NLS system,

$$\mathcal{F}_{nls}(\varphi) := \left((-\hat{\varphi}_1(-n))_{n \in \mathbb{Z}}, (-\hat{\varphi}_2(n))_{n \in \mathbb{Z}}\right),$$

where $\hat{\varphi}_j(n)$ denotes the nth Fourier coefficient of φ_j,

$$\hat{\varphi}_j(n) = \int_0^1 \varphi_j(x) e^{-2\pi i n x} \, dx, \quad \forall n \in \mathbb{Z}, \, j = 1, 2.$$

All these spaces are endowed with the standard norms. So e.g., $h^N \equiv h^N(\mathbb{Z}, \mathbb{C})$ is the Hilbert space of complex valued sequences $z = (z_n)_{n \in \mathbb{Z}}$ with

$$\|z\|_N := \left(\sum_{n \in \mathbb{Z}} \langle n \rangle^{2N} |z_n|^2\right)^{1/2} < \infty, \quad \langle n \rangle := 1 \vee |n|$$

and

$$h_r^N := \left\{(z, w) \in h_c^N \mid w_n = \overline{z}_n \, \forall n \in \mathbb{Z}\right\}.$$

According to [32], the system (5) admits a Lax pair representation (cf. [22])

$$\partial_t L(\varphi) = P(\varphi) L(\varphi) - L(\varphi) P(\varphi)$$

where $L(\varphi)$ is the first order differential operator

$$L(\varphi) := i \begin{pmatrix} 1 & 0 \\ 0 & -1 \end{pmatrix} \partial_x + \begin{pmatrix} 0 & \varphi_1 \\ \varphi_2 & 0 \end{pmatrix},$$

referred to as Zakharov-Shabat operator (ZS operator), and $P(\varphi)$ is a certain second order differential operator. As a consequence, the spectrum $\mathrm{Spec}_p^+ L(\varphi)$ of the (unbounded) operator $L(\varphi)$, considered with domain H_c^1, is invariant with respect to the NLS flow (5). Since the resolvent of $L(\varphi)$ is compact, $\mathrm{Spec}_p^+ L(\varphi)$ is discrete. Hence $\mathrm{Spec}_p^+ L(\varphi)$ being invariant means that the periodic eigenvalues of $L(\varphi)$ are first integrals of (5). We point out that for any

$\varphi \in L_c^2 \setminus L_r^2$, $L(\varphi)$ is *not* selfadjoint and hence eigenvalues might be complex valued and their geometric multiplicities strictly smaller than their algebraic ones. By a slight abuse of terminology, we refer to these eigenvalues of $L(\varphi)$ as periodic eigenvalues of φ. They are candidates for the components of the version of the moment map in our infinite dimensional setup. Unfortunately, due to multiplicities, they do not have the regularity properties, required for our purposes. Instead we will use the regularized determinant of $L(\varphi) - \lambda$.

To continue we need first to introduce some more notation. For any $\varphi \in L_c^2$, denote by $M(x, \lambda, \varphi) \in \mathbb{C}^{2 \times 2}$ the fundamental solution of $L(\varphi)$,

$$L(\varphi) M(x, \lambda, \varphi) = \lambda M(x, \lambda, \varphi), \quad x \in \mathbb{R}, \lambda \in \mathbb{C}$$

and by $\Delta(\lambda, \varphi)$ the discriminant of $L(\varphi)$,

$$\Delta_\lambda(\varphi) \equiv \Delta(\lambda, \varphi) = \operatorname{tr} M(1, \lambda, \varphi)$$

where $\operatorname{tr} M(1, \lambda, \varphi)$ denotes the trace of $M(1, \lambda, \varphi)$. The following results are well known (cf e.g. [13, Theorem 4.1, Lemma 8.3]).

Proposition 2.1 (i) *The discriminant* $\Delta : \mathbb{C} \times L_c^2 \to \mathbb{C}$, $(\lambda, \varphi) \mapsto \Delta(\lambda, \varphi)$ *is analytic and compact.* (ii) *The Poisson bracket of* Δ_λ *and* Δ_μ *vanishes,*

$$\{\Delta_\lambda, \Delta_\mu\}(\varphi) = 0, \quad \forall \lambda, \mu \in \mathbb{C}, \ \forall \varphi \in L_c^2.$$

Remark 2.2 (i) In our infinite dimensional version of the Arnold-Liouville theorem, the functionals Δ_λ, $\lambda \in \mathbb{C}$, will play the role of the components of the moment map in the Arnold-Liouville Theorem, as reviewed in Theorem 1.1.

(ii) For any $\lambda \in \mathbb{C}$, $\Delta_\lambda - 2$ is up to the phase factor $e^{i\lambda}$ the ζ-regularized determinant of $L(\varphi) - \lambda$ – see Appendix A

(iii) The periodic eigenvalues of $L(\varphi)$ coincide with the roots of $\Delta(\varphi) - 2$. In addition, the algebraic multiplicity of any periodic eigenvalue λ of $L(\varphi)$ equals the multiplicity of λ as a root of $\Delta(\cdot, \varphi) - 2$ at λ (cf [14, Appendix C]).

We now focus our attention to the phase space iL_r^2 of the focusing NLS equation. It is customary to denote the level set of $(\Delta_\lambda)_{\lambda \in \mathbb{C}}$ in iL_r^2, containing the element $\psi \in iL_r^2$, by $\operatorname{Iso}(\psi)$,

$$\operatorname{Iso}(\psi) := \{\varphi \in iL_r^2 \mid \Delta_\lambda(\varphi) = \Delta_\lambda(\psi) \ \forall \lambda \in \mathbb{C}\}$$

and to refer to it as the isospectral set of ψ. We write $\operatorname{Iso}_o(\psi)$ for the connected component of $\operatorname{Iso}(\psi)$, containing ψ. We say that a subset $W \subseteq iL_r^2$ is *saturated* if for any $\psi \in W$, $\operatorname{Iso}_o(\psi) \subseteq W$. In Section 3 we prove that the isospectral sets satisfy the following properties.

Proposition 2.3 (i) *For any* $\psi \in i H_r^N$, $N \geq 0$, Iso(ψ) *is a subset of* $i H_r^N$.
(ii) *For any* $\psi \in i H_r^N$, $N \geq 0$, Iso(ψ) *is compact in* $i H_r^N$.

Remark 2.4 By Proposition 3.1, $\int_0^1 \varphi_1(x)\varphi_2(x)\,dx$ is a spectral invariant on L_c^2. Using that for $\varphi \in i L_r^2$, $\int_0^1 \varphi_1(x)\varphi_2(x)\,dx = -\int_0^1 |\varphi_1(x)|^2\,dx$ one can show that for any $\psi \in i L_r^2$ of sufficiently small norm, Iso(ψ) is connected. See Section 3 for more details.

It turns out that we also have to consider the operator $L(\varphi)$ with domain $H^1(\mathbb{R}/2\mathbb{Z}, \mathbb{C}) \times H^1(\mathbb{R}/2\mathbb{Z}, \mathbb{C})$, or said in an informal way, on the interval $[0, 2]$ with periodic boundary conditions. The spectrum of this operator is again discrete and we denote it by $\mathrm{Spec}_p L(\varphi)$. Clearly, $\mathrm{Spec}_p^+ L(\varphi) \subseteq \mathrm{Spec}_p L(\varphi)$. Furthermore, taking multiplicities into account, the set $\mathrm{Spec}_p L(\varphi)$ coincides with the set of roots of $\Delta^2(\cdot, \varphi) - 4$, referred to as characteristic function of $L(\varphi)$ (cf [14, Appendix C]). We say that $\mathrm{Spec}_p L(\varphi)$ is *simple* if each root of $\Delta^2(\cdot, \varphi) - 4$ is simple. Finally, to state our main result, we need to introduce the action variables and their level sets. For any element $(z, w) = ((z_n)_{n\in\mathbb{Z}}, (w_n)_{n\in\mathbb{Z}})$ in $i\ell_r^2 \equiv i h_r^0$, the nth action $I_n \equiv I_n(z, w)$ is defined by $I_n = z_n w_n$, $n \in \mathbb{Z}$. Since $w_n = -\bar{z}_n$, it follows that

$$I_n(z, w) = -|z_n|^2 \leq 0, \quad \forall n \in \mathbb{Z}, \qquad I(z, w) := (I_n(z, w))_{n\in\mathbb{Z}} \in \ell_-^1,$$

where ℓ_-^1 denotes the negative quadrant $\ell^1(\mathbb{Z}, \mathbb{R}_{\leq 0})$ of the ℓ^1−sequence space $\ell^1(\mathbb{Z}, \mathbb{R})$. For any $I \in \ell_-^1$, we denote by $\mathrm{Tor}(I) \subseteq i\ell_r^2$ the compact torus with $\#\{n \in \mathbb{Z} \mid I_n \neq 0\}$ "degrees of freedom",

$$\mathrm{Tor}(I) := \left\{ ((z_n)_{n\in\mathbb{Z}}, (w_n)_{n\in\mathbb{Z}}) \in i\ell_r^2 \,\middle|\, z_n w_n = I_n \, \forall n \in \mathbb{Z} \right\}.$$

We are now ready to state the aforementioned version of the Arnold-Liouville theorem in infinite dimension for $(\Delta_\lambda)_{\lambda\in\mathbb{C}}$.

Theorem 2.5 *Let* $\psi \in i L_r^2$ *and assume that* $\mathrm{Spec}_p L(\psi)$ *is simple. Then there exist an open, saturated neighborhood W of* $\mathrm{Iso}_o(\psi)$ *in* $i L_r^2$ *and a real analytic diffeomorphism*

$$\Phi : W \to \Phi(W) \subseteq i\ell_r^2, \quad \varphi \mapsto \left((z_n(\varphi))_{n\in\mathbb{Z}}, (w_n(\varphi))_{n\in\mathbb{Z}} \right),$$

so that $z_n(\varphi)$, $w_n(\varphi)$, $n \in \mathbb{Z}$, *are complex Birkhoff coordinates for* $(\Delta_\lambda)_{\lambda\in\mathbb{C}}$ *on W, meaning that the following holds:*

(NF1) The coordinate functions $z_n(\varphi)$, $w_n(\varphi)$, $n \in \mathbb{Z}$, *are canonical, i.e.,*

$$\{z_n, w_n\}(\varphi) = -i, \quad \forall n \in \mathbb{Z}, \varphi \in W,$$

whereas the Poisson brackets between all other coordinate functions vanish.

(NF2) For any $\varphi \in W$,

$$\Phi\big(\mathrm{Iso}_o(\varphi)\big) = \mathrm{Tor}\big(I(\varphi)\big), \quad I(\varphi) = \big(I_n(\varphi)\big)_{n \in \mathbb{Z}},$$

where for any $n \in \mathbb{Z}$, $I_n(\varphi) = z_n(\varphi)w_n(\varphi)$.
(NF3) For any integer $N \geq 1$, $\Phi(W \cap iH_r^N) \subseteq ih_r^N$ and

$$\Phi : W \cap iH_r^N \to \Phi\big(W \cap iH_r^N\big) \subseteq ih_r^N$$

is a real analytic diffeomorphism onto its image.

Comments about Theorem 2.5: (i) The Birkhoff coordinates are constructed in terms of action and angle variables. In case of a finite gap potential, the angle variables are defined by a real valued expression involving the Abel map of the spectral curve, associated to the finite gap potential. The question if these expressions are real valued has been a longstanding issue, raised by experts in the field in connection with special solutions of the focusing NLS equation, given in terms of theta functions.

(ii) The open neighborhood $W \subseteq iL_r^2$ in Theorem 2.5 is chosen in such a way that for any $\varphi \in W$, every multiple root of the characteristic function $\Delta^2(\cdot, \varphi) - 4$ is real and of multiplicity two whereas all simple roots are non-real and appear in complex conjugate pairs. Hence each root of $\Delta^2(\cdot, \varphi) - 4$ has multiplicity at most two.

(iii) According to [15], there exists a saturated neighborhood W_0 of 0 in iL_r^2 and a real analytic diffeomorphism

$$\Phi : W_0 \to \Phi(W_0) \subseteq i\ell_r^2, \quad \varphi \mapsto \big((z_n(\varphi))_{n \in \mathbb{Z}}, (w_n(\varphi))_{n \in \mathbb{Z}}\big),$$

so that $\big(z_n(\varphi)\big)_{n \in \mathbb{Z}}, \big(w_n(\varphi)\big)_{n \in \mathbb{Z}}$ are Birkhoff coordinates for $(\Delta_\lambda)_{\lambda \in \mathbb{C}}$ on W_0 in the sense of Theorem 2.5. Note that $\Delta^2(\cdot, 0) - 4 = -4\sin^2(\lambda)$ (cf [13, Theorem 4.1]), hence its roots are $n\pi$, $n \in \mathbb{Z}$, each having muliplicity two.

(iv) Informally, Theorem 2.5 means that $z_n(\varphi), w_n(\varphi), n \in \mathbb{Z}$, can be thought of as (locally defined) nonlinear Fourier coefficients of $\varphi = (\varphi_1, \varphi_2)$. By [15], the differential of the Birkhoff map at $\varphi = 0$ is given by the linear map $(\widehat{\varphi}_1, \widehat{\varphi}_2) \mapsto (\widehat{z}, \widehat{w})$ where $\widehat{z} = (\widehat{z}_n)_{n \in \mathbb{Z}}, \widehat{w} = (\widehat{w}_n)_{n \in \mathbb{Z}}$ and

$$\widehat{z}_n = -\int_0^1 \widehat{\varphi}_1(x)e^{2\pi inx}dx, \quad \widehat{w}_n = -\int_0^1 \widehat{\varphi}_2(x)e^{-2\pi inx}dx.$$

For the proof of Theorem 2.5 we refer to [21]. In that paper, one can also find a brief outline of the proof at the end of the introduction.

We now discuss three immediate applications of Theorem 2.5. In contrast to the classical Arnold-Liouville theorem, the neighborhood W of Theorem 2.5 contains level sets of $(\Delta_\lambda)_{\lambda \in \mathbb{C}}$ of various dimensions. As a first application of Theorem 2.5 we have the following result in this regard.

Corollary 2.6 *For any integer $N \geq 0$, the union of the finite dimensional level sets of $(\Delta_\lambda)_{\lambda \in \mathbb{C}}$ in W is dense in $W \cap i H_r^N$.*

The second application of Theorem 2.5 concerns the focusing NLS and the focusing mKdV equations. Refining the notion of the local Poisson algebra, introduced in Section 1, we define for any integer $N \geq 0$,

$$\mathcal{A}_{W \cap i H_r^N} := \left\{ H : W \cap i H_r^N \to \mathbb{C} \,\middle|\, H \text{ is real analytic, } (PA)_N \text{ is satisfied} \right\},$$

where

$(PA)_N$ For any $\lambda \in \mathbb{C}$ the Poisson bracket $\{H, \Delta_\lambda\}$ is well defined on $W \cap i H_r^N$ and $\{H, \Delta_\lambda\} = 0$ on $W \cap i H_r^N$.

Corollary 2.7 *For any $H \in \mathcal{A}_{W \cap i H_r^N}$, $N \geq 0$, the Hamiltonian $H \circ \Phi^{-1}$ is in Birkhoff normal form, meaning that $H \circ \Phi^{-1}$ is a function of the actions alone. In fact, $H \circ \Phi^{-1}$ is real analytic as a function of the actions.*

Prominent Hamiltonians in the local Poisson algebras are the ones in the NLS hierarchy, $\mathcal{H}_n$, $n \geq 1$. They come up in the asymptotic expansion of the discriminant (cf e.g. [13, Theorem 4.8]). Up to an irrelevant choice of constant factors, the first four are given by $\mathcal{H}_1(\varphi) = \int_0^1 \varphi_1 \varphi_2 \, dx$, $\mathcal{H}_2(\varphi) = \int_0^1 \varphi_2 \partial_x \varphi_1 \, dx$, the NLS Hamiltonian $\mathcal{H}_3(\varphi) = \int_0^1 \left(\partial_x \varphi_1 \cdot \partial_x \varphi_2 + \varphi_1^2 \varphi_2^2 \right) dx$, and the mKdV Hamiltonian,

$$\mathcal{H}_4(\varphi) = \int_0^1 \left(\partial_x^2 \varphi_1 \cdot \partial_x \varphi_2 + 3 \varphi_1 \partial_x \varphi_1 \cdot \varphi_2^2 \right) dx \,.$$

More generally, for any integer $k \geq 1$,

$$\mathcal{H}_{2k+1} = \int_0^1 \left(\partial_x^k \varphi_1 \cdot \partial_x^k \varphi_2 + q_{2k+1} \right) dx \,,$$

$$\mathcal{H}_{2k} = \int_0^1 \left(\partial_x^k \varphi_1 \cdot \partial_x^{k-1} \varphi_2 + q_{2k} \right) dx \tag{7}$$

where q_{2k+1} and q_{2k} are certain polynomials in φ_1, φ_2, and their derivatives up to order $k - 1$. Note that $\mathcal{H}_{2k}$ and $\mathcal{H}_{2k+1}$ are well defined on H_c^N. In Section 3 we prove the following

Proposition 2.8 *For any integer $N \geq 0$ and any $\lambda \in \mathbb{C}$, the Poisson brackets $\{\mathcal{H}_{2N}, \Delta_\lambda\}$ and $\{\mathcal{H}_{2N+1}, \Delta_\lambda\}$ are well defined on H_c^N in terms of*

the pairing between H_c^N and its dual H_c^{-N} and vanish identically. Hence for any neighborhood W as in Theorem 2.5, $\mathcal{H}_{2N}$ and $\mathcal{H}_{2N+1}$ are elements in the local Poisson algebra $A_{W \cap i H_r^N}$.

Application to the focusing NLS equation: According to [3], the focusing NLS equation (3) is globally well-posed on $i H_r^s$ for any $s \geq 0$. For initial data in $W \cap i H_r^N$, $N \geq 1$, with W given as in Theorem 2.5, we can show that the corresponding solution has additional properties by using that in this case it can be expressed in terms of the Birkhoff coordinates on W. (Arguing as in [19], [16], this result could be generalized to more general Sobolev spaces.) Indeed, since by Proposition 2.8, the NLS Hamiltonian $\mathcal{H}_3$ is in $A_{W \cap i H_r^1}$, Corollary 2.7 applies to the focusing NLS equation. When expressed in the Birkhoff coordinates of Theorem 2.5, this equation reads

$$\dot{z}_n = -i \omega_n z_n, \qquad \dot{w}_n = i \omega_n w_n, \qquad n \in \mathbb{Z},$$

where $\omega_n \equiv \omega_n(I) := \partial_{I_n}(\mathcal{H}_{NLS} \circ \Phi^{-1})$ are the NLS frequencies and $I_n = z_n w_n$ are the actions of the initial data φ. The corresponding solution $(z(t), w(t)) = \big((z_n(t))_{n \in \mathbb{Z}}, (w_n(t))_{n \in \mathbb{Z}} \big)$ can thus be found by quadrature,

$$z_n(t) = z_n(0) e^{-i \omega_n t}, \qquad w_n(t) = w_n(0) e^{i \omega_n t}, \qquad \forall t \in \mathbb{R}, \, n \in \mathbb{Z}, \qquad (8)$$

where $(z(0), w(0)) = \Phi(\varphi)$. We claim that $t \mapsto \mathcal{S}_\varphi(\cdot, t) := \Phi^{-1}(z(t), w(t))$ is the solution of (3) with initial data φ, found in [3]. To see that this is the case note that by Corollary 2.6, φ can be approximated in W by finite gap potentials. Being C^∞−smooth, they give rise to a sequence of C^∞−smooth solutions of (3). One can show that for any $T > 0$, they converge in $C([-T, T], W \cap i H_r^N)$ to $\mathcal{S}_\varphi$, implying that $\mathcal{S}_\varphi(x, t)$ coincides with the solution found in [3]. Note that $\mathcal{S}_\varphi(x, t)$ is almost periodic in time and evolves in the compact subset $\text{Iso}_o(\varphi) \subseteq i H_r^N$. Furthermore, it follows from the construction of Birkhoff coordinates that the approximating finite gap solutions above can be represented in terms of theta functions.

Application to the focusing mKdV equation: According to [4], [6], the focusing mKdV equation (4) is globally well posed on $i H_r^s$ for any $s \geq 1/2$. In case the initial data is in $W \cap i H_r^N$, $N \geq 2$, with W given as in Theorem 2.5, we can show that the corresponding solution has additional properties by using that in this case it can be expressed in terms of the Birkhoff coordinates on W. (Arguing as in [19], [16], this result could be generalized to more general Sobolev spaces.) Indeed, by Proposition 2.8, the mKdV Hamiltonian $\mathcal{H}_4$ is in $A_{W \cap i H_r^2}$. The corresponding Hamiltonian vector field is given by

$$X_{\mathcal{H}_4}(\varphi) = -i \Big(-\partial_x^3 \varphi_1 + 6 \varphi_1 \varphi_2 \partial_x \varphi_1, \, -\partial_x^3 \varphi_2 + 6 \varphi_1 \varphi_2 \partial_x \varphi_2 \Big).$$

When restricted to the real subspace of $i H_r^N$, $N \geq 2$,

$$\{\varphi = i(v, v) \in i H_r^N \mid v \text{ real valued}\} \cong H^N(\mathbb{T}, \mathbb{R}),$$

the first component of the Hamiltonian PDE $\partial_t \varphi = X_{\mathcal{H}_4}(\varphi)$ yields the focusing mKdV equation (4), $\partial_t v = -\partial_x^3 v - 6v^2 \partial_x v$. Arguing as in the case of the focusing NLS equation, one concludes that any solution of (4) with initial data $v \in H^N(\mathbb{T}, \mathbb{R})$, $N \geq 2$, so that $i(v, v) \in W$ with W as in Theorem 2.5, can be obtained by quadrature, when expressed in complex Birkhoff coordinates, and hence is globally defined and almost periodic in time.

The third application of Theorem 2.5 concerns a geometric property of $\mathrm{Iso}_o(\varphi)$, $\varphi \in W$. Since $\Phi(\mathrm{Iso}_o(\varphi)) = \mathrm{Tor}(I(\varphi))$ it follows that in the case $\{n \in \mathbb{Z} \mid I_n(\varphi) \neq 0\}$ is *finite*, $\mathrm{Iso}_o(\varphi)$ is a finite dimensional smooth submanifold of $i L_r^2$. Actually, it is a submanifold of $i H_r^N$ for any $N \geq 0$. In case $\{n \in \mathbb{Z} \mid I_n(\varphi) \neq 0\}$ is *infinite*, $\mathrm{Tor}(I(\varphi))$, being a compact subset of $i h_r^0$, is not a manifold. Nevertheless, the space $i h_r^0$ is a direct sum of two closed subspaces, which by a slight abuse of terminology are referred to as tangent and normal space of $\mathrm{Tor}(I(\varphi))$ at $\Phi(\varphi)$. Here the word normal refers to the symplectic structure on $i h_r^0$. The tangent space is the closed $\mathbb{R}-$subspace of $i h_r^0$, spanned by the Hamiltonian vector fields $X_n(\varphi)$ with Hamiltonian $I_n(\varphi)$ for n with $I_n(\varphi) \neq 0$,

$$X_n(\varphi) = \left((-i z_n(\varphi) \delta_{nk})_{k \in \mathbb{Z}}, (i w_n(\varphi) \delta_{nk})_{k \in \mathbb{Z}}\right)$$

and the normal space is the closed $\mathbb{R}-$subspace of $i h_r^0$, spanned by the gradients of $I_n(\varphi)$ for n with $I_n(\varphi) \neq 0$,

$$Y_n(\varphi) = \left((w_n(\varphi) \delta_{nk})_{k \in \mathbb{Z}}, (z_n(\varphi) \delta_{nk})_{k \in \mathbb{Z}}\right),$$

together with the elements $e_n^{(1)}$, $e_n^{(2)} \in i h_r^0$ for $n \in \mathbb{Z}$ with $I_n(\varphi) = 0$, given by

$$e_n^{(1)} = \left((\delta_{nk})_{k \in \mathbb{Z}}, -(\delta_{nk})_{k \in \mathbb{Z}}\right), \quad e_n^{(2)} = \left((i \delta_{nk})_{k \in \mathbb{Z}}, (i \delta_{nk})_{k \in \mathbb{Z}}\right).$$

Since $d\Phi(\varphi)$ is invertible, a corresponding result holds for $\mathrm{Iso}_o(\varphi)$. More precisely, Theorem 2.5 directly yields the following

Corollary 2.9 *For any* $\varphi \in W \cap i H_r^N$, $N \geq 1$, *the real Hilbert space* $i H_r^N$ *splits as a direct sum* $T_\varphi \mathrm{Iso}_o \oplus N_\varphi \mathrm{Iso}_o$ *where* $T_\varphi \mathrm{Iso}_o$ *is the closure of the span of* $d\Phi(\varphi)^{-1}(X_n(\varphi))$, $n \in \mathbb{Z}$ *with* $I_n(\varphi) \neq 0$, *and* $N_\varphi \mathrm{Iso}_o$ *the one of the span of* $d\Phi(\varphi)^{-1}(Y_n(\varphi))$, $n \in \mathbb{Z}$ *with* $I_n(\varphi) \neq 0$, *and* $d\Phi(\varphi)^{-1}(e_n^{(1)})$, $d\Phi(\varphi)^{-1}(e_n^{(2)})$, $n \in \mathbb{Z}$ *with* $I_n(\varphi) = 0$.

The following result, proved in Section 4, addresses the question of how restrictive the assumption of $\mathrm{Spec}_p L(\psi)$ being simple is. Let

$$\mathcal{T} := \{\psi \in i L_r^2 \mid \mathrm{Spec}_p L(\psi) \text{ simple}\}.$$

Recall that a subset A of a complete metric space X is said to be *residual* if it is the intersection of countably many open dense subsets. By Baire's theorem, A is dense in X.

Theorem 2.10 *For any integer $N \geq 0$, the set $\mathcal{T} \cap i H_r^N$ is residual in $i H_r^N$. As a consequence, for any integer $N \geq 0$, the set $\mathcal{B}_N$ of elements $\varphi \in i H_r^N$ with the property that $\mathrm{Iso}_0(\varphi)$ admits local complex Birkhoff coordinates in the sense of Theorem 2.5, is open and dense in $i H_r^N$.*

We say that $\varphi \in i L_r^2$ is a *finite gap potential* if the characteristic function $\Delta^2(\cdot, \varphi) - 4$ has finitely many simple roots (cf e.g. [14]) and that it is a *regular finite gap potential* if $\mathrm{Iso}_0(\varphi)$ is a finite dimensional torus and a neighborhood of $\mathrm{Iso}_0(\varphi)$ admits local complex Birkhoff coordinates in the sense of Theorem 2.5. An immediate consequence of Theorem 2.10 is the following improvement of Corollary 2.6, which answers a longstanding question, raised by experts in the field.

Corollary 2.11 *Any regular finite gap potential in $i L_r^2$ is C^∞−smooth. For any integer $N \geq 0$, the set of such potentials is dense in $i H_r^N$.*

Remark 2.12   As already mentioned above, the solution of the focusing NLS equation for any given regular finite gap potential $q \in i L_r^2$ as initial data, is C^∞−smooth. We expect that by the construction of Birkhoff coordinates, it can be expressed in terms of theta functions. It then follows from the wellposedness results of the focusing NLS equation, established in [3], and Corollary 2.11 that any solution of [3] in $i H_r^N$, $N \geq 1$, can be approximated by solutions, which evolve in the space of regular finite gap potentials.

In view of the application of Theorem 2.5 to the focusing mKdV equation, it is of interest to know if a result corresponding to Theorem 2.10 holds on the scale of phase spaces $H^N(\mathbb{T}, \mathbb{R})$, $N \geq 0$, of this equation. This is indeed the case. Let

$$\widetilde{\mathcal{T}} := \left\{ v \in L^2(\mathbb{T}, \mathbb{R}) \mid \mathrm{Spec}_p L(i(v, v)) \text{ simple} \right\}.$$

At the end of Section 4 we show that by the arguments used in the proof of Theorem 2.10 it follows that for any integer $N \geq 0$, $\widetilde{\mathcal{T}} \cap H^N(\mathbb{T}, \mathbb{R})$ is residual in $H^N(\mathbb{T}, \mathbb{R})$ (cf Theorem 4.2). Hence the result corresponding to Corollary 2.11 also holds in the case of the focusing mKdV equation.

Related work: In the seventies and the eighties, several groups of scientists made pioneering contributions to the development of the theory of integrable PDEs. In the periodic and quasi-periodic setup, deep connections between various equations of this type and complex geometry as well as spectral theory were discovered, allowing to establish the existence of families of finite dimensional invariant tori on which these equations can be linearized. In this way, classes of solutions were found which can be represented in terms of theta functions, referred as finite band solutions. See e.g. [2, 7, 23, 11, 30] and references therein. Further developments of these connections allowed to treat more general classes of solutions. In particular, it was established that many integrable PDEs admit invariant tori with infinitely many degrees of freedom. See e.g. [10, 26] as well as subsequent works [17, 13, 19, 27] and references therein.

3 Proofs of Proposition 2.3 and Proposition 2.8

The main purpose of this section is to show Proposition 2.3 and Proposition 2.8. Key ingredients in the proofs are results on spectral invariants of the ZS operator. We say that a functional $\mathcal{F} : H_c^N \to \mathbb{C}$ is a spectral invariant of the ZS operator on a subspace E of H_c^N, $N \geq 0$, if for any $\psi \in E$,

$$\mathcal{F}(\varphi) = \mathcal{F}(\psi), \qquad \forall\, \varphi \in E_\psi := \{\varphi \in E \mid \Delta_\lambda(\varphi) = \Delta_\lambda(\psi)\ \forall \lambda \in \mathbb{C}\}.$$

We begin by proving Proposition 2.8.

Proof of Proposition 2.8. The Hamiltonians $\mathcal{H}_k$, $k \geq 1$, of the NLS hierarchy appear in the expansion of $\Delta(\lambda)$ (cf [13, Theorem 4.8], [14, Proposition 2.2]). From this expansion one concludes that $\mathcal{H}_k$, $1 \leq k \leq N$ are spectral invariants of the ZS operator on H_c^N. Furthermore, by [13, Theorem 4.1], the L^2–gradient $\partial \Delta_\lambda := (\partial_{\varphi_1} \Delta_\lambda, \partial_{\varphi_2} \Delta_\lambda)$ of $\Delta \equiv \Delta_\lambda$ on L_c^2 is given by

$$i\partial \Delta = \mathring{m}_2 M_1 * M_1 + (\mathring{m}_4 - \mathring{m}_1)M_1 * M_2 - \mathring{m}_3 M_2 * M_2$$

where

$$M_1 \equiv M_1(x, \lambda) := \begin{pmatrix} m_1(x, \lambda) \\ m_3(x, \lambda) \end{pmatrix}, \quad M_2 \equiv M_2(x, \lambda) := \begin{pmatrix} m_2(x, \lambda) \\ m_4(x, \lambda) \end{pmatrix}$$

are the two columns of the fundamental matrix $M(x, \lambda) \equiv M(x, \lambda, \varphi)$ of $L(\varphi)$,

$$\mathring{m}_j \equiv \mathring{m}_j(\lambda) := m_j(1, \lambda), \quad 1 \leq j \leq 4,$$

and the $*$ product of two vectors in $\mathbb{C}^2$ is defined by

$$a * b := \begin{pmatrix} a_2 b_2 \\ a_1 b_1 \end{pmatrix}, \quad \forall\, a = \begin{pmatrix} a_1 \\ a_2 \end{pmatrix}, \ b = \begin{pmatrix} b_1 \\ b_2 \end{pmatrix} \in \mathbb{C}^2.$$

By [13, Theorem 1.1], $M(x, \lambda, \varphi)$ is continuous in x for any $\lambda \in \mathbb{C}$, $\varphi \in L_c^2$ and one verifies that

$$\partial \Delta_\lambda(x + 1, \varphi) = \partial \Delta_\lambda(x, \varphi), \quad \forall x \in \mathbb{R}.$$

Rewriting $L(\varphi)M(x,\lambda) = \lambda M(x,\lambda)$ as

$$i\begin{pmatrix} 1 & 0 \\ 0 & -1 \end{pmatrix}\partial_x M = -\begin{pmatrix} 0 & \varphi_1 \\ \varphi_2 & 0 \end{pmatrix}M + \lambda M,$$

and taking derivatives with respect to x of the latter identity, one concludes that for any $\varphi \in H_c^N$, $\lambda \in \mathbb{C}$, $M(x,\lambda,\varphi)$ is N times continuously differentiable with respect to x and $\partial_x^{N+1}M(x,\lambda,\varphi)$ is locally L^2 integrable. Since $\partial\Delta_\lambda$ is one periodic with respect to the x variable it then follows that $\partial\Delta_\lambda(\cdot,\varphi) \in H_c^{N+1}$ for any $\lambda \in \mathbb{C}$, $\varphi \in H_c^N$. As a consequence, for any $\lambda \in \mathbb{C}$, the Hamiltonian vector field $-i\partial\Delta_\lambda$ defines locally a flow on H_c^N

$$t \mapsto \mathcal{S}_\lambda^t(\varphi) \in H_c^N, \quad \mathcal{S}_\lambda^0(\varphi) = \varphi, \quad \varphi \in H_c^N,$$

for t in an open interval containing 0, which can be chosen locally uniformly in φ. Since $\mathrm{Iso}_o(\varphi)$ denotes the connected component of the level set of $(\Delta_\mu)_{\mu\in\mathbb{C}}$, containing φ, and $\{\Delta_\lambda, \Delta_\mu\} = 0$ for any $\mu \in \mathbb{C}$ by Proposition 2.1 one concludes that for any $\varphi \in H_c^N$, the flow of $-i\partial\Delta_\lambda$ leaves $\mathrm{Iso}_o(\varphi)$ invariant. Furthermore, since on H_c^N, $\mathcal{H}_k$, $1 \le k \le N$, are analytic and spectral invariants of the ZS operator one has for any $1 \le k \le N$,

$$\frac{d}{dt}\big|_{t=0}\mathcal{H}_k\big(\mathcal{S}_\lambda^t(\varphi)\big) = 0, \quad \forall \varphi \in H_c^N, \lambda \in \mathbb{C}.$$

Note that $\partial\mathcal{H}_k$, $1 \le k \le N$, is in the dual H_c^{-N} of H_c^N whereas by the considerations above, $\partial\Delta_\lambda \in H_c^{N+1}$. In view of the definition of the Poisson bracket we thus have shown that for any $1 \le k \le N$,

$$\{\mathcal{H}_k, \Delta_\lambda\}(\varphi) = \frac{d}{dt}\big|_{t=0}\mathcal{H}_k\big(\mathcal{S}_\lambda^t(\varphi)\big) = 0, \quad \forall \varphi \in H_c^N, \lambda \in \mathbb{C}.$$

Hence once we have shown that $\{\mathcal{H}_{2N}, \Delta_\lambda\}(\varphi)$ and $\{\mathcal{H}_{2N+1}, \Delta_\lambda\}(\varphi)$ are well defined, continuous functionals on H_c^N, it follows from the density of H_c^{2N+1} in H_c^N that

$$\{\mathcal{H}_{2N}, \Delta_\lambda\}(\varphi) = 0, \quad \{\mathcal{H}_{2N+1}, \Delta_\lambda\}(\varphi) = 0, \quad \forall \varphi \in H_c^N, \lambda \in \mathbb{C}.$$

To see that this is indeed the case, note that by the definition of the NLS hierarchy,

$$\mathcal{H}_{2N+1}(\varphi) = \int_0^1 \big(\partial_x^N\varphi_1 \cdot \partial_x^N\varphi_2 + q_{2N+1}\big)\, dx$$

where q_{2N+1} is a polynomial in φ_1, φ_2 and its derivatives up to order $N-1$. It is well defined on H_c^N and its L^2–gradient $\partial\mathcal{H}_{2N+1}(\varphi)$ has the leading term

$$\Big((-1)^N\partial_x^N(\partial_x^N\varphi_1), (-1)^N\partial_x^N(\partial_x^N\varphi_2)\Big) \in H_c^{-N}$$

whereas all other terms in the components of $\partial\mathcal{H}_{2N+1}(\varphi)$ are of the form $(-1)^j\partial_x^j p$ where $0 \le j \le N-1$ and p is a polynomial in φ_1, φ_2, and its derivatives up to order $N - 1$. Hence by the dual pairing of H_c^N and H_c^{-N}, the Poisson bracket $\{\mathcal{H}_{2N+1}, \Delta_\lambda\}$ is a well defined, continuous functional on H_c^N for any $\lambda \in \mathbb{C}$. In the same way one shows that this is also the case for $\{\mathcal{H}_{2N}, \Delta_\lambda\}$. $\square$

Next we prove item (i) of Proposition 2.3.

Proof of Proposition 2.3 (i). Assume that $\psi \in i H_r^N$ where N is an integer with $N \geq 1$. Then for any $\varphi \in \mathrm{Iso}(L(\psi))$, $\Delta^2(\lambda, \varphi) - 4 = \Delta^2(\lambda, \psi) - 4$, $\forall \lambda \in \mathbb{C}$. Since $\mathrm{Spec}_p(L(\varphi))$ coincides with the set of roots of $\Delta^2(\lambda, \varphi) - 4$ (cf [13, Theorem 6.1]) and the algebraic multiplicity of any eigenvalue $\lambda \in \mathrm{Spec}_p(L(\varphi))$ equals its multiplicity as a root of $\Delta^2(\lambda, \varphi) - 4$ (cf [14, Appendix C]) it then follows that $\mathrm{Spec}_p(L(\varphi)) = \mathrm{Spec}_p(L(\psi))$. By the characterization of the regularity of a potential $\varphi \in i L_r^2$ in terms of the asymptotics of the eigenvalues in $\mathrm{Spec}_p(L(\varphi))$, established in [18, Theorem 1.1, Theorem 1.2], it then follows that $\varphi \in i H_r^N$. $\square$

To prove Proposition 2.3 (ii), we first need to establish the following auxilary result about the Hamiltonians $\mathcal{H}_k$, $k \geq 1$, in the NLS hierarchy, which is of interest in itself.

Proposition 3.1 (i) *The Hamiltonian* $\mathcal{H}_1(\varphi) = \int_0^1 \varphi_1 \varphi_2 \, dx$ *is a spectral invariant of the ZS operator on* L_c^2. (ii) *For any integer* $N \geq 1$, *the Hamiltonians* $\mathcal{H}_{2N}, \mathcal{H}_{2N+1}$ *are spectral invariants of the ZS operator on* H_c^N.

Remark 3.2   It is well know that $\mathcal{H}_1$ is a spectral invariant of the ZS operator on L_r^2 (cf e.g. [13, Corollary 13.5]). Hence, in particular, Proposition 3.1 extends this result to $i L_r^2$.

Proof of Proposition 3.1. (i) Let $\psi \in L_c^2$ be an arbitrary potential. According to the Counting Lemma (cf [13, Lemma 6.3]) there exists an open, pathwise connected neighborhood $V \subseteq L_c^2$ of the interval $[0, \psi] = \{s\psi \mid 0 \leq s \leq 1\}$ and an integer $R \geq 0$ so that for any $\varphi \in V$, $\Delta^2(\lambda, \varphi) - 4$ has exactly two roots $\lambda_n^-(\varphi), \lambda_n^+(\varphi)$ (counted with multiplicities) in the disc $D_n = \{\lambda \in \mathbb{C} \mid |\lambda - n\pi| < \pi/6\}$ for any $n \in \mathbb{Z}$ with $|n| > R$ and a set $\Lambda_R \equiv \Lambda_R(\varphi)$ of $4R + 2$ additional roots that lie in the disc $B_R = \{\lambda \in \mathbb{C} \mid |\lambda| < R\pi + \pi/6\}$. By [13, Theorem 13.4], for $\varphi \in V$ sufficiently close to 0, $\mathcal{H}_1(\varphi) = \sum_{n \in \mathbb{Z}} I_n(\varphi)$ where for any $n \in \mathbb{Z}$, $I_n(\varphi)$ is the nth action variable of φ, given by the contour integral

$$I_n(\varphi) = \frac{1}{\pi} \int_{\Gamma_n} \frac{\lambda \dot{\Delta}(\lambda, \varphi)}{\sqrt[c]{\Delta^2(\lambda, \varphi) - 4}} \, d\lambda.$$

Here Γ_n, $n \in \mathbb{Z}$, are the contours defined in [13], $\sqrt[c]{\Delta^2(\lambda, \varphi) - 4}$ is the canonical root of $\Delta^2(\lambda, \varphi) - 4$, also defined in [13], and $\dot{\Delta}(\lambda, \varphi) \equiv \partial_\lambda \Delta(\lambda, \varphi)$. By Cauchy's theorem,

$$\sum_{|n| \leq R} I_n = \frac{1}{\pi} \int_{\partial B_R} \frac{\lambda \dot{\Delta}(\lambda, \varphi)}{\sqrt[c]{\Delta^2(\lambda, \varphi) - 4}} d\lambda$$

and for $|n| > R$, we might choose the circle ∂D_n as contour Γ_n. Hence for $\varphi \in V$ sufficiently close to 0,

$$\mathcal{H}_1(\varphi) = \frac{1}{\pi} \int_{\partial B_R} \frac{\lambda \dot{\Delta}(\lambda, \varphi)}{\sqrt[c]{\Delta^2(\lambda, \varphi) - 4}} d\lambda + \sum_{|n| > R} \frac{1}{\pi} \int_{\partial D_n} \frac{\lambda \dot{\Delta}(\lambda, \varphi)}{\sqrt[c]{\Delta^2(\lambda, \varphi) - 4}} d\lambda .$$

$$(9)$$

By Proposition 2.1 and the definition of R, one sees that the canonical root $\sqrt[c]{\Delta^2(\lambda, \varphi) - 4}$ extends analytically to ∂B_R and ∂D_n, $|n| > R$, yielding that the contour integrals $\int_{\partial B_R} \frac{\lambda \dot{\Delta}(\lambda, \varphi)}{\sqrt[c]{\Delta^2(\lambda, \varphi) - 4}} d\lambda$ and $\int_{\partial D_n} \frac{\lambda \dot{\Delta}(\lambda, \varphi)}{\sqrt[c]{\Delta^2(\lambda, \varphi) - 4}} d\lambda$, $|n| > R$, are well defined and analytic on V. By standard estimates (cf [13, Theorem 13.3 and its proof]), $\left| \int_{\partial D_n} \frac{\lambda \dot{\Delta}(\lambda, \varphi)}{\sqrt[c]{\Delta^2(\lambda, \varphi) - 4}} d\lambda \right|$ can be estimated by $C|\lambda_n^+(\varphi) - \lambda_n^-(\varphi)|^2$ where the constant $C > 0$ can be chosen locally uniformly on V. It then follows from the asymptotics $\lambda_n^\pm(\varphi) = n\pi + \ell_n^2$ that $\sum_{|n| > R} \frac{1}{\pi} \int_{\partial D_n} \frac{\lambda \dot{\Delta}(\lambda, \varphi)}{\sqrt[c]{\Delta^2(\lambda, \varphi) - 4}} d\lambda$ is analytic on V as well. Here $\lambda_n^\pm(\varphi) = n\pi + \ell_n^2$ means that $(\lambda_n^\pm(\varphi) - n\pi)_{|n| > R}$ is a ℓ^2–sequence. In fact $(\lambda_n^\pm(\varphi) - n\pi)_{|n| > R}$ is bounded locally uniformly in φ on V (cf [13, Proposition 6.7]). Hence the identity (9) holds on all of V. Since the right hand side of (9) is defined in terms of $\Delta(\lambda, \varphi)$, this shows that $\mathcal{H}_1$ is a spectral invariant for the ZS operator on L_c^2.

(ii) First we consider $\mathcal{H}_{2N+1}$ for $N \geq 1$. Let ψ be an arbitrary potential in H_c^N and choose V, R, and B_R as in the proof of item (i). (Actually, instead of V one could also choose the open ball of radius $\|\psi\|_{H_c^N} + 1$ in H_c^N (cf [13, Lemma 6.3]).) By increasing $R > 0$, if necessary, we can assume that for any $\varphi \in V \cap H_c^N$ and $|n| > R$, $\dot{\Delta}(\cdot, \varphi)$ has precisely one root in the disc D_n, which we denote by $\dot{\lambda}_n \equiv \dot{\lambda}_n(\varphi)$, and that all other roots of $\dot{\Delta}(\cdot, \varphi)$ are contained in the disc B_R (cf [13]). By [13, Theorem 13.6], for $\varphi \in V \cap H_c^N$ sufficiently close to 0,

$$\mathcal{H}_{2N+1}(\varphi) = \frac{4^N}{2N+1} \sum_{n \in \mathbb{Z}} J_{2N+1,n}(\varphi) \tag{10}$$

where $J_{2N+1,n}(\varphi)$, $n \in \mathbb{Z}$, denote the action variables of φ at the level $2N + 1$,

$$J_{2N+1,n}(\varphi) = \frac{1}{\pi} \int_{\Gamma_n} \frac{\lambda^{2N+1} \dot{\Delta}(\lambda, \varphi)}{\sqrt[c]{\Delta^2(\lambda, \varphi) - 4}} d\lambda .$$

As in item (i), Γ_n, $n \in \mathbb{Z}$, are the contours and $\sqrt[c]{\Delta^2(\lambda, \varphi) - 4}$ is the canonical root of $\Delta^2(\lambda, \varphi) - 4$, defined in [13]. (In fact, in [13, Theorem 13.6], the identity (10) is proved for $\varphi \in H_r^{2N}$. Inspecting its proof one sees that it holds for φ in a neighborhood of H_r^N in H_c^N.) To conclude we argue as in the proof of item (i). By Cauchy's theorem, for any φ in $V \cap H_c^N$ sufficiently close to 0,

$$\sum_{|n| \le R} J_{2N+1,n}(\varphi) = \frac{1}{\pi} \int_{\partial B_R} \frac{\lambda^{2N+1} \dot{\Delta}(\lambda, \varphi)}{\sqrt[c]{\Delta^2(\lambda, \varphi) - 4}} \, d\lambda$$

and if $|n| > R$, by choosing again the circle ∂D_n as contour Γ_n,

$$\int_{\Gamma_n} \frac{\lambda^{2N+1} \dot{\Delta}(\lambda, \varphi)}{\sqrt[c]{\Delta^2(\lambda, \varphi) - 4}} \, d\lambda = \int_{\partial D_n} \frac{\lambda^{2N+1} \dot{\Delta}(\lambda, \varphi)}{\sqrt[c]{\Delta^2(\lambda, \varphi) - 4}} \, d\lambda \, .$$

Hence for $\varphi \in V \cap H_c^N$ suffficiently close to 0, (10) yields

$$\begin{aligned}
\mathcal{H}_{2N+1}(\varphi) &= \frac{4^N}{2N+1} \frac{1}{\pi} \int_{\partial B_R} \frac{\lambda^{2N+1} \dot{\Delta}(\lambda, \varphi)}{\sqrt[c]{\Delta^2(\lambda, \varphi) - 4}} \, d\lambda \\
&\quad + \frac{4^N}{2N+1} \sum_{|n|>R} \frac{1}{\pi} \int_{\partial D_n} \frac{\lambda^{2N+1} \dot{\Delta}(\lambda, \varphi)}{\sqrt[c]{\Delta^2(\lambda, \varphi) - 4}} \, d\lambda \, .
\end{aligned} \qquad (11)$$

As in the proof of item (i), one sees that

$$\int_{\partial B_R} \frac{\lambda^{2N+1} \dot{\Delta}(\lambda, \varphi)}{\sqrt[c]{\Delta^2(\lambda, \varphi) - 4}} \, d\lambda \quad \text{and} \quad \int_{\partial D_n} \frac{\lambda^{2N+1} \dot{\Delta}(\lambda, \varphi)}{\sqrt[c]{\Delta^2(\lambda, \varphi) - 4}} \, d\lambda,$$

$|n| > R$, are well defined, analytic function on $V \cap H_c^N$. Since

$$\int_{\partial D_n} \frac{\dot{\Delta}(\lambda, \varphi)}{\sqrt[c]{\Delta^2(\lambda, \varphi) - 4}} \, d\lambda = 0$$

one has for any $|n| > R$,

$$\int_{\partial D_n} \frac{\lambda^{2N+1} \dot{\Delta}(\lambda, \varphi)}{\sqrt[c]{\Delta^2(\lambda, \varphi) - 4}} \, d\lambda = \int_{\partial D_n} \frac{\left(\lambda^{2N+1} - (\dot{\lambda}_n)^{2N+1}\right) \dot{\Delta}(\lambda, \varphi)}{\sqrt[c]{\Delta^2(\lambda, \varphi) - 4}} \, d\lambda$$

where $\dot{\lambda}_n \equiv \dot{\lambda}_n(\varphi)$ is the unique root of $\dot{\Delta}(\lambda, \varphi)$ in D_n. Using that

$$\lambda^{2N+1} - (\dot{\lambda}_n)^{2N+1} = (2N+1)(\dot{\lambda}_n)^{2N}(\lambda - \dot{\lambda}_n) + \binom{2N+1}{2}(\lambda - \dot{\lambda}_n)^2 + \ldots$$

one obtains a polynomial in $\dot{\lambda}_n$ of degree $2N$,

$$\int_{\partial D_n} \frac{\lambda^{2N+1} \dot{\Delta}(\lambda, \varphi)}{\sqrt[c]{\Delta^2(\lambda, \varphi) - 4}} \, d\lambda = (2N+1)(\dot{\lambda}_n)^{2N} I_n + \ldots$$

where $\ldots$ stands for terms of lower order in $\dot{\lambda}_n$. Using that $\dot{\lambda}_n = n\pi + \ell_n^2$ (cf [13, Lemma 6.9]) one then concludes as in the proof of [13, Theorem 13.6] that for any $\varphi \in V \cap H_c^N$ and $|n| > R$

$$\left| \int_{\partial D_n} \frac{\lambda^{2N+1} \dot{\Delta}(\lambda, \varphi)}{\sqrt[c]{\Delta^2(\lambda, \varphi) - 4}} \, d\lambda \right| \leq C |n|^{2N} |\lambda_n^+(\varphi) - \lambda_n^-(\varphi)|^2$$

where the constant $C > 0$ can be chosen locally uniformly on $V \cap H_c^N$. Since for $\varphi \in V \cap H_c^N$, one has $\sum_{|n|>R} |n|^{2N} |\lambda_n^+(\varphi) - \lambda_n^-(\varphi)|^2 \leq C'$ for some constant C' which again can be chosen locally uniformly on $V \cap H_c^N$, one sees that the infinite sum

$$\sum_{|n|>R} \frac{1}{\pi} \int_{\partial D_n} \frac{\lambda^{2N+1} \dot{\Delta}(\lambda, \varphi)}{\sqrt[c]{\Delta^2(\lambda, \varphi) - 4}} \, d\lambda$$

is a well defined, analytic function on $V \cap H_c^N$ and hence identity (11) holds on all of $V \cap H_c^N$. This proves that $\mathcal{H}_{2N+1}$ is a spectral invariant of the ZS operator on H_c^N. In the same way, one shows that $\mathcal{H}_{2N}$ is such a spectral invariant. $\quad \square$

With these preparations made, we are ready to prove Proposition 2.3 (ii).

Proof of Proposition 2.3 (ii) in the case $N = 0$. Assume that $\psi \in iL_r^2$ and that $(\varphi_j)_{j \geq 1}$ is a sequence in $\mathrm{Iso}(\psi)$. By Proposition 3.1(i), $\mathcal{H}_1(\varphi_j) = \mathcal{H}_1(\psi)$ for any $j \geq 1$. Since for any φ in iL_r^2, $\mathcal{H}_1(\varphi) = -\int_0^1 |\varphi_1|^2 dx$, it follows that $(\varphi_j)_{j \geq 1}$ has a weakly convergent subsequence $(\varphi_{j_k})_{k \geq 1}$. Denote its limit by $\varphi \in iL_r^2$. Recall that for any $\lambda \in \mathbb{C}$, Δ_λ is compact (cf Proposition 2.1(i)) and hence

$$\lim_{k \to \infty} \Delta_\lambda(\varphi_{j_k}) = \Delta_\lambda(\varphi), \qquad \forall \lambda \in \mathbb{C},$$

implying that $\varphi \in \mathrm{Iso}(\psi)$. Applying Proposition 3.1(i) once more one concludes that $\mathcal{H}_1(\varphi) = \mathcal{H}_1(\psi)$. Together with the weak convergence of $(\varphi_{j_k})_{k \geq 1}$ this yields that $\lim_{k \to \infty} \varphi_{j_k} = \varphi$ strongly in iL_r^2. $\quad \square$

Proof of Proposition 2.3 (ii) in the case $N \geq 1$. We argue similarly as in the case $N = 0$. Assume that $\psi \in iH_r^N$ for some integer $N \geq 1$ and that $(\varphi_j)_{j \geq 1}$ is a sequence in $\mathrm{Iso}(\psi)$. Recall that by definition (7), for any $\varphi \in iH_r^N$, $\mathcal{H}_{2N+1}(\varphi)$ is of the form

$$\mathcal{H}_{2N+1}(\varphi) = \int_0^1 \left(-|\partial_x^N \varphi_1|^2 + q_{2N+1}(\varphi_1, \varphi_2) \right) dx$$

where $q_{2N+1}(\varphi_1, \varphi_2)$ is a polynomial in φ_1, φ_2 and its derivatives up to order $N - 1$. Since $\mathcal{H}_1(\varphi) = -\int_0^1 |\varphi_1|^2 dx$, it then follows from Proposition 3.1(ii) that $\|\varphi_j\|_{H_c^N} \lesssim \|\psi\|_{H_c^N}$, implying that $(\varphi_j)_{j \geq 1}$ has a subsequence $(\varphi_{j_k})_{k \geq 1}$ which weakly converges in iH_r^N. Denote its limit by $\varphi \in iH_r^N$. By Rellich's embedding theorem, $\lim_{k\to\infty} \varphi_{j_k} = \varphi$ strongly in iH_r^{N-1}. In particular it follows that

$$\lim_{k\to\infty} \Delta_\lambda(\varphi_{j_k}) = \Delta_\lambda(\psi), \qquad \forall \lambda \in \mathbb{C},$$

implying that $\varphi \in \text{Iso}(\psi)$. Hence by Proposition 3.1(ii), $\mathcal{H}_{2N+1}(\varphi) = \mathcal{H}_{2N+1}(\psi)$ Using the strong convergence of φ_{j_k} in iH_r^{N-1} and the specific form of $\mathcal{H}_{2N+1}(\varphi)$ once more, one concludes that $\lim_{k\to\infty} \|\varphi_{j_k}\|_{H_c^N} = \|\varphi\|_{H_c^N}$, yielding that the sequence $(\varphi_{j_k})_{k\geq 1}$ strongly converges to φ. □

4 Proof of Theorem 2.10

In order to prove Theorem 2.10 we first need to make some preliminary considerations and introduce some additional notation. We say that two complex numbers $a, b \in \mathbb{C}$ are lexicographically ordered, $a \preceq b$, if [Re(a) < Re(b)] or [Re(a) = Re(b) and Im(a) ≤ Im(b)]. Furthermore, for any $n \in \mathbb{Z}$ and any integer $R \geq 0$, introduce the discs

$$D_n = \{\lambda \in \mathbb{C} \,|\, |\lambda - n\pi| < \pi/6\}, \qquad B_R = \{\lambda \in \mathbb{C} \,|\, |\lambda| < R\pi + \pi/6\}.$$

As discussed in the proof of Proposition 3.1(i), the Counting Lemma (cf [13, Lemma 6.3]) implies that for any $\psi \in iL_r^2$, there exists an open, pathwise connected neighborhood $V_\psi \subseteq L_c^2$ of the interval $[0, \psi] = \{s\psi \,|\, 0 \leq s \leq 1\}$ and an integer $R_\psi \geq 0$ so that for any $\varphi \in V_\psi$, $\Delta^2(\lambda, \varphi) - 4$ has exactly two roots (counted with multiplicities) in the disc D_n for any integer n with $|n| > R_\psi$ and a set $\Lambda_{R_\psi} \equiv \Lambda_{R_\psi}(\varphi)$ of $4R_\psi + 2$ (counted with multiplicities) additional roots that lie in the disc B_{R_ψ}. The two roots in D_n, $|n| > R_\psi$, are denoted by $\lambda_n^-(\varphi)$, $\lambda_n^+(\varphi)$ and are lexicographically ordered, $\lambda_n^-(\varphi) \preceq \lambda_n^+(\varphi)$. They satisfy $\Delta(\lambda_n^\pm(\varphi), \varphi) = (-1)^n 2$ and have the asymptotics $n\pi + \ell_n^2$ as $|n| \to \infty$ (cf [13, Proposition 6.7]). More precisely, $(\lambda^\pm - n\pi)_{|n|>R_\psi}$ is bounded in ℓ^2, locally uniformly in φ on V_ψ. Furthermore, as already mentioned in the proof of Proposition 2.3 (i) in Section 3, $\text{Spec}_p L(\varphi)$ coincides with the set of roots of $\Delta^2(\lambda, \varphi) - 4$ (with multiplicities). For a potential $\varphi \in iL_r^2$, $\text{Spec}_p L(\varphi)$ has additional properties. By [14, Proposition 2.6], any real root of $\Delta^2(\lambda, \varphi) - 4$ has, when considered as an eigenvalue in $\text{Spec}_p L(\varphi)$, geometric multiplicity two and its algebraic multiplicity is even whereas for a root λ of $\Delta^2(\lambda, \varphi) - 4$ in $\mathbb{C} \setminus \mathbb{R}$, its complex conjugate $\bar{\lambda}$ is also a root of $\Delta^2(\lambda, \varphi) - 4$ and, when considered as an eigenvalue in $\text{Spec}_p L(\varphi)$, has the same algebraic and geometric multiplicity as λ. As a consequence, for any $\varphi \in W_\psi := V_\psi \cap iL_r^2$,

$$\overline{\Lambda_{R_\psi}(\varphi)} = \Lambda_{R_\psi}(\varphi) \quad \text{and} \quad \lambda_n^-(\varphi) = \overline{\lambda_n^+(\varphi)}, \quad \text{Im}\,\lambda_n^+(\varphi) \geq 0, \quad \forall |n| > R_\psi.$$

Proof of Theorem 2.10. First we consider the case $N = 0$. It is to prove that $\mathcal{T}$ is residual in iL_r^2. For any $\psi \in iL_r^2$ and $R \geq R_\psi$, define

$$\mathcal{S}_{\psi,R} := \left\{ \varphi \in W_\psi \mid \lambda \text{ simple root of } \Delta^2(\cdot, \varphi) - 4 \ \forall \lambda \in \Lambda_R(\varphi) \right\},$$

where $W_\psi = V_\psi \cap iL_r^2$ and W_ψ are given as above. Without loss of generality we assume that W_ψ is path connected. Clearly, $\mathcal{S}_{\psi,R}$ is open in W_ψ for any integer $R \geq R_\psi$. We claim that for any integer $R \geq R_\psi$, $\mathcal{S}_{\psi,R}$ is also dense in W_ψ. To this end introduce for any such R the map

$$\mathcal{Q}_R : \mathbb{C} \times W_\psi \to \mathbb{C}, \quad (\lambda, \varphi) \mapsto \mathcal{Q}_R(\lambda) \equiv \mathcal{Q}_R(\lambda, \varphi) = \prod_{\eta \in \Lambda_R(\varphi)} (\eta - \lambda).$$

Note that for any $\varphi \in W_\psi$, $\mathcal{Q}_R(\cdot, \varphi)$ is a polynomial in λ of degree $4R + 2$. Then (cf [14, Theorem 1.2])

$$\mathcal{S}_{\psi,R} = \left\{ \varphi \in W_\psi \mid \mathcal{D}_R(\varphi) \neq 0 \right\}$$

where $\mathcal{D}_R(\varphi)$ denotes the discriminant of $\mathcal{Q}_R(\cdot, \varphi)$, i.e., the resultant of $\mathcal{Q}_R(\cdot, \varphi)$ and its derivative $\partial_\lambda \mathcal{Q}_R(\cdot, \varphi)$. (Indeed, the key property of $\mathcal{D}_R(\varphi)$ is that it vanishes at an element $\varphi \in W_\psi$ iff $\mathcal{Q}_R(\cdot, \varphi)$ has at least one multiple zero.) The discriminant $\mathcal{D}_R(\varphi)$ is given by

$$\mathcal{D}_R(\varphi) = \det \begin{pmatrix} a_0 & \cdots & a_{4R} & a_{4R+1} & a_{4R+2} & 0 & \cdots & 0 \\ \vdots & & \vdots & \vdots & \vdots & \vdots & & \vdots \\ 0 & \cdots & a_0 & a_1 & a_2 & a_3 & \cdots & a_{4R+2} \\ b_0 & \cdots & b_{4R} & b_{4R+1} & 0 & 0 & \cdots & 0 \\ \vdots & & \vdots & \vdots & \vdots & \vdots & & \vdots \\ 0 & \cdots & 0 & b_0 & b_1 & b_2 & \cdots & b_{4R+1} \end{pmatrix}$$

where $(a_0, a_1, \ldots, a_{4R+2})$ is the coefficient vector of the polynomial $\mathcal{Q}_R(\cdot, \varphi)$ and repeated $4R + 1$ times whereas $(b_0, b_1, \ldots, b_{4R+1})$ is the one of $\partial_\lambda \mathcal{Q}_R(\cdot \cdot \varphi)$, and repeated $4R + 2$ times. Note that $\mathcal{D}_R(\varphi)$ is real valued on W_ψ. Indeed, since for any $\varphi \in W_\psi$, $\Lambda_R(\varphi)$ is invariant under complex conjugation, the coefficients of $\mathcal{Q}_R(\cdot, \varphi)$ and the ones of $\partial_\lambda \mathcal{Q}_R(\cdot, \varphi)$ are real valued, implying that $\mathcal{D}_R$ is real valued on W_ψ. Furthermore, $\mathcal{D}_R$ is a real analytic function on W_ψ. Indeed, the coefficients of the polynomials $\mathcal{Q}_R(\cdot, \varphi)$ are symmetric polynomials in the roots $\eta \in \Lambda_R(\varphi)$ of $\Delta^2(\cdot, \varphi) - 4$ and hence can be written as polynomials in

$$s_n(\varphi) := \sum_{\eta \in \Lambda_R(\varphi)} \eta^n, \quad 1 \leq n \leq 4R + 2.$$

Therefore it suffices to show that each s_n is real analytic on W_ψ. To see it, note that by the argument principle, s_n is given by

$$s_n = \frac{1}{2\pi i} \int_{\partial B_R} \lambda^n \frac{\partial_\lambda(\Delta^2(\lambda, \varphi) - 4)}{\Delta^2(\lambda, \varphi) - 4} \, d\lambda$$

and hence is real analytic on W_ψ. Here we used that by Proposition 2.1(i), $\Delta^2(\lambda, \varphi) - 4$ is analytic on $\mathbb{C} \times L_c^2$ and since R is an integer with $R \geq R_\psi$, $\Delta^2(\lambda, \varphi) - 4$ does not vanish for $(\lambda, \varphi) \in \partial B_R \times W_\psi$.

Recall that W_ψ is assumed to be connected. To show that $\mathcal{S}_{\psi,R}$ is dense in W_ψ, it then suffices to prove that it is not empty. To this end we consider elements on W_ψ near 0. They can be parametrized by Birkhoff coordinates. By [15, Theorem 1.1 and formula (3.8)], for any element $\varphi \in W_\psi$ near 0 with complex Birkhoff coordinates $\left((z_k(\varphi))_{k\in\mathbb{Z}}, (w_k(\varphi))_{k\in\mathbb{Z}}\right)$ so that $I_n(\varphi) = z_n(\varphi)w_n(\varphi) \neq 0$ for any $|n| \leq 2R+1$, all the roots of $\Delta^2(\cdot, \varphi)-4$ in $\Lambda_R(\varphi)$ are simple. Hence any such φ is in $\mathcal{S}_{\psi,R}$ and by [15, Theorem 1.1], one concludes that $\mathcal{S}_{\psi,R} \neq \emptyset$.

Altogether we thus have shown that $\mathcal{S}_{\psi,R}$ is open and dense in W_ψ. Clearly, for any integers $R_\psi \leq R < R'$, $\mathcal{S}_{\psi,R'} \subseteq \mathcal{S}_{\psi,R}$ and

$$\mathcal{T} \cap W_\psi = \bigcap_{R \geq R_\psi, R \in \mathbb{Z}} \mathcal{S}_{\psi,R},$$

i.e., $\mathcal{T} \cap W_\psi$ is an intersection of countably many open dense subsets of W_ψ. It then follows from Lemma 4.1 below that $\mathcal{T}$ is residual. In the same way one shows that $\mathcal{T} \cap i H_r^N$ is residual in $i H_r^N$ for any integer $N \geq 1$. In fact, for $N \geq 1$, the neighborhood V_ψ can be chosen be the open ball in H_c^N of radius $\|\psi\|_{H^N} + 1$, centered at 0 since in this case, the integer R_ψ can be chosen so that it only depends on the H_c^1 norm $\|\psi\|_{H_c^1}$ of ψ (cf [13, Lemma 6.4]). $\square$

It remains to prove the lemma used in the proof of Theorem 2.10.

Lemma 4.1 *Assume that X is a separable, complete metric space (hence satisfies the second axiom of countability) and Y is a subset of X with the property that for any $x \in X$, there exists an open neighborhood V_x of x in X so that $Y \cap V_x$ is residual in V_x, meaning that there exist closed subsets $Z_{x,n} \subseteq X$, $n \geq 1$, so that $V_x \setminus Z_{x,n}$ is dense in V_x for any $n \geq 1$ and $Y \cap V_x = \bigcap_{n\geq 1} V_x \setminus Z_{x,n}$. Then Y is residual in X.*

Proof. For any $x \in X$, choose an open neighborhood U_x of x in X so that $\overline{U_x} \subseteq V_x$. Since X satisfies the second axiom of countability there exists a sequence $(x_j)_{j\geq 1}$ in X so that $U_j \equiv U_{x_j}$, $j \geq 1$, covers X (Theorem of Lindelöf). Let $V_j := V_{x_j}$. By assumption, for any $j \geq 1$ there exist closed subsets $Z_{j,n} \subseteq X$, $n \geq 1$, so that $V_j \setminus Z_{j,n}$ is dense in V_j for any $n \geq 1$ and

$$Y \cap V_j = \bigcap_{n \geq 1} V_j \setminus Z_{j,n}. \tag{12}$$

Note that $W_{j,n} := \overline{U_j} \cap Z_{j,n}$, $j, n \geq 1$, are closed subsets of X, are contained in V_j, and have the property that $X \setminus W_{j,n}$, $j, n \geq 1$, are dense in X since

$$\overline{X \setminus W_{j,n}} = \overline{X \setminus V_j} \cup \overline{V_j \setminus W_{j,n}} \supseteq (X \setminus V_j) \cup \overline{V_j \setminus Z_{j,n}} = X.$$

We claim that $Y = \bigcap_{j,n \geq 1} X \setminus W_{j,n}$. To see that $Y \subseteq \bigcap_{j,n \geq 1} X \setminus W_{j,n}$, assume that $x \in Y$. Choose any $j \geq 1$. If $x \in V_j$, then $x \in Y \cap V_j$. Since by (12)

$$Y \cap V_j = \bigcap_{n \geq 1} V_j \setminus Z_{j,n} \subseteq \bigcap_{n \geq 1} V_j \setminus W_{j,n} \subseteq \bigcap_{n \geq 1} X \setminus W_{j,n}$$

it then follows that $x \in \bigcap_{n \geq 1} X \setminus W_{j,n}$. If $x \notin V_j$, then $x \notin W_{j,n} \ \forall n \geq 1$ and hence $x \in \bigcap_{n \geq 1} X \setminus W_{j,n}$. Altogether we have shown that $Y \subseteq \bigcap_{j,n \geq 1} X \setminus W_{j,n}$.

Conversely, to see that $\bigcap_{j,n \geq 1} X \setminus W_{j,n} \subseteq Y$, assume that $x \in \bigcap_{j,n \geq 1} X \setminus W_{j,n}$. Since U_j, $j \geq 1$, covers X there exists $k \geq 1$ so that $x \in U_k$. Hence $x \in U_k \setminus W_{k,n}$ for any $n \geq 1$. By the definition of $W_{k,n}$ one has $U_k \setminus W_{k,n} = U_k \setminus Z_{k,n}$ and hence by (12)

$$x \in \bigcap_{n \geq 1} U_k \setminus Z_{k,n} \subseteq \bigcap_{n \geq 1} V_k \setminus Z_{k,n} = Y \cap V_k \subseteq Y.$$

Altogether we thus have proved that $Y = \bigcap_{j,n \geq 1} X \setminus W_{j,n}$, implying that Y is residual in X. $\qquad\qquad\square$

We finish this section by proving the result mentioned at the end of Section 2, concerning the scale of phase spaces of the focusing mKdV equation. Let us introduce for any integer $N \geq 0$ the real Hilbert space

$$E_r^N := \left\{ (v, v) \,\middle|\, v \in H^N(\mathbb{T}, \mathbb{R}) \right\}.$$

The result described at the end of Section 2 can then be formulated as follows.

Theorem 4.2 *For any integer $N \geq 0$, $\mathcal{T} \cap i E_r^N$ is residual in $i E_r^N$.*

To prove Theorem 4.2, we first need to establish a preliminary result. Consider the linear isometric involution

$$T : L_c^2 \to L_c^2, \quad (\varphi_1, \varphi_2) \mapsto (\varphi_2, \varphi_1).$$

Note that for any $(\varphi_1, \varphi_2) \in i L_r^2$ one has $\varphi_2 = -\overline{\varphi_1}$, hence $T(\varphi_1, \varphi_2) = -(\overline{\varphi_1}, \overline{\varphi_2})$. By [13] there exists a neighborhood V_0 of 0 in L_c^2 and a canonical diffeomorphism $\Phi^{nls} : V_0 \to \Phi^{nls}(V_0) \subseteq h_c^0$, mapping $\varphi \in V_0$ to its complex Birkhoff coordinates $((z_k(\varphi))_{k \in \mathbb{Z}}, (w_k(\varphi))_{k \in \mathbb{Z}})$. Without loss of generality, we can assume that the neighborhood V_0 has been chosen so that $T(V_0) = V_0$. Furthermore, introduce the linear isometric involution

$$S : \ell_c^2 \to \ell_c^2, \quad \left((z_k)_{k\in\mathbb{Z}}, (w_k)_{k\in\mathbb{Z}}\right) \mapsto \left((\check{z}_k)_{k\in\mathbb{Z}}, (\check{w}_k)_{k\in\mathbb{Z}}\right)$$

where $\check{z}_k = w_{-k}$ and $\check{w}_k = z_{-k}$ for any $k \in \mathbb{Z}$.

Lemma 4.3 *For any $\varphi \in V_0$, $\Phi^{nls}(T\varphi) = S\left(\Phi^{nls}(\varphi)\right)$.*

Proof of Lemma 4.3. Since Φ^{nls} is real analytic it suffices to prove the claimed identity for $\varphi \in L_r^2 \cap V_0$. By [12, Theorem 1.2], the identity indeed holds on $L_r^2 \cap V_0$. $\qquad\square$

Proof of Theorem 4.2. First let us consider the case $N = 0$. The same arguments as in the proof of Theorem 2.10 yield the claimed result. It only remains to prove that there exists $\varphi \in iE_r^0$ near 0 so that $\mathrm{Spec}_p L(\varphi)$ is simple. First note that any $\varphi \in iE_r^0$ satisfies $T(\varphi) = \varphi$ and hence by Lemma 4.3, $S(\Phi^{nls}(\varphi)) = \Phi^{nls}(\varphi)$, where Φ^{nls} is the Birkhoff map, defined in the neighborhood V_0 of 0 in L_c^2 as above. It follows that for any $k \in \mathbb{Z}$, $z_{-k}(\varphi) = z_k(\varphi)$ and $w_{-k}(\varphi) = w_k(\varphi)$, implying that

$$I_{-k}(\varphi) = z_{-k}(\varphi)w_{-k}(\varphi) = z_k(\varphi)w_k(\varphi) = I_k(\varphi).$$

As a consequence, any element $\varphi \in iE_r^0$ with $z_k(\varphi) \neq 0 \; \forall k \in \mathbb{Z}$ has the property that $I_k(\varphi) \neq 0$
$\forall k \in \mathbb{Z}$. For such an element φ, $\mathrm{Spec}_p L(\varphi)$ is simple (cf [15, Theorem 1.1 and formula (3.8)]). The case where N is an integer $N \geq 1$ is proved in the same way since for any $N \geq 1$, $\Phi^{nls}(iH_r^N \cap V_0) \subseteq ih_r^N$ and $\Phi^{nls} : iH_r^N \cap V_0 \to \Phi^{nls}(iH_r^N \cap V_0) \subseteq iH_r^N$ is a real analytic diffeomorphism. $\qquad\square$

Appendix A. Regularized Determinant of $L(\varphi) - \lambda$

In this appendix we compute the ζ–regularized determinant of $L(\varphi) - \lambda$ for a C^∞–smooth potential, $\varphi \in C^\infty(\mathbb{T}, \mathbb{C}) \times C^\infty(\mathbb{T}, \mathbb{C})$, and $\lambda \in \mathbb{C} \setminus \mathrm{Spec}_p(L(\varphi))$ by applying the formula, derived in [5, Theorem 1(B)], for such determinants of differential operators with smooth coefficients, acting on a vector bundle over $\mathbb{T} = \mathbb{R}/\mathbb{Z}$. Using the notation introduced in [5], we introduce for $\lambda \in \mathbb{C}$,

$$A(\lambda) := L(\varphi) - \lambda = A_1 D + A_0(x, \lambda), \quad D = \frac{1}{i}\partial_x,$$

where

$$A_1 = \begin{pmatrix} -1 & 0 \\ 0 & 1 \end{pmatrix}, \quad A_0(x, \lambda) = \begin{pmatrix} -\lambda & \varphi_1(x) \\ \varphi_2(x) & -\lambda \end{pmatrix}.$$

The differential operator $A(\lambda)$ is of order $n = 1$ and acts on the trivial vector bundle $\mathbb{T} \times \mathbb{C}^2$ of rank $N = 2$. According to [5, Theorem 1(B)], the ζ−regularized determinant of $A(\lambda)$ is expressed in terms of the monodromy matrix $P_{A(\lambda)} = M(1, \lambda, \varphi)$ where $M(x, \lambda, \varphi)$ denotes the fundamental solution of $L(\varphi)$, $L(\varphi)M(x, \lambda, \varphi) = \lambda M(x, \lambda, \varphi)$. In addition, the formula involves the two factors $R(A(\lambda))$ and $S_\theta(A(\lambda))$, defined in terms of the two coefficient matrices A_1 and $A_0(x, \lambda)$. Since $\mathrm{tr}(A_1^{-1} A_0(x, \lambda)) \equiv 0$, one has by [5, Theorem 1(B)]

$$R\big(A(\lambda)\big) = \exp\left(\frac{i}{2} \int_0^1 \mathrm{tr}(A_1^{-1} A_0(x, \lambda))\, dx\right) = 1\,.$$

To define the factor $S_\theta(A(\lambda))$, note that the principal symbol of $L(\varphi)$ is given by $\sigma_L(\xi) = A_1\xi$, $\xi \in \mathbb{R}$, and hence is independent of x. Since the spectrum of $\sigma_L(\xi)$ consists of the two eigenvalues ξ and $-\xi$, any ray $R_\theta = \{re^{i\theta} \mid r > 0\}$ with $\theta \in [0, 2\pi) \setminus \{0, \pi\}$ has empty intersection with the spectrum of $\sigma_L(\xi)$, $\xi \in \mathbb{R}$. We choose $\theta = \pi/2$ as principal angle. The projections $\Pi_{\pi/2}^{\pm}$ are then given by (cf [5])

$$\Pi_{\pi/2}^+ : \mathbb{C}^2 \to \mathbb{C} \times \{0\}\,, \qquad \Pi_{\pi/2}^- : \mathbb{C}^2 \to \{0\} \times \mathbb{C}$$

and the bundle isomorphism $\Gamma_{\pi/2}$ is defined by $\Gamma_{\pi/2} = \Pi_{\pi/2}^+ - \Pi_{\pi/2}^- = -A_1$. It follows that $\Gamma_{\pi/2}A_1^{-1} = -\mathrm{Id}_{2\times 2}$ and

$$\det(\Gamma_{\pi/2}) = -1\,, \qquad \mathrm{tr}\left(\Gamma_{\pi/2}A_1^{-1} A_0(x, \lambda)\right) = 2\lambda.$$

Hence according to [5],

$$S_{\pi/2}\big(A(\lambda)\big) = \det(\Gamma_{\pi/2}) \exp\left(\frac{i}{2} \int_0^1 \mathrm{tr}\left(\Gamma_{\pi/2}A_1^{-1} A_0(x, \lambda)\right) dx\right) = -e^{i\lambda}\,.$$

Altogether, the ζ-regularized determinant of $L(\varphi) - \lambda$ for values of λ in $\mathbb{C} \setminus \mathrm{Spec}_p(L(\varphi))$ with principal angle $\theta = \pi/2$ is given by

$$\begin{aligned}
\det_{\pi/2}(A(\lambda)) &= i^4 S_{\pi/2}(A(\lambda)) R(A(\lambda)) \det(\mathrm{Id}_{2\times 2} - P_{A(\lambda)}) \\
&= -e^{i\lambda} \det(\mathrm{Id}_{2\times 2} - M(1, \lambda, \varphi)) = e^{i\lambda}(\Delta(\lambda, \varphi) - 2)\,.
\end{aligned}$$

Since $\det_{\pi/2}(A(\lambda))$ is defined to be zero if $A(\lambda)$ is not injective, i.e., if λ is a periodic eigenvalue of $L(\varphi)$, one concludes that for any $\lambda \in \mathbb{C}$, one has

$$\det_{\pi/2}(L(\varphi) - \lambda) = e^{i\lambda}(\Delta(\lambda, \varphi) - 2)\,.$$

Since by Proposition 2.1, $e^{i\lambda}(\Delta(\lambda, \varphi) - 2)$ is an analytic function on $\mathbb{C} \times L_c^2$, one concludes that $\det_{\pi/2}(L(\varphi) - \lambda)$ analytically extends to $\mathbb{C} \times L_c^2$.

Acknowledgment. The authors gratefully acknowledge the support and hospitality of the FIM at ETH Zurich and of the Mathematics Departments of the Northeastern University and the University of Zurich.

References

[1] Arnold, V., *Mathematical methods of classical mechanics*, Graduate Texts in Mathematics, **60**, Springer, 1989

[2] Belokolos, E., Bobenko, A., Enolskii, V., Its, A., Matveev, V., *Algebro-Geometric Approach to Nonlinear Integrable Equations*, Springer Series in Nonlinear Dynamics, Springer, 1994

[3] Bourgain, J., *Fourier transform restriction phenomena for certain lattice subsets and applications to nonlinear evolution equation. I. Schrödinger equations*, Geom. Funct. Anal., **3**(1993), no. 2, 107–156

[4] Bourgain, J., *Fourier transform restriction phenomena for certain lattice subsets and applications to nonlinear evolution equation. II. The KdV equation*, Geom. Funct. Anal., **3**(1993), no. 3, 209–262

[5] Burghelea, D., Friedlander, L., Kappeler, T.,*On the determinant of elliptic differential operators and finite difference operators in vector bundles over S^1*, Commun. Math. Phys. **138**(1991), 1–18

[6] Colliander, J., Keel, M., Staffilani, G., Takaoka, H., Tao, T., *Sharp global well-posedness for KdV and modified KdV on $\mathbb{R}$ and $\mathbb{T}$*, J. Amer. Math. Soc., **16**(2003), no. 3, 705–749.

[7] Dubrovin, B., Krichever, I., Novikov, S., *Integrable Systems. I*, in Arnold, V., Novikov, S. (Eds), *Dynamical Systems IV*, Encyclopedia of Mathematical Sciences, Springer 1990

[8] Duistermaat, J., *On global action-angle variables*, Comm. Pure Appl. Math., **33**(1980), 687–706

[9] Eliasson, H., *Normal forms for Hamiltonian systems with Poisson commuting integrals – elliptic case*, Comment. Math. Helv., **65**(1990), no. 1, 4–35

[10] Flaschka, H., McLaughlin, D., *Canonically conjugate variables for the Korteweg-de Vries equation and the Toda lattice with periodic boundary conditions*, Progr. Theoret. Phys., **55**(1976), no. 2, 438–456

[11] Gesztesy, F., Holden, H., *Soliton equations and their algebro-geometric solutions*, Cambridge Studies in Advanced Mathematics, **79**, Cambridge University Press, Cambridge, 2003

[12] Grébert, B., Kappeler, T., Symmetries of the nonlinear Schrödinger equation, Bull. Soc. math. France **130**(2002), no. 4, 603–618

[13] Grébert, B., Kappeler, T., *The defocusing NLS and its normal form*, EMS Series of Lectures in Mathematics, EMS, Zürich, 2014

[14] Kappeler, T., Lohrmann, P., Topalov, P., *Generic non-selfadjoint Zakharov-Shabat operators*, Math. Ann., **359**(2014), no. 1-2, 427–470

[15] Kappeler, T., Lohrmann, P., Topalov, P., Zung, N., *Birkhoff coordinates for the focusing NLS equation*, Comm. Math. Phys., **285**(2009), no. 3, 1087–1107

[16] Kappeler, T., Molnar, J., *On the well-posedness of the defocusing mKdV equation below L^2*, SIAM J. Math. Anal., **49**(2017), no. 3, 2191–2219

[17] Kappeler, T., Pöschel, *KdV&KAM*, **45**, Ergeb. der Math. und ihrer Grenzgeb., Springer, 2003

[18] T. Kappeler, F. Serier, P. Topalov, *On the characterization of the smoothness of skew-adjoint potentials in periodic Dirac operators*, J. of Funct. Anal., **256**(2009), 2069–2112

[19] Kappeler, T., Topalov, P., *Global well-posedness of KdV in $H^{-1}(\mathbb{T}, \mathbb{R})$*, Duke Journal of Mathematics, **135**(2006), no. 2, 327–360

[20] Kappeler, T., Topalov, P., *On the non-existence of local Birkhoff coordinates for the focusing NLS equation*, arXiv:1807.02455, to appear in Fields Institute Communications

[21] Kappeler, T., Topalov, P., *Arnold-Liouville theorem for integrable PDEs: a case study of the focusing NLS equation*, in preparation

[22] Lax, P., *Integrals of nonlinear equations of evolution and solitary waves*, Comm. Pure Appl. Math., **21**(1968), 467–490

[23] Lax, P., *Periodic solutions of the KdV equation*, Comm. Pure Appl. Math., **28**(1975), 141–188

[24] Li, Y., McLaughlin, D., *Morse and Melnikov functions for NLS PDEs*, Comm. Math. Phys., **162**(1994), no. 1, 175–214

[25] Liouville, J., *Note sur l'intégration des équations différentielles de la Dynamique*, présenté au Bureau des Longitudes le 29 juin 1853, J. Math. Pures Appl., **20**(1855), 137–138

[26] McKean, H., Trubowitz, E., *Hill's operator and hyperelliptic function theory in the presence of infinitely many branch points*, Comm. Pure Appl. Math., **29**(1976), no. 2, 143–226

[27] McKean, H., Vaninsky, K., *Action-angle variables for the cubic Schrödinger equation*, Comm. Pure Appl. Math., **50**(1997), no. 6, 489–562

[28] Mineur, H., *Réduction des systèmes mécaniques à n degrés de liberté admettant n intégrales premières uniformes en involution aux systèmes à variables séparées*, Journal de Math., **15**(1935), 385–389

[29] Moser, J., Zehnder, E., *Notes on dynamical systems*, Courant Lecture Notes in Mathematics, **12**, AMS, 2005

[30] Novikov, S., Manakov, S., Pitaevski, L., Zakharov, V., *Theory of solitons. The inverse scattering method*, Contemporary Soviet Mathematics, Consultants Bureau (Plenum), 1984

[31] Vey, J., *Sur certains systèmes dynamiques séparables*, Amer. J. Math., **100**(1978), no. 3, 591–614

[32] Zakharov, V., Shabat, A., *A scheme for integrating nonlinear equations of mathematical physics by the method of the inverse scattering problem I*, Functional Anal. Appl., **8**(1974), 226–235

[33] Zung, N., *Convergence versus integrability in Birkhoff normal form*, Ann. of Math., **161**(2005), no. 1, 141–156

[34] Zung, N., *A note on focus-focus singularities*, Differential Geom. Appl., **7**(1997), no. 2, 123–130

Thomas Kappeler
Department of Mathematics
University of Zurich, 190 Winterthurerstrasse, 8057 Zurich, Switzerland
Email address: thomas.kappeler@math.uzh.ch

Peter Topalov
Department of Mathematics, Northeastern University, 360 Huntington Ave., 02115 Boston, MA, USA
Email address: p.topalov@northeastern.edu

12

Commuting Hamiltonian Flows of Curves in Real Space Forms

Albert Chern, Felix Knöppel, Franz Pedit and Ulrich Pinkall

Abstract. Starting from the vortex filament flow introduced in 1906 by Da Rios, there is a hierarchy of commuting geometric flows on space curves. The traditional approach relates those flows to the nonlinear Schrödinger hierarchy satisfied by the complex curvature function of the space curve. Rather than working with this infinitesimal invariant, we describe the flows directly as vector fields on the manifold of space curves. This manifold carries a canonical symplectic form introduced by Marsden and Weinstein. Our flows are precisely the symplectic gradients of a natural hierarchy of invariants, beginning with length, total torsion, and elastic energy. There are a number of advantages to our geometric approach. For instance, the real part of the spectral curve is geometrically realized as the motion of the monodromy axis when varying total torsion. This insight provides a new explicit formula for the hierarchy of Hamiltonians. We also interpret the complex spectral curve in terms of curves in hyperbolic space and Darboux transforms. Furthermore, we complete the hierarchy of Hamiltonians by adding area and volume. These allow for the characterization of elastic curves as solutions to an isoperimetric problem: elastica are the critical points of length while fixing area and volume.

1 Introduction

The study of curves and surfaces in differential geometry and geometric analysis has given rise to a number of important global problems, which play a pivotal role in the development of those subjects. A historical example worth mentioning is Euler's study and classification of elastic planar curves, which are the critical points of the bending energy $\int \kappa^2$, the averaged squared curvature of the curve. Variational calculus, conserved quantities, and geometry combine beautifully to provide a complete solution of the problem. Euler presumably did not know that the bending energy belongs to an infinite hierarchy of commuting energy functionals on the space of planar curves, whose commuting symplectic gradients are avatars of the modified Korteweg–de Vries (mKdV) hierarchy. In fact, a more complete picture arises when considering curves in 3-space, in which case the flows are a geometric manifestation of the non-linear Schrödinger hierarchy—of which mKdV is a reduction.

There is evidence that an analogous structure is present on the space of surfaces with abelian fundamental groups in 3- and 4-space. The energy functional in question is the Willmore energy $\int H^2$, averaging the squared mean curvature of the surface, whose critical points are Willmore surfaces. Again, this functional is part of an infinite hierarchy of commuting functionals related to the Davey–Stewartson hierarchy in mathematical physics. The existence of such hierarchies plays a significant role in classification problems in surface geometry including minimal, constant mean curvature, and Willmore surfaces.

The equations mentioned, (m)KdV, non-linear Schrödinger, Davey–Stewartson (and some of its reductions like modified Novikov–Veselov, sinh-Gordon, etc.) have their origins in mathematical physics and serve as prime models for *infinite dimensional integrable systems*. Characterizing features of these equations include soliton solutions, a transformation theory—Darboux transforms—with permutability properties, Lax pair descriptions on loop algebras, dressing transformations, reformulations as families of flat connections—zero curvature descriptions, and explicit solutions arising from linear flows on Jacobians of finite genus spectral curves. The latter provide a stratification of the hierarchy by finite dimensional classical integrable systems giving credence to the terminology.

This note will provide a geometric inroad into these various facets of infinite dimensional integrable systems by discussing in detail the simplest example, namely the space $\mathcal{M}$ of curves γ in $\mathbb{R}^3$. The advantage of this example lies in its technical simplicity without loosing any of the conceptual complexity of the theory. At various junctures we shall encounter loop algebras, zero curvature equations, Jacobians etc. as useful concepts, techniques, and possible directions for further exploration. Rather than working with closed curves, many of the classically relevant examples require curves with monodromy given by an orientation preserving Euclidean motion $h(p) = Ap + a$. Such curves $\gamma : \tilde{M} \to \mathbb{R}^3$ are equivariant in the sense that $\tau^* \gamma = A\gamma + a$, where $\tilde{M} \cong \mathbb{R}$ and τ is a fixed diffeomorphism (translation) with $M = \tilde{M}/\tau \cong S^1$.

The space $\mathcal{M}$ of curves with a fixed monodromy has the structure of a presymplectic manifold thanks to the Marsden–Weinstein 2-form $\sigma \in \Omega^2(\mathcal{M}, \mathbb{R})$ given by equation (2). This 2-form is degenerate along the tangent spaces to the reparametrization orbits of diffeomorphisms on $\tilde{M}$ compatible with the translation τ. The L^2-metric $\langle\,,\,\rangle$ gives $\mathcal{M}$ the additional structure of a Riemannian manifold and the two structures relate via $\sigma(\xi, \eta) = \langle T \times \xi, \eta \rangle$ for $\xi, \eta \in T\mathcal{M}$, and $T \in \Gamma(T\mathcal{M})$ the vector field whose value at $\gamma \in \mathcal{M}$

is the unit length tangent $T(\gamma) = \gamma'$. The Riemannian and pre-symplectic structures allow us to construct variational G_E and symplectic Y_E gradients to a given energy functional $E\colon \mathcal{M} \to \mathbb{R}$. In this article we focus on the Hamiltonian aspects of the space $\mathcal{M}$ and show that $\mathcal{M}$ can be viewed as a phase space for an infinite dimensional integrable system given by a hierarchy of commuting Hamiltonians $E_k \in C^\infty(\mathcal{M}, \mathbb{R})$ and corresponding commuting symplectic vector fields $Y_k \in \Gamma(T\mathcal{M})$.

Having introduced the space $\mathcal{M}$ in Section 2, we start Section 3 with a number of classical energy functionals whose integrants, due to their geometric nature, are all defined over $M \cong S^1$: the length functional E_1, the total torsion functional E_2, measuring the total turning angle of a parallel section in the normal bundle of a curve $\gamma \in \mathcal{M}$ over a period of τ, and the bending, or elastic, energy E_3, which measures the total squared curvature of a curve $\gamma \in \mathcal{M}$ over a period of τ. Their variational and symplectic gradients are well known (see Table 12.1 and the historical Section 7), and the behavior of some of their flows has been studied from geometric analytic and Hamiltonian aspects. There are two more Hamiltonians, E_{-1} and E_{-2}, whose significance in our context seems to be new: they arise, at least in the case of closed curves, from the flux of the infinitesimal translation and rotation vector fields on $\mathbb{R}^3$ through any surface spanned by the curve $\gamma \in \mathcal{M}$. These functionals measure a certain projected enclosed area of γ and the volume of a solid torus generated by revolving γ, respectively. Since E_{-1} and E_{-2} are given by infinitesimal isometries, these functionals are preserved by the symplectic flows of E_k for $1 \leq k \leq 3$. As an example, the vortex filament flow $Y_1 = \gamma' \times \gamma''$ preserves the projected area and volume of the initial curve as shown in Figure 12.3. The functionals E_{-1} and E_{-2} also provide new isoperimetric characterizations of the critical points of the elastic energy E_3 constrained by length and total torsion, the so-called *Euler elastica*. For instance, an Euler elastic curve can also be described as being critical for total torsion under length and enclosed area constraints, or critical for length under enclosed area and volume constraints.

Inspecting the gradients G_k and Y_k of the functionals E_k for $-2 \leq k \leq 3$ in Table 12.1, where we also included the vacuum $E_0 = 0$, one notices the pattern

$$G_{k+1} = -Y_k'$$

along curves $\gamma \in \mathcal{M}$. Combined with the fact that $G_k = T \times Y_k$, we obtain the recursion

$$Y_{k+1}' + T \times Y_k = 0, \quad Y_0 = T \tag{1}$$

which can be solved explicitly and yields an infinite sequence [41, 26] of vector fields $Y_k \in \Gamma(T\mathcal{M})$. Since the vortex filament flow $Y_1 = \gamma' \times \gamma''$ corresponds to the non-linear Schrödinger flow [18] of the associated complex curvature function ψ of the curve γ, it is reasonable to expect the flows Y_k to be avatars of the non-linear Schrödinger hierarchy. We regard this as supporting evidence that the vector fields Y_k commute on $\mathcal{M}$ and are symplectic for a hierarchy of energy functionals E_k on $\mathcal{M}$.

Our discussion of the commutativity of the Y_k in Section 4 is based on the geometry of its generating function

$$Y = \sum_{k \geq 0} Y_k \lambda^{-k}$$

which, pointwise, lies in the loop Lie algebra $\Lambda \mathbb{R}^3$ of formal Laurent series with coefficients in the Lie algebra $(\mathbb{R}^3, \times)$. This loop Lie algebra has the decomposition

$$\Lambda \mathbb{R}^3 = \Lambda^+ \mathbb{R}^3 \oplus \Lambda^- \mathbb{R}^3$$

into Lie subalgebras given by positive and non-negative frequencies in λ and Y is contained in $\Lambda^- \mathbb{R}^3$. The recursion relation (1) for Y_k corresponds to the lowest order non-trivial Lax flow

$$V_0(Y) = Y \times (\lambda Y)_+$$

for the generating loop Y, where $(\ )_+$ denotes projection along the decomposition of $\Lambda \mathbb{R}^3$. The main theorem in this section relates the evolution of a curve $\gamma \in \mathcal{M}$ under the flows Y_k to the evolution of the generating loop Y by the higher order Lie theoretic flows

$$V_k(Y) = Y \times (\lambda^{k+1} Y)_+$$

which, as gradients of invariant functions on $\Lambda \mathbb{R}^3$, are known to commute [16]. This observation eventually implies the commutativity of the vector fields Y_k on $\mathcal{M}$. It appears that our approach to commutativity is new. Related results in the literature [26] are usually concerned with the induced flows on the space of complex curvature functions ψ, rather than with the flows on the geometric objects, the curves $\gamma \in \mathcal{M}$, per se.

The connection of our flows Y_k to commuting vector fields V_k on the loop Lie algebra $\Lambda \mathbb{R}^3$ provides us with a rich solution theory for the dynamics generated by Y_k. A particularly well studied type of solutions are the finite gap curves $\gamma \in \mathcal{M}$. These arise from invariant finite dimensional subspaces $\Lambda_d \subset \Lambda^- \mathbb{R}^3$ of polynomials of degree d in λ^{-1}. The flows V_k are non-trivial only for $k = 0, \ldots, d-1$, and linearize on the (extended) Jacobian $\mathrm{Jac}(\Sigma)$ of a finite genus $g_\Sigma = d - 1$ algebraic curve Σ, the *spectral curve* of the flows V_k. Thus, finite gap solutions can in principle be explicitly parametrized by theta functions on Σ.

A non-trivial problem is to single out those finite gap solutions which give rise to curves $\gamma \in \mathcal{M}$ with prescribed monodromy. It seems feasible that our setting of monodromy preserving flows Y_k could contribute to this problem. A related question, which may be within reach of our approach, is how to approximate a given curve $\gamma \in \mathcal{M}$ by a finite gap solution of low spectral genus g_Σ.

In addition to their algebro-geometric significance, finite gap solutions on Λ_d have a variational characterization as stationary solutions $\gamma \in \mathcal{M}$ to the putative functional E_{d+1} constrained by the lower order Hamiltonians E_k, $1 \leq k \leq d$. For example, Euler elastica γ correspond to elliptic spectral curves and hence can be explicitly parametrized by elliptic functions.

In Section 5 we derive the complete list of Hamiltonians E_k for which Y_k are symplectic. In contrast to previous work [27], where those functionals are calculated from the non-linear Schrödinger hierarchy [14], we work on the space of curves $\mathcal{M}$ directly. The basic ingredient comes from the geometry behind the generating loop $Y = \sum_{k \geq 0} Y_k \lambda^{-k}$ of the flows Y_k, namely the associated family of curves γ_λ for $\lambda \in \mathbb{R}$, see Figure 12.4. These curves osculate the original curve γ at some chosen base point $x_0 \in \tilde{M}$ to second order, and tend to the straight line through $\gamma(x_0)$ in direction of $\gamma'(x_0)$ as $\lambda \to \infty$, see Figure 12.5. The rotation monodromy A_λ of the curves γ_λ, based at $x_0 \in \tilde{M}$, has axis $Y_\lambda(x_0)$ and we show that its rotation angle θ_λ is the Hamiltonian for the generating loop Y, that is,

$$\tfrac{1}{\lambda^2} d\theta = \sigma(Y, -) \, .$$

Applying the Gauß–Bonnet Theorem to the sector traced out by the tangent image T_λ of γ_λ over a period of the translation τ on $\tilde{M}$, we derive in Theorem 5.4 an explicit expression for the angle function $\frac{1}{\lambda^2}\theta = \sum_{k \geq 0} E_k \lambda^{-k}$: for example,

$$E_6 = \int_M \left(-\tfrac{1}{2} \det(\gamma', \gamma''', \gamma'''') + \tfrac{7}{8} |\gamma''|^2 \det(\gamma', \gamma'', \gamma''') \right) \, dx \, .$$

In Section 6 we discuss the associated family γ_λ for complex values of the spectral parameter $\lambda \in \mathbb{C}$. In this case the curves γ_λ can be seen as curves with monodromy in hyperbolic 3-space H^3. Furthermore, the two fixed points on the sphere at infinity of H^3 of the hyperbolic monodromy of γ_λ, based at $x_0 \in \tilde{M}$, define two Darboux transforms $\eta_\pm \in \mathcal{M}$ of the original curve $\gamma \in \mathcal{M}$, see Figure 12.7. These two fixed points define a hyper-elliptic curve over $\lambda \in \mathbb{C}$, which, at least for finite gap curves γ, is biholomorphic to the spectral curve Σ discussed in Section 4. We have thus realized the algebro-geometric spectral curve Σ of a finite gap curve $\gamma \in \mathcal{M}$ in terms of a family of Darboux transforms $\eta \in \mathcal{M}$ parametrized by Σ. In other words, the spectral curve Σ has a canonical realization in Euclidean space $\mathbb{R}^3$ as pictured in Figures 12.7–12.9.

There is a vast amount of literature pertaining to our discussion, for which we have included a final historical section. Here the reader can find a chronological exposition, with references to the relevant literature, of much of the background material.

Acknowledgments: this paper arose from a lecture course the third author gave during his stay at the Yau Institute at Tsinghua University in Beijing during Spring 2018. The development of the material was also supported by SFB Transregio 109 Discretization in Geometry and Dynamics at Technical University Berlin. Software support for the images was provided by SideFX.

2 Symplectic Geometry of the Space of Curves

Let $\tilde{M} \cong \mathbb{R}$ with a fixed orientation and let $\tau \in \mathrm{Diff}(\tilde{M})$ denote a translation so that $M := \tilde{M}/\tau \cong S^1$. A smooth immersion $\gamma : \tilde{M} \to N$ into an oriented Riemannian manifold N is called a *curve with monodromy* if

$$\tau^* \gamma = h \circ \gamma$$

for an orientation preserving isometry h of N, see Figure 12.1. The space of curves $\mathcal{M} = \mathcal{M}_h$ of a given monodromy h is a smooth Fréchet manifold, whose tangent spaces

$$T_\gamma \mathcal{M} = \Gamma_\tau(\gamma^* TN)$$

are given by equivariant vector fields ξ along γ, that is

$$\tau^* \xi = h \, \xi \, .$$

We make no notational distinction between the action of an isometry on N and its derived action on TN. It will be helpful to think of equivariant sections ξ of $\gamma^* TN$ as sections of the τ and h twisted bundle

$$\gamma^* TN_{|M} := \tilde{M} \times \gamma^* TN/(\tau, h) \, .$$

The latter is a real Riemannian vector bundle over M with a metric connection, and splits orthogonally into tangential and normal, to γ, subbundles

$$\gamma^* TN_{|M} = T_\gamma M \oplus \perp_\gamma M \, .$$

Moreover, the tangent space $T_\gamma \mathcal{M} = \Gamma(\gamma^* TN_{|M})$ can now be viewed as the space of sections of the bundle $\gamma^* TN_{|M}$ over M.

The group of orientation preserving diffeomorphisms $\mathrm{Diff}_\tau(\tilde{M})$ commuting with τ acts on $\mathcal{M}$ from the right by reparametrization. We now assume that N is 3-dimensional with volume form $\det \in \Omega^3(N, \mathbb{R})$. Then $\mathcal{M}$ carries a

Figure 12.1 An elastic curve in $\mathbb{R}^3$ with monodromy.

presymplectic structure [30], the *Marsden–Weinstein 2-form* $\sigma \in \Omega^2(\mathcal{M}, \mathbb{R})$, whose value at $\gamma \in \mathcal{M}$ is given by

$$\sigma_\gamma(\xi, \eta) = \int_M \det(d\gamma, \xi, \eta). \tag{2}$$

Here $\xi, \eta \in T_\gamma \mathcal{M} = \Gamma(\gamma^* T N_{|M})$ and $d\gamma \in \Omega^1(M, T_\gamma M)$, therefore the integrant is a well-defined 1-form on $M = \tilde{M}/\tau$. The Marsden–Weinstein 2-form σ is degenerate exactly along the tangent spaces to the orbits of $\mathrm{Diff}_\tau(\tilde{M})$. Thus, σ descends to a symplectic structure on the space $\mathcal{M} / \mathrm{Diff}_\tau(\tilde{M})$ of unparametrized curves, which is singular at $\gamma \in \mathcal{M}$ where $\mathrm{Diff}_\tau(\tilde{M})$ does not act freely. Since all our Hamiltonians will be geometric, and therefore invariant under $\mathrm{Diff}_\tau(\tilde{M})$, they also are defined on the space of unparametrized curves $\mathcal{M} / \mathrm{Diff}_\tau(\tilde{M})$. Nevertheless, we always will carry out calculations on the smooth manifold $\mathcal{M}$ and remind ourselves that results have to be viewed modulo the action of $\mathrm{Diff}_\tau(\tilde{M})$.

In addition to its presymplectic structure, $\mathcal{M}$ carries a Riemannian structure given by the variational L^2-metric

$$\langle \xi, \eta \rangle_\gamma = \int_M (\xi, \eta) dx$$

for $\xi, \eta \in T_\gamma \mathcal{M}$. Here $dx \in \Omega^1(M, \mathbb{R})$ denotes the volume form of the induced metric $dx^2 = (d\gamma, d\gamma)$ on $\tilde{M}$ which descends to M. We emphasize that the 1-form dx depends on the curve γ, in other words, we think of

$$dx: \mathcal{M} \to \Omega^1(M, \mathbb{R}).$$

In classical terms x is the arclength parameter for γ on $\tilde{M}$. Derivatives, including covariant derivatives, along a curve $\gamma \in \mathcal{M}$ with respect to the dual vector field $\frac{d}{dx} \in \Gamma(TM)$ to dx will often be notated by $(\ )'$. For instance, $d\gamma = \gamma' dx$ and $T := \gamma'$ is the unit tangent vector field along γ; if ξ is a section of a bundle with connection ∇ over M, then $\nabla \xi = \xi' dx$.

To avoid overbearing notation, we will not distinguish between F and $F(\gamma)$ for maps F defined on $\mathcal{M}$. For example, $T = \gamma'$ can refer to the unit tangent vector field along γ or the vector field $T \in \Gamma(T\,\mathcal{M})$, whose value at γ is given by $T(\gamma) = \gamma'$.

The vector cross product, defined by the volume form $\det \in \Omega^3(N, \mathbb{R})$, relates the Riemannian metric and the Marsden–Weinstein 2-form on $\mathcal{M}$ by

$$\sigma(\xi, \eta) = \langle T \times \xi, \eta \rangle . \tag{3}$$

Given a Hamiltonian $E \colon \mathcal{M} \to \mathbb{R}$, its *variational gradient* $G_E \in \Gamma(T\,\mathcal{M})$ is given by

$$\langle G_E, \ \rangle = dE .$$

The solutions of the Euler–Lagrange equation $G_E = 0$ are the critical points for the variational problem defined by E. We call a vector field $Y_E \in \Gamma(T\,\mathcal{M})$ on $\mathcal{M}$ a *symplectic gradient* if

$$\sigma(Y_E, \ ) = dE .$$

Note that due to the degeneracy of the Marsden–Weinstein 2-form, Y_E is defined only up to the tangent spaces to the orbits of $\mathrm{Diff}_\tau(\tilde{M})$ given by $C^\infty(\mathcal{M}, \mathbb{R})T \subset T\,\mathcal{M}$. Equation (3) relates the variational and symplectic gradients by

$$G_E = T \times Y_E . \tag{4}$$

A Hamiltonian E is invariant under $\mathrm{Diff}_\tau(\tilde{M})$ if and only if $G_E \in \Gamma(\perp_\gamma M)$ is a normal vector field along $\gamma \in \mathcal{M}$. In this situation, relation (4) can be reversed to

$$Y_E \equiv -T \times G_E \quad \mathrm{mod} \ C^\infty(\mathcal{M}, \mathbb{R})T , \tag{5}$$

and provides a symplectic gradient Y_E for the invariant Hamiltonian E.

There are two primary reasons why we work on the space $\mathcal{M} = \mathcal{M}_h$ of curves with monodromy, rather than just with closed curves whose monodromy $h = \mathrm{id}$ is trivial: first, a number of historically relevant examples, such a elastic curves, give rise to curves with monodromy; second, curves with monodromy appear naturally in the *associated family* of curves, which provides the setup to calculate the commuting hierarchy of Hamiltonians on $\mathcal{M}$ and, at the same time, gives a geometric realization of the *spectral curve*.

3 Energy Functionals on the Space of Curves

Calculating the variational and symplectic gradients for a number of classical Hamiltonians $E \in C^\infty(\mathcal{M}, \mathbb{R})$, we will observe a pattern leading to a recursion relation for an infinite hierarchy of flows. All our functionals $E \colon \mathcal{M} \to \mathbb{R}$ will be what physicists call "local", that is, they factorize via

$$\mathcal{M} \longrightarrow \Omega^1(M, \mathbb{R}) \xrightarrow{\int_M} \mathbb{R}$$

where the values of the first map, producing the integrant, involve only a finite order jet of $\gamma \in \mathcal{M}$.

Given a curve $\gamma \in \mathcal{M}$, we choose a variation $\gamma_t \in \mathcal{M}$ and denote by $(\ )^{\cdot}$ any kind of derivative with respect to t at $t = 0$. The variational gradient G_E is then read off from

$$\dot{E} = dE_\gamma(\dot{\gamma}) = \langle G_E, \dot{\gamma} \rangle .$$

3.1 Length Functional

We begin with the most elementary Hamiltonian, the length functional $E_1 = \int_M dx$, obtained by integrating the volume form dx over M.

Theorem 3.1 *The variational and symplectic gradients of the length functional E_1 are given by $G_1 = -\gamma''$ and $Y_1 \equiv \gamma' \times \gamma''$ mod $C^\infty(\mathcal{M}, \mathbb{R})T$. The latter gives rise to the* **vortex filament flow** *[11] on $\mathcal{M}$, see Figure 12.2.*

Proof. Since $d\gamma_t = T_t dx_t$, we have the formula

$$(d\gamma)^{\cdot} = \nabla\dot{\gamma} = \dot{\gamma}'dx = \dot{T}dx + Td\dot{x} \tag{6}$$

which will be reused a number of times. Taking inner product with T, and recalling $(T, \dot{T}) = 0$ due to $(T, T) = 1$, gives

$$d\dot{x} = (\nabla\dot{\gamma}, T) = d(\dot{\gamma}, T) - (\dot{\gamma}, T')dx .$$

Since $d(\dot{\gamma}, T)$ is exact on M, we can apply Stokes' Theorem and obtain

$$\dot{E}_1 = \int_M d\dot{x} = -\int_M (\dot{\gamma}, T')dx = \langle -\gamma'', \dot{\gamma} \rangle .$$

Hence $G_1 = -\gamma''$ and from (5) the symplectic gradient becomes $Y_1 \equiv \gamma' \times \gamma''$. □

Figure 12.2 A closed curve in $\mathbb{R}^3$ evolving according to the vortex filament flow Y_1.

3.2 Torsion Functional

The next functional to consider is the *total torsion* of a curve $\gamma \in \mathcal{M}$, which turns out to be the angle

$$E_2 = \alpha \quad \mod 2\pi$$

of the inverse holonomy of the oriented rank 2 Riemannian bundle $\perp_\gamma M$ over M. It compares the monodromy of the curve γ with the holonomy of its normal bundle over a period of the curve. The following formula for the total torsion angle, justifying the nomenclature, will be useful for calculating its variational gradient:

Lemma 3.2 *Let $\gamma \in \mathcal{M}$ and $v \in T_\gamma M$ be a unit length normal vector field along γ. Then the total torsion can be expressed by*

$$\alpha = \int_M (\nabla v, T \times v).$$

Proof. We may regard $\perp_\gamma M$ as a unitary complex line bundle over M with complex structure $T \times (\ )$, and Hermitian metric $(\ ,\ )$ and unitary connection ∇ induced by the corresponding Riemannian structures on N. Then

$$\nabla v = T \times v \ \omega$$

for the connection 1-form $\omega \in \Omega^1(M, \mathbb{R})$ with respect to the section v. Since $\xi = v \exp(-i \int \omega)$ is a unit length parallel section in the normal bundle of γ over $\tilde{M}$, the monodromy of $\perp_\gamma M$ is given by $\exp(-i \int_M \omega)$, which proves the claim.   □

Even though the total torsion functional E_2 is only defined modulo 2π, it has a well defined variational gradient.

Theorem 3.3 *The variational gradient of the total torsion E_2 is given by*

$$G_2 = -(\gamma' \times \gamma''' + *R(T)T)$$

where $*$ denotes the Hodge star operator on the 3-dimensional oriented Riemannian manifold N and $R \in \Omega^2(N, \mathbf{so}(TN))$ its Riemannian curvature 2-form. In particular, if N is a 3-dimensional space form, the variational and symplectic gradients of the total torsion have the simple expression

$$G_2 = -\gamma' \times \gamma''' \quad and \quad Y_2 \equiv -\gamma''' \quad \mod C^\infty(M, \mathbb{R})T.$$

The flow generated by Y_2 is called the helicity filament flow *[21] on $\mathcal{M}$.*

Proof. Let $\gamma_t \in \mathcal{M}$ be a variation of γ. Using the characterization of Lemma 3.2, we calculate

$$(\nabla v, T \times v)\dot{} = ((\nabla v)\dot{}, T \times v) + (\nabla v, \dot{T} \times v) + (\nabla v, T \times \dot{v}).$$

The last term $(\nabla v, T \times \dot{v}) = \det(T, \dot{v}, \nabla v) = 0$, since all entries are perpendicular to v and $v^{\perp}$ is 2-dimensional. In the first term, we commute $(\dot{\ })$ with ∇ and accrue a curvature term

$$(\nabla v)\dot{\ } = \nabla \dot{v} + R(\dot{\gamma}, d\gamma)v.$$

Therefore, using the symmetries of the Riemannian curvature R, we arrive at

$$(\nabla v, T \times v)\dot{\ } = (\nabla \dot{v}, T \times v) + (\nabla v, \dot{T} \times v) - (R(v, T \times v)d\gamma, \dot{\gamma}). \quad (7)$$

The curvature term can easily be rewritten as

$$R(v, T \times v)d\gamma = *R(T)T \, dx$$

using the Hodge star $*$ on N and the relation $d\gamma = \gamma' \, dx$. It remains to unravel the first two terms of the right hand side of equation (7) above:

$$(\nabla \dot{v}, T \times v) + (\nabla v, \dot{T} \times v) = d(\dot{v}, T \times v) - (\dot{v}, \nabla T \times v) + (\nabla v, \dot{T} \times v)$$

where we omitted the term $(\dot{v}, T \times \nabla v)$, which again vanishes due to all entries being perpendicular to v. Since the terms $\nabla T \times v$ and $\dot{T} \times v$ are tangential, we can rewrite

$$(\dot{v}, \nabla T \times v) = (\dot{v}, T)(T, \nabla T \times v) = (v, \dot{T})(\nabla T, T \times v)$$

and likewise for $(\nabla v, \dot{T} \times v)$, where we used properties of the cross product. Combining the above formulas yields

$$(\nabla \dot{v}, T \times v) + (\nabla v, \dot{T} \times v) = (\dot{T}, -v(\nabla T, T \times v) + T \times v(v, \nabla T))$$
$$= (\dot{T}, T \times \nabla T) + d(\dot{v}, T \times v).$$

Next we replace $\dot{T}$ by (6) and notice that $T \times \nabla T$ is normal, so that

$$(\dot{T}, T \times \nabla T) = (\dot{\gamma}', T \times \nabla T)$$
$$= d(\dot{\gamma}, T \times T') - (\dot{\gamma}, T \times T'')dx.$$

We now note that both 1-forms, $d(\dot{v}, T \times v)$ and $d(\dot{\gamma}, T \times T')$, appearing in the above expressions are exact on M. Therefore, inserting the collected formulas into (7) and applying Stokes' Theorem yields

$$\dot{E}_2 = \int_M (\nabla v, T \times v)\dot{\ } = \int_M (\dot{\gamma}, -T \times T'' - *R(T)T)dx$$
$$= \langle -T \times T'' - *R(T)T, \dot{\gamma} \rangle,$$

which gives the stated result for the variational gradient G_2 after putting $T = \gamma'$. The vanishing of $*R(T)T$ for a space of constant curvature follows from the special form of the curvature tensor R, since T is perpendicular to v and $T \times v$. The expression for the symplectic gradient Y_2 comes from (5). $\qquad \square$

3.3 Elastic Energy

The next functional to be discussed is the *bending* or *elastic energy*

$$E_3 = \tfrac{1}{2} \int_M |\gamma''|^2 dx \,,$$

the total squared length of the curvature $\gamma'' \in \Gamma(\perp_\gamma M)$ of a curve $\gamma \in \mathcal{M}$. This functional has been pivotal in the development of many areas of mathematics, including variational calculus, geometric analysis, and integrable systems. In comparison, the previously discussed total torsion appears to be less familiar, presumably because it is trivial for planar curves which, historically, were the first to be studied.

Theorem 3.4 *The variational gradient of the elastic energy E_3 is given by*

$$G_3 = \left(\gamma''' + \tfrac{3}{2}|\gamma''|^2\gamma'\right)' - R(\gamma', \gamma'')\gamma' \,,$$

with $R \in \Omega^2(N, \mathbf{so}(TN))$ the Riemannian curvature. In case N has constant curvature K, the variational gradient simplifies to

$$G_3 = \left(\gamma''' + \left(\tfrac{3}{2}|\gamma''|^2 - K\right)\gamma'\right)' \,.$$

Proof. Let $\gamma_t \in \mathcal{M}$ be a variation of γ. Replacing $\gamma' = T$ in the integrand of E_3, we obtain

$$\tfrac{1}{2}(|T'|^2 dx)^{\boldsymbol{\cdot}} = ((T')^{\boldsymbol{\cdot}}, T')dx + \tfrac{1}{2}|T'|^2 d\dot{x}$$

where $d\dot{x} = (\dot{\gamma}', T)dx$ from (6). To evaluate the first term on the right hand side, we take the t-derivative of $\nabla T = T'dx$ to calculate $(T')^{\boldsymbol{\cdot}}$. This gives

$$(\nabla T)^{\boldsymbol{\cdot}} = (T')^{\boldsymbol{\cdot}} dx + T' d\dot{x}$$

and, commuting ∇ and t-derivatives,

$$(\nabla T)^{\boldsymbol{\cdot}} = \nabla \dot{T} + R(\dot{\gamma}, d\gamma)T$$

accruing a curvature term. Together with our formula for $d\dot{x}$, this provides the expression

$$(T')^{\boldsymbol{\cdot}} = \dot{T}' + R(\dot{\gamma}, T)T - T'(\dot{\gamma}', T) \,.$$

Therefore,

$$\tfrac{1}{2}(|T'|^2 dx)^{\boldsymbol{\cdot}} = (\nabla \dot{T}, T') - \tfrac{1}{2}|T'|^2(\nabla \dot{\gamma}, T) - (R(T, \nabla T)T, \dot{\gamma}) \,,$$

where we used the symmetries of the Riemannian curvature R and $\nabla T = T'dx$. It remains to calculate

$$(\nabla \dot{T}, T') = d(\dot{T}, T') - (\dot{T}dx, T'') \equiv -(\nabla \dot{\gamma}, T'') + (\nabla \dot{\gamma}, T)(T, T'')$$

$$= -d(\dot{\gamma}, T'') + (\dot{\gamma}, \nabla T'') - (\nabla \dot{\gamma}, T)|T'|^2 \,,$$

where we used (6), the unit length of T, and $\equiv$ indicates equality modulo exact 1-forms on M. Inserting this last expression yields

$$\tfrac{1}{2}(|T'|^2 dx)^{\cdot} = (\dot{\gamma}, \nabla T'') - \tfrac{3}{2}|T'|^2(\nabla \dot{\gamma}, T) - (R(T, \nabla T)T, \dot{\gamma})$$
$$\equiv (\nabla(T'' + \tfrac{3}{2}|T'|^2 T) - R(T, \nabla T)T, \dot{\gamma}).$$

Finally, we uses use Stokes' Theorem and obtain

$$\dot{E}_3 = \int_M \left(\left(T'' + \tfrac{3}{2}|T'|^2 T \right)' - R(T, T')T, \dot{\gamma} \right) dx$$
$$= \left\langle \left(T'' + \tfrac{3}{2}|T'|^2 T \right)' - R(T, T')T, \dot{\gamma} \right\rangle,$$

which gives the stated result for the variational gradient G_3. The expression of G_3 for a space form N derives from the special form of the curvature tensor R in that case. $\qquad\square$

3.4 Flux Functional

There is another, somewhat less known, Hamiltonian which will play a role in isoperimetric characterizations of critical points of the functionals E_k. If $V \in \Gamma(TN)$ is a volume preserving vector field on N, for example a Killing field, then $i_V \det \in \Omega^2(N, \mathbb{R})$ is a closed 2-form which has a primitive $\alpha_V \in \Omega^1(N, \mathbb{R})$, provided that N has no second cohomology. We additionally assume that we can choose the primitive α_V invariant under the monodromy h, that is $h^*\alpha_V = \alpha_V$. In particular, this will imply that the vector field V is h-related to itself. Then the primitive α_V along γ, that is, the 1-from $\gamma^*\alpha_V \in \Omega^1(M, \mathbb{R})$, descends to M and we obtain the *flux functional*

$$E_V = \int_M \gamma^*\alpha_V .$$

It is worth noting that on closed curves the flux functional is indeed the flux of the vector field V through any disk spanned by γ. The primitive α_V of $i_V \det$ is defined only up to a closed 1-form and the value of E_V depends on this choice. When calculating the variational gradient G_V of E_V only homotopic curves (of the same monodromy) play a role, along which the value of E_V is independent of the choice of primitive by Stokes' Theorem.

Theorem 3.5 *The variational and symplectic gradients of the flux functional E_V are given by*

$$G_V = \gamma' \times (V \circ \gamma) \quad and \quad Y_V \equiv V \circ \gamma \mod C^\infty(M, \mathbb{R})T .$$

The Hamiltonian flow of E_V is given by moving an initial curve γ along the flow of the vector field V on N.

Proof. If $\gamma_t \in \mathcal{M}$ is a variation of γ, then

$$(\gamma_t^* \alpha_V)^{\cdot} = d i_{\dot{\gamma}} \alpha_V + i_{\frac{d}{dt}} (\gamma^* d \alpha_V) = d i_{\dot{\gamma}} \alpha_V + \det(V \circ \gamma, \dot{\gamma}, d\gamma),$$

where we used the Cartan formula for the Lie derivative and $i_V \det = d\alpha_V$. Since

$$\det(V \circ \gamma, \dot{\gamma}, d\gamma) = (d\gamma \times V \circ \gamma, \dot{\gamma})$$

Stokes' Theorem gives us

$$(G_V)^{\cdot} = \int_M (\gamma' \times V \circ \gamma, \dot{\gamma}) dx = \langle \gamma' \times V \circ \gamma, \dot{\gamma} \rangle,$$

and we read off the formula for the variational gradient G_V. Applying (5), we obtain the symplectic gradient $Y_V = V \circ \gamma$. □

As an instructive example, we look at the special case $N = \mathbb{R}^3$, where there are two obvious choices for V. For unit length $v \in \mathbb{R}^3$, we have the infinitesimal translation $V(p) = v$ and infinitesimal rotation $V(p) = v \times p$ giving rise to two more Hamiltonians E_V. The monodromy h for the space of curves $\mathcal{M} = \mathcal{M}_h$ in $\mathbb{R}^3$ is an orientation preserving Euclidean motion $h = Ap + a$. The condition that V is h-related to itself means that v is an eigenvector of the rotational monodromy A. To simplify notation, we choose the origin of $\mathbb{R}^3$ to lie on the axis $\mathbb{R}v$. We can easily find the h-invariant primitives $\alpha_V \in \Omega^1(\mathbb{R}^3, \mathbb{R})$ of $i_V \det$: for an infinitesimal translation $\alpha_V = \det(v, p, dp)$ and for an infinitesimal rotation $\alpha_V = \frac{1}{2}|p - (p, v)v|^2(v, dp)$.

Theorem 3.6 *Let $N = \mathbb{R}^3$ and let $V(p) = v$ be an infinitesimal translation with unit length $v \in \mathbb{R}^3$ an eigenvector of the rotational part of the monodromy $h = Ap + a$. Then the flux functional becomes*

$$E_{-1} = \frac{1}{2} \int_M (\gamma \times d\gamma, v).$$

For a closed curve the vector $\frac{1}{2} \int_M \gamma \times d\gamma$, in reference to Kepler, is called the area vector of γ and thus we call E_{-1} the area functional. Indeed, E_{-1} measures the signed enclosed area of the orthogonal projection of a closed curve γ onto the plane $v^\perp$.

The variational and symplectic gradients of E_{-1} are given by

$$G_{-1} = \gamma' \times v \quad \text{and} \quad Y_{-1} \equiv v \mod C^\infty(\mathcal{M}, \mathbb{R})T.$$

The proof follows immediately from Theorem 3.5.

Theorem 3.7 *Let $N = \mathbb{R}^3$ and let $V(p) = v \times p$ be an infinitesimal rotation with unit length eigenvector $v \in \mathbb{R}^3$ of the rotational part of the monodromy $h = Ap + a$. Then the flux functional becomes*

$$E_{-2} = \tfrac{1}{2} \int_M (|\gamma - (\gamma, v)v|^2 d\gamma, v) .$$

For a closed curve the vector $\tfrac{1}{2} \int_M (|\gamma - (\gamma, v)v|^2 d\gamma$ is called the volume vector *of γ and thus we call E_{-2} the* volume functional. *Indeed, E_{-2} measures the volume of the solid torus of revolution generated by rotating the closed curve γ around the axis $\mathbb{R}v$.*

The variational and symplectic gradients of E_{-2} are given by

$$G_{-2} = \gamma' \times (v \times \gamma) \quad and \quad Y_{-2} \equiv v \times \gamma \quad \mod C^{\infty}(\mathcal{M}, \mathbb{R})T .$$

Again, the proof follows from Theorem 3.5 and an elementary calculation involving surfaces of revolution.

It will be convenient to add the *vacuum*, the trivial Hamiltonian $E_0 = 0$ with trivial variational gradient $G_0 = 0$ and symplectic gradient $Y_0 = \gamma'$. The Hamiltonian flow generated by Y_0 on $\mathcal{M}$ is reparametrization and thus descends to the trivial flow on the quotient $\mathcal{M}/\operatorname{Diff}_\tau(\mathcal{M})$.

Figure 12.3 A closed curve γ moving according to vortex filament flow and the full tori obtained from γ by rotation a disk bounded by γ around a fixed axis. All these full tori have the same volume.

3.5 Hierarchy of Flows

One of the reasons to list those functionals, besides their historical significance, is the curious pattern

$$G_{k+1} = -Y_k' \tag{8}$$

their gradients seem to exhibit for curves in Euclidean 3-space $\mathbb{R}^3$, at least for $-2 \leq k \leq 1$. The pattern breaks at $k = 2$ since G_3, as a variational gradient of the geometrically given energy functional E_3, is normal to γ but $Y_2' = -\gamma^{(iv)}$ is not. This suggests that our initial choice of Y_2 should be augmented by a tangential component in order to have Y_2' normal. For any vector field $Y \in \Gamma(T \mathcal{M})$, we have $d\dot{x} = (\nabla \dot{\gamma}, T)$ along its flow $\dot{\gamma} = Y(\gamma)$. Therefore, the condition that Y' is normal to γ is equivalent to the fact that all curves γ_t of the flow have the same induced volume form $dx_t = dx$, that is, the same parametrization. In the case of Y_2 this leads to the tangential correction term $-\frac{3}{2}|\gamma''|^2\gamma'$. Therefore, by combining (4) with the recursion equation (8), we arrive at the recursion

$$Y_k' + T \times Y_{k+1} = 0, \quad Y_0 = T \tag{9}$$

defining an infinite hierarchy of vector fields $Y_k \in \Gamma(T \mathcal{M})$.

Theorem 3.8 ([26, 41]) *The recursion* (9) *has the explicit solution*

$$Y_{k+1} = T \times Y_k' - \frac{1}{2}\sum_{i=1}^{k}(Y_i, Y_{k-i+1})T, \quad k \geq 0.$$

In particular, Y_k is a homogeneous differential polynomial of degree k in T, and thus a differential polynomial of degree $k + 1$ in γ. Moreover, if γ is contained in a plane $E \subset \mathbb{R}^3$, then $Y_{2k}(\gamma): \tilde{M} \to E$ is planar, whereas $Y_{2k+1}(\gamma): \tilde{M} \to E^\perp$ is normal. In other words, the sub-hierarchy of even flows preserves planar curves.

Proof. Our prescription for the flows Y_k demands a tangential correction to (5) of the form

$$Y_k = -T \times G_k + f_k T = T \times Y_{k-1}' + f_k T$$

in order for (8) to hold. Since $f_k = (Y_k, T)$ the first few f_k can easily be read off: $f_0 = 1$, $f_1 = 0$, and $f_2 = -\frac{1}{2}|T'|^2$. To calculate a general expression, we first note that

$$(G_k, Y_l) = -(G_k, T \times G_l) = (T \times G_k, G_l) = -(Y_k, G_l)$$

where we used (5). On the other hand, using (8), we obtain

$$(G_k, Y_l) = (-Y'_{k-1}, Y_l) = -(Y_{k-1}, Y_l)' + (Y_{k-1}, Y'_l)$$
$$= -(Y_{k-1}, Y_l)' - (Y_{k-1}, G_{l+1})$$
$$= -(Y_{k-1}, Y_l)' + (G_{k-1}, Y_{l+1}).$$

Recalling $G_0 = 0$, this relation telescopes to

$$(G_k, Y_l) = -\sum_{i=1}^{k} (Y_{k-i}, Y_{l+i-1})', \tag{10}$$

which, together with (8), (4) and $T = Y_0$, implies

$$f'_k = (Y_k, T)' = (Y'_k, T) + (Y_k, T') = (Y_k, Y'_0) = -(Y_k, G_1) = (G_k, Y_1)$$
$$= -\sum_{i=1}^{k-1} (Y_{k-i}, Y_i)' - (Y_0, Y_k)' = -\sum_{i=1}^{k-1} (Y_{k-i}, Y_i)' - f'_k.$$

This shows that (up to a constant of integration) $f_k = -\frac{1}{2}\sum_{i=1}^{k-1}(Y_{k-i}, Y_i)$. The remaining statements of the theorem follow from simple induction arguments.

□

At this stage it is worth to discuss some of the immediate consequences of the last theorem.

Corollary 3.9 *If Y_k are symplectic gradients for Hamiltonians $E_k \in C^\infty(\mathcal{M}, \mathbb{R})$, then E_k is constant along every flow Y_l for $k, l \geq -2$.*

Proof. The relation (10) implies that

$$dE_k(Y_l) = \int_M (G_k, Y_l)dx = 0$$

at least for $k, l \geq 0$. One can easily derive an analogous relation to (10) for $-2 \leq k, l \leq 0$, which proves the claim. □

As an example, we apply this result to the vortex filament flow $\dot{\gamma} = \gamma' \times \gamma''$ of a closed initial curve γ_0. Corollary 3.9 then implies that the enclosed volumes $E_{-2}(\gamma_t) = E_{-2}(\gamma_0)$ of the solid tori of revolution around the axis $\mathbb{R}\upsilon$ generated by γ_t are equal for all times t in the existence interval of the flow, see Figure 12.3. Similarly, the projected areas $E_{-1}(\gamma_t)$ of the curves γ_t are all equal.

Corollary 3.9 also provides evidence as to why the flows Y_k should commute: assuming Y_k are symplectic flows to geometric, that is, parametrization invariant, Hamiltonians E_k, they descend to the quotient $\mathcal{M}/\text{Diff}_\tau(\tilde{M})$. The latter is symplectic, at least away from its singularities. By standard symplectic techniques Corollary 3.9 then implies that the flows on the (non-singular part of the) quotient commute. This would at least show that the flows Y_k on $\mathcal{M}$ commute modulo T. More is true though, as we shall discuss in Section 4, where we relate the flows Y_k to Lax flows on a loop algebra to show that they genuinely commute.

3.6 Constrained Critical Curves in $\mathbb{R}^3$

We continue to denote by $E_k \colon \mathcal{M} \to \mathbb{R}$ the putative Hamiltonians for the flows $Y_k \in \Gamma(T\mathcal{M})$.

Definition 3.10 *A curve $\gamma \in \mathcal{M}$ is critical for the functional E_{k+1} under the constraints $E_1 \ldots, E_k$, if the constrained Euler–Lagrange equation*

$$G_{k+1} = \sum_{i=1}^{k} c_i G_i$$

holds for some constants, the Lagrange multipliers, $c_i \in \mathbb{R}$.

The recursion relation (8) implies that the condition for an E_{k+1} critical constrained curve is equivalent to

$$Y_k = \sum_{i=0}^{k-1} c_i Y_i + v \tag{11}$$

for some $c_i \in \mathbb{R}$, where $v \in \mathbb{R}^3$ necessarily is an eigenvector of the rotational part of the monodromy of γ. In terms of flows this says that Y_k evolves an initial curve γ, modulo lower order flows, by a pure translation in the direction of the monodromy axis. Since $G_k = T \times Y_k$ by (4), the above is equivalent to

Table 12.1. *The variational and symplectic gradients of the listed functionals for curves in $\mathbb{R}^3$.*

Functional	Variational gradient	Symplectic gradient
$E_{-2} = \frac{1}{2}\int_M (\lvert \gamma$ $- (\gamma, v)v\rvert^2 d\gamma, v)$	$G_{-2} = \gamma' \times (v \times \gamma)$	$Y_{-2} = v \times \gamma$
$E_{-1} = \frac{1}{2}\int_M (\gamma \times d\gamma, v)$	$G_{-1} = \gamma' \times v$	$Y_{-1} = v$
$E_0 = 0$	$G_0 = 0$	$Y_0 = \gamma'$
$E_1 = \int_M dx$	$G_1 = -\gamma''$	$Y_1 = \gamma' \times \gamma''$
$E_2 = \alpha$	$G_2 = -\gamma' \times \gamma'''$	$Y_2 = -\gamma''' - \frac{3}{2}\lvert\gamma''\rvert^2\gamma'$
$E_3 = \frac{1}{2}\int_M \lvert\gamma''\rvert^2 dx$	$G_3 = \left(\gamma''' + \frac{3}{2}\lvert\gamma''\rvert^2\gamma'\right)'$	$Y_3 = -\gamma' \times \gamma''''$ $- \frac{3}{2}\lvert\gamma''\rvert^2\gamma' \times \gamma''$ $+ \det(\gamma', \gamma'', \gamma''')\gamma'$

$$G_k = \sum_{i=0}^{k-1} c_i G_i + T \times v = \sum_{i=-1}^{k-1} c_i G_i .$$

This last says that γ is E_k critical under the constraints $E_{-1}, \ldots, E_{k-1}$, where we remember $E_0 = 0$. Applying this construction once more gives the equivalent condition

$$G_{k-1} = \sum_{i=-2}^{k-2} c_i G_i ,$$

that is to say, γ is E_{k-1} critical under the constraints $E_2, \ldots, E_{k-2}$.

Theorem 3.11 *The following statements for a curve $\gamma \in \mathcal{M}$ are equivalent:*

(1) γ is E_{k+1} critical under constraints $E_1, \ldots, E_k$.

(2) γ is E_k critical under constraints $E_{-1}, \ldots, E_{k-1}$.

(3) γ is E_{k-1} critical under constraints $E_{-2}, \ldots, E_{k-2}$.

(4) The flow $\dot{\gamma} = Y_k$ evolves, modulo a linear combination of lower order flows, by a pure translation in the direction of the axis of the mondoromy of the initial curve γ.

(5) The flow $\dot{\gamma} = Y_{k-1}$ evolves, modulo a linear combination of lower order flows, by a Euclidean motion with axis along the monodromy of the initial curve γ.

From this observation, we obtain a striking isoperimetric characterization of *Euler elastica*, that is, curves critical for the elastic energy E_3 constrained by the length E_1 and total torsion E_2.

Corollary 3.12 *The following conditions on a curve $\gamma \in \mathcal{M}$ are equivalent:*

(1) γ is an Euler elastic curve.

(2) γ is critical for total torsion with length and area E_{-1} constraints.

(3) γ is critical for length with area and volume E_{-2} constraints.

(4) The helicity filament flow $\dot{\gamma} = Y_2$ evolves, modulo the vortex filament flow and reparametrization, by a pure translation in the direction of the axis of the mondoromy of the initial curve γ.

(5) The vortex filament flow $\dot{\gamma} = Y_1$ evolves, modulo reparametrization, by a Euclidean motion with axis along the monodromy of the initial curve γ.

4 Lax Flows on Loop Algebras

We have constructed a hierarchy of vector fields Y_k on $\mathcal{M}$ which are symplectic for $-2 \leq k \leq 3$ with respect to explicitly given Hamiltonians $E_{-2}, \ldots, E_3$, see Table 12.1. In order for our hierarchy Y_k to qualify as an "infinite dimensional integrable system", the least we need to verify is

(1) The vector fields Y_k commute, that is $[Y_i, Y_j]_{\mathcal{M}} = 0$, for which Corollary 3.9 provides evidence.

(2) There is a hierarchy of Hamiltonians $E_k \in C^\infty(\mathcal{M}, \mathbb{R})$ for which Y_k are the symplectic gradients.

To this end it will be helpful to rewrite the recursion (9) in terms of the generating function

$$Y = \sum_{k \geq 0} Y_k \lambda^{-k}$$

of the hierarchy $Y_k \in \Gamma(T\mathcal{M})$ as

$$Y' + \lambda T \times Y = 0. \tag{12}$$

This last is a Lax pair equation on the loop Lie algebra

$$\Lambda \mathbb{R}^3 = \mathbb{R}^3((\lambda^{-1}))$$

of formal Laurent series with poles at $\lambda = \infty$. The Lie algebra structure on $\Lambda \mathbb{R}^3$ is the pointwise structure induced from the Lie algebra $(\mathbb{R}^3, \times)$. From the construction of Y_k in Theorem 3.8, we see that

$$(Y, Y) = \sum_{k,l \geq 0} (Y_k, Y_l) \lambda^{-k-l} = 1. \tag{13}$$

One way to address the commutativity of the flows Y_k is to relate them to commuting vector fields on $\Lambda \mathbb{R}^3$ obtained by standard methods, such as Adler–Kostant–Symes or, more generally, R-matrix theory.

Without digressing too far, we summarize the gist of these constructions [16]: given a Lie algebra $\mathfrak{g}$, a vector field $X : \mathfrak{g} \to \mathfrak{g}$ is ad-invariant if

$$dX_\xi[\xi, \eta] = [X(\xi), \eta], \quad \xi, \eta \in \mathfrak{g}. \tag{14}$$

Standard examples of ad-invariant vector fields are gradients (with respect to an invariant inner product) of Ad-invariant functions if $\mathfrak{g}$ is the Lie algebra of a Lie group G. If $X, \tilde{X}$ are ad-invariant vector fields, then (14) implies

$$[\xi, X(\xi)] = 0 \quad \text{and} \quad [X(\xi), \tilde{X}(\xi)] = 0$$

for $\xi \in \mathfrak{g}$. These are just a restatement of the fact that gradients of Ad-invariant functions commute with respect to the Lie–Poisson structure on $\mathfrak{g}$. One further ingredient is the so-called R-matrix, that is, an endomorphism $R \in \text{End}(\mathfrak{g})$ satisfying the Yang–Baxter equation

$$R([R\xi, \eta] + [\xi, R\eta]) - [R\xi, R\eta] = c[\xi, \eta]$$

for some constant c. This relation is a sufficient condition for $[\xi, \eta]_R = [R\xi, \eta] + [\xi, R\eta]$ to satisfy the Jacobi identity. Thus, $[\ ,\ ]_R$ defines a second Lie algebra structure on $\mathfrak{g}$. A typical example of an R-matrix (relating to the

Adler–Kostant–Symes method) arises from a decomposition $\mathfrak{g} = \mathfrak{g}_+ \oplus \mathfrak{g}_-$ into Lie subalgebras: $R = \frac{1}{2}(\pi_+ - \pi_-)$ with $\pi_\pm$ the projections along the decomposition. Whereas the Hamiltonian dynamics of Ad-invariant functions with respect to the original Lie-Poisson structure on $\mathfrak{g}$ is trivial, the introduction of an R-matrix renders this dynamics non-trivial for the new Lie–Poisson structure while keeping the original commutativity. Given an ad-invariant vector field X on $\mathfrak{g}$ and a constant μ, then

$$V(\xi) = [\xi, (R + \mu \operatorname{id})X(\xi)]$$

is the corresponding "Hamiltonian" vector field for the Lie–Poisson structure corresponding to $[\,,\,]_R$. We collect the basic integrability properties of the vector fields V, which are verified by elementary calculations using the properties of ad-invariant vector fields.

Lemma 4.1 *Let X, $\tilde{X}$ be ad-invariant vector fields on the Lie algebra $\mathfrak{g}$.*

(1) The corresponding Hamiltonian vector fields V, $\tilde{V}$ on $\mathfrak{g}$ commute, that is, their vector field Lie bracket $[V, \tilde{V}]_{C^\infty} = 0$.

(2) Let $\xi : D \to \mathfrak{g}$ be a solution to

$$d\xi = V(\xi)dt + \tilde{V}(\xi)d\tilde{t} \tag{15}$$

over a 2-dimensional domain D. Then the $\mathfrak{g}$-valued 1-form

$$\beta = (R + \mu \operatorname{id})X(\xi)dt + (R + \tilde{\mu} \operatorname{id})\tilde{X}(\xi)d\tilde{t}$$

on D satisfies the Maurer–Cartan equation $d\beta + \frac{1}{2}[\beta \wedge \beta] = 0$.

If G is a Lie group for $\mathfrak{g}$, the ODE

$$dF = F\beta$$

has a solution $F : D \to G$ by (ii) and $\xi = \operatorname{Ad} F^{-1}\eta$ solves (15) for any initial condition $\eta \in \mathfrak{g}$. Thus, Ad-invariant functions are constant along solutions to (15), which is referred to as "isospectrality" of the flow. Specifically, if an Ad-invariant inner product on $\mathfrak{g}$ is positive definite and the flow (15) is contained in a finite dimensional subspace of $\mathfrak{g}$, then the flow is complete.

With this at hand, the recursion equation (12) for the generating loop $Y = \sum_{k \geq 0} Y_k \lambda^{-k}$ can be seen as the first in a hierarchy of commuting flows on the loop algebra $\Lambda \mathbb{R}^3$. We have the splitting

$$\Lambda \mathbb{R}^3 = \Lambda^+ \mathbb{R}^3 \oplus \Lambda^- \mathbb{R}^3$$

into the Lie subalgebras $\Lambda^+\mathbb{R}^3 = \mathbb{R}^3[\lambda]_0$ of polynomials without constant term and power series $\Lambda^-\mathbb{R}^3 = \mathbb{R}^3[[\lambda^{-1}]]$ in λ^{-1} with R-matrix $R = \frac{1}{2}(\pi_+ - \pi_-)$. For $k \in \mathbb{Z}$ the vector fields

$$X_k(\xi) = \lambda^{k+1}\xi$$

on $\Lambda\mathbb{R}^3$ are ad-invariant with corresponding commuting vector fields

$$V_k(\xi) = \xi \times \left(R + \tfrac{1}{2}\,\mathrm{id}\right) X_k(\xi) = \xi \times (\lambda^{k+1}\xi)_+$$

as shown in Lemma 4.1. For $\xi \in \Lambda^-\mathbb{R}^3$ we have

$$\xi \times V_k(\xi) = \xi \times (\lambda^{k+1}\xi)_+ = -\xi \times (\lambda^{k+1}\xi)_- ,$$

so that the Lie subalgebra $\Lambda^-\mathbb{R}^3$ is preserved under the flows V_k. In particular, the recursion equation (12) can be rewritten as

$$Y' + (\lambda Y)_+ \times Y = 0, \tag{16}$$

which is the lowest flow V_0 of the hierarchy V_k restricted to $\Lambda^-\mathbb{R}^3$. The Lie theoretic flows V_k are related to the geometric flows Y_k as follows:

Theorem 4.2 *If $\gamma \in \mathcal{M}$ evolves by Y_k, that is, $\dot{\gamma} = Y_k(\gamma)$, then the generating loop $Y = \sum_{k \geq 0} Y_k \lambda^{-k}$ evolves by V_k, that is, $\dot{Y} = V_k(Y)$.*

Proof. We derive an inhomogeneous ODE for $\dot{Y}$ which then can be solved by variation of the constants. The vector fields V_0 and V_k commute, thus

$$\dot{Y}' = (Y')^{\cdot} = \dot{Y} \times \lambda Y_0 + Y \times \lambda \dot{Y}_0$$

due to (16) and $Y_0 = (\lambda Y)_+$. From the explicit form of Y_k given in Theorem 3.8 it follows that the the reparametrization flow Y_0 commutes with all Y_k. Therefore,

$$\dot{Y}_0 = \dot{T} = (\gamma')^{\cdot} = \dot{\gamma}' = Y_k' = Y_{k+1} \times Y_0$$

and we obtain the inhomogeneous ODE

$$\dot{Y}' + \lambda Y_0 \times \dot{Y} = (\lambda Y_0 \times Y_{k+1}) \times Y .$$

Since Y is a solution to the homogeneous equation, we make the variation of the constants Ansatz

$$\dot{Y} = [Y, Z].$$

Inserting this into the inhomogeneous ODE and using Jacobi identity yields

$$(Z' + \lambda Y_0 \times Z - \lambda Y_0 \times Y_{k+1}) \times Y = 0,$$

which is easily seen to be solved by $Z = (\lambda^{k+1}Y)_+$. Thus, the general solution of the inhomogeneous ODE is $\dot{Y} = [Y, (\lambda^{k+1}Y)_+] + cY$ for some t-independent c. But $(Y, Y) = 1$ by (13), hence $c = 0$ and $\dot{Y} = V_k$ as stated. $\qquad\square$

We are now in a position to show the commutativity of the geometric flows Y_k.

Theorem 4.3 *The flows Y_k commute on $\mathcal{M}$.*

Proof. Considering the flows $\dot{\gamma} = Y_k$ and $\overset{\circ}{\gamma} = Y_l$, we need to verify that

$$\overset{\circ}{Y}_k = \dot{Y}_l .$$

The previous theorem implies that the generating loop $Y = \sum_{k \geq 0} Y_k \lambda^{-k}$ satisfies

$$\dot{Y} = V_k \quad \text{and} \quad \overset{\circ}{Y} = V_l .$$

Thus it suffices to show that the λ^{-l} coefficient of $V_k = Y \times (\lambda^{k+1}Y)_+$ is equal to the λ^{-k} coefficient of $V_l = Y \times (\lambda^{l+1}Y)_+$. Unravelling this condition we need to check the equality

$$\sum_{\substack{i+j=k+l+1 \\ 0 \leq j \leq k}} Y_i \times Y_j = \sum_{\substack{i+j=k+l+1 \\ 0 \leq j \leq l}} Y_i \times Y_j$$

which indeed holds due to the skew symmetry of the vector cross product. $\qquad\square$

At this stage it is worth noting that we have provided a dictionary translating the geometrically defined flows Y_k on $\mathcal{M}$ into the purely Lie theoretic defined flows V_k on $\Lambda \mathbb{R}^3$. There is a rich solution theory of such equations and a particularly well studied class of solutions are the "finite gap solutions", which arise from finite dimensional algebro-geometric integrable systems.

In much of what follows, we will need to complexify the loop Lie algebra $\Lambda \mathbb{R}^3$ to $\Lambda \mathbb{C}^3$ allowing for complex values of the loop parameter λ. Depending on the context, the Lie algebra $\mathbb{R}^3 = \mathbf{su}_2$ will be identified with the Lie algebra of the special unitary group $\mathbf{SU}_2$. Likewise, $\mathbb{C}^3 = \mathbf{sl}_2(\mathbb{C})$ will be identified with the Lie algebra of the special linear group $\mathbf{SL}_2(\mathbb{C})$. The flows V_k can be extended to $\Lambda \mathbb{C}^3$ and satisfy the reality condition $V_k(\bar{\xi}) = \overline{V_k(\xi)}$. In particular, the vector fields V_k preserve $\Lambda \mathbb{R}^3 \subset \Lambda \mathbb{C}^3$.

Lemma 4.4 *The subspaces*

$$\Lambda_d = \left\{ \xi \in \Lambda^- \mathbb{C}^3 ; \xi = \sum_{k=0}^{d} \xi_k \lambda^{-k} \right\} \subset \Lambda \mathbb{C}^3$$

of Laurent polynomials of degree $d \in \mathbf{N}$ are invariant under the commuting flows V_k acting non-trivially only for $0 \leq k \leq d - 1$ on Λ_d. In particular, for real initial data $\eta \in \Lambda_d^{\mathbb{R}}$ the flow

$$d\xi = \sum_{k=0}^{d-1} V_k(\xi) dt_k \tag{17}$$

has a real global solution $\xi \colon \mathbb{R}^d \to \Lambda_d^{\mathbb{R}}$ with $(\xi, \xi) = (\eta, \eta)$.

Proof. Since

$$V_k = \xi \times (\lambda^{k+1}\xi)_+ = -\xi \times (\lambda^{k+1}\xi)_-$$

we see that V_k are tangent to Λ_d. The invariant inner product on $\Lambda \mathbb{R}^3$ restricts to a positive definite inner product on the finite dimensional space $\Lambda_d^{\mathbb{R}}$ which V_k preserves by Lemma 4.1. In particular, the flows evolve on a compact finite dimensional sphere and thus are complete. $\qquad\square$

A fundamental invariant of the dynamics of V_k is the hyperelliptic *spectral curve*

$$\Sigma = \{(\lambda, \mu) \in \mathbb{C}^\times \times \mathbb{C}; \ \mu^2 + (\xi, \xi) = 0\}$$

which only depends on the initial condition $\eta \in \Lambda_d^{\mathbb{R}}$ of a solution $\xi : \mathbb{R}^d \to \Lambda_d^{\mathbb{R}}$ of the flow (17). The spectral curve Σ completes to a genus $d - 1$ curve equipped with a real structure covering complex conjugation in λ. Since $(\xi, \xi) = \mathrm{tr}(\xi^2)$, the spectral curve encodes the constant eigenvalues of the "isospectral" solution ξ. Alternatively, one could consider for each $t \in \mathbb{R}^d$ the eigenline curve

$$\Sigma_t = \{(\lambda, \mathbb{C}v); \ \xi_\lambda(t)v = \mu_\lambda v\} \subset \mathbb{C} \times \mathbb{CP}^1, \tag{18}$$

which can be shown to be biholomorphic to Σ. Pulling pack the canonical bundle over $\mathbb{CP}^1$ under $\Sigma_t \subset \mathbb{C} \times \mathbb{CP}^1$ gives a family of holomorphic line bundles $L_t \to \Sigma$. The dynamics of the ODE (17) is encoded in the eigenline bundle flow $L : \mathbb{R}^d \to \mathrm{Jac}(\Sigma)$ of ξ in the Jacobian $\mathrm{Jac}(\Sigma)$ of Σ. The flow of line bundles L_t is known to be linear and thus the vector fields V_k linearize on $\mathrm{Jac}(\Sigma)$. This exhibits the finite gap theory as a classical integrable system with explicit formulas for solutions in terms of theta functions on the spectral curve Σ.

Returning to our geometric picture, finite gap solutions have a purely variational description when viewed on the space of curves $\mathcal{M}$ with fixed monodromy. Let $E_k \in \mathbb{C}^\infty(\mathcal{M}, \mathbb{R})$ be the putative Hamiltonians for the vector fields Y_k.

Theorem 4.5 *The following statements on a curve $\gamma \in \mathcal{M}$ are equivalent:*

(1) γ is an E_{d+1} critical curve constrained by $E_1, \dots, E_d$.
(2) The lowest order flow

$$\xi' = V_0(\xi)$$

admits a solution $\xi = \sum_{k=0}^d \xi_k \lambda^{-k}$ on $\Lambda_d^{\mathbb{R}}$ with $(\xi, \xi) = 1$.

The relation between the curve γ and the solution ξ is given by $\gamma' = \xi_0$.

Proof. From (11) we know that γ is E_{d+1} constrained critical if and only if

$$Y_{d+1} = \sum_{k=0}^{d} c_k Y_k$$

for constants $c_i \in \mathbb{R}$. Setting

$$\xi_k = Y_k - c_d Y_{k-1} - \cdots - c_{d-k+1} Y_0 \qquad (19)$$

for $k = 0, \ldots, d$ provides a Laurent polynomial solution to $\xi' = V_0(\xi)$ due to the recursion relation $Y_k' + Y_0 \times Y_{k+1} = 0$ for Y_k in (8). From (13) we know that $(Y, Y) = 1$ and we also have $\xi_0 = Y_0 = T = \gamma'$. To show the converse, we start from a Laurent polynomial solution $\xi = \sum_{k=0}^{d} \xi_k \lambda^{-k}$ and verify the relations (19) inductively starting from $Y_0 = \xi_0$. Since $\xi_{d+1} = 0$, this gives the desired constrained criticality relation on $Y_{d+1} = \sum_{k=0}^{d} c_k Y_k$. $\qquad \square$

Theorem 4.5 provides a fairly explicit recipe [9, 10], [31] for the construction of all E_{d+1} critical constrained curves $\gamma \in \mathcal{M}$ with a given monodromy, together with their isospectral deformations by the higher order flows Y_k for $k = 1, \ldots, d - 1$. Since the Lax flows $d\xi = \sum_{k=0}^{d-1} V_k(\xi) dt_k$ are linear flows on the Jacobian of the spectral curve Σ, the solution ξ can in principle be expressed in terms of theta functions on Σ. The curve γ is then obtained from $\xi_0 = \gamma'$ with $t_0 = x$ and the "higher times" t_k accounting for the isospectral deformations. Periodicity conditions, such as the required monodromy of the resulting curve γ, can be discussed via the Abel–Jacobi map. A classical example is given by Euler elastica, that is, curves $\gamma \in \mathcal{M}$ with a given monodromy critical for the bending energy E_3 under length and torsion constraints: from our discussion we know that they can be computed explicitly from elliptic spectral curves.

5 Associated Family of Curves

The remaining question to address is why the commuting hierarchy Y_k of vector fields on $\mathcal{M}$ are the symplectic gradients of Hamiltonians $E_k: \mathcal{M} \to \mathbb{R}$. This requires a closer look at the geometry behind the generating loop $Y = \sum_{k \geq 0} Y_k \lambda^{-k}$ where $Y_0 = T = \gamma'$. Lemma 4.1 implies that the generating loop Y, which satisfies the Lax equation

$$Y' + \lambda T \times Y = 0,$$

along a curve $\gamma \in \mathcal{M}$ is given by

$$Y = F^{-1} Y(x_0) F, \qquad (20)$$

where $F: \tilde{M} \times \mathbb{C} \to \mathbf{SL}_2(\mathbb{C})$ solves the linear equation

$$F' = F\lambda T, \quad F(x_0) = \mathbb{1}. \tag{21}$$

Here $x_0 \in \tilde{M}$ denotes an arbitrarily chosen base point. We remind the reader that $(\ )'$ indicates derivative with respect to the, via the curve γ, induced volume 1-form dx on $\tilde{M}$. Even though Y is only a formal loop, F is smooth on $\tilde{M}$, holomorphic in $\lambda \in \mathbb{C}$, and takes values in $\mathbf{SU}_2$ for real values $\lambda \in \mathbb{R}$. In other words, we can think of $F: \tilde{M} \to \Lambda^+\mathbf{SL}_2(\mathbb{C})$ as a map into the loop group of holomorphic loops into $\mathbf{SL}_2(\mathbb{C})$ equipped with the reality condition $F_{\bar{\lambda}} = (F_\lambda{}^*)^{-1}$. Note that (21) implies that $F_0 = \mathbb{1}$ identically on $\tilde{M}$, which is to say that F maps into the based (at $\lambda = 0$) loop group.

Definition 5.1 *Let $\gamma \in \mathcal{M}$ be a curve with monodromy. The associated family of curves $\gamma_\lambda: \tilde{M} \to \mathbb{R}^3$ is given by*

$$d\gamma_\lambda = F_\lambda \, d\gamma \, F_\lambda^{-1}, \quad \gamma_\lambda(x_0) = \gamma(x_0)$$

for real $\lambda \in \mathbb{R}$.

We note that the associated curves γ_λ induce the same volume 1-form $dx_\lambda = dx$ on $\tilde{M}$ as the original curve γ, so that $(\ )'$ is defined without ambiguity. Since $F_{\lambda=0} = \mathbb{1}$, the associated family γ_λ is a deformation of the initial curve $\gamma = \gamma_0$ osculating to first order at $\gamma(x_0)$. In fact, γ_λ osculates to second order as can be seen by calculating the complex curvatures $\psi_\lambda: \tilde{M} \to \mathbb{C}$ of the curves γ_λ, which agree at $x_0 \in \tilde{M}$ as shown in Figure 12.4. The complex curvature $\psi: \tilde{M} \to \mathbb{C}$ of a curve $\gamma \in \mathcal{M}$ is given by $\gamma'' = \psi\xi$ for a unit length parallel section $\xi \in \Gamma(\perp_\gamma M)$ of the normal bundle, a unitary complex line bundle. It has been shown [18] that the complex curvature evolves under the non-linear Schrödinger equation, if the corresponding curve evolves under the vortex filament flow Y_1.

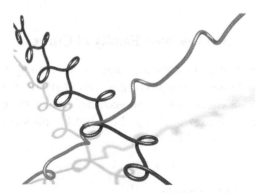

Figure 12.4 A curve γ (dark grey) together with a member γ_λ (light grey) of its associated family for real λ. The two curves osculate to second order at $\gamma(x_0)$.

In general, the curves γ_λ have changing monodromies and thus are not contained in the original $\mathcal{M}$. As we shall see, the dependency of these monodromies on the spectral parameter λ gives a geometric realization of the spectral curve and, at the same time, provides the Hamiltonians E_k for the flows Y_k.

Lemma 5.2 *Let $\gamma \in \mathcal{M}$ be a curve with monodromy $\tau^*\gamma = A\gamma A^{-1} + a$ where $A \in \mathbf{SU}_2$ and $a \in \mathbb{R}^3$. Then we have:*

(1) The associated family γ_λ can be expressed via the Sym formula

$$\gamma_\lambda - \gamma(x_0) = \frac{dF_\lambda}{d\lambda} F_\lambda^{-1}.$$

(2) The rotation monodromy $\tilde{A}_\lambda$ of γ_λ is given by

$$\tilde{A} = (\tau^*F)(x_0)A \in \Lambda^+\mathbf{SU}_2$$

and due to $F_0 = \mathbb{1}$, we have $\tilde{A}_0 = A$. The translation monodromy of γ_λ becomes

$$a_\lambda = \frac{d\tilde{A}_\lambda}{d\lambda}\tilde{A}_\lambda^{-1} - \tilde{A}_\lambda\,\gamma(x_0)\tilde{A}_\lambda^{-1}.$$

(3) The total torsion E_2 of γ_λ is given by

$$E_2(\gamma_\lambda) = E_2(\gamma) + \lambda E_1(\gamma)$$

with $E_1(\gamma)$ the length of γ over a period of τ.

Proof. To check the Sym formula, we calculate

$$\gamma_\lambda' = \left(\frac{dF_\lambda}{d\lambda}F_\lambda^{-1}\right)' = \frac{dF_\lambda'}{d\lambda}F_\lambda^{-1} = F_\lambda T F_\lambda^{-1}.$$

Since $F(x_0) = \mathbb{1}$ we also have $\frac{dF(x_0)}{d\lambda}F^{-1}(x_0) = 0$, which verifies $\gamma_\lambda(x_0) = \gamma(x_0)$. From the Sym formula we read off the monodromy of the curve γ_λ as stated. To compute the formula for $\tilde{A}$, we observe that the ODE (21) has τ^*FA as a solution since γ has rotational monodromy $\tau^*T = ATA^{-1}$. Therefore, $\tau^*FA = \tilde{A}F$ for some $\tilde{A} \in \Lambda^+\mathbf{SU}_2$ and evaluation at $x_0 \in \tilde{M}$ gives $\tilde{A} = (\tau^*F)(x_0)A$. For the total torsion of γ_λ we use its characterization in Lemma 3.2: if $v \in T_\gamma\mathcal{M}_{(A,a)}$ is a normal vector field along γ, that is, a section on $\perp_\gamma M$, then $v_\lambda = F_\lambda v F_\lambda^{-1} \in T_{\gamma_\lambda}\mathcal{M}_{(\tilde{A}_\lambda, a_\lambda)}$ is a normal field along γ_λ. Therefore,

$$E_2(\gamma_\lambda) = \int_M (\nabla v_\lambda, T_\lambda \times v_\lambda) = \int_M (\nabla T + \lambda T \times v dx, T \times v)$$

$$= E_2(\gamma) + \lambda E_1(\gamma)$$

where we used $\nabla T_\lambda = F_\lambda(\nabla T + \lambda T \times v dx)F_\lambda^{-1}$ and the invariance of the inner product. $\square$

Since the flows Y_k are genuine vector fields on $\mathcal{M}$, they satisfy $\tau^* Y_k = A Y_k A^{-1}$ along $\gamma \in \mathcal{M}$. Thus, the generating loop $Y = \sum_{k \geq 0} Y_k \lambda^{-k}$ along a curve $\gamma \in \mathcal{M}$ is also invariant under the monodromy $\tau^* Y = A Y A^{-1}$. Therefore, (20) implies

$$Y(x_0) = \tau^*(Y(x_0)) = \tau^*(FYF^{-1}) = \tilde{A} Y(x_0) \tilde{A}^{-1} \tag{22}$$

which shows that the axis of the rotation monodromy $\tilde{A}$ is in the direction of $Y(x_0)$. Ignoring for the moment that $Y(x_0) \in \Lambda^- \mathbb{R}^3$ is only a formal Laurent series, we can express the monodromy

$$\tilde{A} = \exp\left(-\frac{1}{2}\theta Y(x_0)\right)$$

by its unit length axis and rotation angle θ. Note that $\tilde{A} = \tilde{A}(x_0)$ depends on the base point $x_0 \in \tilde{M}$ by Lemma 5.2. In particular, as one moves the base point

$$\tilde{A}(x) = F(x)^{-1} \tilde{A} F(x), \quad x \in \tilde{M} \tag{23}$$

changes by conjugation. This has the important implication that the monodromy angle θ is independent of $x_0 \in \tilde{M}$. Therefore, θ depends only on the curve $\gamma \in \mathcal{M}$ and we obtain for each real $\lambda \in \mathbb{R}$ a (modulo 2π) well defined function

$$\theta_\lambda \colon \mathcal{M} \to \mathbb{R}.$$

The monodromy angle function θ of the associated family turns out to be the Hamiltonian for the generating loop $Y = \sum_{k \geq 0} Y_k \lambda^{-k}$.

Theorem 5.3 *Let $\gamma_t \in \mathcal{M}$ be a variation of $\gamma = \gamma_0$ with rotational monodromy $A \in \mathrm{SU}_2$, then*

$$\frac{1}{\lambda^2}\dot{\theta} = \frac{1}{\lambda^2} d\theta_\gamma(\dot{\gamma}) = \sigma(Y, \dot{\gamma}).$$

In other words, the generating loop Y is the symplectic vector field for the Hamiltonian $\lambda^{-2}\theta$.

Before continuing, we need to remedy the problem that the generating loop Y of the genuine vector fields Y_k is only a formal series. From (22) we know that $Y(x_0)$ is in the direction of the axis of the rotation $\tilde{A}$, which is given by the trace-free part of $\tilde{A}$. Since $\tilde{A} \in \Lambda^+ \mathrm{SU}_2$ is a real analytic loop, we may choose a unit length real analytic map $\tilde{Y}(x_0) \colon \mathbb{R} \to S^2 \subset \mathbb{R}^3$ spanning the axis of $\tilde{A}$, that is

$$\mathbb{R}\tilde{Y}(x_0)_\lambda = \mathbb{R}\left(\tilde{A}_\lambda - \tfrac{1}{2}\operatorname{tr}\tilde{A}_\lambda \mathbb{1}\right).$$

From (23) we deduce $\tilde{Y} = F^{-1}\tilde{Y}(x_0)F$ and therefore $\tilde{Y}$ satisfies the same Lax equation (12) as the formal loop Y. It can be shown by asymptotic analysis [14, Chapter I.3] that $\tilde{Y}_\lambda \to T$ as $\lambda \to \infty$ in the smooth topology in $\Gamma(\gamma^* T\mathbb{R}^3_{|M})$.

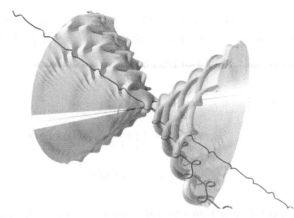

Figure 12.5 The curves γ_λ in the associated family of an elastic curve sweep out a surface that is close to the quadratic cone swept out by their monodromy axes. The intersection of this cone with a sphere centered at $\gamma(x_0)$ is a quartic curve in bijective algebraic correspondence with the real part of the spectral curve Σ, which is elliptic in this case. As the spectral parameter $\lambda \to \infty$, the curves γ_λ straighten, as can be seen when approaching the white sector.

Since $T_\lambda = F_\lambda T F_\lambda^{-1}$, we also obtain that $T_\lambda \to T(x_0)$ as $\lambda \to \infty$. In particular, the curves γ_λ in the associated family straighten out and tend to the line with tangent $T(x_0)$, as can be seen in Figure 12.5. Knowing that $\tilde{Y}$ is smooth at $\lambda = \infty$, it has the Taylor series expansion $\tilde{Y} = \sum_{k \geq 0} \tilde{Y}_k \lambda^{-k}$ and satisfies the recursion equation (12) with the same initialization $\tilde{Y}_0 = Y_0 = T$. But then the explicit construction of the solutions to this recursion in Theorem 3.8 shows that $Y_k = \tilde{Y}_k$ for $k \geq 0$. In other words, instead of working with Y in the construction of the angle function θ, we need to work with $\tilde{Y}$, and likewise in any of the arguments to come. We will still keep writing Y, but think $\tilde{Y}$, as not to stray too far from our geometric intuition. With this being said, we come to the proof of Theorem 5.3.

Proof. We first derive an expression for the variation $\dot{\tilde{A}}$ of the monodromy. Expressing $\dot{F} = HF$ the equations $F' = F(\lambda T)$, $F(x_0) = \mathbb{1}$ imply that

$$H' dx = F\lambda(\dot{T} dx + T d\dot{x})F^{-1}, \quad H(x_0) = 0$$

which, after inserting (6), results in

$$H' = F\lambda\dot{\gamma}' F^{-1}, \quad H(x_0) = 0.$$

Now $\tilde{A} = (\tau^* F)(x_0)A$ and therefore

$$\dot{\tilde{A}} = (\tau^* \dot{F})(x_0)A = (\tau^* H)(x_0)(\tau^* F)(x_0)A = (\tau^* H)(x_0)\tilde{A}$$

$$= \int_I F\lambda\dot{\gamma}' F^{-1}\tilde{A} dx = \int_I F\lambda\dot{\gamma}' \tilde{A}(x)F^{-1} dx,$$

where we used (23) and the fundamental domain $I \subset \tilde{M}$ for τ has oriented boundary $\partial I = \{x_0\} \cup \{\tau(x_0)\}$. Since $\tilde{A}(x) = \exp\left(-\frac{1}{2}\theta Y(x)\right) = \cos\left(\frac{1}{2}\theta\right)\mathbb{1} - \sin\left(\frac{1}{2}\theta\right) Y(x)$ with $Y(x)$ trace free, we arrive at

$$-\dot{\theta}\sin\left(\tfrac{1}{2}\theta\right) = \operatorname{tr}\dot{\tilde{A}} = \int_I \operatorname{tr}(\lambda\dot{\gamma}'\tilde{A}(x))dx$$

$$= \int_I \operatorname{tr}\left(\lambda\dot{\gamma}'\left(\cos\left(\tfrac{1}{2}\theta\right)\mathbb{1} - \sin\left(\tfrac{1}{2}\theta\right)Y\right)\right)dx$$

$$= -\sin\left(\tfrac{1}{2}\theta\right)\int_M (\lambda\dot{\gamma}', Y)dx\,.$$

Finally, we cancel $\sin\left(\frac{1}{2}\theta\right)$ on both sides, apply Stokes' Theorem, and use (12) to obtain

$$\dot{\theta} = \int_M (\lambda\dot{\gamma}', Y)dx = \int_M d(\lambda\dot{\gamma}, Y) - (\lambda\dot{\gamma}, Y')dx = \int_M (\lambda\dot{\gamma}, \lambda\,T\times Y)dx$$
$$= \sigma(\lambda^2 Y, \dot{\gamma})\,.$$

$\square$

It remains to calculate an explicit expression for the Taylor expansion

$$\frac{1}{\lambda^2}\theta = \sum_{k\geq 0} E_k \lambda^{-k} \tag{24}$$

of the monodromy angle θ at $\lambda = \infty$ to obtain concrete formulas for the hierarchy of Hamiltonians $E_k \colon \mathcal{M} \to \mathbb{R}$ from Theorem 5.3. Starting with a curve $\gamma \in \mathcal{M}$ with rotation monodromy A, we consider its associated family of curves γ_λ with rotation monodromy $\tilde{A}_\lambda$ at the base point $x_0 \in \tilde{M}$. The tangent image $T_\lambda \colon \tilde{M} \to S^2 \subset \mathbb{R}^3$ defines the following sector on S^2, see Figure 12.6:

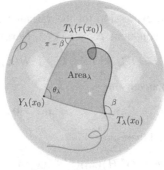

Figure 12.6 The monodromy angle θ_λ can be computed in terms of the area of the sector enclosed by the tangent image T_λ.

(1) The monodromy axis point $Y_\lambda(x_0)$ connected to $T_\lambda(x_0)$ by a geodesic arc, then the curve T_λ traversed from x_0 to $\tau(x_0)$ along a fundamental domain $I \subset \tilde{M}$, and the geodesic arc connecting $T_\lambda(\tau(x_0))$ back to the axis point $Y_\lambda(x_0)$. Since $Y_\lambda \to T$ tends to T and $T_\lambda \to T(x_0)$ tends to $T(x_0)$ for large $\lambda \in \mathbb{R}$, there is no ambiguity in this prescription.

(2) The exterior angles of this sector are $\pi - \theta_\lambda$ at $Y_\lambda(x_0)$, some angle β at $T_\lambda(x_0)$, and the angle $\pi - \beta$ at $T_\lambda(\tau(x_0))$, since T_λ has rotation monodromy around the axis $Y_\lambda(x_0)$ with angle θ_λ.

Parallel transport in the normal bundle $\perp_{\gamma_\lambda} \tilde{M}$ along γ_λ is the same as parallel transport with respect to the Levi-Civita connection on S^2 along the tangent image T_λ. Applying the Gauß–Bonnet Theorem to the parallel transport around our sector of area Area_λ gives the relation

$$E_2(\gamma_\lambda) + (\pi - \theta_\lambda) + \beta + (\pi - \beta) + \text{Area}_\lambda = 2\pi .$$

where we used Lemma 3.2. Applying Lemma 5.2 (iii), this unravels to

$$\theta_\lambda = \lambda E_1(\gamma) + E_2(\gamma) + \text{Area}_\lambda .$$

Now the area of a spherical sector of the type described above is given by

$$\text{Area}_\lambda = \int_I \frac{\det(Y_\lambda(x_0), T_\lambda, T_\lambda')}{1 + (Y_\lambda(x_0), T_\lambda)} \, dx$$

which we can further simplify. From Definition 5.1 of the associated family, we have $T_\lambda = F_\lambda T F_\lambda^{-1}$ and the generating loop $Y = \sum_{k \geq 0} Y_k \lambda^{-k}$, as a solution to the Lax equation (12), satisfies $Y = F^{-1} Y(x_0) F$, which leads to

$$\text{Area}_\lambda = \int_M \frac{\det(Y, T, T')}{1 + (Y, T)} \, dx .$$

To calculate $(Y, T) = \sum_{k \geq 0}(Y_k, T)\lambda^{-k}$, we need the tangential components of Y_k, which can be read off from the proof of Theorem 3.8:

$$(T, Y_0) = 1, \quad (T, Y_1) = 0, \quad (T, Y_k) = -\frac{1}{2} \sum_{i=1}^{k-1} (Y_{k-i}, Y_i) .$$

Expressing $\frac{1}{1+(T,Y)}$ via the geometric series, Theorem 5.3 gives rise to explicit formulas for the Hamiltonians E_k.

Theorem 5.4 *The Hamiltonians E_k for the commuting flows Y_k are given by the generating function*

$$\sum_{k \geq 0} E_k \lambda^{-k} = 0 + E_1 \lambda^{-1} + E_2 \lambda^{-2}$$

$$+ \frac{\lambda^{-2}}{2} \int_M \sum_{i \geq 1, j \geq 0} (Y_i, Y_1) \lambda^{-i} (-1)^j \left(\sum_{l \geq 2} \frac{1}{2} (T, Y_l) \lambda^{-l} \right)^j \, dx .$$

For instance,

- $E_4 = -\frac{1}{2} \int_M \det(\gamma', \gamma'', \gamma''') dx,$
- $E_5 = \int_M \left(\frac{1}{2} |\gamma'''|^2 - \frac{5}{8} |\gamma''|^4 \right) dx,$
- $E_6 = \int_M \left(-\frac{1}{2} \det(\gamma', \gamma''', \gamma'''') + \frac{7}{8} |\gamma''|^2 \det(\gamma', \gamma'', \gamma''') \right) dx.$

6 Darboux Transforms and the Spectral Curve Revisited

The geometry of the associated family γ_λ of a curve $\gamma \in \mathcal{M}$ played a pivotal role in our discussions so far. For instance, from (18) the real part of the spectral curve Σ is given by the eigenlines of the monodromy A_λ for real spectral parameter $\lambda \in \mathbb{R}$. A natural question to ask is whether there are associated curves γ_λ for complex spectral parameters $\lambda \in \mathbb{C}$. This is indeed the case, provided that we allow the curves γ_λ to live as curves with monodromy in hyperbolic 3-space H^3. The two fixed points of their monodromies on the sphere at infinity of H^3 exhibit the spectral curve Σ as a hyperelliptic branched cover over the complex λ-plane. Moreover, these fixed points give rise to periodic Darboux transforms $\eta \in \mathcal{M}$ of the original curve γ as curves in $\mathbb{R}^3$. Thus, we have associated to a curve $\gamma \in \mathcal{M}$ of a given monodromy h a Riemann surface Σ worth of curves $\eta \in \mathcal{M}$, the Darboux transforms of γ of the same monodromy h. Given $\gamma \in \mathcal{M}$ and using the notation from the previous section, we have a solution $F_\lambda \colon \tilde{M} \to \mathbf{SL}_2(\mathbb{C})$ of $F' = F(\lambda T)$, $F(x_0) = \mathbb{1}$, for all $\lambda \in \mathbb{C}$. Using the description of hyperbolic space H^3 as the set of all hermitian 2×2-matrices with determinant one and positive trace, we define the associated family

$$\gamma_\lambda \colon \tilde{M} \to H^3, \qquad \gamma_\lambda = F_\lambda F_\lambda^*$$

for non-real values $\lambda \in \mathbb{C} \setminus \mathbb{R}$. Since $\tau^* F A = \tilde{A} F$ with $\tilde{A} \in \Lambda^+ \mathbf{SL}_2(\mathbb{C})$ for non-real λ, the curves γ_λ in hyperbolic space H^3 have monodromy

$$\tau^* \gamma_\lambda = \tilde{A}_\lambda \gamma_\lambda \tilde{A}_\lambda^*.$$

Furthermore, the curves γ_λ have constant speed $2\mathrm{Im}(\lambda)$, as can be seen from

$$\gamma_\lambda' = F(\lambda T + \bar{\lambda} T^*) F^* = 2\mathrm{Im}(\lambda) i F T F^*.$$

Stereographic projection

$$\pi \colon H^3 \to \mathbf{su}_2 = \mathbb{R}^3, \qquad \pi(p) = i \frac{\mathrm{tr}\, p\, \mathbb{1} - 2p}{2 + \mathrm{tr}\, p}$$

realizes H^3 as the Poincare model B^3, the unit ball centered at the origin. Since $d_{\mathbb{1}} \pi(T) = \frac{T}{2}$, we rescale B^3 by $\frac{1}{\mathrm{Im}\lambda}$ and center it at $\gamma(x_0)$. Then the curve

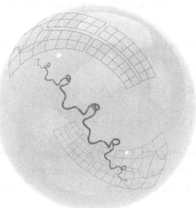

Figure 12.7 A curve γ (dark grey) together with a member γ_λ (light grey) of its associated family for complex λ in hyperbolic 3-space.

Figure 12.8 The part of the spectral curve that lies over a rectangular domain in the λ-plane.

$$\hat{\gamma}_\lambda = \gamma(x_0) + \frac{1}{\mathrm{Im}\lambda} \pi \circ \gamma_\lambda$$

in $\mathbb{R}^3$ touches γ to first order at $\gamma(x_0)$, see Figure 12.7.

Viewed as hyperbolic motions, the monodromy matrices $\tilde{A}_\lambda$ have two fixed points on the sphere at infinity. Stereographic projection π realizes these fixed points as two unit vectors $S_\pm \in S^2$. The set of all pairs $(\lambda, S_\pm) \in \mathbb{C} \times S^2 = \mathbb{C} \times \mathbb{CP}^1$ is biholomorphic to the eigenline spectral curve (18). This means that as λ varies, the pairs (S_+, S_-) of points on S^2 trace out a Euclidean image of the spectral curve, see Figure 12.8.

It turns out that a slightly different scaling of the hyperbolic bubbles around the points $\gamma(x)$, which agrees with the above in case $\lambda \in i\mathbb{R}$ is purely

Figure 12.9 As the base point x_0 moves along a curve γ (medium grey), the curves (light grey) in its associated family for fixed complex λ spiral towards points that trace out Darboux transforms $\eta_\pm$ (dark grey) of γ.

imaginary, is closely related to *Darboux transforms* in the sense of [20, 32] of $\gamma \in \mathcal{M}$, see Figure 12.9.

Theorem 6.1 *Let $\gamma \in \mathcal{M}$ then the two curves $\eta_\pm \colon \tilde{M} \to \mathbb{R}^3$ given by*

$$\eta_\pm = \gamma + \frac{\mathrm{Im}(\lambda)}{|\lambda|^2} S_\pm$$

are Darboux transforms of γ having the same monodromy as γ, that is $\eta_\pm \in \mathcal{M}$. In particular, $\eta_\pm$ have constant distance to γ and induce the same arclength on $\tilde{M}$ as γ. All Hamiltonians E_k, $k \geq -2$, satisfy

$$E_k(\eta_\pm) = E_k(\gamma).$$

Proof. For real λ, as we move the base point along $\tilde{M}$, equation (23) implies that $S_\pm$ satisfy the differential equation

$$S'_\pm = -\lambda T \times S_\pm.$$

Viewing $S \mapsto T \times S$ as a vector field on S^2, we see that an imaginary part of λ adds a multiple of the same vector field rotated by $\pi/2$:

$$S'_\pm = -\mathrm{Re}(\lambda)T \times S_\pm - \mathrm{Im}(\lambda)T \times (T \times S_\pm).$$

According to equation (25) of [32] (note that the letters S and T are interchanged there) this is precisely the equation needed for

$$\eta_\pm := \gamma(x_0) + \frac{\mathrm{Im}(\lambda)}{|\lambda|^2} S_\pm$$

to be Darboux transforms of γ. Darboux transforms exhibit the so-called Bianchi permutability: Darboux transforming γ with parameter λ followed by a Darboux transform with parameter μ has the same result as the same procedure with the roles of λ and μ interchanged. This was proved in a discrete setting in [32] and follows by continuum limit for the smooth case. The differential equation that determines Darboux transforms is the same as the one that determines the monodromies of the curves in the associated family. Therefore, as a consequence of permutability, $\eta_\pm$ has the same monodromy angle function θ as γ. Thus, by (24) we have $E_k(\eta_\pm) = E_k(\gamma)$. $\square$

For a finite gap curve $\gamma \in \mathcal{M}$ the hierarchy of flows corresponds to the osculating flag of the Abel–Jacobi embedding of the spectral curve Σ at the point over $\lambda = \infty$. On the other hand, Darboux transformations correspond to translations along secants of the Abel image of Σ in its Jacobian $\text{Jac}(\Sigma)$.

7 History of Elastic Curves and Hamiltonian Curve Flows

For the early history of elastic space curves we follow [39], see also [29, 40]. In 1691 Jakob Bernoulli posed the problem of determining the shape of bent beams [4]. It was his nephew Daniel Bernoulli who, in 1742, realized in a letter to Euler [3] that this problem amounts to minimizing $\int \kappa^2$ for the curve that describes the beam. Euler then classified in 1744 all planar elastic curves [13, 8]. Lagrange started in 1811 to investigate elastic space curves, but he ignored the gradient of total torsion that in general has to be part of the variational functional. This was pointed out in 1844 by Binét [5], who wrote the complete Euler–Lagrange equations and was able to solve them up to quadratures. Then in 1859 Kirchhoff [24] discovered that these Euler–Lagrange equations can be interpreted as the equations of motion of the Lagrange top, a fact that became famous as the *Kirchhoff analogy*. Finally, in 1885 Hermite solved the equations explicitly [19] in terms of elliptic functions.

In 1906 Max Born wrote his Ph.D. thesis on the stability of elastic curves [7]. More recently, elastic curves in spaces other than $\mathbb{R}^3$ have been explored [37]. The gradient flows of their variational functionals have been studied from the viewpoint of Geometric Analysis [12]. Planar critical points of elastic energy under an area constraint were also investigated [2, 15].

The history of Hamiltonian curve flows begins in 1906 with the discovery of the vortex filament equation by Da Rios [11], who was a Ph.D. student of Levi-Civita. In 1932 Levi-Civita described what nowadays would be called the one-soliton solution [28]. The corresponding curves are indeed elastic loops already found by Euler, but Da Rios and Levi-Civita were seemingly not aware of this fact. For further details on the history see [35] and also [34]. Under the name of *localized induction approximation* this equation is a standard tool in Fluid Dynamics [36]. Rigorous estimates indicate that this approximation to the 3D Euler equation is valid even over a short time [22]. Recently it became possible to study knotted vortex filaments experimentally [25]. The symplectic form on the space of curves was found by Marsden and Weinstein [30] in 1983, see also Chapter VI.3 of [1]. Codimension two submanifolds in higher dimensional ambient spaces can be treated similarly [23].

The relationship between vortex filaments and the theory of integrable systems was discovered in 1972 by Hasimoto. He showed the equivalence of the vortex filament flow with the nonlinear Schrödinger equation [18]. The non-linear Schrödinger hierarchy then led to the discovery of the hierarchy of flows for space curves [27, 41, 26]. This made it possible to study in detail [17, 9, 10] those curves in $\mathbb{R}^3$ corresponding to finite gap solutions of the nonlinear Schrödinger equation [33]. Since every other of the Hamiltonian curve flows preserves the planarity of curves, this allows for a self-contained approach to the mKdV hierarchy of flows on plane curves [31].

In magnetohydrodynamics vortex filaments can carry a trapped magnetic field, in which case the total torsion becomes part of the Hamiltonian. As discussed above, the resulting flow is the helicity filament flow [21].

Darboux transformations with purely imaginary λ have been studied under the name *bicycle transformations* [38, 6]. Those with general complex λ were investigated in [32] and in [20], where they were called Bäcklund transformations.

References

[1] ARNOLD, V. I., AND KHESIN, B. A. *Topological Methods in Hydrodynamics*, vol. 125. Springer, 1999.

[2] ARREAGA, G., CAPOVILLA, R., CHRYSSOMALAKOS, C., AND GUVEN, J. Area-constrained planar elastica. *Phys. Rev. E 65* (Feb 2002), 031801.

[3] BERNOULLI, D. The 26th letter to Euler. In *Correspondence Mathématique et Physique*, vol. 2. 1742.

[4] BERNOULLI, J. Quadratura curvae, e cujus evolutione describitur inflexae laminae curvatura. In *Die Werke von Jakob Bernoulli*. 1691, pp. 223–227.

[5] BINET, J. Mémoire sur l'integration des équations de la courbe élastique à double courbure. *Compte Rendu* (1844).

[6] BOR, G., LEVI, M., PERLINE, R., AND TABACHNIKOV, S. Tire tracks and integrable curve evolution. *arXiv preprint arXiv:1705.06314* (2017).

[7] BORN, M. *Untersuchungen über die Stabilität der elastischen Linie in Ebene und Raum, unter verschiedenen Grenzbedingungen.* PhD thesis, Universität Göttingen, 1906.

[8] BRYANT, R., AND GRIFFITHS, P. Reduction for constrained variational problems and $\int \kappa^2/2\, ds$. *American Journal of Mathematics 108*, 3 (1986), 525–570.

[9] CALINI, A. Recent developments in integrable curve dynamics. *Geometric Approaches to Differential Equations 15* (2000), 56–99.

[10] CALINI, A. Integrable dynamics of knotted vortex filaments. In *Proceedings of the Fifth International Conference on Geometry, Integrability and Quantization* (2004), pp. 11–50.

[11] DA RIOS, L. S. Sul moto d'un liquido indefinito con un filetto vorticoso di forma qualunque. *Rendiconti del Circolo Matematico di Palermo (1884-1940) 22*, 1 (1906), 117–135.

[12] DZIUK, G., KUWERT, E., AND SCHÄTZLE, R. Evolution of elastic curves in $\mathbb{R}^n$: Existence and computation. *SIAM Journal on Mathematical Analysis 33*, 5 (2002), 1228–1245.

[13] EULER, L. *Methodus inveniendi lineas curvas maximi minimive proprietate gaudentes sive solutio problematis isoperimetrici latissimo sensu accepti.* Springer, 1952.

[14] FADDEEV, L., AND TAKHTAJAN, L. *Hamiltonian Methods in the Theory of Solitons.* Springer, 1987.

[15] FERONE, V., KAWOHL, B., AND NITSCH, C. The elastica problem under area constraint. *Mathematische Annalen 365*, 3-4 (2016), 987–1015.

[16] FERUS, D., PEDIT, F., PINKALL, U., AND STERLING, L. Minimal tori in S^4. *Journal für die reine und angewandte Mathematik (Crelles Journal) 429* (1992), 1–48.

[17] GRINEVICH, P. G., AND SCHMIDT, M. U. Closed curves in $\mathbb{R}^3$ and the nonlinear Schrödinger equation. In *Nonlinearity, Integrability And All That: Twenty Years After NEEDS'79.* World Scientific, 2000, pp. 139–145.

[18] HASIMOTO, H. A soliton on a vortex filament. *Journal of Fluid Mechanics 51*, 3 (1972), 477–485.

[19] HERMITE, C. *Sur quelques applications des fonctions elliptiques.* 1885.

[20] HOFFMANN, T. Discrete Hashimoto surfaces and a doubly discrete smoke-ring flow. In *Discrete Differential Geometry.* Springer, 2008, pp. 95–115.

[21] HOLM, D. D., AND STECHMANN, S. N. Hasimoto transformation and vortex soliton motion driven by fluid helicity. *arXiv preprint nlin/0409040* (2004).

[22] JERRARD, R. L., AND SEIS, C. On the vortex filament conjecture for Euler flows. *Archive for Rational Mechanics and Analysis 224*, 1 (2017), 135–172.

[23] KHESIN, B. The vortex filament equation in any dimension. *Procedia IUTAM 7* (2013), 135–140.

[24] KIRCHHOFF, G. Über das Gleichgewicht und die Bewegung eines unendlich dünnen elastischen Stabes. *Journal für die reine und angewandte Mathematik 56* (1859), 285–313.

[25] KLECKNER, D., AND IRVINE, W. T. Creation and dynamics of knotted vortices. *Nature Physics 9*, 4 (2013), 253–258.

[26] LANGER, J. Recursion in curve geometry. *The New York Journal of Mathematics 5*, 25–51 (1999).

[27] LANGER, J., AND PERLINE, R. Poisson geometry of the filament equation. *Journal of Nonlinear Science 1*, 1 (1991), 71–93.

[28] LEVI-CIVITA, T. Attrazione newtoniana dei tubi sottili e vortici filiformi (newtonian attraction of slender tubes and filiform vortices). *Annali R. Scuola Norm. Sup. Pisa 1* (1932), 1–33.

[29] LEVIEN, R. The elastica: a mathematical history. Tech. rep., University of California, 2008.

[30] MARSDEN, J., AND WEINSTEIN, A. Coadjoint orbits, vortices, and Clebsch variables for incompressible fluids. *Physica D: Nonlinear Phenomena 7*, 1 (1983), 305–323.

[31] MATSUTANI, S., AND PREVIATO, E. From Eulers elastica to the mKdV hierarchy, through the Faber polynomials. *Journal of Mathematical Physics 57*, 8 (2016), 081519.

[32] PINKALL, U., SPRINGBORN, B., AND WEISSMANN, S. A new doubly discrete analogue of smoke ring flow and the real time simulation of fluid flow. *Journal of Physics A 40*, 42 (2007).

[33] PREVIATO, E. Hyperelliptic quasi-periodic and soliton solutions of the nonlinear Schrödinger equation. *Duke Mathematical Journal 52*, 2 (1985), 329–377.

[34] RICCA, R. L. Rediscovery of Da Rios equations. *Nature 352* (1991), 561–562.

[35] RICCA, R. L. The contributions of Da Rios and Levi-Civita to asymptotic potential theory and vortex filament dynamics. *Fluid Dynamics Research 18*, 5 (1996), 245–268.

[36] SAFFMAN, P. G. *Vortex Dynamics*. Cambridge University Press, 1992.

[37] SINGER, D. A. Lectures on elastic curves and rods. In *AIP Conference Proceedings* (2008), vol. 1002, pp. 3–32.

[38] TABACHNIKOV, S. On the bicycle transformation and the filament equation: Results and conjectures. *Journal of Geometry and Physics 115* (2017), 116–123.

[39] TJADEN, E. *Einfache elastische Kurven*. PhD thesis, TU Berlin, 1991.

[40] TRUESDELL, C. The influence of elasticity on analysis: the classic heritage. *Bulletin of the American Mathematical Society 9*, 3 (1983), 293–310.

[41] YASUI, Y., AND SASAKI, N. Differential geometry of the vortex filament equation. *Journal of Geometry and Physics 28*, 1 (1998), 195–207.

Albert Chern
Institute of Mathematics, MA 8-4, Technical University Berlin, Strasse des 17. Juni 136, 10623 Berlin, GERMANY
E-mail address: chern@math.tu-berlin.de

Felix Knöppel
Institute of Mathematics, MA 8-4, Technical University Berlin, Strasse des 17. Juni 136, 10623 Berlin, GERMANY
E-mail address: knoeppel@math.tu-berlin.de

Franz Pedit
Department of Mathematics and Statistics, University of Massachusetts, Amherst, MA 01003, USA
E-mail address: pedit@math.umass.edu

Ulrich Pinkall
Institute of Mathematics, MA 8-4, Technical University Berlin, Strasse des 17. Juni 136, 10623 Berlin, GERMANY
E-mail address: pinkall@math.tu-berlin.de

13

The Kowalewski Top Revisited

F. Magri

Abstract. The paper is a commentary of one section of the celebrated paper by Sophie Kowalewski on the motion of a rigid body with a fixed point. Its purpose is to show that the results of Kowalewski may be recovered by using the separability conditions obtained by Tullio Levi Civita in 1904.

1 Introduction

The present paper is a commentary of Sec. 2 of the paper *Sur le problème de la rotation d' un corps solide autour d' un point fixe* written by Sophie Kowalewski in 1889 [1]. This celebrated paper is composed of seven sections. Each contains a relevant result.

Sec. 1 deals with the integrability of the Euler-Poisson equations of Mechanics, that is with the integrability of the equations of a rigid body with a fixed point in motion under the action of gravity. Starting from the remark that the classical cases of Euler and Lagrange are solved by means of elliptic functions, which are meromorphic functions in the complex plane, Kowalewski sets the question of determining all the rigid bodies for which the solutions of the Euler-Poisson equations are meromorphic functions of the complex time. She finds a case missed by Euler and Lagrange (henceforth known as the Kowalewski top), and she claims that there are no other cases [2]. This claim raised the objections of Liapunov and Markov [4], due to some gap in Kowalewski's proof , but now it is commonly accepted as true. The original idea of Kowalewski has become one of the main techniques to detect the integrability of the equations of motion of a dynamical system, and even accepted as one of the possible definitions of the notion of integrable system [17].

In Sec. 2 Kowalewski demonstrates that the six equations of motion of her top may be reduced to a system of two first- order differential equations having the form of Euler's equations for a hyperelliptic curve. In this section Kowalewski introduces the main tools needed to perform the reduction. They are:

- The quartic integral k^2 of Kowalewski.
- The biquadratic function $R(x_1, x_2)$ of Kowalewski.
- The fifth-order polynomial $R_1(s)$ of Kowalewski.

Let us explain intuitively their role. The integral k^2 (combined with the classical integrals of the energy, of the vertical component of the angular momentum, and of the weight) serves to define a two-dimensional foliation. The motion of the top takes place on the leaves of this foliation, and therefore from an analytical point of view the top is a system with two degrees of freedom. The biquadratic function $R(x_1, x_2)$ serves to introduce a special system of coordinates s_1 and s_2 on the leaves of the foliation. They are defined by

$$ s := \frac{1}{2}w + \frac{1}{2}l_1, $$

where $3l_1$ is the energy of the top, and w is a solution of the quadratic equation

$$ w^2 - 2\frac{R(x_1, x_2)}{(x_1 - x_2)^2}w - \frac{R_1(x_1, x_2)}{(x_1 - x_2)^2} = 0 $$

called *the fundamental equation* by Golubev [7]. The second coefficient of this equation is the function of $R(x_1, x_2)$ defined by

$$ (x_1 - x_2)^2 R_1(x_1, x_2) := R(x_1, x_1)R(x_2, x_2) - R(x_1, x_2)^2. $$

The fifth-order polynomial $R_1(s)$ serves to write explicitly the reduced equations of motion on the leaves of the two-dimensional invariant foliation. The main result of Sec. 2 is the proof that in the coordinates s_1 and s_2 the Euler-Poisson equations of the Kowalewski top assume the form

$$ 0 = \frac{ds_1}{\sqrt{R_1(s_1)}} + \frac{ds_2}{\sqrt{R_1(s_2)}} $$

$$ dt = \frac{s_1 ds_1}{\sqrt{R_1(s_1)}} + \frac{s_2 ds_2}{\sqrt{R_1(s_2)}}, $$

that is the form of Euler's equations related to the hyperelliptic curve $y^2 = R_1(s)$.

In the remaining five sections, finally, Kowalewski solves these equations and computes the angular velocity of the body as a function of the time in terms of θ-functions associated with the hyperelliptic curve. Her impressive computations were completed a few years later by Kotter [3].

As this brief description may suggest, the most critical point in the process of Kowalewski's solution is the choice of the coordinates s_1 and s_2. It is really a difficult problem to explain why the fundamental equation works so well. A noticeable advance in this direction was a remark by A. Weil on an old theorem by Euler, concerning the solution of Euler's equation on a general elliptic curve [8]. As noticed in [9], the fundamental equation of Kowalewski coincides with the "canonical equation" of Euler related to the elliptic curve $y^2 = R(x, x)$ (see also [16]). This remark leads us to interpret the change of variables of Kowalewski as an addition formula for elliptic functions, and points out an algebro-geometric origin of the fundamental equation. The algebro-geometric viewpoint has been widely developed in the past by M. Adler, P. van Moerbeke, P. Vanhaecke, L. Haine and collaborators, giving rise to the beautiful theory of algebraically completely integrable systems (see [10] and the references collected therein).

The viewpoint of the present paper is different. According to the analysis pursued in the paper, the origin of the fundamental equation of Kowalewski must be found in the geometry of the leaves of the two-dimensional invariant foliation of the top. In the paper we prove that these leaves carry a remarkable differential geometric structure. We identify it and we prove that its occurrence is neither specific to Kowalewski's top or occasional. It is a characteristic trait of separable systems, which is implied by the separability conditions discovered by Tullio Levi Civita in 1904 [6]. They claim that a Hamilton-Jacobi equation is separable in a set of canonical coordinated (u_j, v_j) if and only if the Hamiltonian function verifies the equations

$$\frac{\partial H}{\partial u_k}\frac{\partial H}{\partial u_j}\frac{\partial^2 H}{\partial v_k \partial v_j} - \frac{\partial H}{\partial u_k}\frac{\partial H}{\partial v_j}\frac{\partial^2 H}{\partial v_k \partial u_j} - \frac{\partial H}{\partial v_k}\frac{\partial H}{\partial u_j}\frac{\partial^2 H}{\partial u_k \partial v_j}$$
$$+ \frac{\partial H}{\partial v_k}\frac{\partial H}{\partial v_j}\frac{\partial^2 H}{\partial u_k \partial u_j} = 0.$$

In the last section we use the information acquired from the study of these equations to construct the fundamental equation of Kowalewski in the case of the top.

The paper is organized as follows. Sec. 2 is a brief survey of the second section of Kowalewski's work. Sec. 3 is an initial exploration of the differential geometry of the leaves of the two-dimensional invariant foliation of the top. A threefold geometric structure is identified and described by a 2-form, a tensor field of type $(1, 1)$, and a metric. Its study leads to the discovery of several new remarkable identities verified by the function $R(x_1, x_2)$ of Kowalewski. These identities are explained in the next two sections. In particular: Sec. 4 presents a new criterion of separability, called the dual form of Levi Civita's conditions; Sec. 5 presents a new algorithm for the search of separation coordinates, called the method of Kowalewski's conditions. When applied to Kowalewski's top in Sec. 6, this method quickly gives the fundamental equation of Kowalewski.

2 Excerpts from Kowalewski's Paper

In the second section of her paper, Kowalewski consider the equations of motion of a rigid body with a fixed point under the action of gravity. She assumes that the principal moments of inertia satisfy the relations

$$A = B = 2\,C = 2\,I,$$

and that the center of mass of the body belongs to the equatorial plane, that is to the plane orthogonal to the axis of material symmetry passing through the fixed point. She considers as dynamical variables the three components (p, q, r) of the angular velocity of the body, and the direction cosines $(\gamma, \gamma', \gamma'')$ of the invariable direction of the gravity with respect to the principal axes of inertia. Furthermore, she exploits the freedom in the choice of the inertial axes and in the choice of the units of measure to set $I = 1$, and to attribute the values

$$x_0 = 1 \quad y_0 = 0 \quad z_0 = 0$$

to the coordinates of the center of mass in the moving reference frame. In place of these variables we prefer to use the components of the angular momentum

$$L_1 = 2pI \quad L_2 = 2qI \quad L_3 = rI$$

and the components of the moment of the weight with respect to the fixed point

$$y_1 = c_0\gamma \quad y_2 = c_0\gamma' \quad y_3 = c_0\gamma''$$

where $c_0 = Mgx_0$. Accordingly we write the equations of motion of the rigid body in the form

$$\dot{L}_1 = \frac{1}{2}L_2L_3$$

$$\dot{L}_2 = -\frac{1}{2}(L_3L_1 + y_3)$$

$$\dot{L}_3 = y_2$$

$$\dot{y}_1 = L_3y_2 - \frac{1}{2}L_2y_3$$

$$\dot{y}_2 = \frac{1}{2}L_1y_3 - L_3y_1$$

$$\dot{y}_3 = \frac{1}{2}L_2y_1 - \frac{1}{2}L_1y_2. \tag{1}$$

A first property worthy of attention is the homogeneity of the equations of motion. If one assigns degree 1 to the components of the angular momentum and degree 2 to the components of the moment of the weight, one can easily notice that

$$deg(\dot{L}_j) = 2 \quad deg(\dot{y}_j) = 3 \quad j = 1, 2, 3.$$

This means that the vector field defined by the equations of motion is homogeneous of degree 1. A second property concerns the integrals of motion. As noticed by Kowalewski, the above equations admit four and not only three integrals of motion. These integrals may be chosen in many different ways. In the present paper we choose

$$c_1 = y_1^2 + y_2^2 + y_3^2$$
$$c_2 = L_1 y_1 + L_2 y_2 + L_3 y_3$$
$$h_1 = \frac{1}{4}(L_1^2 + L_2^2 + 2L_3^2) - y_1$$
$$h_2 = \frac{1}{32}(L_1^2 - L_2^2 + 4y_1)^2 + \frac{1}{32}(2L_1 L_2 + 4y_2)^2, \tag{2}$$

and we notice that these integrals are related to those chosen by Kowalewski by the relations

$$h_1 = 3l_1 \quad h_2 = \frac{1}{2}k^2 \quad c_2 = l.$$

For c_1 Kowalewski fixes the value $c_1 = c_0^2$, while we keep its value arbitrary. This will allow us to readily control the degree of homogeneity of certain expressions containing the integrals of motion which will appear afterwards, by noticing that

$$deg(h_1) = 2 \quad deg(c_2) = 3 \quad deg(h_2) = deg(c_1) = 4.$$

The final scope of Kowalewski in this section is to infer the form of the equations of motion restricted to the level surfaces of the integrals of motion. To this end she starts to study a convenient parametrization of these surfaces. We skip this part of the paper, slightly departing from the line traced by Kowalewski, preferring to pass directly to the components of the dynamical vector field. We notice that the components $\dot{L}_1$ and $\dot{L}_2$ depend linearly on the coordinates L_3 and y_3, so that their squares depend linearly on L_3^2, $y_3 L_3$, and y_3^2. On the level surfaces of the integrals of motion the last variables are polynomial functions of (L_1, L_2, y_1, y_2), and therefore they can be easily eliminated. If one replaces the variables y_1 and y_2 by the variables (of the same degree) $e_1 = \frac{1}{4}(L_1^2 - L_2^2) + y_1$ and $e_2 = \frac{1}{2}L_1 L_2 + y_2$, the process of elimination gives:

$$8\dot{L_1}^2 = (4e_1 - 2L_1^2 + 4h_1)L_2^2$$
$$8\dot{L_1}\dot{L_2} = (4e_2L_2 - 4h_1L_1 - 4c_2)L_2 + L_1L_2(L_1^2 - L_2^2))$$
$$8\dot{L_2}^2 = -4e_1L_2^2 + 4h_1L_1^2 + 8c_2L_1 + 8c_1 - 1/2(L_1^2 - L_2^2)^2 - 1/2(e_1^2 + e_2^2).$$

We stress that we have worked so far in the real domain. Kowalewski works, instead, in the complex domain. Her complex variables are used almost universally in the literature on Kowalewski's top, and a special role is attributed to them. I believe that this role has been overestimated, and my point of view is that one should privilege a tensorial viewpoint where all the coordinates are on the same footing. Nevertheless, it is mandatory at this point to pass to Kowalewski's complex coordinates, in order to set the connection with her paper and to see the function $R(x_1, x_2)$ to appear . Kowalewski uses the variables $x_1 = \frac{1}{2}(L_1 + L_2i)$ and $x_2 = \frac{1}{2}(L_1 - L_2i)$, and the auxiliary functions $\xi_1 = e_1 + e_2i$ and $\xi_2 = e_1 - e_2i$. Accordingly she obtains the expressions

$$-4\dot{x_1}^2 = R(x_1) + (x_1 - x_2)^2\xi_1$$
$$4\dot{x_1}\dot{x_2} = R(x_1, x_2)$$
$$-4\dot{x_2}^2 = R(x_2) + (x_1 - x_2)^2\xi_2.$$

In her notations the functions $R(x_1, x_2)$, $R(x_1)$, and $R(x_2)$ are defined by

$$R(x_1, x_2) = -x_1^2x_2^2 + 2h_1x_1x_2 + c_2(x_1 + x_2) + c_1 - 2h_2$$
$$R(x_1) = -x_1^4 + 2h_1x_1^2 + 2c_2x_1 + c_1 - 2h_2$$
$$R(x_2) = -x_2^4 + 2h_1x_2^2 + 2c_2x_2 + c_1 - 2h_2.$$

To eliminate the variables ξ_1 and ξ_2 from the equations of motion for x_1 and x_2, one disposes of two relations: the first is $\xi_1\xi_2 = 2h_2$; the second follows from $\dot{x_1}^2\dot{x_2}^2 - (\dot{x_1}\dot{x_2})^2 = 0$. Together they imply the linear constraint

$$R(x_2)\xi_1 + R(x_1)\xi_2 + R_1(x_1, x_2) + 2h_2 = 0$$

where the function $R_1(x_1, x_2)$ appears for the first time. From $\dot{x_1}^2\dot{x_2}^2 - (\dot{x_1}\dot{x_2})^2 = 0$ one readily sees that

$$R_1(x_1, x_2)(x_1 - x_2)^2 = R(x_1)R(x_2) - R(x_1, x_2)^2.$$

This equation emphasizes the prominent role of the function $R(x_1, x_2)$, which generates all the other relevant functions of the theory of the top. Therefore, to understand Kowalewski's paper one has to clarify the specific role of this function.

The elimination of the variables ξ_1 and ξ_2 leads to a system of implicit equations for $\dot{x_1}$ and $\dot{x_2}$ sufficiently complicated to give up any hope of further progress. For this reason Kowalewski chooses another route, by introducing a new set of variables. More or less at the middle of the section she writes:

"Au lieu des deux variables x_1 et x_2 introduisons maintenant deux variables nouvelles s_1 et s_2, définies par les équations

$$s_1 = \frac{R(x_1, x_2)}{2(x_1 - x_2)^2} - \frac{\sqrt{R(x_1)}\sqrt{R(x_2)}}{2(x_1 - x_2)^2} + \frac{1}{2}l_1$$

$$s_2 = \frac{R(x_1, x_2) + \sqrt{R(x_1)}\sqrt{R(x_2)}}{2(x_1 - x_2)^2} + \frac{1}{2}l_1; \tag{3}$$

s_1 et s_2 sont, comme on le voit, les deux racines de l'équation algébrique de second degré

$$(x_1 - x_2)^2 \left(s - \frac{1}{2}l_1\right)^2 - R(x_1, x_2)\left(s - \frac{1}{2}l_1\right) - \frac{1}{4}R_1(x_1, x_2) = 0." \tag{4}$$

She does not provide motivations for her choice, but proceeds to prove a remarkable property of this change of variables. Let x_1 and x_2 be two arbitrary functions of the time, and let $\dot{x}_1$ and $\dot{x}_2$ be their time derivatives. Let furthermore S_1 and S_2 be the functions of s defined by

$$S_1 = 4s_1^3 - g_2 s_1 - g_3 \qquad S_2 = 4s_2^3 - g_2 s_2 - g_3$$

with $g_2 = k^2 - c_0^2 + 3l_1^2$ and $g_3 = l_1(k^2 - c_0^2 - l_1^2) + l^2 c_0^2$. Then Kowalewski proves that

$$\frac{ds_1}{\sqrt{S_1}} = +\frac{dx_1}{\sqrt{R(x_1)}} + \frac{dx_2}{\sqrt{R(x_2)}}$$

$$\frac{ds_2}{\sqrt{S_1}} = -\frac{dx_1}{\sqrt{R(x_1)}} + \frac{dx_2}{\sqrt{R(x_2)}}. \tag{5}$$

This equation expresses a property of the Jacobian of the transformation from the coordinates (x_1, x_2) to the coordinates (s_1, s_2) and therefore, at the end, a property of the function $R(x_1, x_2)$ which defines the transformation. This property has attracted the attention of A. Weil and directed the further studies towards the algebro-geometric explanation of the change of variables of Kowalewski. In the last section we shall see that this property has a nice differential-geometric interpretation as well.

Joining together the two results obtained so far (the implicit equations for $\dot{x}_1$ and $\dot{x}_2$, and the property of the Jacobian of the transformation), Kowalewski arrives at her main result. After some computations, she writes:

"d'où il suit

$$0 = \frac{ds_1}{\sqrt{R_1(s_1)}} + \frac{ds_2}{\sqrt{R_1(s_2)}}$$

$$dt = \frac{s_1 ds_1}{\sqrt{R_1(s_1)}} + \frac{s_2 ds_2}{\sqrt{R_1(s_2)}}, \tag{6}$$

$R_1(s)$ étant un polynome du cinquième degré, et les racines de l'équation $R_1(s) = 0$ étant toutes différentes entre elles, les équations différentielles nous conduisent aux fonctions ultraelliptiques".

It is this result of Kowalewski that we intend to analyse in the following sections. Before we remember that the same result has attracted the attention of many authors interested in the Lax representation of dynamical systems. Their goal is to recover the hyperelliptic curve $y^2 = R_1(s)$ as the spectral curve of a suitable Lax matrix with spectral parameter, in order to relate the method of solution of Kowalewski to the process of linearization of the flow on the Jacobian of the spectral curve explained in ([12]) and ([13]). Lax matrices with the required property have been presented in [14]. Other Lax matrices have been constructed for some generalizations of the top, called the Kowalewski tops in two-fields (see ([15]).

3 Exploring the Geometry of the Kowaleski Top

In this section we start to look at the Kowalewski top from the perspective of differential geometry. The aim is to identify three differential geometric structures attached to the top that the algebro-geometric approach of Kowalewski has left in a cone of shadow. The search of these structures is guided by the study of Levi Civita's conditions worked out in the next section. In this section we adopt an informal approach, letting the geometric structures emerge directly from the example.

The first step is to introduce into the picture a second vector field. It is the vector field X_{h_2} associated with the quartic integral of motion discovered by Kowalewski. It is constructed by means of the Hamiltonian structure of the equations of motion of the rigid body (see, for instance, [11]). Its equations are:

$$L_1' = e_1 \dot{L}_1 + e_2 \dot{L}_2$$
$$L_2' = e_2 \dot{L}_1 - e_1 \dot{L}_2$$
$$L_3' = e_1 y_2 - e_2 y_1$$
$$y_1' = \frac{1}{2} y_3 (e_1 L_2 - e_2 L_1)$$
$$y_2' = \frac{1}{2} y_3 (e_1 L_1 + e_2 L_2)$$
$$y_3' = \frac{1}{2} y_1 (e_2 L_1 - e_1 L_2) - \frac{1}{2} y_2 (e_1 L_1 + e_2 L_2). \tag{7}$$

They show that X_{h_2} is a homogeneous vector field of degree 3. The above equations have been written in a semi-explicit form in order to stress one point: the first two components of X_{h_2} belong to the ideal generated by the first two components of X_{h_1}. In other words, the components of the two vector fields satisfy two syzygies, and the coefficients of the syzygies are the functions e_1 and e_2 already encountered before. We shall see, in the last section, that this seemingly odd property is relevant for understanding Kowalewski's paper. It is also worth noticing that this is the only point where the Hamiltonian structure of the equations of motion of the rigid body plays a role. It will never be used afterwards. We stress this point to emphasize that the Hamiltonian scheme of the theory of integrable systems is less important for the solution of the equations of motion than commonly admitted. The basic scheme may be described as follows. A certain number of commuting vector fields are given on a manifold, together with a complementary number of functions which are constant along the vector fields. The level surfaces of the functions define a foliation to which the vector fields are tangent. If the leaves of the foliation carry the additional geometric structure described below the vector fields may be solved by separation of variables. There is no need of any Hamiltonian structure. We shall presently exhibit three remarkable geometric structures attached to the two-dimensional foliation of the Kowalewski top. They are described by a 2-form ω, a tensor field M of type $(1, 1)$, and a metric g respectively. All are related in a surprising way to the function $R(x_1, x_2)$ of Kowalewski.

The 2 form ω The search for the integrals of motion of a dynamical system, described by a vector field X, may be viewed as the search for the 1-forms α which annihilates X and which admits an integrating factor:

$$f\alpha = dI.$$

The potential I of α is the sought integral. This process may be extended to foliations. Let us consider on the phase space of the top a class of homogeneous 2-forms ω, and let us look for those among them that annihilate the Kowalewski foliation: $\omega(X_{h_1}, X_{h_2}) = 0$. The minimal degree of homogeneity worth of attention is 3. The corresponding forms are:

$$\omega = \sum_{j<k,l} a^l_{jk} L_l dL_j \wedge dL_k + \sum_{jk} b_{jk} dL_j \wedge dy_k,$$

where the coefficients a^l_{jk} and b_{jk} are constant. Evaluating these forms on the vector fields of Kowalewski, one readily checks that only four of them annihilate the foliation. Three are trivial and will not be specified. The fourth is

$$\omega = \frac{1}{2}(L_1 dL_2 - L_2 dL_1) \wedge dL_3 + dL_2 \wedge dy_3.$$

The main point is that this 2-form admits the integrating factor $f = L_2^{-2}$, so that

$$f\omega = d\alpha.$$

The potential α is not uniquely defined, but a convenient choice is

$$\alpha = \frac{\dot{L}_1 dL_1 + \dot{L}_2 dL_2}{L_2^2}. \tag{8}$$

This form is a homogeneous potential of ω that belongs to the span of dL_1 and dL_2. The existence of this 1-form is the first information provided by the differential-geometric analysis of the top.

We shall now show that the 1-form α allows to discover a very subtle *differential identity* satisfied by the functions $R(x_1, x_2)$ and $R_1(x_1, x_2)$ used by Kowalewski to build her separation polynomial. To arrive at this identity we need to evaluate the components of α on the basis of vector fields X_{h_1} and X_{h_2}. According to Eq. 8, they are

$$\alpha(X_{h_1}) = \frac{\dot{L}_1{}^2 + \dot{L}_2{}^2}{L_2^2}$$

$$\alpha(X_{h_2}) = \frac{\dot{L}_1 L_1' + \dot{L}_2 L_2'}{L_2^2}.$$

If one evaluates these components in the coordinate system used by Kowalewski, one discovers that

$$\alpha(X_{h_1}) = -\frac{R(x_1, x_2)}{(x1 - x2)^2}$$

$$\alpha(X_{h_2}) = -\frac{1}{2}\frac{R_1(x_1, x_2)}{(x1 - x2)^2} + h_2.$$

So the 1-form α gives the coefficients of the fundamental equation of Kowalewski up to some irrelevant additive constant. More importantly, it provides a new information on these coefficients.

Lemma 3.1 *The functions $R(x_1, x_2)$ and $R_1(x_1, x_2)$ of Kowalewski verify the differential constraint*

$$X_{h_1}\left(\frac{1}{2}\frac{R_1(x_1, x_2)}{(x1 - x2)^2}\right) - X_{h_2}\left(\frac{R(x_1, x_2)}{(x1 - x2)^2}\right) = 0. \tag{9}$$

Proof. Let us denote by $A = \alpha(X))$ and $B = \alpha(Y)$ the values of α on any pair of commuting vector fields tangent to the two-dimensional foliation. Since α is the potential of a 2-form vanishing on the foliation $d\alpha(X, Y) = 0$. Since the vector fields X and Y commute, the Palais formula for the exterior differential of 1-forms gives $X(B) - Y(A) = 0$. The lemma is proved by choosing $X = X_{h_1}$ and $Y = X_{h_2}$.   □

The tensor field M The second remarkable geometric structure defined on the foliation of the Kowalewski top is a tensor field of type $(1, 1)$, henceforth denoted by M. It is generated by the fundamental equation of Kowalewski.

To explain the definition of this tensor field, we need to make a digression on the relation between second-order polynomials and torsionless tensor fields. Let (F, G) be an ordered pair of functions on the phase space of the top, which we identify as the coefficients of the second-order polynomial $Q(u) = u^2 + Fu + G$. Let moreover M be a tensor field of type $(1, 1)$ on the leaves of the Kowalewski foliation, which we specify by its components on the basis X_{h_1} and X_{h_2}:

$$M X_{h_1} = m_1 X_{h_1} + m_2 X_{h_2}$$
$$M X_{h_2} = m_3 X_{h_1} + m_4 X_{h_2}.$$

Let's assume that $\dot{F} G' - F' \dot{G} \neq 0$ almost everywhere on the phase space of the top (the symbols $\dot{F}$ and F' denoting the derivatives of the function F along the vector fields X_{h_1} and X_{h_2} respectively). Then we claim that the functions F and G generate a torsionless tensor field M on the leaves of the foliation.

Lemma 3.2 *There is a unique torsionless tensor field M, on the leaves of the foliation of the top, which admits the polynomial $Q(u)$ as its characteristic polynomial. Its components are given by*

$$m_1 = \frac{\dot{G} G' + G \dot{F} F' - F \dot{F} G'}{\dot{F} G' - F' \dot{G}}$$

$$m_2 = \frac{-\dot{G}^2 + F \dot{F} \dot{G} - G \dot{F}^2}{\dot{F} G' - F' \dot{G}}$$

$$m_3 = \frac{G'^2 + F F' G' + G \dot{F}^2}{\dot{F} G' - F' \dot{G}}$$

$$m_4 = \frac{-\dot{G} G' + F F' G' - G \dot{F} F'}{\dot{F} G' - F' \dot{G}}$$

Proof. If one computes the Nijenhuis torsion of the tensor field M according to the standard procedure, one finds that M is torsionless if and only if its components satisfy the pair of differential equations

$$X_{h_1}(m_1 m_4 - m_2 m_3) + m_2(m_1' + m_4') - m_4(\dot{m}_1 + \dot{m}_4) = 0$$
$$-X_{h_2}(m_1 m_4 - m_2 m_3) + m_1(m_1' + m_4') - m_3(\dot{m}_1 + \dot{m}_4) = 0.$$

By adding the conditions

$$m_1 + m_4 = -F$$
$$m_1 m_4 - m_2 m_3 = G,$$

which express that $Q(u)$ is the characteristic polynomial of M, one obtains a system of four linear algebraic equations for the coefficients of M. The determinant of the coefficients of the linear system is different from zero owing to the inequality satisfied by the functions F and G. Hence the system has a unique solution. This solution is given by the formulas written above. $\qquad\square$

We use the recipe provided by this Lemma to work out the tensor field M associated with the quadratic polynomial defining the fundamental equation of Kowalewski. The result is :

$$m_1 = \frac{R(x_1, x_2)}{(x1 - x2)^2} + \frac{1}{6}h_1$$

$$m_2 = \frac{1}{2}$$

$$m_3 = \frac{1}{2}\frac{R_1(x_1, x_2)}{(x1 - x2)^2}$$

$$m_4 = \frac{1}{6}h_1.$$

Therefore also the tensor field M is closely related to the coefficients $R(x_1, x_2)$ and $R_1(x_1, x_2)$ of the fundamental equation. What does it mean? This question will be answered in Sec. 5. Here we simply show a noticeable consequence of this occurrence.

Proposition 3.3 *The vector fields X_{h_1} and X_{h_2} commute with respect to the deformed commutator defined by the tensor field M associated with the fundamental equation of Kowalewski:*

$$[X_{h_1}, X_{h_2}]_M = 0.$$

Proof. Let us recall that every torsionless tensor field of type $(1, 1)$ defines a deformed commutator on vector fields

$$[X, Y]_M := [MX, Y] + [X, MY] - M[X, Y].$$

By expanding the vector equation $[X_{h_1}, X_{h_2}]_M = 0$ on the basis X_{h_1} and X_{h_2}, one finds that the equation is true if and only if the components of M satisfy the first-order differential constraints

$$X_{h_1}(m_3) = X_{h_2}(m_1)$$
$$X_{h_1}(m_4) = X_{h_2}(m_2).$$

The second equation is manifestly verified since the components m_2 and m_4 are constant. The first equation coincides with the differential constraint discovered in the study of the 1-form α. This Proposition explains the geometric meaning of Lemma 3.1. $\qquad\square$

The metric g The third remarkable geometric structure defined on the foliation of Kowalewski's top is a metric g. In the coordinates (L_1, L_2), it has the particularly nice form

$$g := \frac{dL_1 \otimes dL_1 + dL_2 \otimes dL_2}{L_2^2}. \tag{10}$$

It is characterized by the following two properties:

- It maps the vector field X_{h_1} into the 1-form α: $g(X_{h_1}, X) = \alpha(X)$.
- It makes the tensor field M symmetric: $g(MX, Y) = g(MY, X)$.

It also displays a strict connection with the coefficients of the fundamental equation. Indeed its components on the basis X_{h_1} and X_{h_2} are

$$g(X_{h_1}, X_{h_1}) = -\frac{R(x_1, x_2)}{(x1 - x2)^2}$$

$$g(X_{h_1}, X_{h_2}) = -\frac{1}{2}\frac{R_1(x_1, x_2)}{(x1 - x2)^2} + h_2$$

$$g(X_{h_2}, X_{h_2}) = -2h_2\frac{R(x_1, x_2)}{(x1 - x2)^2}.$$

I guess that this metric has a relevant role in the theory of the Kowalewski top. Probably it points out a connection with the theory of Killing tensors but this connection is unclear to me. Therefore, I restrict myself to signal the existence of this metric without elaborating on it.

Summing up, one can say that the investigation of the two-dimensional invariant foliation of the top according to the procedures of differential geometry reveals a threefold unsuspected role of the basic function $R(x_1, x_2)$ of Kowalewski.

4 Levi Civita and Beyond

In this and the next section we develop a new point of view on the process of separation of variables. It was inspired by the outcomes of the previous investigation of Kowalewski's top, and it aims to provide a coherent interpretation of these outcomes. Two are the major achievements supplied by the new point of view: the first is a new criterion of separability, called the dual form of the Levi Civita conditions of 1904; the second is a new algorithm for the search of separation coordinates. It fits quite well with the work of Kowalewski, and for this reason it is called the method of Kowalewski's conditions. In this section we deal with the criterion.

To quickly explain the content of this criterion, let us consider a Liouville integrable system on a symplectic manifold of dimension $2n$. The system is assigned by giving n independent functions $(H_1, \ldots, H_n)$ which are in involution. These functions are known in a system of canonical coordinates (x_j, y_j) fixed in advance. By assumption they do not verify the Levi Civita separability conditions in these coordinates. The problem of the search for separation coordinates is the problem of finding a new system of canonical coordinates

$$u_j = \Phi_j(x_1, \ldots, x_n, y_1, \ldots, y_n)$$
$$v_j = \Psi_j(x_1, \ldots, x_n, y_1, \ldots, y_n)$$

such that the functions H_a verify Levi Civita's separability conditions when written in the new coordinates. To imply this property the functions Φ_j must be quite special. In this section we prove that these functions necessarily obey a system of n second-order differential constraints. These constraints have the form of n polynomial equations on the first and second derivatives of the functions Φ_j along the Hamiltonian vector fields associated with the functions H_a. They are called the dual form of the Levi Civita conditions because they work as Levi Civita's conditions but in the opposite sense: while the Levi Civita conditions demand to consider the derivatives of the functions H_a with respect to the coordinates u_j, the new criterion of separability demands to consider the derivatives of the coordinates u_j along the Hamiltonian vector fields associated with the functions H_a. The proof of the existence of the new differential constraints rests on the following geometric interpretation of Levi Civita's separability conditions.

Proposition 4.1 *Consider a separable Liouville integrable system, that is a Liouville integrable system plus a system of separation coordinates* (u_j, v_j). *Assume that the functions* H_a *depend on all the momenta* v_j, *so that the derivatives of the functions* H_a *with respect to* v_j *do not vanish. (This assumption is common within the theory of Levi Civita conditions). Hence, the Lagrangian foliation of the system is equipped with an infinite number of tensors fields* M *of type* $(1, 1)$ *which satisfy the following two conditions*

- *The Nijenhuis torsion of* M *vanishes.*
- *The Hamiltonian vector fields associated with the functions* H_a *commute in pair with respect to the deformed commutator defined by* M:

$$[X_{H_a}, X_{H_b}]_M = 0. \tag{11}$$

Proof. As is known, a set of function H_a which are both in involution and separable (in the sense that they satisfy the Levi Civita separability conditions) are necessarily in separated involution. This means that they satisfy the stronger involutivity conditions

$$\frac{\partial H_a}{\partial u_j}\frac{\partial H_b}{\partial v_j} - \frac{\partial H_a}{\partial v_j}\frac{\partial H_b}{\partial u_j} = 0$$

in the separation coordinates. This property has several important consequences. The main is that one can introduce the auxiliary functions

$$\lambda_j := \frac{\frac{\partial H_a}{\partial u_j}}{\frac{\partial H_a}{\partial v_j}}$$

which do not depend on the choice of the function H_a. Related to them there are the vector fields

$$X_j := \frac{\partial}{\partial u_j} - \lambda_j \frac{\partial}{\partial v_j}.$$

By the condition of separated involution, these vector fields are tangent to the leaves of the Lagrangian foliation. By the linear independence of the vector fields $\frac{\partial}{\partial u_j}$, they form a basis in the tangent space to the leaves of the foliation. By the Levi Civita conditions they commute. So the final result is that the leaves of the Lagrangian foliation associated with a separable Liouville integrable system admit a distinguished basis of commuting vector fields X_j.

Let us use this basis to define the tensor field M by assigning its eigenvalues and its eigenvectors. We agree that M admits the vector fields X_j as eigenvectors, and that the corresponding eigenvalues μ_j are functions of the conjugate canonical coordinates (u_j, v_j) associated with X_j. By this agreement

$$X_k(\mu_j) = 0 \qquad for \quad k \neq j.$$

We claim that this tensor field satisfies the two conditions stated in the Lemma. The proof is a simple computation. For the first condition (the vanishing of the torsion of M), one has to evaluate the torsion on the basis formed by the vector fields X_j by using the equation $M X_j = \mu_j X_j$. One readily recognizes that all the terms vanish owing to the assumption made on the eigenvalues. For the second condition, one preliminarily expands the Hamiltonian vector field X_{H_a} on the basis X_j :

$$X_{H_a} = \sum_j \frac{\partial H_a}{\partial v_j} X_j.$$

Then one evaluates the deformed commutator $[X_{H_a}, X_{H_b}]_M$. By using the spectral definition of the tensor field M once again, one finds that the deformed commutator is the sum of three groups of terms: two of them vanish because of the condition of separated involution; the third vanishes because of the assumption on the eigenvalues. □

Let us make three comments. The first concerns the arbitrariness in the definition of the tensor field M. One may notice that this arbitrariness reflects a similar degree of arbitrariness in the definition of the separation coordinates. Indeed, it is clear that any canonical transformation like

$$\bar{u}_j = \mu_j(u_j, v_j)$$
$$\bar{v}_j = v_j(u_j, v_j)$$

produces another system of separation coordinates. One may set up a 1:1 correspondence among separation coordinates and tensor fields M by imposing the additional constraint

$$\mu_j := u_j. \tag{12}$$

Under this condition, the tensor field M gives a faithful representation of the coordinate system. This condition will be constantly enforced henceforth. The second comment is that the above result shows that the example of the Kowalewski top is not isolated. All separable Liouville integrable systems share with the Kowalewski top the property that the leaves of their Lagrangian foliation carry a tensor field M. The third comment is about the intrinsic character of the geometric conditions satisfied by the tensor field M. They are tensorial and can be written in an arbitrary coordinate system. This property is remarkable, since we started from Levi Civita's conditions, which are nontensorial. In a sense, the Proposition provides a tensorial formulation of the nontensorial Levi Civita conditions.

The two conditions on M (the vanishing of the torsion and the vanishing of the deformed commutator) impose severe restrictions on the eigenvalues of M. The next step is to work out these restrictions explicitly. Although the argument which follows is completely general, we shall confine ourselves to the case $n = 2$ for brevity, and we agree to denote by the symbols f_p and f_s the derivatives of any function f along the vector fields X_{H_1} and X_{H_2} spanning the two-dimensional invariant foliation.

As in the example of the top, we expand the tensor field M on the basis of the Hamiltonian vector fields

$$M X_{H_1} = m_1 X_{H_1} + m_2 X_{H_2}$$
$$M X_{H_2} = m_3 X_{H_1} + m_4 X_{H_2},$$

and we notice that the vanishing of the torsion of M entails that the components (m_1, m_2, m_3, m_4) can be computed as functions both of the eigenvalues of M and of the coefficients of its characteric polynomial. If we call u_1 and u_2 the eigenvalues of M, and F and G the coefficients of its characteristic polynomial, we find

$$m_1 = \frac{u_1 u_{1p} u_{2s} - u_2 u_{1s} u_{2p}}{u_{1p} u_{2s} - u_{1s} u_{2p}}$$

$$m_2 = \frac{u_2 u_{1p} u_{2p} - u_1 u_{1p} u_{2p}}{u_{1p} u_{2s} - u_{1s} u_{2p}}$$

$$m_3 = \frac{u_1 u_{1s} u_{2s} - u_2 u_{1s} u_{2s}}{u_{1p} u_{2s} - u_{1s} u_{2p}}$$

$$m_4 = \frac{u_2 u_{1p} u_{2s} - u_1 u_{1s} u_{2p}}{u_{1p} u_{2s} - u_{1s} u_{2p}}$$

and

$$m_1 = \frac{+GpGs + GFpFs - FFpGs}{FpGs - FsGp}$$

$$m_2 = \frac{-Gp^2 + FFpGp - GFp^2}{FpGs - FsGp}$$

$$m_3 = \frac{+Gp^2 + FFsGs + GFp^2}{FpGs - FsGp}$$

$$m_4 = \frac{-GpGs + FFsGs - GFpFs}{FpGs - FsGp}.$$

The second representation has been proved in Lemma 3.2; the first representation follows from the second one, by setting $F = -(u_1 + u_2)$ and $G = u_1 u_2$.

Let us now recall that the vanishing of the deformed commutator $[X_{H_1}, X_{H_2}]_M = 0$ requires

$$m_{3p} - m_{1s} = 0$$

$$m_{4p} - m_{2s} = 0.$$

By inserting the above representations of the coefficients of M into these constraints, one obtains the desired conditions on the eigenvalues of M and on the coefficients of its characteristic polynomial. We shall specify them in a moment. Before we interpret these conditions from the standpoint of the theory of separation of variables. Imagine a two-dimensional separable Liouville integrable system, defined by a pair of functions H_1 and H_2. Let us assume that the separation coordinates are defined as the roots of a quadratic equation $Q(u) = u^2 + Fu + G = 0$, as in the example of Kowalewski. According to the geometric interpretation of the Levi Civita separability conditions, the leaves of the Lagrangian foliation of the separable system are endowed with a tensor field M whose eigenvalues are the separation coordinates and whose characteristic polynomial is $Q(u)$. This tensor field satisfies the two conditions studied above. Hence the separation coordinates verify the constraints following from these conditions. The conditions just obtained provide, therefore, a new criterion of separability, written either on the roots or on the coefficients of the polynomial which defines the separation coordinates.

Criterion The separation coordinates u_1 and u_2 of a two-dimensional separable Liouville integrable system satisfy the dual Levi Civita separability conditions

$$u_{1p}u_{2p}u_{1ss} - (u_{1p}u_{2s} + u_{1s}u_{2p})u_{1ps} + u_{1s}u_{2s}u_{1pp} = 0$$

$$u_{1p}u_{2p}u_{2ss} - (u_{1p}u_{2s} + u_{1s}u_{2p})u_{2ps} + u_{1s}u_{2s}u_{2pp} = 0. \tag{13}$$

The coefficients of the separating polynomial of a two-dimensional separable Liouville integrable system satisfy the dual Levi Civita separability conditions

$$(-G_s^2 + FF_sG_s - GF_s^2)F_{pp}$$
$$+ (2G_pG_s - FF_pG_s - FF_sG_p + 2GF_pF_s)F_{ps}$$
$$+ (-G_p^2 + FF_pG_p + GF_p^2)F_{ss} = 0$$
$$(-G_s^2 + FF_sG_s - GF_s^2)G_{pp}$$
$$+ (2G_pG_s - FF_pG_s - FF_sG_p + 2GF_pF_s)G_{ps}$$
$$+ (-G_p^2 + FF_pG_p + GF_p^2)G_{ss} = ((F_pG_s - F_sG_p))^2. \tag{14}$$

As clarified by the argument leading to the criterion, the criterion has a large domain of application. For instance, the first equation is verified by the coordinates s_1 and s_2 of Kowalewski, while the second equation is verified by the coefficients of her fundamental equation. It can, therefore, be considered a suitable point of departure for the search of separation coordinates. Nevertheless, it does not answer two important questions: How does one solve these conditions? Are these conditions sufficient as well? Both questions will be addressed in the next section. Before, we add the remark that there is a second approach to the dual Levi Civita separability conditions which avoids the use of the Hamiltonian setting. It is based on the theory developed by Staeckel [5], but its exposition goes outsides the limit of the present paper.

5 The Method of Kowalewski's Conditions

The idea that inspires the method of Kowalewski's conditions is to lower the order of the differential equations satisfied by the separation coordinates. It rests on the remark that there is a special class of solutions of the dual Levi Civita conditions which are selected by a system of *first-order* differential equations. These first-order equations are the Kowalewski conditions.

Let us recall the geometrical setting. It consists of three elements: a two-dimensional foliation, the equations of the foliation (that is a set of independent functions which are constant on the leaves of the foliation), and a pair of commuting vector fields tangent to the foliation. To be definite, we assume that there are two commuting vector fields X_1 and X_2 and four functions (H_1, H_2, H_3, H_4) as in the example of the top. The vector fields are not required to be Hamiltonian. The derivatives of the function f along X_1 and X_2 are denoted by f_p and f_s as before.

Definition 5.1 (Kowalewski's conditions) *Let T and D be two functions such that $T_p D_s - T_s D_p \neq 0$. They are said to verify the Kowalewski conditions if*

$$T_s - D_p = 0$$
$$D_s - TD_p + DT_p = 0.$$

Furthermore, let A and B be another pair of functions, with $A \neq 0$. They are said to verify the auxiliary system attached to the solution (T, D) of the Kowalewski conditions if

$$A_s - B_p = 0$$
$$B_s - TB_p + DA_p = 0.$$

There are four reasons to consider the Kowalewski conditions and the related auxiliary system. We now present them in the form of four separate claims, each dealing with a different aspect of the method of Kowalewski's conditions. To state them, we agree to denote by $Q(w) := w^2 + Tw + D$ the quadratic polynomial associated with the solution (T, D) of the Kowalewski conditions; by w_1 and w_2 its roots; by L the unique torsionless tensor field of type $(1, 1)$, on the leaves of the two-dimensional foliation, having $Q(w)$ as characteristic polynomial. Before this tensor field was called M, but from now on, for clarity, we shall call L the tensor field associated with the solutions of Kowalewski's conditions, and M the tensor field associated with the solutions of the dual Levi Civita conditions.

The first claim justifies our interest in the Kowalewski conditions.

Proposition 5.2 *Any solution of the Kowalewski condition is a solution of the dual Levi Civita conditions.*

Proof. Consider the Kowalewski conditions, and derive them along the vector fields X_1 and X_2. The result is a system of four polynomial relations among T, D and their first- and second-order derivatives. Add these four relations to Kowalewski's equations themselves. This gives a set of six polynomial relations. Consider now the dual Levi Civita conditions on the coefficients T and D of the polynomial $Q(w)$, and regard them as a pair of polynomial relations on T, D, and their first- and second-order derivatives. The proof is completed by noticing that the last two polynomials belong to the ideal generated by the six polynomials engendered by Kowalewski's conditions.

□

The second claim points out the distinctive properties of the solutions of Kowalewski's conditions.

Proposition 5.3 *The roots w_1 and w_2, and the tensor field L associated with any solution of Kowalewski's condition enjoy the following properties:*

- *The roots w_1 and w_2 verify the first-order constraints (called the second form of the Kowalewski conditions)*

$$w_{1s} + w_2 w_{1p} = 0$$
$$w_{2s} + w_1 w_{2p} = 0.$$

- *The roots w_1 and w_2 verify the dual Levi Civita conditions.*
- *The tensor field L has the simplified form*

$$LX_1 = -TX_1 + X_2$$
$$LX_2 = -DX_1.$$

Proof. The first property is proved by inserting the relations $T = -(w_1 + w_2)$ and $D = w_1 w_2$ into the Kowalewski conditions. The second property follows from the first property: it is proved by the same technique used to prove the previous Proposition. The third property is proved by evaluating the components of L as rational functions of T, D, and their first derivatives. Modulo the Kowalewski conditions, these rational functions take the simplified form shown in the Lemma.   □

The third claim explains why Kowalewski's conditions are interesting for the search of separation coordinates.

Proposition 5.4 *The roots w_1 and w_2 associated with any solution (T, D) of Kowalewski's conditions are separation coordinates for the vector fields X_1 and X_2 tangent to the foliations.*

Proof. This Proposition has been proved in [17]. We repeat here briefly the argument leading to the conclusion. The conditions $w_{1s} + w_2 w_{1p} = 0$ and $w_{2s} + w_1 w_{2p} = 0$ imply the validity of the following expansions:

$$\psi_1 \frac{\partial}{\partial w_1} = X_2 + w_1 X_1$$

$$\psi_2 \frac{\partial}{\partial w_2} = X_2 + w_2 X_1.$$

The vector fields X_1 and X_2 commute as well as the vector fields $\frac{\partial}{\partial w_1}$ and $\frac{\partial}{\partial w_2}$. This property of commutativity entails that the function ψ_1 depends only on w_1, and that the function ψ_2 depends only on w_2. To close the proof it is sufficient to write the previous vector expansions in components. For the vector fields X_1 and X_2 one obtains the equations

$$\frac{w_{1p}}{\psi_1} + \frac{w_{2p}}{\psi_2} = 0$$

$$\frac{w_1 w_{1p}}{\psi_1} + \frac{w_2 w_{2p}}{\psi_2} = 1,$$

and

$$\frac{w_{1s}}{\psi_1} + \frac{w_{2s}}{\psi_2} = 1$$

$$\frac{w_1 w_{1s}}{\psi_1} + \frac{w_2 w_{2s}}{\psi_2} = 0$$

respectively. Assuming that ψ_1 and ψ_2 are algebraic functions of w_1 and w_2 respectively, one recognizes in these equations the Euler equations associated with the algebraic curves defining ψ_1 and ψ_2. This is the mechanism of separation of variables induced by the Kowalewski conditions. $\qquad\square$

The fourth claim, finally, gives a geometrical interpretation of the Kowalewski conditions.

Proposition 5.5 *Let K be the unique tensor field of type $(1, 1)$ on the leaves of the two-dimensional foliation characterized by the following two properties:*

- *It maps the vector field X_1 into X_2: $K X_1 = X_2$.*
- *The functions T and D, solutions of Kowalewski's conditions, are the trace and the determinant of K: $T := tr K$ and $D := det K$:*

Then the differentials of the trace T and of the determinant D satisfy the recursion relation

$$K dT = dD.$$

Proof. Consider the tensor field $K = L + T Id$. Since $-T$ and D are the trace and the determinant of L according to Lemma 5.3, it follows that T and D are the trace and the determinant of K. Since $L X_1 = -T X_1 + X_2$ it is obvious that $K X_1 = X_2$. To prove the last property, let us evaluate the 1-form $K dT - dD$ on the vector fields X_1 and X_2. One obtains:

$$K dT(X_1) - dD(X_1) = dT(K X_1) - dD(X_1) = dT(X_2) - dD(X_1)$$
$$= T_s - D_p.$$
$$K dT(X_2) - dD(X_2) = dT(K X_2) - dD(X_2) = dT(T X_2 - D X_1) - dD(X_2)$$
$$= T T_s - D T p - D_s.$$

These identities prove the claim. $\qquad\square$

To complete the discussion, we add (without proof) three more claims which show the role of the auxiliary system, and clarify the relations between the dual Levi Civita conditions (of the previous section) and the Kowalewski conditions (of this section).

The first claim shows how to construct solutions of the dual Levi Civita conditions from solutions of the Kowalewski conditions.

Proposition 5.6 *Let (T, D) and (A, B) be any solution of the Kowalewski conditions and of the related auxiliary system. Then the functions*

$$F = AT - 2B$$
$$G = A^2 D - ABT + B^2$$

solve the dual Levi Civita conditions. Furthermore, any solution of the dual Levi Civita conditions may be represented in this way.

The second claim shows the opposite relation.

Proposition 5.7 *Let (F, G) be any solution of the dual Levi Civita conditions, and let (m_1, m_2, m_3, m_4) be components of the tensor field M associated with the quadratic polynomial $Q(u) = u^2 + Fu + G$. Assume that $m_2 \neq 0$. Then the functions*

$$A = m_2$$
$$B = m_4$$
$$T = \frac{(m_4 - m_1)}{m_2}$$
$$D = -\frac{m_3}{m_2}$$

are solutions of the Kowalewski conditions and of the auxiliary system. Furthermore, any solution of the Kowalewski conditions may be obtained in this way.

The third claim, finally, explains the relation between the roots of the polynomials $Q(u) = u^2 + Fu + G$ and $Q(w) = w^2 + Tw + D$.

Proposition 5.8 *Let (F, G) and (T, D, A, B) be two related solutions of the dual Levi Civita conditions and of the Kowalewski conditions:*

$$F = AT - 2B$$
$$G = A^2 D - ABT + B^2$$

Then the roots u_1 and u_2 are related to the roots w_1 and w_2 according to

$$u = Aw + B.$$

Moreover: the root u_1 is a function only of w_1; the root u_2 is a function only of w_2. Hence the roots (u_1, u_2) are separation coordinates for the vector fields X_1 and X_2 as well.

It is worth noticing that according to the last Proposition no other restiction on the coordinates u_1 and u_2 is required in order to guarantee that they are separation coordinates. Hence the dual Levi Civita conditions are necessary and sufficient for two-dimensional invariant foliations.

Collecting the information split among the different claims, one arrives at the present algorithm for the construction of the separation coordinates of a pair of commuting vector fields X_1 and X_2 satisfying the assumptions of the geometric scheme adopted in this paper.

Algorithm To construct separation coordinates for the vector fields X_1 and X_2, the first step is to find a solution (T, D) of Kowalewski's conditions

$$T_s - D_p = 0$$
$$D_s - TD_p + DT_p = 0. \tag{15}$$

Then the roots w_1 and w_2 of the quadratic equation

$$Q(w) = w^2 + Tw + D = 0 \tag{16}$$

are a first system of such coordinates. In this system of coordinates the equations of motion have the form

$$\frac{w_{1p}}{\psi_1} + \frac{w_{2p}}{\psi_2} = 0$$
$$\frac{w_1 w_{1p}}{\psi_1} + \frac{w_2 w_{2p}}{\psi_2} = 1,$$

and

$$\frac{w_{1s}}{\psi_1} + \frac{w_{2s}}{\psi_2} = 1$$
$$\frac{w_1 w_{1s}}{\psi_1} + \frac{w_2 w_{2s}}{\psi_2} = 0$$

respectively. If the functions ψ_1 and ψ_2 are algebraic functions defined by the polynomial equations

$$P_1(w_1, \psi_1) = 0 \qquad P_2(w_2, \psi_2) = 0$$

the above equations of motion are Euler type equations related to the algebraic curves defined by the above polynomials. Successively, one may look at the solutions of the auxiliary system

$$A_s - B_p = 0$$
$$B_s - TB_p + DA_p = 0. \tag{17}$$

associated to the solution found at the previous step. With any solution of the auxiliary system one may construct a new system of separation coordinates u_1 and u_2 according to

$$u = Aw + B. \tag{18}$$

The new coordinates are the roots of the quadratic equation

$$Q(u) = u^2 + Fu + G = 0 \tag{19}$$

whose coefficients are related to the solutions of Kowalewski's conditions according to

$$F = AT - 2B$$
$$G = A^2 D - ABT + B^2 \tag{20}$$

This algorithm will be applied, in the next section, to the Kowalewski top. The aim is to show that the few remarks on the geometry of the top made in Sec. 3 lead quickly to the fundamental equation of Kowalewski.

6 Rediscovering the Coordinates of Kowalewski

According to Propositions 5.4 and 5.5 of the previous section, the problem of constructing a pair of separation coordinates w_1 and w_2 for the top is equivalent to the problem of constructing a tensor field K, on the leaves of the two-dimensional invariant foliation, such that

$$K X_{h_1} = X_{h_2}$$
$$K d(tr(K)) = d(det(K)),$$

We now show how to construct this tensor field.

We use two pieces of information readily available from the study of the vector fields X_{h_1} and X_{h_2}. The first is provided by the syzygies

$$L_1' = e_1 \dot{L}_1 + e_2 \dot{L}_2$$
$$L_2' = e_2 \dot{L}_1 - e_1 \dot{L}_2$$

They tell us that the tensor field K defined by

$$K dL_1 = e_1 dL_1 + e_2 dL_2$$
$$K dL_2 = e_2 dL_1 - e_1 dL_2$$

is symmetric and verifies the first condition: $KX_{h_1} = X_{h_2}$. The second is provided by the 1-form α. It tells us that the functions

$$\alpha(X_{h_1}) = \frac{\dot{L_1}^2 + \dot{L_2}^2}{L_2^2}$$

$$\alpha(X_{h_2}) = \frac{\dot{L_1}L_1' + \dot{L_2}L_2'}{L_2^2}.$$

verify the first Kowalewski condition

$$X_{h_1}\left(\frac{\dot{L_1}L_1' + \dot{L_2}L_2'}{L_2^2}\right) - X_{h_2}\left(\frac{\dot{L_1}^2 + \dot{L_2}^2}{L_2^2}\right) = 0.$$

This information can be elaborated as follows. First let us notice that the most general symmetric tensor field K satisfying the condition $KX_{h_1} = X_{h_2}$ is:

$$KdL_1 = e_1 dL_1 + e_2 dL_2 + f(\dot{L_2}^2 dL_1 - \dot{L_1}\dot{L_2} dL_2)$$

$$KdL_2 = e_2 dL_1 - e_1 dL_2 + f(-\dot{L_1}\dot{L_2} dL_1 + \dot{L_1}^2 dL_2),$$

where f is an arbitrary function. Then let us notice that the trace and the determinant of this tensor field

$$T = f(\dot{L_1}^2 + \dot{L_2}^2)$$

$$D = f(\dot{L_1}L_1' + \dot{L_2}L_2') - (e_1^2 + e_2^2),$$

coincide with the components of the 1-form α (up to an irrelevant additive constant) if we choose $f = L_2^{-2}$. Let us make this choice. We know that the functions T and D verify already the first Kowalewski condition. It remains only to check the second condition. This can be easily done. Then, we can claim that the roots of the quadratic polynomial $Q(w) = w^2 + Tw + D$ are separation coordinates for the top, without computing the coordinates explicitly , and without writing the equations of motion in the new coordinates. We know this property *a priori* as a consequence of the Kowalewski conditions.

The study of the tensor field K provides another interesting result. Let us pass to the coordinates x_1 and x_2 of Kowalewski, and let us represent the tensor field K in these coordinates. The result is

$$Kdx_1 = -\frac{R(x_1, x_2)}{(x_1 - x2)^2}dx_1 + \frac{R(x_1)}{(x_1 - x_2)^2}dx_2$$

$$Kdx_2 = \frac{R(x_2)}{(x_1 - x_2)^2}dx_1 - \frac{R(x_1, x_2)}{(x_1 - x_2)^2}dx_2.$$

This is a key formula. It reveals that the function $R(x_1, x_2)$ and its allied functions are the components of the tensor field K. Therefore, all these functions are united into a single geometric object. Moreover it becomes clear that the fundamental equation of Kowalewski is just the characteristic equation of the tensor field K. These outcomes are a concrete illustration of the claim that the separation coordinates are strictly related to a geometric structure possessed by the leaves of the invariant foliation of the top.

As a final remark, let us remember that the tensor field K satisfies the condition: $KdT = dD$. This is a differential condition which must pass to the function $R(x_1, x_2)$ defining the components of K. One readily proves that this function must satisfy the partial differential equations

$$\frac{1}{2}\frac{\partial R(x_1, x_2)}{\partial x_2} + \frac{1}{4}\frac{dR(x_1)}{dx_1} + R(x_1, x_2) - R(x_1) = 0$$

$$\frac{1}{2}\frac{\partial R(x_1, x_2)}{\partial x_1} + \frac{1}{4}\frac{dR(x_2)}{dx_2} - R(x_1, x_2) + R(x_1) = 0$$

They are the last identities we want to emphasize. They entail the basic identities (5), used by Kowalewski to linearize the flow of the top. These identities have thus received two different types of interpretations. From the algebro-geometric standpoint they are the trace of an addition formula for elliptic functions. From the differential geometric standpoint they are the condition KdT = dD. This merging of the two points of view is an interesting fact that deserves further study.

Acknowledgements. This paper is dedicated to Emma Previato on the occasion of her 65th birthday. I like to notice that Emma has studied in Padova, as did Tullio Levi Civita, and that she has done important work in the field of algebraic geometry, as did Sophie Kowalewski.

References

[1] Kowalevski S., *Sur le problème de la rotation d'un corps solide autour d'un point fixe* , Act. Math., 12, (1889), 177–232.

[2] Kowalevski S., *Sur une propriété du système d'équations différentielles définissant la rotation d'un corps solide autour d'un point fixe*, Act. Math., 14, (1890).

[3] Kotter F., *Sur le cas traité par M.me Kowalevski de rotation d'un corps solide autour d'un point fixe*, Act. Math., 17, (1892), 27–263.

[4] Lyapunov A., *On a property of the differential equations of the problem of motion of a rigid body having a fixed point*, Kharkhov Math. Soc. Trans, 4,(1894).

[5] Staeckel P., *Uber die Integration der Hamilton-Jacobischen Differentialgleichung mittelst separation der Variabeln*, Habilitationsschrift., Halle, (1891).

[6] Levi Civita T., *Sull'integrazione della equazione di Hamilton-Jacobi per separazione di variabili*, Math. Annalen., 59, (1904), 383–397.

[7] Golubev V.V., *Lectures on integration of the equations of motion of a rigid body about a fixed point*, Israel program for scientific translations, Haifa, (1960).

[8] Weil A., *Euler and the Jacobians of elliptic curves*, Arhitmetic and Geometry: Volume dedicater to I.R.Shafarevich,M.Artin ed. (1983), 353–359.

[9] Horozov E. and van Moerbeke P., *The full geometry of Kowalevski's top and (1,2)- abelian surfaces* , Comm. Pure and Appl. Math., 42, (1989), 357–407.

[10] Adler M., van Moerbeke P. and Vanhaecke P., *Algebraic integrability, Painlevé geometry and Lie algebras*, Springer, (2013).

[11] Jurdjevic V., *Integrable Hamiltonian systems on Lie groups: Kowalewski type*, Annals of Math., 150, (1999), 605–644.

[12] Griffiths P. A., *Linearizing flow and a cohomological interpretation of Lax equations*, Amer. J. of Math., 107, (1985), 1445–1483.

[13] Sklyanin E. K., *Separation of variables- New trends*, Progr. Theor. Phys., 118, (1995), 35–60.

[14] Adler M., van Moerbeke P., *The Kowalewski and Henon-Heiles Motions as Manakow Geodesic Flows on SO(4)- a Two-Dimensional family of Lax Pairs*, Commun. Math. Phys., 113 , (1988),659–700.

[15] Belokolos E.D.,Bobenko A.I., Enol'ski V.Z., Its A.R. and Matveev V.B., *Algebro-geometric approach to nonlinear integrable equations*, Springer , Nonlinear Dynamics, (1994).

[16] Audin M. and Silhol R., *Variétés abéliennes réelles et toupie de Kowalevski*, Compos. Math., 87, (1993), 153–229.

[17] Audin M., *Two notions of integrability* Differential equations and quantum groups, www-irma-u-strasbg.fr, (2006).

[18] Magri F. and Skripnik T., *The Clebsch system*, arXiv Math: 1512.04872, (2015).

F. Magri
Dipartimento di Matematica ed Applicazioni, Universita' di Milano Bicocca,
20125 Milano, Italy

14

The Calogero-Françoise Integrable System: Algebraic Geometry, Higgs Fields, and the Inverse Problem

Steven Rayan, Thomas Stanley and Jacek Szmigielski

Dedicated to Emma Previato on the occasion of her 65th birthday.

Abstract. We review the Calogero-Françoise integrable system, which is a generalization of the Camassa-Holm system. We express solutions as (twisted) Higgs bundles, in the sense of Hitchin, over the projective line. We use this point of view to (a) establish a general answer to the question of linearization of isospectral flow and (b) demonstrate, in the case of two particles, the dynamical meaning of the theta divisor of the spectral curve in terms of mechanical collisions. Lastly, we outline the solution to the inverse problem for CF flows using Stieltjes' continued fractions.

1 Introduction

The idea of viewing certain non-linear problems as arising from *isospectral* deformations of linear operators goes back to P. D. Lax, who, in [25], connected the existence of infinitely-many integrals of motion for the Korteweg-de Vries equation (KdV) with an isospectral deformation of a linear operator L_u parametrized by a solution u to the KdV equation. More concretely, introducing the one-dimensional (in x) Schrödinger operator $L_u = -D^2 + u(x, t)$ he observed that the KdV equation $u_t + uu_x + u_{xxx} = 0$ was equivalent to the operator equation

$$\dot{L}_u = [P, L_u], \tag{1}$$

where P is a certain third-order differential operator depending on u and its x derivative. This line of research was taken up by J. Moser, especially in the context of Hamiltonian systems with finitely many degrees of freedom [31, 32]. Some of these systems, like the finite Toda lattice or an n-dimensional rigid body [28], had a finite dimensional Lax pair and the dynamical problem despite being isospectral from the outset had no a priori relation to complex geometry even though a deeper analysis in each case was unequivocally pointing to the existence of such a connection. This connection was established within a Lie-theoretic context in [2] leading to many years of fruitful interaction between Lie theory (mostly Kac-Moody Lie algebras) and the theory of integrable systems. One of the decisive contributions to this theme was E. Previato's paper with M. Adams and J. Harnad [1], in some sense complementing the work of J. Moser [32].

In the mid-1970s, yet another class of integrable, finite-dimensional systems was obtained from reductions of Lax integrable PDEs, i.e. famous finite-zone potentials [33, 15, 26, 30, 24], and led to the appearance of invariant *spectral curves*. This had become a dominant research direction for many years to come and Emma beautifully reviewed this vast area in her 1993 lecture notes [34] placing emphasis on the old paper of Bourchnal and Chaundy [10].

The present paper is about a different occurrence of spectral curves, also due to the reduction from a PDE given by a Lax pair equation, but the reduction is in smoothness. We now turn to describing schematically the situation, leaving the details to Section 2. The Camassa-Holm equation [12] (CH)

$$m_t + 2mu_x + um_x = 0, \quad m = u - u_{xx}.$$

was invented as a model for nonlinear water waves with nonlinear dispersion. It has a Lax pair

$$L = -D^2 + \frac{1}{4} - \lambda m, \qquad P = \left(\frac{1}{2\lambda} + u\right) D + \frac{u_x}{2},$$

from which it is clear that the central object in this endeavour is m, while u should be thought of as a potential producing m. Even though the Lax equation has to be slightly modified (see Section 3), the computation is elementary for smooth m. However, this is not so if m is non-smooth, for example if m is a discrete measure, because then the Lax equation involves a multiplication of distributions with overlapping singular supports and this leads to certain subtle phenomena (see i.e. [13]).

It is the presence of spectral curves, which in this setting arise out of the reduction from a smooth m to a discrete measure m, that brings algebraic geometry into play. To bring to bear this aspect fully, we recall that the work of Adams-Harnad-Previato [1] is part of a sequence of results in the 1980s and early 1990s that translate classical integrable systems theory into the framework of complex algebraic geometry. At the centre of this theme is the *Hitchin system*, discovered in [20] as an algebraically completely integrable Hamiltonian system defined on an enlargement of the cotangent bundle of the moduli space of stable holomorphic bundles on a fixed Riemann surface X of genus $g \geq 2$. The entire system has a modular interpretation, namely as the moduli space of stable "Higgs bundles" on X, which consist of holomorphic vector bundles together with 1-form-valued maps called "Higgs fields". Higgs bundles themselves arise as solutions to a dimensional reduction of the self-dual Yang-Mills equations in four dimensions, as in [19]. The Hamiltonians for this system have a wonderfully explicit description in terms of characteristic data of the Higgs field.

Versions of the Hitchin system arise in lower genus, too. To accommodate surfaces X with $g = 0$ or $g = 1$, one can make one of two (related) modifications. On the one hand, the Higgs field can be allowed to take values in a line bundle other than the bundle of 1-forms, leading to an integrable system studied in genus 0 by P. Griffiths [18] and A. Beauville [6] and in arbitrary genus by E. Markman [29]. Retroactively, we refer to this as a *twisted Hitchin system*, as it is formally a moduli space of Higgs bundles but with Higgs fields that have been twisted to take values in a line bundle of one's choosing. The Hamiltonians have the same description as in the original Hitchin system but, generally speaking, the resulting integrable system is *superintegrable*: it contains more Poisson-commuting Hamiltonians than are necessary. In more geometric terms, the total space of the system is a torus fibration in which the base typically has dimension larger than that of the fibre. The other modification is to puncture X at finitely-many points and to allow the Higgs field to develop poles at these points, as in [36, 9, 8]. Typically, one asks for the residues of the Higgs field at the poles to satisfy a certain Lie-theoretic condition, such as being semisimple. This scheme has the virtue of preserving certain desirable properties of the original Hitchin system, such as the existence of a holomorphic symplectic form. (In contrast, the twisted Hitchin systems generally fail to be globally symplectic and possess a family of degenerate Poisson structures that depend on a choice of divisor, as in [29].)

A folklore belief is that every completely integrable system should be realizable as a Hitchin system of some kind, for some choice of Riemann surface X. If true, this has the advantage of providing a systematic origin for spectral curves, namely as branched covers of the Riemann surface X. A natural question is: when and how can a particular integrable system be identified with a Hitchin system? In some sense, the original Hitchin systems for $g \geq 2$ give rise to integrable systems of KdV / KP-type (for instance, [23]). Classically-known integrable systems tend to feature integrability in terms of elliptic integrals and hence involve the projective line $\mathbb{P}^1$ and elliptic curves. For example, geodesic flow on the ellipsoid and Nahm's equations are twisted Hitchin systems on $X = \mathbb{P}^1$, as described in [22]. Here, the Lax pair integrability can be expressed explicitly in terms of Laurent series in an affine chart on the $\mathbb{P}^1$.

In this article, we ask this question for the Calogero-Françoise integrable system, which arises as a generalization of the Camassa-Holm dynamics. We demonstrate how one can fit the CF integrable system into a twisted Hitchin system on $X = \mathbb{P}^1$, with Higgs fields taking values in the line bundle $\mathcal{O}(d)$, whose transition function is z^d in the local coordinate. One nice feature of this identification is that the theta divisor in the Jacobian of the spectral curve can be interpreted as a dynamical collision locus. We demonstrate this explicitly in the case $d = 2$. Along the theta divisor, we also see a transition to a Hitchin system with poles of order 1, capturing the singular dynamics algebro-geometrically. Finally, we examine the inverse problem for CF from the point of view of continued fractions, in the sense of Stieltjes.

We hope that the mix of integrable systems theory and complex algebraic geometry in this article reflects some of the spirit of E. Previato's groundbreaking work over the past several decades.

Acknowledgements

We are grateful to P. Boalch for useful discussions concerning Hitchin systems and Lax integrability. The first and third named authors acknowledge the support of Discovery Grants from the Natural Sciences and Engineering Research Council of Canada (NSERC). The second named author was supported by the NSERC USRA program.

2 Calogero-Françoise Hamiltonian System

The main reference for this section is [5]. We nevertheless present the main aspects of the setup to introduce notation and the main dynamical objects. F. Calogero and J.-P. Françoise introduced in [11] a family of completely integrable Hamiltonian systems with Hamiltonian

$$H(x_1, \ldots, x_d, m_1, \ldots, m_d)) = \frac{1}{2} \sum_{j,k=1}^{d} m_j m_k G_{v,\beta}(x_j - x_k) \qquad (2)$$

where

$$G_{v,\beta}(x) = \frac{\beta_-}{2v} e^{-2v|x|} + \frac{\beta_+}{2v} e^{2v|x|}, \qquad (3)$$

and $\{x_1, \ldots, x_d\}$ and $\{m_1, \ldots, m_d\}$ are canonical positions and momenta respectively. For future use we will define β as a 2×2 diagonal matrix $\mathrm{diag}(\beta_-, \beta_+)$.

The authors of [11] constructed explicitly d Hamiltonians $\{H_j, j = 1, \ldots, d\}$ and directly showed that they were in involution, i.e. $\{H_j, H_k\} = 0$, with respect to the canonical Poisson bracket. The special case $\beta_+ = 0, \beta_- = 1$ was used as a motivating example and we briefly describe now this special case. In 1993, R. Camassa and D. Holm [12] proposed what would turn out to be one of the most studied nonlinear partial differential equations of the last three decades, namely

$$m_t + 2mu_x + um_x = 0, \quad m = u - u_{xx}. \tag{4}$$

The equation was originally derived from the Hamiltonian for Euler's equation in the shallow water approximation. One of the outstanding properties of the resulting equation is that it captures some aspects of "slope-steepening" and the breakdown of regularity of solutions, while at the same time it exhibits numerous intriguing aspects of Lax integrability, the connections to continued fractions of Stieltjes' type being one. One feature that stands out in the present context is the existence of non-smooth solitons, dubbed *peakons*. These are obtained from the *peakon ansatz*

$$u(x, t) = \sum_{j=1}^{d} m_j(t) e^{-|x - x_j(t)|} \tag{5}$$

for which m becomes a finite sum of weighted Dirac measures

$$m = 2 \sum_{j=1}^{d} m_j(t) \delta_{x_j(t)}, \tag{6}$$

and subsequently, upon substituting into (4), one ends up with the systems of ODEs for positions x_j and momenta m_j

$$\dot{x}_j = u(x_j), \qquad \dot{m}_j = -m_j \langle u_x \rangle (x_j), \tag{7}$$

where $\langle f \rangle (x_j)$ denotes the arithmetic average of the right and left limits of f at x_j. Moreover, the peakon equation (7) is Hamiltonian with respect to the canonical Poisson bracket and Hamiltonian

$$H(x_1, \ldots, x_d, m_1, \ldots, m_d) = \frac{1}{2} \sum_{i,j=1}^{d} m_i m_j e^{-|x_i - x_j|}. \tag{8}$$

The CH equation (4) has a Lax pair

$$L = -D^2 + \frac{1}{4} - \lambda m, \qquad P = \left(\frac{1}{2\lambda} + u \right) D + \frac{u_x}{2} \tag{9}$$

whose compatibility indeed yields (4) (see, however, the discussion in Section 3).

It was shown in [3, 4] that peakon equations can be explicitly integrated using classical results of analysis including the Stieltjes' continued fractions and the moment problem. The CH peakon Hamiltonian (8) was the starting point for the analysis in [11] and clearly the Hamiltonian (2) is a natural generalization of (8). The fact that this generalization fits in with the CH equation (4) was proven in [5]. We will review the analysis based on that paper with due attention to the emergence of a spectral curve and associated Riemann surface, both of which were absent from the analysis in [11] and were only in the background in [5].

3 CF Flows: the Peakon Side

Given a measure $m \in \mathscr{M}(\mathbf{R})$ and $\lambda \in \mathbf{C}$ we form the operator pencil

$$L(\lambda) = D^2 - v^2 - 2v\lambda m. \tag{10}$$

We introduce another operator

$$B(\lambda) = \left(\frac{1}{2v\lambda} - u\right) D + \tfrac{1}{2}u_x, \tag{11}$$

and observe that

$$[B(\lambda), L(\lambda)] = 2v\lambda(2u_x m + mu_x) - \left(m + \tfrac{1}{2}u_{xx} - 2v^2 u\right)_x + \bmod L(\lambda), \tag{12}$$

where $\bmod L(\lambda)$ means that this part vanishes on the kernel of $L(\lambda)$, from which we conclude that the operator equation (valid identically in λ)

$$\dot{L}(\lambda) = [B(\lambda), L(\lambda)] + \bmod L(\lambda) \tag{13}$$

implies

$$m_t = (mD + Dm)u, \qquad [D(4v^2 - D^2)]u = 2m_x. \tag{14}$$

Remark 3.1 In his 1976 paper S. Manakov [27] introduced a generalization of the Lax formalism. The main new aspect amounted to replacing the standard Lax pair formulation $\dot{L} = [B, L]$ with what would become known as a *Manakov triple formulation* by postulating the existence of another operator, say, C such that a generalized Lax equation $\dot{L} = [B, L] + CL$ holds. Needless to say the CH equation and in fact many other integrable equations have since been identified as satisfying some form of the Manakov triple formalism. In the CH case, $C = -2u_x$.

The main reason for reviewing this derivation, despite its obvious affinity with the CH Lax pair (9), is to emphasize the second equation determining how the potential u is related to the measure m. Upon one integration we get $(4v^2 - D^2)u = 2m + c$ for some c, resulting in shifting $u \to u + \frac{c}{4v^2}$ which can easily be absorbed by the Gallilean transformation $t = t', x = x' + ct$. Thus we could assume that $c = 0$ and that's precisely what was done in [5]. Yet, in this paper we will choose a particular constant c, specified later, to fit more naturally with other developments. We note that $\frac{1}{2}G_{v,\beta}$ in (3) is the most general even fundamental solution of $(4\nu^2 - D^2)$, provided $\beta_- - \beta_+ = 1$. Moreover, any rescaling of $G_{\nu,\beta}$ results in a rescaling of $H(x_1, \ldots, x_d, m_1, \ldots, m_d)$ which, in turn, can be compensated by changing the time scale. So the assumption $\beta_- - \beta_+ = 1$ causes no loss of generality and will be in force for the remainder of the paper.

We will concentrate from this point onward on the peakon sector whose definition we record to fix notation

$$u(x) = \sum_{j=1}^{d} m_j G_{\nu,\beta}(x - x_j) - C, \qquad m = \sum_{j=1}^{d} m_j \delta_{x_j}. \qquad (15)$$

Moreover, we assign the labels to positions in an increasing order $x_1 < x_2 < \cdots < x_d$. In the CH peakon case ($\beta_+ = 0$), the positivity of masses (momenta) m_j is crucial for the global existence of solutions [4] so we make the same assumption that $m_j > 0$ until further notice. We note that the evolution equation (14) has to be interpreted in the sense of distributions. In particular

$$m_t = \sum_{j=1}^{d} \left(\dot{m}_j \delta_{x_j} - \dot{x}_j m_j \delta_{x_j}^{(1)} \right)$$

while the term $(mD + Dm)u$ has to be properly defined since the singular supports of m and u coincide. The regularization consistent with Lax integrability turns out to be to assign the average value to u_x at any point x_j. Thus $u_x \delta_{x_j} \overset{def}{=} \langle u_x \rangle (x_j) \delta_{x_j}$. This point is explained in [4]. The resulting CF Hamiltonian system has the same form as (7) except for the definition of u which is now given by equation (15).

The presence of masses at x_j divides $\mathbf{R}$ into intervals $I_j = (x_{j-1}, x_j)$ with the proviso that $x_0 = -\infty$, $x_{d+1} = +\infty$. We will need the asymptotic behaviour of u on I_1, I_{d+1} which follows trivially from the definition of u.

Lemma 3.2 *Let us set $M_\pm = \sum_{j=1}^{d} m_j e^{\pm 2\nu x_j}$. Then the asymptotic behaviour of u is given by*

$$u(x) = \frac{\beta_-}{2\nu} M_- e^{2\nu x} + \frac{\beta_+}{2\nu} M_+ e^{-2\nu x} - C, \qquad x \in I_1, \qquad (16)$$

$$u(x) = \frac{\beta_-}{2\nu} M_+ e^{-2\nu x} + \frac{\beta_+}{2\nu} M_- e^{2\nu x} - C, \qquad x \in I_{d+1}. \qquad (17)$$

Remark 3.3 The main difference between the CF scenario and the original peakon CH case is the presence of a non-vanishing, actually exponentially growing, tail at $\pm\infty$. Eventually, this has a real impact on the type of algebraic curve to which the problem is associated.

3.1 Forward Problem

We concentrate now on solving $L(\lambda)\Phi = 0$ for $L(\lambda), m$ given by (10), (15) respectively. The mathematics involved is elementary, but one gets an interesting insight into the emergence of an underlying finite dimensional dynamical system. The singular support of m consists of positions $x_1 < x_2 < \cdots < x_d$. On the complement, that is on the union of open intervals $\bigcup_{j=1}^{d+1} I_j$, we are solving

$$(D^2 - v^2)\Phi = 0.$$

Let us denote by Φ_j the restriction of Φ to I_j. Then on each I_j, we have

$$\Phi_j = a_j e^{vx} + b_j e^{-vx}, \qquad 1 \le j \le d+1;$$

however, while crossing the right endpoint x_j, we have

$$\Phi_j(x_j) = \Phi_{j+1}(x_j), \qquad [D\Phi](x_j) = (D\Phi_{j+1} - D\Phi_j)(x_j) = 2v\lambda\Phi_j(x_j).$$

If we use as a basis $\{e^{vx}, e^{-vx}\}$, then Φ_j can be identified with $\begin{bmatrix} a_j \\ b_j \end{bmatrix}$ and the last equation can be written

$$\Phi_{j+1} = T_j\Phi_j, \qquad T_j = I + \lambda m_j \begin{bmatrix} 1 & e^{-2vx_j} \\ -e^{2vx_j} & -1 \end{bmatrix}, \qquad (18)$$

where I denotes the 2×2 identity matrix.

We now define the transition matrix

$$T = T_d \dots T_1, \qquad (19)$$

which maps $\Phi_1 \rightarrow \Phi_{d+1}$. In the next step we want to determine the time evolution of T.

Lemma 3.4 *Let*

$$B_- = \begin{bmatrix} \frac{1}{2\lambda} + C & \beta_- M_- \\ -\beta_+ M_+ & -\frac{1}{2\lambda} - C \end{bmatrix},$$

$$B_+ = \begin{bmatrix} \frac{1}{2\lambda} + C & \beta_+ M_- \\ -\beta_- M_+ & -\frac{1}{2\lambda} - C \end{bmatrix}.$$

Then

$$\dot{T} = B_+ T - T B_-. \qquad (20)$$

Proof. The proof can be found in [5] but for the sake of completeness and to give the reader a sense of the origin of the geometric underpinnings of the CF system we present an economical version of the argument. We start by observing that the generalized Lax Equation 13 implies that

$$L(\lambda) \left(\frac{\partial}{\partial t} - B(\lambda) \right) \Phi = \left(\frac{\partial}{\partial t} - B(\lambda) \right) L(\lambda) \Phi = 0. \tag{21}$$

Suppose we denote by $\hat{\Phi}$ the fundamental solution normalized to be $\hat{\Phi}_1 = I$ in I_1 (corresponding to $e^{\nu x}$ and $e^{-\nu x}$ as linearly independent solutions). Then (21) implies that

$$\left(\frac{\partial}{\partial t} - B(\lambda) \right) \hat{\Phi} = \hat{\Phi} C(t) \tag{22}$$

where $C(t)$ is a 2×2 matrix whose entries do not depend on x. Let us denote by B_j the restriction of $B(\lambda)$ to I_j. Then $\hat{\Phi}_j = T_{j-1} \cdots T_1 \hat{\Phi}_1$, and

$$\left(\frac{\partial}{\partial t} - B_j \right) \hat{\Phi}_j = \hat{\Phi}_j C(t) = T_{j-1} \cdots T_1 \hat{\Phi}_1 C(t) = T_{j-1} \cdots T_1 \left(\frac{\partial}{\partial t} - B_1 \right) \Phi_1$$

Hence $\left(\frac{\partial}{\partial t} - B_j \right) \hat{\Phi}_j = T_{j-1} \cdots T_1 \left(\frac{\partial}{\partial t} - B_1 \right) \hat{\Phi}_1$. In particular, for $j = d + 1$ and denoting $B_{d+1} = B_+$ and $B_1 = B_-$, we obtain

$$\left(\frac{\partial}{\partial t} - B_+ \right) T = -T B_1, \tag{23}$$

since $\hat{\Phi}_1 = I$. Finally, the computation of the matrix of B_- and B_+ in the basis $\{e^{\nu x}, e^{-\nu x}\}$ follows readily from Lemma 3.2. □

We observe that B_- and B_+ are very closely related: they differ only by the placement of β_- and β_+, which are interchanged by conjugating with the diagonal matrix β, namely,

$$B_- \beta = \beta B_+ \tag{24}$$

This leads us to an interesting corollary.

Corollary 3.5 *Let* $A(z) = z^d T \left(\lambda = \frac{1}{z} \right) \beta$ *and set* $B(z) = B_+ \left(\lambda = \frac{1}{z} \right)$. *Then* $A(z)$ *satisfies the Lax equation*

$$\dot{A}(z) = [B(z), A(z)]. \tag{25}$$

We observe that now both $A(z)$ and $B(z)$ are matrix valued polynomials in z of degrees d and 1 respectively. Clearly, we can associate to $A(z)$ a *spectral curve*

$$\boxed{C(w, z) = \det \left(wI - A(z) \right) = 0,}$$

or more succinctly

$$C(w, z) = \{(w, z) : w^2 = \text{tr}(A(z))w - z^{2d} \det \beta\}. \tag{26}$$

We remark that the compactification of the affine curve $C(w, z) = 0$, which we will denote by Y, is a 2-fold cover of $\mathbb{P}^1$ on which w is single valued.

3.2 Higgs Fields and Properties of A and B

The first observation about A is that it differs only by a multiplication by $z^d \beta$ from $T(\lambda = 1/z)$ in Equation 19. Thus A is a matrix valued polynomial of degree d in z, $A = \sum_{j=0}^{d} A_j z^j$. By factoring z^d this defines a holomorphic function around ∞ with a local parameter $\tilde{z} = \frac{1}{z}$.

As a matrix, each $A(z)$ acts by left multiplication of course on the vector space $V = \mathbb{C}^2$. The passage from linear algebra to geometry occurs by viewing $z \in \mathbb{P}^1$ as parametrizing a family of such vector spaces, and so the whole object A acts by left multiplication on the vector bundle $E = \mathbb{P}^1 \times \mathbb{C}^2$. This is a rank-2 vector bundle with trivial holomorphic structure. Typically, the isomorphism class of holomorphic line bundles on $\mathbb{P}^1$ with transition function z^n is denoted by $\mathscr{O}(n)$. The first Chern class of $\mathscr{O}(n)$, identified with an integer via Poincaré duality, is n. This integer is called the *degree* of the line bundle and we write $\deg \mathscr{O}(n) = n$. The degree tells us the number of times that a generic holomorphic section vanishes. The now-classical splitting theorem of Birkhoff and Grothendieck says that every holomorphic bundle on $\mathbb{P}^1$ decomposes uniquely (up to ordering) as a sum of holomorphic line bundles. In the case of our E, this is $E = \mathscr{O} \oplus \mathscr{O}$, where $\mathscr{O} := \mathscr{O}(0) \cong \mathbb{P}^1 \times \mathbb{C}$ is the trivial line bundle (or "structure sheaf") on $\mathbb{P}^1$. The degree is additive with respect to both tensor products and direct sums of line bundles. Hence, $\deg E = 0$.

In this framework, A is a holomorphic map from E to itself, tensored by $\mathscr{O}(d)$:

$$\boxed{A \in H^0(\mathbb{P}^1, \text{End}(E) \otimes \mathscr{O}(d))}$$

The data of A is the *Higgs field* in our setup; together, (E, A) is the *Higgs bundle*. We refer to the dimension of the fibre of E as the *rank* of E. In this case, the rank of E is $r = 2$. We also ought to remark that the pair (E, ϕ) is a "twisted" Higgs bundle relative to the formulation of Hitchin [19], as the Higgs bundles coming from gauge theory would be $\mathscr{O}(-2)$-valued on $\mathbb{P}^1$ while our d is strictly positive.

We will now compute the coefficients A_j for the Higgs field in terms of the original data $\{x_1, \ldots, x_d, m_1, \ldots, m_d\}$. This step is not relevant to the geometry of the problem but it is crucial if one actually wants to solve the original peakon equation $\dot{x}_j = u(x_j)$, $\dot{m}_j = -m_j \langle u_x \rangle (x_j)$.

Recall that

$$m = \sum_{j=1}^{d} m_j \delta_{x_j}, \quad x_1 < x_2 < \cdots < x_d,$$

and (see (18))

$$T_j = I + \lambda m_j \begin{bmatrix} 1 & e^{-2vx_j} \\ -e^{2vx_j} & -1 \end{bmatrix}.$$

The first elementary observation is that if we set $X_j = \begin{bmatrix} 1 & e^{-2vx_j} \\ -e^{2vx_j} & -1 \end{bmatrix}$ then

$$X_j = \begin{bmatrix} 1 \\ -e^{2vx_j} \end{bmatrix} \begin{bmatrix} 1 & e^{-2vx_j} \end{bmatrix}. \tag{27}$$

Definition 3.6 *The binomial coefficient $\binom{S}{j}$ denotes the collection of j-element subsets of the set S, and $[d]$ denotes the integer interval $\{1, 2, 3, \ldots, d\}$. Moreover, the elements of a set $I \in \binom{[d]}{j}$ are labeled in increasing order: $I = \{i_1 < i_2 < \cdots < i_j\}$. Finally, given a collection of matrices $\{a_1, a_2, \ldots, a_d\}$ we denote the ordered product of matrices labeled by the multi-index I as $a_I \overset{def}{=} a_{i_j} a_{i_{j-1}} \cdots a_{i_1}$.*

Lemma 3.7 *Given a multi-index $I \in \binom{[d]}{j}$ and the set of matrices $\{X_1, X_2, \ldots, X_d\}$ as in Equation 27 then*

$$X_I = \left(\prod_{k=1}^{j-1} \left(1 - e^{-2v(x_{i_{k+1}} - x_{i_k})}\right) \right) \begin{bmatrix} 1 & e^{-2vx_{i_1}} \\ -e^{2vx_{i_j}} & -e^{2v(x_{i_j} - x_{i_1})} \end{bmatrix} \tag{28}$$

with the proviso that the empty product is taken to be 1 when $j = 1$.

Proof. It suffices to observe that

$$X_j X_i = \begin{bmatrix} 1 \\ -e^{2vx_j} \end{bmatrix} \begin{bmatrix} 1 & e^{-2vx_j} \end{bmatrix} \begin{bmatrix} 1 \\ -e^{2vx_i} \end{bmatrix} \begin{bmatrix} 1 & e^{-2vx_i} \end{bmatrix}$$

$$= (1 - e^{-2v(x_j - x_i)}) \begin{bmatrix} 1 \\ -e^{2vx_j} \end{bmatrix} \begin{bmatrix} 1 & e^{-2vx_i} \end{bmatrix}, \quad i < j,$$

and then proceed by induction on the number of terms. $\qquad\square$

Given a multi-index $I \in \binom{[d]}{j}$ we denote

$$\prod_{k=1}^{j-1} \left(1 - e^{-2v(x_{i_{k+1}} - x_{i_k})}\right) \overset{def}{=} f_I. \tag{29}$$

Theorem 3.8

$$T = I + \sum_{j=1}^{d} \lambda^j \sum_{I \in \binom{[d]}{j}} f_I m_I \begin{bmatrix} 1 & e^{-2vx_{i_1}} \\ -e^{2vx_{i_j}} & -e^{2v(x_{i_j}-x_{i_1})} \end{bmatrix} \tag{30}$$

In particular,

$$T = \lambda^d f_{[d]} m_{[d]} \begin{bmatrix} 1 & e^{-2vx_1} \\ -e^{2vx_d} & -e^{2v(x_d-x_1)} \end{bmatrix} + O(\lambda^{d-1}), \qquad \lambda \to \infty, \tag{31}$$

$$A(z) = \beta z^d + \sum_{j=1}^{d} z^{d-j} \sum_{I \in \binom{[d]}{j}} f_I m_I \begin{bmatrix} \beta_- & \beta_+ e^{-2x_{i_1}} \\ -\beta_- e^{2x_{i_j}} & -\beta_+ e^{2(x_{i_j}-x_{i_1})} \end{bmatrix}, \tag{32}$$

$$\mathrm{tr} A(z) = f_{[d]} m_{[d]} (\beta_- - \beta_+ e^{2v(x_d-x_1)}) + O(z), \qquad z \to 0, \tag{33}$$

$$A(z) = \beta z^d + \begin{bmatrix} \beta_- M & \beta_+ M_- \\ -\beta_- M_+ & -\beta_+ M \end{bmatrix} z^{d-1} + O(z^{d-2}), \qquad z \to \infty, \tag{34}$$

where $M = \sum_{j=1}^{d} m_j$ *is the total mass (momentum) and* $M_\pm = \sum_{j=1}^{d} e^{\pm 2vx_j} m_j$ *(see Lemma 3.2).*

In view of the invariance of $\mathrm{tr} A(z)$ we immediately have:

Corollary 3.9 *Under the CF flow*

1. *the total mass (momentum) M is conserved;*
2. *if at $t = 0$ all masses $m_j(0)$ have the same sign then they can not collide, meaning, $x_i(t) \neq x_j(t), i \neq j$ for all times.*

Proof. We start off by noting that individual masses cannot change signs. Indeed the equation of motion $\dot{m}_j = -m_j \langle u_x \rangle (x_j)$ implies that $m_j(t)$ has the same sign as $m_j(0)$. The first statement follows immediately from the invariance of $\mathrm{tr} A(z)$, since $\mathrm{tr} A(z) = (\beta_- + \beta_+) z^d + M z^{d-1} + O(z^{d-2})$. For the second claim it suffices to prove that the neighbours cannot collide. First, we observe that $m_{[d]}$ will remain bounded away from zero if masses have the same sign. Suppose now the constant of motion $f_{[d]} m_{[d]} (\beta_- - \beta_+ e^{2v(x_d-x_1)})$ is not zero at $t = 0$. Then, $f_{[d]} = \prod_{k=1}^{d-1} \left(1 - e^{-2(x_{k+1}-x_k)}\right)$ will remain non-zero if $\beta_- - \beta_+ e^{2v(x_d(0)-x_1(0))} \neq 0$. $\square$

4 Linearization

One of the advantages of the Higgs bundle framework is that the moduli space of Higgs bundles on a Riemann surface X is fibred by tori, each of which is the Jacobian of a spectral curve for a Higgs field. In a sense, all possible spectral curves covering the given X appear in the moduli space. At the same time, the total space of the moduli space is an algebraically completely integrable system [20] that extends, in a canonical way, the phase space structure of the cotangent bundle to the moduli space of bundles on X. Furthermore, explicit Hamiltonians are given by invariants of the Higgs fields — these are the components of the so-called "Hitchin map", which sends a Higgs field A to the coefficients of its characteristic polynomial. In the twisted Higgs setup at genus 0, the relationship to the cotangent bundle of the moduli space of bundles fails to persist, but the integrable system, with the description of its Hamiltonians as characteristic coefficients, does.

The *spectral correspondence* identifies isomorphism classes of Higgs bundles with a fixed characteristic polynomial with isomorphism classes of holomorphic line bundles on the associated spectral curve Y. This correspondence is developed for 1-form-valued Higgs bundles on X of genus $g \geq 2$ by Hitchin [20] and for arbitrary genus and Higgs fields taking values in an arbitrary line bundle $L \to X$ by Beauville-Narasimhan-Ramanan [7] — see also [29, 14]. The essence of the correspondence is that the eigenspaces of a Higgs field A for a holomorphic vector bundle $E \to X$ form a line bundle S on Y, which as a curve is embedded in the total space of L (since A is L-valued and so the eigenvalues are sections of L). If r is the rank of E, then Y will be an r-sheeted cover of X. If π is the projection from the total space of L to X, then the direct image $(\pi|_Y)_* S$ is a vector bundle E' on X, isomorphic to the original vector bundle E. Let z be a local coordinate on X (just as with $X = \mathbb{P}^1$ in the preceding discussion). The map that multiplies sections $s(z)$ of S by $w(z)$, where $w(z)$ is the corresponding point on Y, is the action of eigenvalues on eigenspaces. This map pushes forward to an L-valued endomorphism A' of E whose spectrum is Y. This operation that starts with (E, A) and ends with the isomorphic Higgs bundle (E', A') is, at almost every point of X, diagonalization.

When $X = \mathbb{P}^1$ and $L = \mathcal{O}(d)$, Lax partners B that complement the Higgs fields A can be systematically computed and written down explicitly, as carried out by Hitchin in [22]. The only assumption necessary on A is that it has a smooth, connected spectral curve Y, which is a fairly generic property. By Bertini's theorem on pencils of divisors, the generic characteristic polynomial produces a smooth spectral curve. At the same time, the moduli space consists only of "stable" Higgs bundles, which are Higgs bundles (E, A) with a restriction on which subbundles of E can be preserved by A. Normally, this is imposed to ensure that the moduli space is topologically well-formed. The precise condition for Higgs bundles, called *slope stability*, originates in [20], and in our case is as follows: (E, A) on $\mathbb{P}^1$ is *semistable* if for each nonzero, proper subbundle U with $A(U) \subseteq U \otimes \mathcal{O}(d)$ we have that

$$\frac{\deg(U)}{\mathrm{rank}(U)} \leq \frac{\deg(E)}{\mathrm{rank}(E)}.$$

Generically, A will be *very stable*, meaning that it preserves no subbundle whatsoever (other than E itself and 0). This corresponds to the spectral curve being connected.

Embedding a known integrable system into a Hitchin-type system on $\mathbb{P}^1$ can therefore lead to new insights about linearized flows. Having produced such an embedding for CF — we established the existence of Higgs fields A acting on $E = \mathcal{O} \oplus \mathcal{O}$ that solve the peakon equation — we may now compute the linearization along the Hitchin fibres directly.

Theorem 4.1 *Let* $A(z, t) \in H^0(\mathbb{P}^1, End(E) \otimes \mathcal{O}(d))$ *be a solution to the peakon equation given above, such that its spectral curve Y (constant for all t) is a double cover of $X = \mathbb{P}^1$ that is embedded as a non-singular, connected subvariety in the total space of $\mathcal{O}(d)$. Let $J^{\tilde{g}-1}$ be the Jacobian of degree $\tilde{g} - 1$ line bundles on Y and Θ be the theta divisor and $\tilde{g}$ is the genus of Y. The CF flow as given by Corollary 3.5 is linearized on $J^{\tilde{g}-1} \setminus \Theta$.*

Proof. The problem of linearization for matrix Lax equations with a spectral parameter z has been studied by several authors [22, 18] We will use pertinent to this problem material from N. Hitchin's lectures in [22]. For the Lax equation of the type $\dot{A}(z) = [B(z), A(z)]$ with polynomial matrix valued $A(z), B(z)$ the linearization on $J^{g-1} \setminus \Theta$ happens if and only if $B(z)$ has a specific form dependent on $A(z)$ ([22, Lecture 5]):

$$B(z) = \gamma(z, A(z))^+,$$

where

$$\gamma(z, w) = \sum_{i=1}^{r-1} \frac{b_i(z) w^i}{z^N}.$$

In this formula, r is the degree (in w) of the spectral curve $\det(w - A(z)) = 0$, N is an integer, $b_i(z)$ are polynomials in z, and $^+$ means the projection on the polynomial part. In our case $r = 2$ so $\gamma(z, w) = \frac{b(z) w}{z^N}$. We now choose $b(z) = 1$ and $N = d - 1$ and settle with $B(z) = \left(\frac{A(z)}{z^{d-1}}\right)^+$. Then by Theorem 3.8 we get

$$B(z) = \begin{bmatrix} \beta_- z & 0 \\ 0 & \beta_+ z \end{bmatrix} + \begin{bmatrix} \beta_- M & \beta_+ M_- \\ -\beta_- M_+ & -\beta_+ M \end{bmatrix}.$$

However, $B(z)$ is not unique; any power of $A(z)$ can be added to $B(z)$ without changing the Lax equation, in particular any multiple of the identity can be added with impunity. For example, by writing $\beta_- = \frac{1}{2}(\beta_- + \beta_+) + \frac{1}{2}(\beta_- - \beta_+)$ and the same for β_+, we see that one can take

$$B(z) = \begin{bmatrix} \frac{z}{2} & 0 \\ 0 & -\frac{z}{2} \end{bmatrix} + \begin{bmatrix} \frac{\beta_- + \beta_+}{2} M & \beta_+ M_- \\ -\beta_- M_+ & -\frac{\beta_- + \beta_+}{2} M \end{bmatrix}.$$

It suffices now to set

$$\boxed{C = \frac{\beta_- + \beta_+}{2} M}$$

in Lemma 3.4 to complete the proof.   □

5 Two CF Peakons: Collisions

In this section we will analyze, as a concrete example, the case of $d = 2$ to get a better insight into the global existence of CF flows, but also to demonstrate the special dynamical meaning of the theta divisor $\Theta \subset J^{\tilde{g}-1}$.

We recall that the moduli space of stable twisted Higgs bundles with $d = 2$ on $\mathbb{P}^1$ on a rank 2 holomorphic bundle of degree 0 was studied algebro-geometrically by the first named author in [35]. Over $\mathbb{C}$, there is a 9-dimensional moduli space of such Higgs bundles. The base of the Hitchin fibration is 8-dimensional, reflecting the extreme underdetermined nature of the integrable system: we only need 2 real Hamiltonians, as per the dimension of the fibre, but we actually have a 16-dimensional space of such Hamiltonians available to us.

Theorem 6.1 in [35] characterizes exactly which holomorphic bundles E with these topological invariants admit the structure of a semistable Higgs bundle (E, A). These are precisely $E = \mathcal{O} \oplus \mathcal{O}$ and $E = \mathcal{O}(1) \oplus \mathcal{O}(-1)$. (A rank 2, degree 0 vector bundle $E = \mathcal{O}(k) \oplus \mathcal{O}(-k)$ with $k \geq 2$ will necessarily have a preserved sub-line bundle $\mathcal{O}(k)$ and $k > 0$, which violates stability as posed above.) It is also shown in Proposition 8.1 in the same reference that, for $d = 2$, there is up to isomorphism a unique Higgs field A for the bundle $E = \mathcal{O}(1) \oplus \mathcal{O}(-1)$, once the characteristic coefficients of A have been fixed. These Higgs fields form a section of the Hitchin system here, intersecting each torus in this unique point. In the isospectral problem, where the characteristic coefficients are constant, we thus have a unique Higgs field A corresponding to $E = \mathcal{O}(1) \oplus \mathcal{O}(-1)$ in the Hitchin fibre determined by β_+, β_-, M, ν. All of the remaining Higgs bundles in the fibre are ones with $E = \mathcal{O} \oplus \mathcal{O}$.

This begs the question of the meaning of this unique point. Algebro-geometrically, the meaning is clear: it is the theta divisor Θ, or rather a twist of it. To see this, note that a Higgs bundle (E, A) valued in $\mathcal{O}(2)$ with rank 2 and degree 0 yields, via the spectral correspondence, a degree 2 line bundle S on the spectral curve Y, which is a genus $\tilde{g} = 1$ curve that covers $\mathbb{P}^1$ $2 : 1$. (See again Section 8 of [35].) In general, the genus of the spectral curve can be computed through the Riemann-Hurwitz formula by noting the number of zeros of the determinant of A, which is $2d$. As such, we will have $\tilde{g} = d - 1$ in general. Subsequently, $\deg(S)$ can then be computed via the Grothendieck-Riemann-Roch Theorem. This, in fact, will always be d and so $\deg(S) = 2$ in this case. (It is perhaps useful to point out that the degree of S need not match the degree of E, as ramification in general disturbs such invariants. A precise formula is given in Proposition 4.3 of Chapter 2 of [22].) All in all, the spectral correspondence employs J^2 of Y while the theta divisor is a subvariety of J^0, as $\tilde{g} - 1 = 0$ in this case.

Note that Θ is complex codimension 1 in J^0. In this particular case, it is a single point: the isomorphism class of the trivial line bundle $\mathcal{O}_Y := Y \times \mathbb{C}$. The remaining points in J^0 are holomorphic line bundles of degree 0 for which there is no global nonzero holomorphic section. To bridge the gap between J^0 and J^2, note that when $E = \mathcal{O} \oplus \mathcal{O}$, we have $E \otimes \mathcal{O}(-1) \cong \mathcal{O}(-1) \oplus \mathcal{O}(-1)$, which has no nonzero holomorphic sections. On the other hand, when $E = \mathcal{O}(1) \oplus \mathcal{O}(-1)$, we have $E \otimes \mathcal{O}(-1) \cong \mathcal{O} \oplus \mathcal{O}(-2)$, which has a 1-dimensional space of holomorphic sections. Upstairs on Y, the associated line bundle is $S \otimes \pi^* \mathcal{O}(-1)$, where π is again the projection of $\mathcal{O}(2)$ (restricted to Y). Since S is a degree 2 cover, the degree of S is shifted by $2 \cdot (-1)$ when we twist by $\pi^* \mathcal{O}(-1)$. In other words, we get $E = \mathcal{O}(-1) \oplus \mathcal{O}(-1)$ as the pushforward of any line bundle in J^0 except for if we push forward $\mathcal{O}_Y$ (i.e. the theta divisor), which instead gives us $E = \mathcal{O} \oplus \mathcal{O}(-2)$.

Translating all of this back to J^2, every line bundle in J^2 pushes forward to give $E = \mathcal{O} \oplus \mathcal{O}$ except for the distinguished line bundle $\mathcal{O}_Y \otimes \pi^* \mathcal{O}(1)$, which pushes forward to give $E = \mathcal{O}(1) \oplus \mathcal{O}(-1)$ — and, along with it, a unique Higgs field A for this bundle whose spectrum is Y.

Dynamically, we wish to demonstrate that this unique Higgs bundle structure on $\mathcal{O}(1) \oplus \mathcal{O}(-1)$, and hence the theta divisor itself, is a collision solution, extending the smooth dynamics presented earlier.

Now, recall the transition matrix T for $d = 2$:

$$
T = T_2 T_1 = \left(I + \lambda m_2 \begin{bmatrix} 1 & e^{-2vx_2} \\ -e^{2vx_2} & -1 \end{bmatrix} \right) \left(I + \lambda m_1 \begin{bmatrix} 1 & e^{-2vx_1} \\ -e^{2vx_1} & -1 \end{bmatrix} \right)
$$

$$
= I + \lambda \begin{bmatrix} m_1 + m_2 & m_1 e^{-2vx_1} + m_2 e^{-2vx_2} \\ -(m_1 e^{2vx_1} + m_2 e^{2vx_2}) & -(m_1 + m_2) \end{bmatrix}
$$

$$
+ \lambda^2 m_1 m_2 \begin{bmatrix} 1 - e^{2v(x_1 - x_2)} & e^{-2vx_1} - e^{-2vx_2} \\ e^{2vx_1} - e^{2vx_2} & 1 - e^{2v(x_2 - x_1)} \end{bmatrix}.
$$

Also, recall that by Corollary 3.5 $A(z) = z^2 T\left(\frac{1}{z}\right)\beta$, and $M = m_1 + m_2$. Hence,

$$A(z) = \begin{bmatrix} \beta_-(z^2 + zM + m_1 m_2(1 - e^{2\nu(x_1 - x_2)})) \\ \beta_-(-z(m_1 e^{2\nu x_1} + m_2 e^{2\nu x_2}) + m_1 m_2(e^{2\nu x_1} - e^{2\nu x_2})) \end{bmatrix}$$

$$\begin{matrix} \beta_+(z(m_1 e^{-2\nu x_1} + m_2 e^{-2\nu x_2}) + m_1 m_2(e^{-2\nu x_1} - e^{-2\nu x_2})) \\ \beta_+(z^2 - zM + m_1 m_2(1 - e^{2\nu(x_2 - x_1)})) \end{matrix} \Bigg].$$

Since $\mathrm{Tr}\, A(z)$ must be invariant, i.e. since the expression

$$z^2(\beta_+ + \beta_-) + zM + m_1 m_2(1 - e^{2\nu(x_1 - x_2)})(\beta_- - \beta_+ e^{2\nu(x_2 - x_1)})$$

is time independent, it must follow that both M and

$$\boxed{C_2 = m_1 m_2(1 - e^{2\nu(x_1 - x_2)})(\beta_- - \beta_+ e^{2\nu(x_2 - x_1)})}$$

are invariant. By the relation $\beta_- - \beta_+ = 1$, both β_+ and β_- are also determined by $\mathrm{Tr}\, A(z)$. We recall that by Corollary 3.9 there are no collisions ($x_1 = x_2$) if masses m_1, m_2 are of the same sign. However, when masses have opposite signs the collision will occur exactly as they do in the $\beta_+ = 0$ case [12, 4].

Numerical solutions to the $d = 2$ case show a collision between particles x_1 and x_2 with various initial conditions. The positions until the collision occurs for initial conditions $x_1(0) = 1, x_2(0) = 2, m_1(0) = 5, m_2(0) = -1, \nu = 2, \beta_+ = 0.018$ are shown by the following graph:

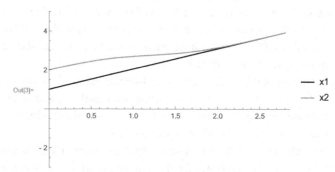

The masses also grow very large near the collision point with the same initial conditions:

Let us briefly describe on a heuristic level the mechanics of such collisions. Suppose the masses have opposite signs and $C_2 < 0$. When $x_2 - x_1$ becomes small the masses (momenta) grow large, one becoming large negative, the other large positive, while preserving the constant $M = m_1 + m_2$. Moreover, at the collision $x_1 \to x_2$, $\beta_- - \beta_+ e^{2\nu(x_2 - x_1)} \to 1$ (as $\beta_- - \beta_+ = 1$), and thus we have that the z dependent off-diagonal terms involve:

$$m_1 m_2(1 - e^{2\nu(x_1 - x_2)}) \to C_2.$$

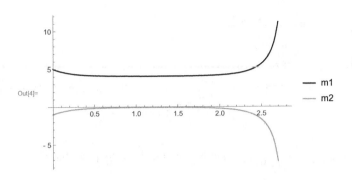

Therefore at the collision, we have:

$$A(z) \rightarrow \begin{bmatrix} \beta_-(z^2 + zM + C_2) & \beta_+(zMe^{-2vx_1} + C_2 e^{-2vx_1}) \\ \beta_-(-zMe^{2vx_1} - C_2 e^{2vx_1}) & \beta_+(z^2 - zM - C_2) \end{bmatrix}$$

$$= \begin{bmatrix} \beta_-(z^2 + zM + C_2) & \beta_+ Me^{-2vx_1}\left(z + \frac{C_2}{M}\right) \\ -\beta_- Me^{2vx_1}\left(z + \frac{C_2}{M}\right) & \beta_+(z^2 - zM - C_2) \end{bmatrix}.$$

We see that the off diagonal terms have the same zeros, so we will conjugate the matrix by $D = \begin{bmatrix} \frac{1}{z + \frac{C_2}{M}} & 0 \\ 0 & z + \frac{C_2}{M} \end{bmatrix}$. This singular automorphism of E has an $\mathcal{O}(-1)$ section and a $\mathcal{O}(1)$ section, and thus has transformed the collision Higgs field A into one for the bundle $E = \mathcal{O}(1) \oplus \mathcal{O}(-1)$. We are now precisely at the (twisted) theta divisor in the fibre, as claimed earlier. The new form of the Higgs field is

$$A' = D^{-1}AD = \begin{bmatrix} \beta_-(z^2 + zM + C_2) & \beta_+ Me^{-2vx_1}\left(z + \frac{C_2}{M}\right)^3 \\ -\frac{\beta_- Me^{2vx_1}}{z + \frac{C_2}{M}} & \beta_+(z^2 - zM - C_2) \end{bmatrix}.$$

Let $P = \begin{bmatrix} a & b \\ 0 & d \end{bmatrix}$ be a gauge transformation of $\mathcal{O}(1) \oplus \mathcal{O}(-1)$, meaning a, d are numbers and b is a polynomial with $\deg b = 2$. Then conjugating A' by P gives:

$$A'' = P^{-1}A'P$$

$$= \begin{bmatrix} a^{-1} & -\frac{b}{ad} \\ 0 & d^{-1} \end{bmatrix} \begin{bmatrix} \beta_-(z^2 + zM + C_2) & \beta_+ Me^{-2vx_1}\left(z + \frac{C_2}{M}\right)^3 \\ -\frac{\beta_- Me^{2vx_1}}{z + \frac{C_2}{M}} & \beta_+(z^2 - zM - C_2) \end{bmatrix} \begin{bmatrix} a & b \\ 0 & d \end{bmatrix}$$

$$
= \begin{bmatrix} \beta_-(z^2 + zM + C_2) + \dfrac{\beta_- M e^{2\nu x_1}}{z + \frac{C_2}{M}}\frac{b}{d} \\[2ex] -\dfrac{\beta_- M e^{2\nu x_1}}{z + \frac{C_2}{M}}\frac{a}{d} \end{bmatrix}
$$

$$
\begin{bmatrix} \frac{b}{a}(\beta_-(z^2 + zM + C_2) + \dfrac{\beta_- M e^{2\nu x_1}}{z + \frac{C_2}{M}}\frac{b}{d} - \beta_+(z^2 - zM - C_2)) + \frac{d}{a}\beta_+ M e^{-2\nu x_1}\left(z + \frac{C_2}{M}\right)^3 \\[2ex] -\dfrac{\beta_- M e^{2\nu x_1}}{z + \frac{C_2}{M}}\frac{b}{d} + \beta_+(z^2 - zM - C_2) \end{bmatrix}.
$$

Choosing $\frac{a}{d} = -\dfrac{1}{\beta_- M e^{2\nu x_1}}$, and $b = \beta_- M a \left(z + \frac{C_2}{M}\right)^2$ we have:

$$
A'' = \begin{bmatrix} \beta_- z^2 & \beta_- M \left(z + \frac{C_2}{M}\right)^2 (-z^2 + \beta_+(zM + C_2)) - \beta_+ \beta_- M^2 \left(z + \frac{C_2}{M}\right)^3 \\[2ex] \dfrac{1}{z + \frac{C_2}{M}} & \beta_+ z^2 + zM + C_2 \end{bmatrix}.
$$

With further simplification we have:

$$
A'' = \begin{bmatrix} \beta_- z^2 & -\beta_- M \left(z + \frac{C_2}{M}\right)^2 z^2 \\[2ex] \dfrac{1}{z + \frac{C_2}{M}} & \beta_+ z^2 + zM + C_2 \end{bmatrix}.
$$

We see that A'' is determined by β_+, β_-, M, C_2, which are all determined by $\operatorname{Tr} A(z)$. Therefore any Higgs fields with the same trace can be conjugated to the same form A'' at the collision $x_1 = x_2$, and so we have the uniqueness of the collision point, up to gauge.

We also note, somewhat in contrast to [35], that the Higgs field A'' has a pole of order 1 at $z = -C_2/M$ on $\mathbb{P}^1$, and so is not purely holomorphic on this chart. This originates in the fact that the off-diagonal terms of the original solution A for $E = \mathcal{O} \oplus \mathcal{O}$ were of degree strictly less than 2. Hence, the collision necessitates not only a change of bundle type but also a change of Higgs field type, that is, to a parabolic Higgs field with an order 1 pole.

Remark 5.1   The case of $d = 2$ was also investigated by N. Hitchin recently in [21] in the context of Nahm's equations, where the singular Hitchin fibres are studied and non-classical conserved quantities are shown to exist.

6 Inverse Problem: Recovering m_j, x_j

Even though we already know by Theorem 4.1 that the dynamics linearize on the Jacobian $J^{\tilde{g}-1}$ of the Riemann surface Y, the actual task is to solve the peakon ODE system (7). We will proceed in the following way. First, we will study the eigenvector mapping

$$
\phi_t : Y \to \mathbb{P}(E) \tag{35}
$$

with $E = \mathcal{O} \oplus \mathcal{O}$, given by the eigenvalue problem:

$$A(z)v = wv. \tag{36}$$

For reasons of symmetry we will shift $w \to w + \frac{1}{2} \mathrm{Tr}\,(A(z))$. This is equivalent to making A traceless, which corresponds to passing to the SL(2, $\mathbb{C}$) Hitchin moduli space, which has fibres of the same dimension $d-1$ over a smaller base. Now, we will construct a meromorphic function W on Y by taking the ratio of two holomorphic sections of the line bundle $\phi_t^* \mathcal{O}_{\mathbb{P}(E)}(1) \in J^0(Y)$, where $\mathcal{O}_{\mathbb{P}(E)}(1)$ is the hyperplane bundle. Recall that the off-diagonal polynomials of A are generically of degree $d-1$ (see Theorem 3.8). Indeed, by expanding (36) with shifted w, we get

$$A_{11}(z)\zeta_1 + A_{12}(z)\zeta_2 = \left(w + \frac{1}{2}\mathrm{Tr}\,A(z)\right)\zeta_1 \tag{37}$$

$$A_{21}(z)\zeta_1 + A_{22}(z)\zeta_2 = \left(w + \frac{1}{2}\mathrm{Tr}\,A(z)\right)\zeta_2, \tag{38}$$

which results in three useful formulas for the *generalized Weyl function* $W = \frac{\zeta_2}{\zeta_1}$:

$$W = \frac{w - \frac{1}{2}\mathrm{Tr}\,(A(z)\sigma_3)}{A_{12}(z)}, \tag{39}$$

$$W = \frac{A_{21}(z)}{w + \frac{1}{2}\mathrm{Tr}\,(A(z)\sigma_3)}, \tag{40}$$

$$W = \frac{A_{21}(z) + A_{22}(z)W}{A_{11}(z) + A_{12}(z)W}, \tag{41}$$

where $\sigma_3 = \begin{bmatrix} 1 & 0 \\ 0 & -1 \end{bmatrix}$. Thus W can be viewed as a ratio of two holomorphic functions on an affine chart on the sphere, one with zeros at the zeros of A_{12}, the other with zeros at the zeros of A_{21}, properly lifted to Y (i.e. if $A_{12}(z_0) = 0$ then the unique lift is $(z_0, w_0 = -\frac{1}{2}\mathrm{Tr}\,(A(z_0)\sigma_3))$). The numbers of zeros and poles are equal, and so we view this as a meromorphic section of a holomorphic line bundle U of degree 0 on Y.

The curious reader may wonder how this squares up with the degree of the line bundle in the preceding section. For $d = 2$, the line bundle S on Y had degree 2. This is again precisely the difference between working in J^0 and J^2, the translation of which was achieved by twisting by the pullback of $\mathcal{O}(1)$, the generator of the Picard group of the projective line. In general, the actual line bundle S that pushes forward to reconstruct the Higgs bundle on $\mathbb{P}^1$ is, by Grothendieck-Riemann-Roch, of degree d and so the two natural Jacobians are J^d and J^0, the former being the home of the actual spectral line bundle S and the latter being the home of the line bundle U whose section is the above ratio. If d is odd, then the passage from J^0 to J^d involves an additional line bundle R

of degree 1 on Y that completes the equivalence $S = U \otimes \pi^* \mathscr{O}(1)^{(d-1)/2} \otimes R$. In the context of the preceding section, collisions of particles occur when zeros of the numerator and denominator of the meromorphic section align. Once all of the zeros upstairs line up with all of those downstairs, we are now at a line bundle with a constant holomorphic section — in other words, the trivial line bundle $\mathscr{O}_Y$. This point will correspond (via twisting to degree $\tilde{g} - 1 = d - 2$) to a line bundle in the theta divisor. In general, the emerging picture is that Θ is stratified by different pairings of collisions, which in turn correspond to pairings of zeros and poles in W.

The three formulas for W above are also quite useful insofar as they reveal different aspects of going to the limit $\beta_+ = 0$. We recall that this limit corresponds to the pure Camassa-Holm (CH) peakons ($\beta_- = 1$) for which the inverse formulas yielding masses and positions exist [3]. To make this connection we observe that, if $\beta_+ \to 0$, then by (41) one automatically obtains

$$W = \frac{A_{21}(z)}{A_{11}(z)} \overset{\text{Corollary 3.5}}{=} \frac{T_{21}(\lambda)}{T_{11}(\lambda)}, \qquad \lambda = \frac{1}{z},$$

which is the desired result. If, on the other hand, one uses (39) then one realizes that in the limit the spectral curve becomes $w = \frac{1}{2} A_{11}(z)$ and the genus drops to 0. Either way one obtains the original Weyl function of the CH peakon problem and $A_{11}(z)$ is the desired spectral invariant.

After shifting w the algebraic curve $C(w, z)$ reads

$$w^2 = P(z), \quad P = \tfrac{1}{4}(\operatorname{Tr} A(z))^2 - \det \beta z^{2d}. \tag{42}$$

It is elementary to check, in view of $\beta_- - \beta_+ = 1$, that $P(z) = z^{2d} + O(z^{2d-1})$, $z \to \infty$. Moreover, in the limit $\beta_+ \to 0$, $P(z)$ becomes a perfect square as we indicated earlier.

Since we will need a bit of information about this surface let us review some elementary facts about this surface. Assuming for simplicity that all roots of $P(z)$ are simple we can write

$$w^2 = \prod_{j=1}^{2d}(z - z_j), \quad z_i \neq z_j, \quad i \neq j. \tag{43}$$

The curve Y resulting from the compactification of $C(w, z) = 0$ has a branch points at z_j. We will define the upper sheet as the one on which $w = +\sqrt{P}$ and the value of the right hand side being defined as positive for large positive values of z, with $w = -\sqrt{P}$ on the lower sheet. Let us denote by C_d the coefficient of z^0 in $\operatorname{Tr} A(z)$. Observe that if $C_d \neq 0$ then $z = 0$ is not a branch point and thus we have two lifts $\pi^{-1}(0)$ on Y to be denoted 0^+ on the upper sheet, 0^- on the lower sheet respectively.

The following theorem is crucial for finding a solution to the inverse problem:

Theorem 6.1 *Suppose* $C_d = \operatorname{Tr} A(0) > 0$, *then on the upper sheet of* Y, *using* z *as a local parameter around* 0^+, *we have*

$$W - \frac{A_{21}(z)}{A_{11}(z)} = O(z^{2d}), \quad z \to 0. \tag{44}$$

The same result holds on the lower sheet if $C_d < 0$.

Proof. Assume $C_d > 0$. Using (40) we see

$$W - \frac{A_{21}(z)}{A_{11}(z)} = \frac{A_{21}(z)}{A_{11}(z)} \frac{\left(-w + \frac{1}{2}\operatorname{Tr} A(z)\right)}{\left(w + \frac{1}{2}\operatorname{Tr}(A(z)\sigma 3)\right)}.$$

Multiplying top and bottom of the right hand side by $\left(w + \frac{1}{2}\operatorname{Tr} A(z)\right)$ and using the equation determining the surface Y we get

$$W - \frac{A_{21}(z)}{A_{11}(z)} = \frac{A_{21}(z)}{A_{11}(z)} \frac{\det \beta z^{2d}}{\left(\left(w + \frac{1}{2}A_{11}(z)\right)^2 - \frac{1}{4}A_{22}^2(z)\right)}.$$

We now observe that by Theorem 3.8 the first term $\frac{A_{21}(z)}{A_{11}(z)}$ has a limit as $z \to 0$. We claim that, under the condition that $C_d > 0$, the denominator $\left(w + \frac{1}{2}A_{11}(z)\right)^2 - \frac{1}{4}A_{22}^2(z))$ has a nonzero limit equal $A_{11}(0)C_d$. That $A_{11}(0) \neq 0$ follows from $C_d = f_{[d]} m_{[d]} (\beta_- - \beta_+ e^{2\nu(x_d - x_1)}) = A_{11}(0)$ $\left(1 - \frac{\beta_+}{\beta_-} e^{2\nu(x_d - x_1)}\right)$. To prove the actual claim we observe that $0^+ = \left(\frac{1}{2}(A_{11}(0) + A_{22}(0)), 0\right)$ hence on the upper sheet

$$\left(w + \frac{1}{2}A_{11}(z)\right)^2 - \frac{1}{4}A_{22}^2(z) \to \left(A_{11}(0) + \frac{1}{2}A_{22}(0)\right)^2 - \frac{1}{4}A_{22}^2(0)$$
$$= A_{11}(0)(A_{11}(0) + A_{22}(0)),$$

as claimed. In the case of $C_d < 0$, we go to the lower sheet where this time $0^- = \left(\frac{1}{2}(A_{11}(0) + A_{22}(0)), 0\right)$ and the rest of the computation goes through identically. $\qquad \square$

For the sake of comparison with [3] we will also state the previous theorem using a different local parametrization $\lambda = \frac{1}{z}$ and write the result in terms of $T_{11}(\lambda)$ and $T_{21}(\lambda)$.

Corollary 6.2 *Suppose* $C_d = \operatorname{Tr} A(0) > 0$, *then on the upper sheet of* Y, *using* $\frac{1}{\lambda}$ *as a local parameter around* 0^+, *we have*

$$W - \frac{T_{21}(\lambda)}{T_{11}(\lambda)} = O\left(\frac{1}{\lambda^{2d}}\right), \quad \lambda \to \infty \qquad (45)$$

The same result holds on the lower sheet if $C_d < 0$.

The next theorem gives a complete solution of the inverse problem of determining m_j, x_j from the knowledge of W:

Theorem 6.3 *Suppose Y is given by Equation 42 and $C_d > 0$. Let W be the generalized Weyl function. Then the first 2d terms of the Taylor expansion of zW around $z = 0$ on the upper sheet determine uniquely the peakon measure m, given by* (6), *and thus the field $A(z)$.*

Likewise, if $C_d < 0$, then we have the analogous result provided the expansion is computed on the lower sheet of Y.

Proof. (Sketch) It is clear that knowing the peakon measure m we can define the field $A(z)$ as in Theorem 3.8. By a minor modification of the arguments in [3, 4] the operator $L(\lambda) = D^2 - \nu^2 - 2\nu\lambda m$ (see Section 3) is unitarily equivalent to $\tilde{L}(\lambda) = D^2 - 2\nu\lambda\tilde{m}$ defined on the finite interval $[-\frac{1}{2\nu}, \frac{1}{2\nu}]$, where the measure $\tilde{m}$ is related to m in a simple way. Thus without any loss of generality we can assume that $L(\lambda)$ in our initial problem is stated for $D^2 - 2\nu\lambda m$. This problem has a very natural interpretation, namely it describes a classical inhomogeneous string of length $\frac{1}{\nu}$ with discrete mass density m (see [17, 16]), and, consequently, the inverse problem can be solved using the formulas originally obtained for Stieltjes' continued fractions [37], as explained in [3], except that the Weyl function in our case is not a rational function on $\mathbf{P}^1$ but a meromorphic function (or section) on Y.

Recall the main premise of Stieltjes' work. Given an asymptotic expansion of a function

$$f(\lambda) = \sum_{i=0}^{\infty} (-1) \frac{c_i}{\lambda^{i+1}}, \quad \lambda \to \infty \qquad (46)$$

one associates to it a sequence $\{f_j, j \in \mathbf{N}\}$ of continued fractions, which are Padé approximants when written as rational functions:

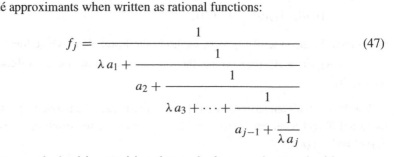

$$f_j = \cfrac{1}{\lambda a_1 + \cfrac{1}{a_2 + \cfrac{1}{\lambda a_3 + \cdots + \cfrac{1}{a_{j-1} + \cfrac{1}{\lambda a_j}}}}} \qquad (47)$$

These are obtained by requiring that each f_j approximates f with an error term being $O\left(\frac{1}{\lambda^{j+1}}\right)$. In other words, the jth approximant recovers j terms of the asymptotic expansion of f. The essential role in determining how the coefficients $a_1, a_2, \cdots$ are constructed from the asymptotic expansion is played by the infinite Hankel matrix H constructed out of the coefficients of the expansion (46) via

$$
H = \begin{bmatrix}
c_0 & c_1 & c_2 & c_3 & \cdots \\
c_1 & c_2 & c_3 & c_4 & \cdots \\
c_2 & c_3 & c_4 & c_5 & \cdots \\
\vdots & \vdots & \vdots & \vdots & \vdots
\end{bmatrix}
$$

and the determinants of the $k \times k$ submatrices of H whose top-left entry is c_ℓ in the top row of H, where $\ell \geq 0, k > 0$. We denote this determinant by Δ_k^ℓ and impose the convention that $\Delta_0^\ell = 1$. Using this notation, the formulas for a_j in (47) read

$$
a_{2k} = \frac{(\Delta_k^0)^2}{\Delta_k^1 \Delta_{k-1}^1}, \qquad a_{2k+1} = \frac{(\Delta_k^1)^2}{\Delta_k^1 \Delta_{k+1}^1}. \tag{48}
$$

We note that in this setup the function f vanishes at $\lambda = \infty$. Thus to compare with our case we might consider $zW = \frac{1}{\lambda} W$. Then Equation 45 reads

$$
\frac{W}{\lambda} - \frac{T_{21}(\lambda)}{\lambda T_{11}(\lambda)} = O\left(\frac{1}{\lambda^{2d+1}}\right), \qquad \lambda \to \infty.
$$

This result shows that the first $2d$ terms in the asymptotic — in fact analytic — expansion of $\frac{W}{\lambda}$ and $\frac{T_{21}(\lambda)}{\lambda T_{11}(\lambda)}$ are identical. However, the full continued fraction expansion of $\frac{T_{21}(\lambda)}{\lambda T_{11}(\lambda)}$ is known (see [3]) to be

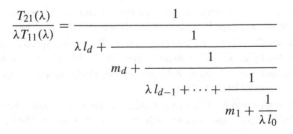

where $l_j = x_{j+1} - x_j$, $j = 0, \ldots, d$ and $x_0 = -\frac{1}{2\nu}$, $x_{d+1} = \frac{1}{2\nu}$. In our case, we have $2d$ terms which give us the first $(2d-1)$-th approximants by Equation 48, while we formally need to go up to the $(2d+1)$-th approximant to determine m_1 and l_0. However, recall that we know that the total length of the string is $\frac{1}{\nu}$; hence, $l_0 + = \frac{1}{\nu} - (l_1 \cdots + l_d)$. Likewise, the total mass M is known from the coefficient of $\operatorname{Tr} A(z)$ at z^{d-1}; hence, $m_1 = M - (m_d + \cdots + m_2)$.

$\square$

Remark 6.4   The inverse problem for the CF equation (7) with u satisfying (15) (in the special case $C = 0$) is given a different treatment in [5]. The solution outlined in the present paper is closer in spirit to the original solution of the inverse problem for the case $\beta_+ = 0$ in [4].

The outstanding problem for either treatment is to characterize the generalized Weyl functions W with the known theta function representation of the flow (see [5, Theorem 6.2]). For example, a function $f(z)$ is a Weyl function for the CH peakon problem ($\beta_+ = 0$) if and only if f is

1. a rational function vanishing at infinity with simple poles on the real axis;
2. all its residues are positive.

It remains an open question what replaces these conditions in the CF case.

References

[1] M. Adams, J. Harnad, and E. Previato. Isospectral hamiltonian flows in finite and infinite dimensions. *Communications in Mathematical Physics*, 117(3):451–500, 1988.

[2] M. Adler and P. van Moerbeke. Completely integrable systems, Euclidean Lie algebras, and curves. *Adv. in Math.*, 38(3):267–317, 1980.

[3] R. Beals, D. H. Sattinger, and J. Szmigielski. Multi-peakons and a theorem of Stieltjes. *Inverse Problems*, 15(1):L1–L4, 1999.

[4] R. Beals, D. H. Sattinger, and J. Szmigielski. Multipeakons and the classical moment problem. *Adv. Math.*, 154(2):229–257, 2000.

[5] R. Beals, D. H. Sattinger, and J. Szmigielski. Periodic peakons and Calogero-Françoise flows. *J. Inst. Math. Jussieu*, 4(1):1–27, 2005.

[6] A. Beauville. Jacobiennes des courbes spectrales et systèmes hamiltoniens complètement intégrables. *Acta Math.*, 164(3-4):211–235, 1990.

[7] A. Beauville, M. S. Narasimhan, and S. Ramanan. Spectral curves and the generalised theta divisor. *J. Reine Angew. Math.*, 398:169–179, 1989.

[8] O. Biquard and P. Boalch. Wild non-abelian Hodge theory on curves. *Compos. Math.*, 140(1):179–204, 2004.

[9] H. U. Boden and K. Yokogawa. Moduli spaces of parabolic Higgs bundles and parabolic $K(D)$ pairs over smooth curves. I. *Internat. J. Math.*, 7(5):573–598, 1996.

[10] J. L. Burchnall and T. W. Chaundy. Commutative Ordinary Differential Operators. *Proc. London Math. Soc. (2)*, 21:420–440, 1923.

[11] F. Calogero and J.-P. Françoise. A completely integrable Hamiltonian system. *J. Math. Phys.*, 37(6):2863–2871, 1996.

[12] R. Camassa and D. D. Holm. An integrable shallow water equation with peaked solitons. *Phys. Rev. Lett.*, 71(11):1661–1664, 1993.

[13] X. Chang and J. Szmigielski. Lax integrability and the peakon problem for the modified Camassa-Holm equation. *Comm. Math. Phys.*, 358(1):295–341, 2018.

[14] R. Donagi and E. Markman. Spectral covers, algebraically completely integrable, Hamiltonian systems, and moduli of bundles. In *Integrable systems and quantum groups (Montecatini Terme, 1993)*, volume 1620 of *Lecture Notes in Math.*, pages 1–119. Springer, Berlin, 1996.

[15] B. A. Dubrovin and S. P. Novikov. A periodic problem for the Korteweg-de Vries and Sturm-Liouville equations. Their connection with algebraic geometry. *Dokl. Akad. Nauk SSSR*, 219:531–534, 1974.

[16] H. Dym and H. P. McKean. *Gaussian processes, function theory, and the inverse spectral problem*. Academic Press [Harcourt Brace Jovanovich Publishers], New York, 1976. Probability and Mathematical Statistics, Vol. 31.

[17] F. R. Gantmacher and M. G. Krein. *Oscillation matrices and kernels and small vibrations of mechanical systems*. AMS Chelsea Publishing, Providence, RI, revised edition, 2002. Translation based on the 1941 Russian original, edited and with a preface by Alex Eremenko.

[18] P. A. Griffiths. Linearizing flows and a cohomological interpretation of Lax equations. *Amer. J. Math.*, 107(6):1445–1484 (1986), 1985.

[19] N. J. Hitchin. The self-duality equations on a Riemann surface. *Proc. London Math. Soc. (3)*, 55(1):59–126, 1987.

[20] N. J. Hitchin. Stable bundles and integrable systems. *Duke Math. J.*, 54(1):91–114, 1987.

[21] N. J. Hitchin. Remarks on Nahm's equations. *ArXiv e-prints 1708.08812*, Aug. 2017.

[22] N. J. Hitchin, G. B. Segal, and R. S. Ward. *Integrable Systems: Twistors, Loop Groups, and Riemann Surfaces*, volume 4 of *Oxford Graduate Texts in Mathematics*. The Clarendon Press, Oxford University Press, New York, 2013.

[23] A. R. Hodge and M. Mulase. Hitchin integrable systems, deformations of spectral curves, and KP-type equations. In *New developments in algebraic geometry, integrable systems and mirror symmetry (RIMS, Kyoto, 2008)*, volume 59 of *Adv. Stud. Pure Math.*, pages 31–77. Math. Soc. Japan, Tokyo, 2010.

[24] A. R. Its and V. B. Matveev. Hill operators with a finite number of lacunae. *Funkcional. Anal. i Priložen.*, 9(1):69–70, 1975.

[25] P. D. Lax. Integrals of nonlinear equations of evolution and solitary waves. *Comm. Pure Appl. Math.*, 21:467–490, 1968.

[26] P. D. Lax. Periodic solutions of the KdV equation. *Comm. Pure Appl. Math.*, 28:141–188, 1975.

[27] S. V. Manakov. The method of the inverse scattering problem, and two-dimensional evolution equations. *Uspehi Mat. Nauk*, 31(5(191)):245–246, 1976.

[28] S. V. Manakov. A remark on the integration of the Eulerian equations of the dynamics of an n-dimensional rigid body. *Funkcional. Anal. i Priložen.*, 10(4):93–94, 1976.

[29] E. Markman. Spectral curves and integrable systems. *Compositio Math.*, 93(3):255–290, 1994.

[30] H. P. McKean and P. van Moerbeke. The spectrum of Hill's equation. *Invent. Math.*, 30(3):217–274, 1975.

[31] J. Moser. Three integrable Hamiltonian systems connected with isospectral deformations. *Advances in Math.*, 16:197–220, 1975.

[32] J. Moser. Geometry of quadrics and spectral theory. In *The Chern Symposium 1979 (Proc. Internat. Sympos., Berkeley, Calif., 1979)*, pages 147–188. Springer, New York-Berlin, 1980.

[33] S. P. Novikov. A periodic problem for the Korteweg-de Vries equation. I. *Funkcional. Anal. i Priložen.*, 8(3):54–66, 1974.

[34] E. Previato. Seventy years of spectral curves: 1923–1993. In *Integrable systems and quantum groups*, pages 419–481. Springer, 1996.

[35] S. Rayan. Co-Higgs bundles on $\mathbb{P}^1$. *New York J. Math.*, 19:925–945, 2013.

[36] C. T. Simpson. Harmonic bundles on noncompact curves. *J. Amer. Math. Soc.*, 3(3):713–770, 1990.

[37] T. J. Stieltjes. Recherches sur les fractions continues. *Ann. Fac. Sci. Toulouse Sci. Math. Sci. Phys.*, 8(4):J1–J122, 1894.

Steven Rayan
Department of Mathematics and Statistics, University of Saskatchewan, 106 Wiggins Road, Saskatoon, Saskatchewan, S7N 5E6, Canada; rayan@math.usask.ca

Thomas Stanley
Department of Mathematics and Statistics, University of Saskatchewan, 106 Wiggins Road, Saskatoon, Saskatchewan, S7N 5E6, Canada; ths808@mail.usask.ca

Jacek Szmigielski
Department of Mathematics and Statistics, University of Saskatchewan, 106 Wiggins Road, Saskatoon, Saskatchewan, S7N 5E6, Canada; szmigiel@math.usask.ca

15

Tropical Markov Dynamics and Cayley Cubic

K. Spalding and A.P. Veselov

Dedicated to Emma Previato on her 65th birthday

Abstract. We study the tropical version of Markov dynamics on the Cayley cubic, introduced by V.E. Adler and one of the authors. We show that this action is semi-conjugated to the standard action of $SL_2(\mathbb{Z})$ on a torus, and thus is ergodic with the Lyapunov exponent and entropy given by the logarithm of the spectral radius of the corresponding matrix.

1 Introduction

In 1880 A.A. Markov [19] discovered a remarkable relation between the theory of binary quadratic forms and the following Diophantine equation known as the *Markov equation*

$$x^2 + y^2 + z^2 = 3xyz. \tag{1}$$

Markov showed that all positive integer solutions can be found from the obvious one $x = y = z = 1$ by applying the symmetry

$$(x, y, z) \to (x, y, 3xy - z) \tag{2}$$

(which is a corollary of the Vieta formula for the Markov equation considered as a quadratic with respect to z) and permutations. The corresponding *Markov numbers*

$$1, 2, 5, 13, 29, 34, 89, 169, 194, 233, 433, 610, 985 \ldots$$

play a very important role in the theory of Diophantine approximations determining the rank of the "most irrational" numbers (see for detail [7]). Many other relations were discovered later, including the theory of Frobenius manifolds and the related Painlevé-VI equation [8], Teichmüller spaces [9] and various problems in algebraic geometry [10, 22].

The growth of Markov numbers was investigated by Don Zagier [26], who used the parallel (going back to Cohn [5]) between the Markov tree and the Euclidean algorithm described by the equation

$$a + b = c \tag{3}$$

with coprime a, b. One can view this parallel as a "tropicalization" (known also as Maslov's "dequantization" [18]): if we write

$$x = e^{\frac{a}{\hbar}}, y = e^{\frac{b}{\hbar}}, z = e^{\frac{c}{\hbar}}$$

and let $\hbar \to 0$, then we have from the Markov equation that

$$a + b = c$$

assuming that a, b are less than c. Similar ideas were used by Andy Hone [12], who studied the growth problem in relation with Halburd's Diophantine approach to integrability [11].

In our paper [23] we used this relation to study the growth of the Markov numbers as functions of the paths on the Markov tree, where the Markov numbers are "naturally growing", see Fig. 15.1. One can view this representation as a version of the Conway topograph [6] for Markov triples.

More precisely, we defined the Lyapunov exponents of the Markov and Euclid trees $\Lambda(\xi), \xi \in \mathbb{R}P^1$ as

$$\Lambda(\xi) = \limsup_{n \to \infty} \frac{\ln(\ln z_n(\xi))}{n} = \limsup_{n \to \infty} \frac{\ln c_n(\xi)}{n} \tag{4}$$

where $z_n(\xi), c_n(\xi)$ are the corresponding numbers along the path γ_ξ on the Markov and Euclid trees respectively (see details in [23]). The function $\Lambda(\xi)$ is $PGL_2(\mathbb{Z})$-invariant and has some interesting properties studied in [23].

In the present paper we consider the tropical version of the integrable case of Markov dynamics on the real surface given by

$$x^2 + y^2 + z^2 = 3xyz + \frac{4}{9}$$

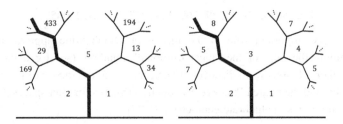

Figure 15.1 Markov and Euclid trees with a path

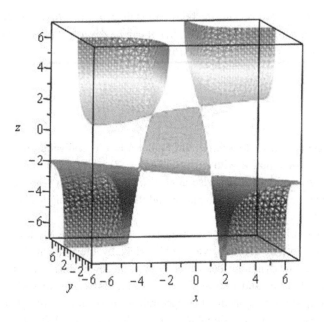

Figure 15.2 Cayley cubic

or, equivalently after scaling the variables by 3:

$$x^2 + y^2 + z^2 = xyz + 4. \tag{5}$$

From the algebro-geometric point of view the relation (5) determines the classical surface known as the *Cayley cubic*. It was studied by Arthur Cayley in [2] and can be characterised as the cubic surface with 4 (which is maximal possible) conical singularities. The real version of the Cayley cubic is shown in Fig. 15.2 prepared using MAPLE.

It has four infinite sheets similar to the Markov surface (1), but in this case the positive sheet (shown in the top right corner on Fig. 15.2) can be parametrized explicitly as

$$x = 2 \cosh a, \, y = 2 \cosh b, z = 2 \cosh c \tag{6}$$

with $c = a + b$ (or in symmetric form $a + b + c = 0$), which is in a good agreement with the tropical arguments above. This observation was used by Zagier [26] to study the growth of Markov numbers and earlier by Mordell [20] for studying the Diophantine properties of this equation.

However, the Cayley cubic has also the middle part with the trigonometric parametrization

$$x = 2 \cos a, \, y = 2 \cos b, z = 2 \cos c \tag{7}$$

with $a+b+c = 0$. It bounds the body called the *spectrahedron*, which appears in a range of applications (see [21, 25]). It can be described as the set of semi-positive symmetric matrices of the form

$$A = \begin{pmatrix} 1 & x/2 & y/2 \\ x/2 & 1 & z/2 \\ y/2 & z/2 & 1 \end{pmatrix}.$$

Indeed one can check that the condition $\det A = 0$ is equivalent to the Cayley relation (5). Note that A with entries given by (7) is the Gram matrix for three unit vectors in $\mathbb{R}^3$ with pairwise angles a, b, c, so geometrically $\det A = 0$ means that these vectors are coplanar and thus one of the angles is the sum of the others.

It is natural to ask if there is a "tropical" analogue of the Markov dynamics on the middle part of the Cayley cubic. Here by the tropical dynamics we simply mean the dynamics determined by the piecewise linear maps.

Vsevolod Adler and one of the authors [1, 24] came up with a natural suggestion, simply replacing the spectrahedron by the regular tetrahedron with the vertices at the singular points, which are $(2, 2, 2)$, $(2, -2, -2)$, $(-2, 2, -2)$, $(-2, -2, 2)$ (see Fig. 15.3).

The corresponding boundary, which we will denote T, is determined by the "tropical" Cayley equation

$$\max\{-u - v - w, -u + v + w, u - v + w, u + v - w\} = 2. \qquad (8)$$

Note that this is *not* an analogue of the Cayley surface in the sense of tropical algebraic geometry [14], so the terminology might be a bit confusing (see more in Concluding remarks).

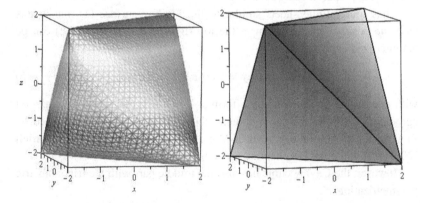

Figure 15.3 Cayley's spectrahedron and tetrahedron

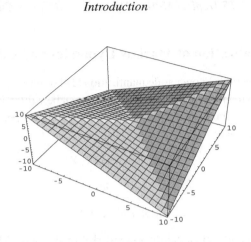

Figure 15.4 Function $f(x, y)$

Following [1] consider the corresponding action of the modular group $PSL_2(\mathbb{Z})$ generated by cyclic permutation of u, v, w and the tropical Vieta involution

$$(u, v, w) \rightarrow (u, v, -w + 2f(u, v)) \tag{9}$$

where $f : \mathbb{R}^2 \rightarrow \mathbb{R}$ is a piecewise linear function defined by

$$f(u, v) = \begin{cases} v & \text{if} & u \geq |v|, \\ u & \text{if} & v \geq |u|, \\ -v & \text{if} & -u \geq |v|, \\ -u & \text{if} & -v \geq |u|. \end{cases} \tag{10}$$

The plot of the function f is shown in Fig. 15.4.

The aim of this paper is to study the properties of this action, which we will call *tropical Cayley-Markov dynamics*. Our main result is the following

Theorem 1.1 *The tropical Cayley-Markov action of a hyperbolic element $A \in SL_2(\mathbb{Z})$ on T is ergodic, with the Lyapunov exponent and entropy given by the logarithm of the spectral radius of A. Their average growth along the path γ_ξ on the planar binary tree is given by the function $\Lambda(\xi)$.*

The proof is by constructing the semi-conjugation of this action with the standard action of $SL_2(\mathbb{Z})$ on a torus, using a natural tropical analogue of the parametrisation (7).

We should mention that the same idea was used by Cantat and Loray [3, 4] to compute the topological entropy of the (generalised) Markov dynamics (see also the important work of Iwasaki and Uehara [15, 16] in this direction).

2 Tropicalization of Markov Dynamics and Cayley Cubic

Tropicalization (also known as dequantization [18] or ultra-discretization [13]) can be applied to any dynamical system which can be written in algebraic form without a minus sign (subtraction-free), by replacing the operation of addition and multiplication by

$$X \oplus Y = \max(X, Y)$$

and

$$X \otimes Y = X + Y$$

respectively. It is clear that this does not work directly for the Markov dynamics in the form (2) because of the minus sign.

However one can consider another Vieta version (cf. Hone [12])

$$(x, y, z) \rightarrow (x, y, (x^2 + y^2)/z), \tag{11}$$

which can be naturally tropicalized as

$$(X, Y, Z) \rightarrow (X, Y, \max(2X, 2Y) - Z). \tag{12}$$

Together with cyclic permutations of X, Y, Z, this generates the action of the modular group $PSL_2(\mathbb{Z})$, which is known to be isomorphic to the free product $\mathbb{Z}_2 * \mathbb{Z}_3$.

It has an invariant

$$\Phi = \max(2X, 2Y, 2Z) - (X + Y + Z),$$

or, equivalently,

$$\Phi = \max(X - Y - Z, Y - X - Z, Z - X - Y), \tag{13}$$

which is the tropical version of the integral

$$F = \frac{x^2 + y^2 + z^2}{xyz},$$

invariant under the Vieta involution (2).

It is easy to see that the tropical equation $\Phi = 0$ for positive integers X, Y, Z defines the Euclidean algorithm and, as explained above, describes the asymptotic growth of the Markov triples in the logarithmic scale.

Let us now turn to the Cayley cubic case

$$x^2 + y^2 + z^2 = xyz + 4.$$

Adding 4 to the right hand side of the equation does not change much the asymptotic behaviour at infinity in the positive octant, and thus the tropicalization, which is the same as in the Markov case. However, it changes

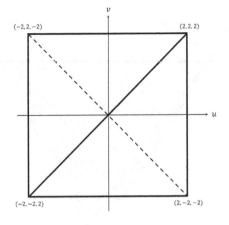

Figure 15.5 Projection of tropical Cayley surface T

the shape of the surface near the origin by adding the part bounded by 4 singular points (see Fig. 15.2 above).

Following [1], replace this part by the surface T of the tetrahedron with the same vertices. The projection of T to the (u, v)-coordinate plane is a 2-to-1 map to the corresponding square (see Fig. 15.5).

One can check directly that the piecewise linear involution (9), (10) swaps the branches of this double cover similarly to the Markov involution (2), which was the motivation for introducing the tropical Cayley-Markov dynamics in [1].

Proposition 2.1 *The function*

$$\Psi = \max\{-u + v + w, u - v + w, u + v - w, -u - v - w\} \qquad (14)$$

is invariant under the tropical Cayley-Markov dynamics (9), (10). *The level set* $\Psi = c, c > 0$ *is the surface of the regular tetrahedron with the vertices* (c, c, c), $(c, -c, -c)$, $(-c, c, -c)$, $(-c, -c, c)$.

The proof is by direct check. Note the difference of (14) with (13), which can be rewritten equivalently as

$$\Phi = -\min(-X + Y + Z, X - Y + Z, X + Y - Z).$$

3 Lyapunov Exponents and Entropy of the Tropical Cayley-Markov Dynamics

Now we would like to study the dynamical properties of the tropical Cayley-Markov action $PSL_2(\mathbb{Z}) = \mathbb{Z}_2 * \mathbb{Z}_3$, where the action of $\mathbb{Z}_2$ is given by (9), (10).

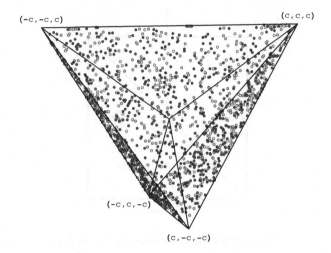

Figure 15.6 The level set of tropical Markov dynamics

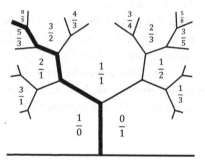

Figure 15.7 The Farey tree with marked "golden" Fibonacci path.

It is easy to see that this action preserves the usual Lebesgue measure on the surface of T.

The numerical calculations [1] showed the ergodic behaviour of the orbits of tropical Cayley-Markov dynamics at the level set $\Psi = c$, see Fig. 15.6.

Now we are ready to prove this and our main Theorem 1.1. For this we need some results from our paper [23].

Let us consider first the *Farey tree*, where at each vertex we have the fractions $\frac{p}{r}, \frac{q}{s}$ and their *Farey mediant* $\frac{p+q}{r+s}$ (see Fig. 15.7). Using the Farey tree we can identify the infinite paths γ on a binary tree with real numbers $\xi \in [0, \infty]$ using the theory of continued fractions. For example, for the golden ratio $\xi = \varphi := \frac{\sqrt{5}+1}{2}$ we have the Fibonacci path shown in bold.

One can use the Farey tree to describe the monoid $SL_2(\mathbb{N})$ consisting of matrices from $SL_2(\mathbb{Z})$ with non-negative entries. Indeed, two neighbouring fractions $\frac{p}{r}, \frac{q}{s}$ can be combined into the matrix

$$A = \begin{pmatrix} p & q \\ r & s \end{pmatrix} \in SL_2(\mathbb{N}). \tag{15}$$

It can be shown [23] that the Lyapunov exponents (4) can be equivalently defined as

$$\Lambda(\xi) = \limsup_{n \to \infty} \frac{\ln \rho(A_n(\xi))}{n}, \tag{16}$$

where $A_n(\xi) \in SL_2(\mathbb{N})$ is attached to the n-th edge along the path γ_ξ and $\rho(A)$ is the *spectral radius* of the matrix A, defined as the maximum of the modulus of its eigenvalues.

Let's introduce the tropical cosine function $\cos_T x$ as the period-2 piecewise linear even function given on the period by

$$\cos_T x = 1 - 2|x|, \quad x \in [-1, 1]$$

(see Fig. 15.8). It is known in Fourier theory as the *even triangle wave function*.

Define the tropical parametrization of T by the following tropical analogue of (7):

$$u = 2\cos_T \varphi, \; v = 2\cos_T \psi, \; w = 2\cos_T \chi, \tag{17}$$

where $\chi = \varphi + \psi$ and $(\phi, \psi) \in T^2 = \mathbb{R}^2/(2\mathbb{Z})^2$.

The corresponding map determines the 2-to-1 folding of the torus T^2 into the surface of the tetrahedron T (see Fig. 15.9).

The key observation now is the following

Proposition 3.1 *The parametrisation (17) semi-conjugates the tropical Cayley-Markov action of A with the standard action of A on the torus T^2.*

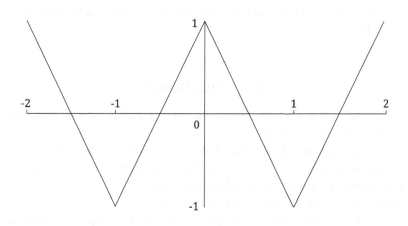

Figure 15.8 Tropical cosine (even triangle wave) function.

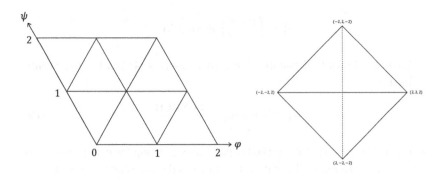

Figure 15.9 Folding of the torus T^2 to the surface of tetrahedron T

Indeed, u and v determine ϕ and ψ by (17) uniquely up to a sign, which means that the two values of the corresponding coordinate w are $w = 2\cos_T(\phi \pm \psi)$. Thus the tropical Cayley-Markov involution corresponds to the linear maps $(\pm\phi, \pm\psi, \pm(\phi + \psi)) \to (\pm\phi, \pm\psi, \pm(\phi - \psi))$, describing the action of $PSL_2(\mathbb{Z})$ on the so-called lax superbases (see [6]).

Note that the surface of the tetrahedron T is the quotient of the torus T^2 by the central symmetry group $\mathbb{Z}_2$, with fixed points corresponding to the vertices of the tetrahedron, so we have the following commutative diagram of the group actions

$$
\begin{array}{ccc}
T^2 & \xrightarrow{SL_2(\mathbb{Z})} & T^2 \\
\downarrow \mathbb{Z}_2 & & \downarrow \mathbb{Z}_2 \\
T & \xrightarrow{PSL_2(\mathbb{Z})} & T.
\end{array}
\tag{18}
$$

Since the action of a hyperbolic element $A \in SL_2(\mathbb{Z})$ on a torus is known to be ergodic with the Lyapunov exponent and entropy given by the natural logarithm of the spectral radius of A (see e.g. [17]), this completes the proof of Theorem 1.1.

4 Concluding Remarks

It is natural to ask how exceptional the case of Cayley-Markov dynamics considered in this chapter is.

Although replacing the spectrahedron by the regular tetrahedron looks like a natural tropicalization (in the sense of piecewise linearization), it cannot be explained by either traditional tropical algebraic geometry [14] or the dequantisation/ultra-discretization procedure [18, 13].

It would be nice therefore to see more similar examples in order to understand if there is any new general procedure here.

Although this is not directly related to Emma Previato's work, we believe that it is in the spirit of her very broad algebro-geometric view on integrable systems. One of us (APV) had enjoyed many years of friendship with Emma, so we are very happy to dedicate this work for her 65th birthday and to wish her the best on this occasion.

5 Acknowledgements

We are very grateful to Vsevolod Adler for many valuable discussions and crucial contribution to our preliminary work [1], to Andy Hone and Masatoshi Noumi, who attracted our attention to the important papers [3, 4, 15, 16].

This work was very much stimulated by the fruitful discussions with Leonid Chekhov, Boris Dubrovin, Andy Hone and Oleg Lisovyi. We are also grateful to Yiru Ye for help with preparation of the figures and to the referee for the helpful comments.

The work of K.S. was supported by the EPSRC as part of PhD study at Loughborough.

References

[1] V.E. Adler and A.P. Veselov *Tropical Markov dynamics*. Unpublished, 2009.

[2] A. Cayley *A Memoir on Cubic Surfaces*. Phil. Trans. Royal Soc. London **159** (1869), 231–326.

[3] S. Cantat *Bers and Hénon, Painlevé and Schrödinger*. Duke Math. Journal **149** (2009), issue 3, 411–457.

[4] S. Cantat and F. Loray *Holomorphic dynamics, Painlevé VI equation and character varieties*. Annales de l'Institut Fourier, Vol. 59 no. 7 (2009), p. 2927–2978.

[5] H. Cohn *Growth types of Fibonacci and Markoff*. Fibonacci Quart. **17** (1979), 178–183.

[6] J.H. Conway *The Sensual (Quadratic) Form*, Carus Mathematical Monographs, Vol.26. Mathematical Association of America, 1997.

[7] T.W. Cusick and M.E. Flahive *The Markoff and Lagrange Spectra*. Math. Surveys and Monographs, **30**. AMS, Providence, Rhode Island, 1989.

[8] B. Dubrovin *Integrable systems and classification of 2-dimensional topological field theories*. Progr. Math., **115**, Birkhäuser, Boston, MA, 1993.

[9] V.V. Fock *Dual Teichmüller spaces*. arxiv:dg-ga/9702018v3, 1997.

[10] P. Hacking, Y. Prokhorov *Smoothable del Pezzo surfaces with quotient singularities*. Compositio Math. **146** (2010), 169–192.

[11] R.G. Halburd *Diophantine integrability*. J. Phys. A: Math. Gen. **38** (2005) L263–L269.

[12] A.N. Hone *Diophantine non-integrability of a third-order recurrence with the Laurent property*. J. Phys. A **39** (2006), no. 12, L171–L177.

[13] R. Inoue, A. Kuniba, T. Takagi *Integrable structure of box-ball systems: crystal, Bethe ansatz, ultradiscretization and tropical geometry.* J. Phys. A **45** (2012), 1–64.

[14] I. Itenberg, G. Mikhalkin, E. Shustin *Tropical Algebraic Geometry* (2nd ed.). Birkhäuser Basel, 2009.

[15] K. Iwasaki *A modular group action on cubic surfaces and the monodromy of the Painlevé VI equation.* Proc. Japan Acad. Ser. A Math. Sci. **78** (2002),131–135.

[16] K. Iwasaki, T. Uehara *An ergodic study of Painlevé-VI.* Math. Ann. **338** (2007), no. 2, 295–345.

[17] A. Katok, B. Hasselblatt *Introduction to the Modern Theory of Dynamical Systems.* Cambridge Univ. Press, 1995.

[18] G.L. Litvinov *The Maslov dequantization, idempotent and tropical mathematics: a very brief introduction.* Contemp. Math. **377**, AMS, Providence, RI, 2005.

[19] A.A. Markoff *Sur les formes binaires indéfininies*, Math. Ann. **17** (1880), 379–399.

[20] L.J. Mordell *On the integer solutions of the equation $x^2 + y^2 + z^2 + 2xyz = n$.* J. London Math. Soc. **28** (1953), 500–510.

[21] M. Ramana, A.J. Goldman *Some geometric results in semidefinite programming.* J. Global Optimization, **7:1** (1995), 33–50.

[22] A.N. Rudakov *The Markov numbers and exceptional bundles on P^2.* Izv. Akad. Nauk SSSR Ser. Mat. **52** (1988), 100–112.

[23] K. Spalding and A.P. Veselov *Lyapunov spectrum of Markov and Euclid trees.* Nonlinearity **30** (2017), 4428–53.

[24] A.P. Veselov *Yang-Baxter and braid dynamics.* Talk at GADUDIS conference, Glasgow, April 2009.

[25] C. Vinzant *What is ... a spectrahedron?* Notices of the AMS, **61:5** (2014), 492–494.

[26] D. Zagier *On the number of Markoff numbers below a given bound.* Math. Comp. **39** (1982), no. 160, 709–723.

K. Spalding

IFLScience Ltd, 6 Kean St, London WC2B 4AS, UK
E-mail address: kathrynspalding7@gmail.com

A.P. Veselov

Department of Mathematical Sciences, Loughborough University, Loughborough LE11 3TU, UK and Moscow State University, Moscow 119899, Russia
E-mail address: A.P.Veselov@lboro.ac.uk

16

Positive One–Point Commuting Difference Operators[*][†]

Gulnara S. Mauleshova and Andrey E. Mironov

Abstract. In this paper we study a new class of rank one commuting difference operators containing a shift operator with only positive degrees. We obtain equations which are equivalent to the commutativity conditions in the case of hyperelliptic spectral curves. Using these equations we construct explicit examples of operators with polynomial and trigonometric coefficients.

1 Introduction and Main Results

Every maximal commutative ring of difference operators is isomorphic to a ring of meromorphic functions on an algebraic spectral curve Γ with poles $p_1, \ldots, p_s$. Such operators are called *s–point operators*. Common eigenfunctions of these operators form a vector bundle of rank l over an affine part of Γ, where l is called the *rank* of the operators. Krichever [1] (see also [2]) found two–point operators of rank one. Krichever and Novikov [3], [4] essentially developed a classification of commuting rings of difference operators, but the classification is not completed. The existence of one–point higher rank operators is proved in [3], [4]. Such operators were studied [5] in the case of hyperelliptic spectral curves. We mention here that in every maximal ring of operators discovered in [1]–[5], there are operators of the form

$$v_m(n)T^m + \cdots + v_0(n) + v_{-1}(n)T^{-1} + \cdots + v_{-k}(n)T^{-k}, \quad m, k > 0,$$

where $T\psi(n) = \psi(n+1)$, i.e. the shift operator is included in the operator with positive as well as negative degrees. In our short note [6] we found a new

[*] Mathematics Subject Classification: 34A30, 37K10
[†] The research was supported by the Russian Foundation for Basic Research (Grant 18-01-00411)
Keywords: commuting difference operators

class of rank one commuting difference operators such that these operators have the form

$$L_m = T^m + u_{m-1}(n)T^{m-1} + \cdots + u_0(n),$$

i.e. T has only positive degrees in L_m. We call such operators *positive operators*.

In this paper we give proofs of the results announced in [6] (Theorems 1.1–1.7), construct new examples (Theorem 1.8, see below), and also find spectral data for one–point higher rank operators (Section 3).

We choose the following spectral data

$$S = \{\Gamma, \gamma_1, \ldots, \gamma_g, q, k^{-1}, K_n\},$$

where Γ is a Riemann surface of genus g, $\gamma = \gamma_1 + \cdots + \gamma_g$ is a non–special divisor on Γ, $q \in \Gamma$ is a fixed point, k^{-1} is a local parameter near q, $K_n \in \Gamma$, $n \in \mathbb{Z}$ is a set of points in general position. If K_n does not depend on n, i.e. $K_n = q^+ \neq q$ then we get the spectral data for two–point operators of rank one (see [1]).

Theorem 1.1 *There exists a unique function* $\psi(n, P)$, $n \in \mathbb{Z}$, $P \in \Gamma$, *which has the following properties:*

1. *The divisor of zeros and poles of* ψ *has the form*

$$\gamma_1(n) + \cdots + \gamma_g(n) + K_1 + \cdots + K_n - \gamma_1 - \cdots - \gamma_g - nq,$$

if $n > 0$ *and has the form*

$$\gamma_1(n) + \cdots + \gamma_g(n) - K_{-1} - \cdots - K_n - \gamma_1 - \cdots - \gamma_g - nq,$$

if $n < 0$.

2. *In the neighborhood of* q, *the function* ψ *has the form*

$$\psi = k^n + O(k^{n-1}).$$

3. $\psi(0, P) = 1$.

The function $\psi(n, P)$ is called the *Baker–Akhiezer function*. For a meromorphic function $g(P)$ on Γ with a unique pole of order m in q, $g = k^m + O(k^{m-1})$ there exists a unique operator such that $L_m\psi(n, P) = g(P)\psi(n, P)$.

Remark 1.2   In the case of the two–dimensional discrete Schrödinger operator, Krichever [7] considered spectral data in which an additional set of points K_n appears similar to our construction.

We consider the hyperelliptic spectral curve Γ given by the equation

$$w^2 = F_g(z) = z^{2g+1} + c_{2g}z^{2g} + \cdots + c_0,$$

we choose $q = \infty$. Let $\psi(n, P)$ be the corresponding Baker–Akhiezer function. There exist commuting operators L_2, L_{2g+1} such that

$$L_2\psi = ((T + U_n)^2 + W_n)\psi = z\psi, \quad L_{2g+1}\psi = w\psi.$$

Theorem 1.3 *The following identity holds:*

$$L_2 - z = (T + U_n + U_{n+1} + \chi_n(P))(T - \chi_n(P)),$$

where

$$\chi = \frac{\psi(n+1, P)}{\psi(n, P)} = \frac{S_n}{Q_n} + \frac{w}{Q_n}, \quad S_n(z) = -U_n z^g + \delta_{g-1}(n)z^{g-1} + \cdots + \delta_0(n),$$

$$Q_n = -\frac{S_{n-1} + S_n}{U_{n-1} + U_n}. \tag{1}$$

Functions U_n, W_n, S_n satisfy the equation

$$F_g(z) = S_n^2 + (z - U_n^2 - W_n)Q_n Q_{n+1}. \tag{2}$$

The equation (2) can be linearized. Namely, if we replace n by $n + 1$ in (2), take a difference with (2) and apply (1), then the right hand side is divisible by Q_{n+1}. So, we obtain the linear equation on S_n.

Corollary 1.4 *Functions $S_n(z)$, U_n, W_n satisfy the equation*

$$(U_n + U_{n+1})(S_n - S_{n+1})$$
$$- (z - U_n^2 - W_n)Q_n + (z - U_{n+1}^2 - W_{n+1})Q_{n+2} = 0. \tag{3}$$

In the case of the elliptic spectral curve Γ given by the equation

$$w^2 = F_1(z) = z^3 + c_2 z^2 + c_1 z + c_0,$$

the equation (2) is exactly solvable.

Corollary 1.5 *Operator*

$$L_2 = (T + U_n)^2 + W_n,$$

where

$$U_n = -\frac{\sqrt{F_1(\gamma_n)} + \sqrt{F_1(\gamma_{n+1})}}{\gamma_n - \gamma_{n+1}}, \quad W_n = -c_2 - \gamma_n - \gamma_{n+1}, \tag{4}$$

γ_n *is an arbitrary functional parameter, commutes with the operator*

$$L_3 = T^3 + (U_n + U_{n+1} + U_{n+2})T^2 + (U_n^2 + U_{n+1}^2 + U_n U_{n+1} + W_n - \gamma_{n+2})T$$
$$+ (\sqrt{F_1(\gamma_n)} + U_n(U_n^2 + W_n - \gamma_n)).$$

Theorem 1.3 allows us to construct explicit examples.

Theorem 1.6 *Operator*

$$L_2 = (T + r_1 \cos(n))^2 + \frac{r_1^2 \sin(g) \sin(g+1)}{2 \cos^2(g + \frac{1}{2})} \cos(2n), \quad r_1 \neq 0$$

commutes with an operator L_{2g+1} of order $2g + 1$.

Theorem 1.7 *Operator*

$$L_2 = (T + \alpha_2 n^2 + \alpha_0)^2 - g(g+1)\alpha_2^2 n^2, \quad \alpha_2 \neq 0$$

commutes with an operator L_{2g+1} of order $2g + 1$.

Theorem 1.8 *Operator*

$$L_2 = (T + \beta_1 a^n)^2 + \frac{a^{2g} + a^{2g+2} - a^{4g+2} - 1}{(a^{2g+1} + 1)^2} \beta_1^2 a^{2n}, \quad \beta_1 \neq 0, \quad a \neq 1,$$

commutes with an operator L_{2g+1} of order $2g + 1$.

In Section 2 we give proofs of Theorems 1.1–1.8. In Section 3 we point out the spectral data for one–point positive higher rank commuting difference operators. In Section 4 we discuss a discretization problem of commuting ordinary differential operators with the help of one-point positive commuting difference operators.

2 Proofs of Theorems 1.1–1.8

In this section we give proofs of Theorems 1.1–1.8.

2.1 Theorem 1.1

Consider the case $n > 0$ (the case $n < 0$ is similar). By Riemann–Roch theorem, the dimension of the space of meromorphic functions on Γ with the pole divisor $\gamma_1 + \cdots + \gamma_g + nq$ is equal to

$$l(\gamma_1 + \cdots + \gamma_g + nq) = n + 1.$$

The requirement that a function from this space vanishes at points $K_1, \ldots, K_n$ extracts a one–dimensional subspace of this space. Condition (2) gives us the unique function. Theorem 1.1 is proved.

For any meromorphic function $g(P)$ on Γ with the unique pole of order m at q there exists a unique operator of the form

$$L_m = T^m + u_{m-1}(n)T^{m-1} + \cdots + u_0(n)$$

such that

$$L_m \psi = g(P)\psi.$$

Indeed, consider the case $n > 0$ (the case $n < 0$ is similar). Let us note that for any $u_j(n)$ the pole divisor of the function

$$\varphi(n, P) = L_m \psi - g(P)\psi$$

has the form

$$\gamma_1 + \cdots + \gamma_g + (m + n - 1)q.$$

Moreover, the function $\varphi(n, P)$ has zeros at points $K_1, \ldots, K_n$. Therefore, we can choose functions $u_{m-1}(n), \ldots, u_0(n)$ such that the function $\varphi(n, P)$ has zeros at points $K_1, \ldots, K_n, K_{n+1}, \ldots, K_{m+n}$. Thus $\varphi(n, P) = 0$. Similarly, for a meromorphic function $f(P)$ with the unique pole of order s at q there exists an operator L_s such that $L_s \psi = f(P)\psi$. Operators L_m and L_s commute.

2.2 Theorem 1.3

We have the identity

$$L_2 \psi = \big((T + U_n)^2 + W_n\big)\psi = z\psi.$$

Set $\chi_n(P) = \frac{\psi(n+1, P)}{\psi(n, P)}$, then $(T - \chi_n(P))\psi(n, P) = 0$.
The operator $L_2 - z$ can be factorized

$$(T + U_n)^2 + W_n - z = (T + \tilde{\chi}_n)(T - \chi_n(P)), \tag{5}$$

where

$$\tilde{\chi}_n = U_n + U_{n+1} + \chi_{n+1}(P),$$

herewith $\chi_n(P)$ satisfies the equation

$$-z + U_n^2 + W_n + \chi_n(P)(U_n + U_{n+1} + \chi_{n+1}(P)) = 0. \tag{6}$$

The equation (6) follows from the direct comparison of the left and right parts in (5). The function $\psi(n, P)$ satisfies (1)–(3), consequently, the function $\chi_n(P) = \frac{\psi(n+1, P)}{\psi(n, P)}$ has the first order pole in q, as well as simple poles in $\gamma_1(n), \ldots, \gamma_g(n)$. Thus $\chi_n(P)$ is a rational function with the pole divisor

$$\big(\chi_n(P)\big)_\infty = \gamma_1(n) + \cdots + \gamma_g(n) + q,$$

and in the neighborhood of q it has the expansion

$$\chi_n(P) = \frac{1}{k} + O(1).$$

Let $\gamma_i(n)$ have coordinates $(\alpha_i(n), \mu_i(n))$. Set

$$Q_n = (z - \alpha_1(n)) \ldots (z - \alpha_g(n)).$$

Then the rational function $\chi_n(P)$ has the form

$$\chi_n(P) = \frac{S_n}{Q_n} + \frac{w}{Q_n}, \tag{7}$$

where S_n is a polynomial of degree g with respect to z

$$S_n = \delta_g(n)z^g + \delta_{g-1}(n)z^{g-1} + \cdots + \delta_0(n).$$

Substituting (7) in (6), we get that the polynomial Q_n has the form

$$Q_n = -\frac{S_{n-1} + S_n}{U_{n-1} + U_n}.$$

Hence $\delta_g(n) = -U_n$. Thus

$$S_n = -U_n z^g + \delta_{g-1}(n)z^{g-1} + \cdots + \delta_0(n) \tag{8}$$

satisfies the equation (2). Theorem 1.3 is proved.

In the proof of theorems 1.6 and 1.7 we use the following observation. Let us assume that U_n, W_n and S_n are even functions in n

$$U_n = U_{-n}, \quad W_n = W_{-n}, \quad S_n = S_{-n}.$$

Let R_n be the left hand side of (3)

$$\begin{aligned}
R_n = {} & S_{n-1}(U_{n+1} + U_{n+2})(z - U_n^2 - W_n) \\
& + S_n(U_{n+1} + U_{n+2})(z + U_n U_{n+1} + U_{n-1}(U_n + U_{n+1}) - W_n) \\
& - S_{n+1}(U_{n-1} + U_n)(z + U_n U_{n+1} + (U_n + U_{n+1})U_{n+2} - W_{n+1}) \\
& - S_{n+2}(U_{n-1} + U_n)(z - U_{n+1}^2 - W_{n+1}).
\end{aligned}$$

Then R_n is skew invariant under the replacement $n \to -n - 1$, i.e. $R_n = -R_{-n-1}$.

2.3 Theorem 1.6

To prove Theorem 1.6 it is enough to show that for

$$U_n = r_1 \cos(n), \qquad W_n = \frac{r_1^2 \sin(g) \sin(g+1)}{2 \cos^2(g + \frac{1}{2})} \cos(2n)$$

there is a polynomial (8) satisfying (3).

In our case the equation (3) has the form

$$R_n = \cos\left(n + \frac{3}{2}\right)\left(S_{n-1}\left(2z - r_1^2\left(2\cos^2(n) + \frac{\sin(g)\sin(g+1)}{\cos^2(g+\frac{1}{2})}\cos(2n)\right)\right)\right.$$

$$+ S_n\left(2z + r_1^2\left(2\cos(1) + \cos(2) + \frac{\cos^2(\frac{1}{2})(2\cos(2g+1)+1)}{\cos^2(g+\frac{1}{2})}\cos(2n)\right)\right)\right)$$

$$- \cos\left(n - \frac{1}{2}\right)\left(S_{n+2}\left(2z - r_1^2\left(2\cos^2(n+1) + \frac{\sin(g)\sin(g+1)}{\cos^2(g+\frac{1}{2})}\cos(2n+2)\right)\right)\right.$$

$$+ S_{n+1}\left(2z + r_1^2\left(2\cos(1) + \cos(2) + \frac{\cos^2(\frac{1}{2})(2\cos(2g+1)+1)}{\cos^2(g+\frac{1}{2})}\cos(2n+2)\right)\right)\right)$$

$$= 0. \tag{9}$$

Let

$$S_n = A_{2g+1}\cos((2g+1)n) + A_{2g-1}\cos((2g-1)n)$$
$$+ \cdots + A_3\cos(3n) + A_1\cos(n),$$

where $A_i = A_i(z)$. Then U_n, W_n and S_n are even functions in n, and, as it was mentioned above, $R_n = -R_{-n-1}$. It is a remarkable fact that after substituting S_n in (9) we obtain

$$R_n = \alpha_{2g+4}\cos((2g+4)n) + \alpha_{2g+2}\cos((2g+2)n)$$
$$+ \cdots + \alpha_4\cos(4n) + \alpha_2\cos(2n)$$
$$= 0,$$

where

$$\alpha_s = 8r_1^2 A_{s-3}\cos^2\left(\frac{1}{2}\right)(\cos(2g+1) - \cos(3-s))\sin\left(\frac{3-s}{2}\right)$$

$$+ A_{s-1}\left(8z\sin\left(\frac{5-3s}{2}\right) - r_1^2\sin\left(\frac{1-3s}{2}\right) - 4r_1^2\sin\left(\frac{5-3s}{2}\right)\right)$$

$$+ 4r_1^2\sin\left(\frac{1-s}{2}\right) + 8z\sin\left(\frac{3-s}{2}\right) + 4r_1^2\sin\left(\frac{5-s}{2}\right) + 2r_1^2\sin\left(\frac{7-s}{2}\right)$$

$$+ 2\sin(1)\sin(2g)(-2(2z + r_1^2(-1 + \cos(1) + \cos(2)))\cos\left(\frac{s}{2}\right)\sin\left(\frac{3}{2}\right)$$

$$+ 2(r_1^2 - 2z))\cos\left(\frac{3s}{2}\right)\sin\left(\frac{5}{2}\right) + 2\cos\left(\frac{3}{2}\right)(2z + r_1^2(1 + 3\cos(1)$$

$$+ \cos(2)))\sin\left(\frac{s}{2}\right) - 2(r_1^2 - 2z))\cos\left(\frac{5}{2}\right)\sin\left(\frac{3s}{2}\right))$$

$$+ 2\cos(1)\cos(2g)(2(2z + r_1^2(-1 + \cos(1) + \cos(2)))\cos\left(\frac{s}{2}\right)\sin\left(\frac{3}{2}\right)$$

$$- 2(r_1^2 - 2z)\cos\left(\frac{3s}{2}\right)\sin\left(\frac{5}{2}\right) - 2\cos\left(\frac{3}{2}\right)(2z + r_1^2(1 + 3\cos(1)$$

$$+ \cos(2)))\sin\left(\frac{s}{2}\right) + 2(r_1^2 - 2z)\cos\left(\frac{5}{2}\right)\sin\left(\frac{3s}{2}\right)) - 3r_1^2\sin\left(\frac{1+s}{2}\right)$$

$$+ r_1^2\sin\left(\frac{3(1+s)}{2}\right) - 2r_1^2\sin\left(\frac{3+s}{2}\right) - r_1^2\sin\left(\frac{5+s}{2}\right)$$

$$+ 2r_1^2\sin\left(\frac{1+3s}{2}\right)) - A_{s+1}\left(2r_1^2\sin\left(\frac{1-3s}{2}\right) + r_1^2\sin\left(\frac{3}{2} - \frac{3s}{2}\right)$$

$$- 3r_1^2\sin\left(\frac{1-s}{2}\right) - 2r_1^2\sin\left(\frac{3-s}{2}\right) - r_1^2\sin\left(\frac{5-s}{2}\right)$$

$$+ 2\cos(1)\cos(2g)(2(2z + r_1^2(-1 + \cos(1) + \cos(2)))\cos\left(\frac{s}{2}\right)\sin\left(\frac{3}{2}\right)$$

$$- 2(r_1^2 - 2z)\cos\left(\frac{3s}{2}\right)\sin\left(\frac{5}{2}\right) + 2\cos\left(\frac{3}{2}\right)(2z + r_1^2(1 + 3\cos(1)$$

$$+ \cos(2)))\sin\left(\frac{s}{2}\right) - 2(r_1^2 - 2z)\cos\left(\frac{5}{2}\right)\sin\left(\frac{3s}{2}\right))$$

$$+ 2\sin(1)\sin(2g)(-2(2z + r_1^2(-1 + \cos(1) + \cos(2)))\cos\left(\frac{s}{2}\right)\sin\left(\frac{3}{2}\right)$$

$$+ 2(r_1^2 - 2z)\cos\left(\frac{3s}{2}\right)\sin\left(\frac{5}{2} - 2\cos\left(\frac{3}{2}\right)(2z + r_1^2(1 + 3\cos(1)$$

$$+ \cos(2)))\sin\left(\frac{s}{2}\right) + 2(r_1^2 - 2z)\cos\left(\frac{5}{2}\right)\sin\left(\frac{3s}{2}\right) + 4r_1^2\sin\left(\frac{1+s}{2}\right)$$

$$+ 8z\sin\left(\frac{3+s}{2}\right) + 4r_1^2\sin\left(\frac{5+s}{2}\right) + 2r_1^2\sin\left(\frac{7+s}{2}\right)$$

$$- r_1^2\sin\left(\frac{1+3s}{2}\right) - 4r_1^2\sin\left(\frac{5+3s}{2}\right) + 8z\sin\left(\frac{5+3s}{2}\right))$$

$$+ 8r_1^2 A_{s+3}\cos^2\left(\frac{1}{2}\right)(-\cos(1 + 2g) + \cos(3 + s))\sin\left(\frac{3+s}{2}\right).$$

From the above formula it follows that $\alpha_{2g+4} = 0$, automatically. So, we have $g + 1$ equations $\alpha_{2g+2} = \alpha_{2g} = \cdots = \alpha_2 = 0$ and $g + 1$ unknown functions $A_1, A_3, \ldots, A_{2g+1}$. From the equations

$$\alpha_{2g+2} = \alpha_{2g} = \cdots = \alpha_4 = 0$$

we express $A_{2g-1}, A_{2g-3}, \ldots, A_3, A_1$ via A_{2g+1}. For example,

$$A_{2g-1} = -A_{2g+1}$$
$$\times \frac{\cos(\frac{1}{2} + g)(4z + r_1^2(2\cos(1) + \cos(2) - 1) - 2(r_1^2 - 2z)\cos(2g + 1))}{8r_1^2 \cos^3(\frac{1}{2})\sin(\frac{1}{2})\sin(\frac{1}{2} - g)}.$$

It turns out that $\alpha_2 = 0$ automatically. Indeed, from

$$R_n + R_{-n-1} = \alpha_{2g+4}(1 + \cos(2g + 4))\cos((2g + 4)n) + \cdots$$
$$+ \alpha_4(1 + \cos(4))\cos(4n) + \alpha_2(1 + \cos(2))\cos(2n)$$
$$- \alpha_{2g+4}\sin(2g + 4)\sin((2g + 4)n) - \cdots - \alpha_4\sin(4)\sin(4n)$$
$$- \alpha_2\sin(2)\sin(2n)$$
$$= \alpha_2(1 + \cos(2))\cos(2n) - \alpha_2\sin(2)\sin(2n)$$
$$= 0.$$

Hence $\alpha_2 = 0$. Next, we choose A_{2g+1} such that coefficient at z^g in $S_n(z)$ is $-U_n$. We find S_n, satisfying the equation (3). Theorem 1.6 is proved.

2.4 Theorem 1.7

Let

$$U_n = a_2 n^2 + a_0, \qquad W_n = -g(g + 1)a_2^2 n^2.$$

In this case the equation (3) has the form

$$R_n = (S_{n-1} + S_n)(2a_0 + a_2(5 + 6n + 2n^2))(a_2^2 g(1 + g)n^2 + z - (a_0 + a_2 n^2)^2)$$
$$+ (S_n - S_{n+1})(2a_0 + a_2 - 2a_2 n + 2a_2 n^2)(2a_0 + a_2 + 2a_2 n + 2a_2 n^2)$$
$$\times (2a_0 + a_2(5 + 6n + 2n^2)) - (S_{n+1} + S_{n+2})(2a_0 + a_2 - 2a_2 n + 2a_2 n^2)$$
$$\times (a_2^2 g(1 + g)(1 + n)^2 - (a_0 + a_2(1 + n)^2)^2 + z)$$
$$= 0. \tag{10}$$

Let

$$S_n = B_{2g+2}n^{2g+2} + B_{2g}n^{2g} + \cdots + B_2 n^2 + B_0, \qquad B_i = B_i(z).$$

Then U_n, W_n and S_n are even functions in n, and, as it was mentioned above, $R_n = -R_{-n-1}$. After substituting S_n in (10) we obtain

$$R_n = \vartheta_{2g+8}(z)n^{2g+8} + \vartheta_{2g+7}(z)n^{2g+7} + \cdots + \vartheta_0(z) = 0,$$

where

$$\vartheta_s = -2B_{s-3}a_2^3(5+2g-s)(s-4)(2g+s-3)$$

$$+ B_{s-2}a_2^3((s-4)(s-2)-4g(g+1))(s-3)(s+1)$$

$$+ B_{s-1}\left(\frac{1}{6}a_2^2(a_2(3(s-3)(8+(s-3)s(8+(s-3)s))\right.$$

$$-4g(s-1)(27+s(5s-22))-4g^2(s-1)(27+s(5s-22)))$$

$$\left.-12a_0(6+s(-29+4g(g+1)-3(s-6)s)))-8a_2(s-3)z\right)$$

$$- B_s\left(\frac{1}{6}a_2(s+1)(6a_0a_2(4g(1+s)-8+4g^2(1+s)-s(2+3(s-3)s))\right.$$

$$+ a_2^2(-(s-2)(-6+(s-3)(s-1)^2s)+2g(6+s(9+4(s-4)s))$$

$$\left.+2g^2(6+s(9+4(s-4)s)))+24(s-2)z\right)$$

$$+ \sum_{m=1}^{2g+4}\left((-1)^m\left(C_{s+m}^m(2a_0+5a_2)(z-a_0^2)+C_{s+m}^{m+1}6a_2(a_0^2-z)\right.\right.$$

$$+ C_{s+m}^{m+2}a_2(5a_2^2g(1+g)+2a_0a_2(-5+g+g^2)-6a_0^2+2z)$$

$$+ C_{s+m}^{m+3}6a_2^2(2a_0-a_2g(1+g))+C_{s+m}^{m+4}a_2^2(a_2(-5+2g(1+g))-6a_0)$$

$$+ C_{s+m}^{m+5}6a_2^3-C_{s+m}^{m+6}2a_2^3)-\left(C_{s+m}^m(2a_0+a_2)(a_2^2(g^2+g+4)+3a_0^2\right.$$

$$+ 10a_0a_2+z)+C_{s+m}^{m+1}2a_2(9a_0^2+2a_2^2+2a_0a_2(g^2+g+4)-z)$$

$$+ C_{s+m}^{m+2}a_2(18a_0^2-a_2^2(g^2+g-2)+2a_0a_2(g^2+g+19)+2z)$$

$$+ C_{s+m}^{m+3}2a_2^2(18a_0+a_2g(1+g))-C_{s+m}^{m+4}a_2^2(18a_0+a_2(15+2g(1+g)))$$

$$+ C_{s+m}^{m+5}18a_2^3+C_{s+m}^{m+6}6a_2^3)+2^m\left(C_{s+m}^m(2a_0+a_2)(a_0^2+2a_0a_2\right.$$

$$- a_2^2(g^2+g-1)-z)-C_{s+m}^{m+1}4a_2(3a_0^2+a_2^2-2a_0a_2(g^2+g-2)+z)$$

$$+ C_{s+m}^{m+2}4a_2(6a_0^2+a_2^2g(g+1)-2a_0a_2(g^2+g-5)-2z)$$

$$- C_{s+m}^{m+3}16a_2^2(a_2g(g+1)-6a_0)+C_{s+m}^{m+4}16a_2^2(6a_0+a_2(5-2g(g+1)))$$

$$\left.\left.+ C_{s+m}^{m+5}192a_2^3+C_{s+m}^{m+6}128a_2^3)\right)B_{s+m},\right.$$

where $0 \le s < 2g+4$, $C_p^k = \frac{p!}{k!(p-k)!}$ at $p \ge k$, $C_p^k = 0$ at $p < k$.
From the above formula it follows that

$$\vartheta_{2g+8} = \vartheta_{2g+7} = \vartheta_{2g+6} = \vartheta_{2g+5} = \vartheta_{2g+4} = 0.$$

So, we have

$$R_n = \vartheta_{2g+3} n^{2g+3} + \vartheta_{2g+2} n^{2g+2} + \cdots + \vartheta_0 = 0.$$

From the condition $\vartheta_{2g+3} = 0$ we express B_{2g} via B_{2g+2}

$$B_{2g} = B_{2g+2} \frac{12 a_0 a_2 (2g^2 + 2g - 1) - a_2^2 g (8g^3 + 20g^2 + 7g - 8) - 12z}{12 a_2^2 (2g - 1)}.$$

We have

$$
\begin{aligned}
R_n + R_{-n-1} = {} & 2\vartheta_{2g+2} n^{2g+2} + (2g+2)\vartheta_{2g+2} n^{2g+1} \\
& + (2\vartheta_{2g} - (2g+1)\vartheta_{2g+1} + (2g+2)(2g+1)\vartheta_{2g+2}) n^{2g} \\
& + \cdots + (\vartheta_{2g+2} + \cdots + 2\vartheta_0) \\
= {} & 0.
\end{aligned}
$$

Hence from the condition $\vartheta_{2g+3} = 0$ it follows that $\vartheta_{2g+2} = 0$. Similarly, from $\vartheta_{2g-(2k-1)} = 0$ we express $B_{2g-(2k+2)}$ via B_{2g+2} and

$$
\begin{aligned}
R_n + R_{-n-1} = {} & 2\vartheta_{2g-2k} n^{2g-2k} + (2g-2k)\vartheta_{2g-2k} n^{2g-(2k+1)} \\
& + \cdots + (\vartheta_{2g+2k} + \cdots + 2\vartheta_0) \\
= {} & 0,
\end{aligned}
$$

hence $\vartheta_{2g-2k} = 0$.

So, from

$$\vartheta_{2g+3} = \vartheta_{2g+1} = \cdots = \vartheta_3 = \vartheta_1 = 0$$

we express $B_{2g}, \ldots, B_2, B_0$ via B_{2g+2}, herewith we have

$$\vartheta_{2g+2} = \vartheta_{2g} = \cdots = \vartheta_2 = \vartheta_0 = 0.$$

Next, we choose B_{2g+2} such that coefficient at z^g in $S_n(z)$ is $-U_n$. We find S_n, satisfying the equation (3). Theorem 1.7 is proved.

Remark 2.1 We checked that for $g = 1, \ldots, 5$ the operator

$$L_2 = (T + \alpha_2 n^2 + \alpha_1 n + \alpha_0)^2 - g(g+1)\alpha_2 n(\alpha_2 n + \alpha_1), \quad \alpha_2 \neq 0$$

commute with L_{2g+1}. Probably this operator L_2 commutes with L_{2g+1} for all $g \in \mathbb{N}, \alpha_2 \neq 0, \alpha_1, \alpha_0$.

2.5 Theorem 1.8

Let

$$U_n = \beta a^n, \qquad W_n = \frac{a^{2g} + a^{2g+2} - a^{4g+2} - 1}{(a^{2g+1} + 1)^2} \beta^2 a^{2n}.$$

The equation (3) has the form

$$S_{n-1}((a+1)(a^{2g+2}+a)^2\beta z - a^{2(n+g+1)}(a+1)^3\beta^3)$$

$$+S_n a(a+1)\beta(a^{2n}(a+1)^2(a^{2g+1}+a^{4g+2}+1)\beta^2 + a(a^{2g+1}+1)^2 z)$$

$$-S_{n+1}(a^{2n+1}(a+1)^3(a^{2g+1}+a^{4g+2}+1)\beta^3 + (a+1)(a^{2g+1}+1)^2\beta z)$$

$$S_{n+2}(a^{2(n+g+1)}(a+1)^3\beta^3 - (a+1)(a^{2g+1}+1)^2\beta z) = 0. \tag{11}$$

Let

$$S_n = G_{2g+1}a^{(2g+1)n} + G_{2g-1}a^{(2g-1)n} + \cdots + G_1 a^n, \qquad G_i = G_i(z).$$

After substituting S_n in (11) we obtain

$$R_n = \mu_{2g+4}a^{(2g+4)n} + \mu_{2g+2}a^{(2g+2)n} + \cdots + \mu_4 a^{4n} = 0,$$

where

$$\mu_p = G_{p-3}\frac{(a+1)^2(a^3-a^p)(a^{2g+4}-a^p)(a^{2g+p}-a^2)\beta^2}{(a^{2g+1}+1)^2}$$

$$+ G_{p-1}a^2(a^2-a^p)(a+a^p)(a^2+a^p)z.$$

From the above formula it follows that $\mu_{2g+4} = 0$. So, we have g equations $\mu_{2g+2} = \mu_{2g} = \cdots = \mu_4 = 0$ and $g+1$ unknown functions $G_1, G_3, \ldots, G_{2g+1}$. From $\mu_p = 0$, we express $G_{2g-1}, G_{2g-3}, \ldots, G_1$ via G_{2g+1}. For example,

$$G_{2g-1} = -G_{2g+1}\frac{a^{1-2g}(a^{2g+1}+1)^3 z}{\beta_1^2(a-1)(a+1)^3(a^{2g}-a)}.$$

Next, we choose G_{2g+1} such that the coefficient at z^g in $S_n(z)$ is $-U_n$. We find S_n, satisfying the equation (3). Theorem 1.8 is proved.

3 One–Point Higher Rank Positive Operators

In this section we introduce spectral data for positive operators of rank $l > 1$.
    We take the following spectral data

$$S = \{\Gamma, q, k^{-1}, K_n, \xi_0(n), \gamma, \alpha\},$$

where Γ is a Riemann surface of genus g, $q \in \Gamma$ is a fixed point, k^{-1} is a local parameter near q, $K_n \in \Gamma, n \in \mathbb{Z}$ is a set of points in general position, $\xi_0(n) = (\xi_0^1(n), \ldots, \xi_0^{l-1}(n), 1)$, at $n > 0$, $\xi_0(n) = (1, \xi_0^2(n), \ldots, \xi_0^l(n))$, at $n < 0$, $\xi_0^j(n)$ is a function in n, for simplicity we assume $\xi_0^j(n) \neq 0$,

$\gamma = \gamma_1 + \cdots + \gamma_{lg}$ is a divisor ($\gamma_j \in \Gamma$ are in general position), α is a set of vectors

$$\alpha_1, \ldots, \alpha_{lg}, \qquad \alpha_j = (\alpha_{j,1}, \ldots, \alpha_{j,l-1}).$$

Pair (γ, α) is called *Tyurin parameters*, (γ, α) defines a semistable holomorphic bundle over Γ with holomorphic sections $\eta_1, \ldots, \eta_l$. Points $\gamma_1, \ldots, \gamma_{lg}$ are points of linear dependence of the sections, herewith

$$\eta_l(\gamma_k) = \sum_{i=1}^{l-1} \alpha_{j,i} \eta_j(\gamma_k), \qquad k = 1, \ldots, lg.$$

Theorem 3.1 *There is a unique vector function* $\psi(n, P) = (\psi_1(n, P), \ldots, \psi_l(n, P))$ *which satisfies the following properties.*

1. *In the neighborhood of point q the function* $\psi(n, P)$ *has the form*

$$\psi(n, P) = \left(\sum_{s=0}^{\infty} \frac{\xi_s(n)}{k^s} \right) \begin{pmatrix} 0 & 1 & 0 & \ldots & 0 & 0 \\ 0 & 0 & 1 & \ldots & 0 & 0 \\ \multicolumn{6}{c}{\ldots\ldots\ldots\ldots\ldots\ldots} \\ 0 & 0 & 0 & \ldots & 0 & 1 \\ k & 0 & 0 & \ldots & 0 & 0 \end{pmatrix}^n, \tag{12}$$

$$\xi_s(n) = (\xi_s^1(n), \ldots, \xi_s^l(n)).$$

2. *The vector function* $\psi(n, P)$ *has gl simple poles in* $\gamma_1, \ldots, \gamma_{lg}$ *such that*

$$Res_{\gamma_i} \psi_j = \alpha_{i,j} Res_{\gamma_i} \psi_l. \tag{13}$$

3. *Let* $n > 0$ *and let* $n = lm + s$, $m \in \mathbb{N}$, $0 < s \le l$, ψ *has simple zeros in* $K_1, \ldots, K_m$

$$\psi(K_p) = 0, \qquad 1 \le p \le m, \tag{14}$$

additionally $\psi_1, \ldots, \psi_s$ *have simple zero in* K_{m+1}

$$\psi_1(K_{m+1}) = 0, \ldots, \psi_s(K_{m+1}) = 0. \tag{15}$$

Let $n < 0$ *and let* $n = -lm - s$, $m \in \mathbb{N}$, $0 < s \le l$, ψ *has simple poles in* $K_{-m}, \ldots, K_{-1}$

$$\psi(K_p) = \infty, \qquad -m \le p \le -1, \tag{16}$$

additionally $\psi_1, \ldots, \psi_s$ *have simple pole in* K_{-m-1}

$$\psi_1(K_{-m-1}) = \infty, \ldots, \psi_s(K_{-m-1}) = \infty. \tag{17}$$

4. $\psi(0, P) = \xi_0(0).$

Proof. Let us consider the case $n > 0$. Let

$$A = \begin{pmatrix} 0 & 1 & 0 & \ldots & 0 & 0 \\ 0 & 0 & 1 & \ldots & 0 & 0 \\ \ldots\ldots\ldots\ldots\ldots\ldots\ldots \\ 0 & 0 & 0 & \ldots & 0 & 1 \\ k & 0 & 0 & \ldots & 0 & 0 \end{pmatrix}.$$

Then

$$A^2 = \begin{pmatrix} 0 & 0 & 1 & 0 & \ldots & 0 & 0 \\ 0 & 0 & 0 & 1 & \ldots & 0 & 0 \\ \ldots\ldots\ldots\ldots\ldots\ldots\ldots\ldots \\ 0 & 0 & 0 & 0 & \ldots & 0 & 1 \\ k & 0 & 0 & 0 & \ldots & 0 & 0 \\ 0 & k & 0 & 0 & \ldots & 0 & 0 \end{pmatrix}, \quad A^3 = \begin{pmatrix} 0 & 0 & 0 & 1 & \ldots & 0 & 0 \\ 0 & 0 & 0 & 0 & \ldots & 0 & 0 \\ \ldots\ldots\ldots\ldots\ldots\ldots\ldots\ldots \\ k & 0 & 0 & 0 & \ldots & 0 & 0 \\ 0 & k & 0 & 0 & \ldots & 0 & 0 \\ 0 & 0 & k & 0 & \ldots & 0 & 0 \end{pmatrix}, \quad \ldots,$$

$$A^l = \begin{pmatrix} k & 0 & 0 & \ldots & 0 & 0 \\ 0 & k & 0 & \ldots & 0 & 0 \\ \ldots\ldots\ldots\ldots\ldots\ldots \\ 0 & 0 & 0 & \ldots & k & 0 \\ 0 & 0 & 0 & \ldots & 0 & k \end{pmatrix}.$$

We have $A^n = k^m A^s$. Hence, the pole divisors of ψ_j are

$$(\psi_j)_\infty = \gamma_1 + \cdots + \gamma_{lg} + q(m+1), \qquad 0 < j \le s$$
$$(\psi_j)_\infty = \gamma_1 + \cdots + \gamma_{lg} + qm, \qquad s < j \le l.$$

The conditions (12)–(15) define the unique function $\psi(n, P)$.

Similarly, when $n < 0$, the pole divisors of ψ_j are

$$(\psi_j)_\infty = \gamma_1 + \cdots + \gamma_{lg} + K_{-1} + \cdots + K_{-m-1}, \qquad 0 < j \le s$$
$$(\psi_j)_\infty = \gamma_1 + \cdots + \gamma_{lg} + K_{-1} + \cdots + K_{-m}, \qquad s < j \le l.$$

The conditions (12)–(13), (16)–(17) define the unique function $\psi(n, P)$. Theorem 3.1 is proved.

For meromorphic functions $f(P), g(P)$ with unique poles in q of orders m, s there are two commuting operators L_{lm}, L_{ns} of orders lm and ns such that

$$L_{lm}\psi = f(P)\psi, \qquad L_{ns}\psi = g(P)\psi.$$

Theorem 3.1 is an analogue of Krichever's Theorem (see [9]) for spectral data of higher rank commuting ordinary differential operators. Higher rank commuting ordinary differential operators were studied for example in [10]–[15] (see another citations in [16]).

We plan to study commuting higher rank positive operators later. Here we demonstrate only that the class of one–point higher rank operators is very interesting. It contains operators with polynomial coefficients.

Example 3.2 *Operator*

$$L_4 = T^4 + 2n^2 T^3 + \frac{3}{2}(n-1)n(n^2 - n - 2)T^2$$

$$+ \frac{1}{2}(n-2)n^2(n^3 - 4n^2 - n + 10)T$$

$$+ \frac{1}{16}(n-3)(n-2)(n-1)n(n(n-3)-4)((n-3)n-6)$$

commutes with the operator

$$L_6 = T^6 + (3n^2 + 6n + 8)T^5 + \frac{1}{4}(n(n+1)(32 + 15n(n+1)) - 6)T^4$$

$$+ \frac{1}{2}n^2(n^2 - 2)(5n^2 + 7)T^3$$

$$+ \frac{1}{16}(n-2)(n-1)n(n+1)((n-1)n(15(n-1)n - 38) - 36)T^2$$

$$+ \frac{1}{16}(n-2)^2 n^2(12 + (n-2)n((n-2)n - 5)(3(n-2)n - 11))T$$

$$+ \frac{1}{64}(n-4)(n-3)(n-2)(n-1)n(n+1)$$

$$\times ((n-3)n - 6)((n-4)(n-3)n(n+1) - 6).$$

The spectral curve of the pair L_4, L_6 is given by the equation

$$w^2 = \frac{1}{4}z^2(32z - 3)^2.$$

Since

$$[T, n] = T, \qquad [x, (-x\partial_x)] = x,$$

the replacement $T \to x$, $n \to (-x\partial_x)$ in L_4, L_6 gives a pair of commuting differential operators with polynomial coefficients. So, we get a commutative subalgebra in the first Weyl algebra $A_1 = \mathbb{C}[x][\partial_x]$.

4 Concluding Remarks

Theorem 1.3 has the following remarkable application.

Let $\wp(x)$, $\zeta(x)$ are Weierstrass functions

$$(\wp'(x))^2 = 4\wp^3(x) + g_2\wp(x) + g_3, \qquad -\zeta'(x) = \wp(x).$$

Set

$$A_1 = -2\zeta(\varepsilon) - \zeta(x - \varepsilon) + \zeta(x + \varepsilon),$$

$$A_2 = -\frac{3}{2}(\zeta(\varepsilon) + \zeta(3\varepsilon) + \zeta(x - 2\varepsilon) - \zeta(x + 2\varepsilon).$$

Further, for odd $g = 2g_1 + 1$ we set

$$A_g = A_1 \prod_{k=1}^{g_1} \left(1 + \frac{\zeta(x - (2k+1)\varepsilon) - \zeta(x + (2k+1)\varepsilon)}{\zeta(\varepsilon) + \zeta((4k+1)\varepsilon)} \right),$$

for even $g = 2g_1$ we set

$$A_g = A_2 \prod_{k=2}^{g_1} \left(1 + \frac{\zeta(x - 2k\varepsilon) - \zeta(x + 2k\varepsilon)}{\zeta(\varepsilon) + \zeta((4k-1)\varepsilon)} \right).$$

Set

$$L_2 = \frac{T_\varepsilon^2}{\varepsilon^2} + A_g(x, \varepsilon)\frac{T_\varepsilon}{\varepsilon} + \wp(\varepsilon),$$

where T_ε is the shift operator, $T_\varepsilon \psi(x) = \psi(x + \varepsilon)$.

- *The operator L_2 commutes with an operator of the form*

$$L_{2g+1} = \sum_{j=0}^{2g+1} B_j(x, \varepsilon)\frac{T_\varepsilon^j}{\varepsilon^j}.$$

There is a decomposition

$$L_2 = \partial_x^2 - g(g+1)\wp(x) + O(\varepsilon).$$

We checked that when $g = 1, 2$, and numerically checked that when $g = 3, \ldots, 6$ the spectral curve of L_2, L_{2g+1} does not depend on the parameter ε and coincides with the spectral curve of the Lamé operator $\partial_x^2 - g(g+1)\wp(x)$. So, we have a remarkable discretization of the Lamé operator preserving all integrable properties, in particular the spectral curve. The proof of the above statement is very long, so we will give it in an another paper.

The above statement allows us to propose the following conjecture:

Let $\tilde{A}$ be a maximal ring of commuting ordinary differential operators, then there is a ring of one–point positive commuting difference operators A (A is isomorphic to $\tilde{A}$), operators from A have the form

$$L_n = \sum_{j=0}^{n} v_j(x, \varepsilon)T_\varepsilon^j,$$

and

$$L_n = \tilde{L}_n + O(\varepsilon),$$

where $\tilde{L}_n \in \tilde{A}$ is an differential operator of order n.

References

[1] I.M. Krichever, *Algebraic curves and non–linear difference equations*, Russian Math. Surveys, **33**:4, (1978) 215–216.

[2] D. Mumford, *An algebro–geometric construction of commuting operators and of solutions to the Toda lattice equation, Korteweg–de Vries equation and related non–linear equations*, Proceedings of the International Symposium on Algebraic Geometry (Kyoto Univ., Kyoto, 1977), Kinokuniya, Tokyo, 1978, 115–153.

[3] I.M. Krichever, S.P. Novikov, *Two–dimensionalized Toda lattice, commuting difference operators, and holomorphic bundles*, Russian Math. Surveys, **58**:3, (2003) 473–510.

[4] I.M. Krichever, S.P. Novikov, *Holomorphic bundles and scalar difference operators: one-point constructions*, Russian Math. Surveys, **55**:1, (2000) 180–181.

[5] G.S. Mauleshova, A.E. Mironov, *Commuting difference operators of rank two*, Russian Math. Surveys, **70**:3, (2015) 557–559.

[6] G.S. Mauleshova, A.E. Mironov, *One–point commuting difference operators of rank 1*, Doklady Mathematics, **93**:1, (2016) 62–64.

[7] I.M. Krichever, *Two–dimensional periodic difference operators and algebraic geometry*, Dokl. Akad. Nauk SSSR, **285**:1, (1985) 31–36.

[8] G.S. Mauleshova, A.E. Mironov, *One–point commuting difference operators of rank 1 and their relation with finite–gap Schrödinger operators*, Doklady Mathematics, **97**:1, (2018) 62–64.

[9] I.M. Krichever, *Commutative rings of ordinary linear differential operators*, Funct. Anal. Appl., **12**:3, (1978) 175–185.

[10] I.M. Krichever, S.P. Novikov, *Holomorphic bundles over algebraic curves and non–linear equations*, Russian Mathematical Surveys, **35**:6, (1980) 53–79.

[11] P.G Grinevich, *Rational solutions for the equation of commutation of differential operators*, Functional Analysis and Its Applications, **16**:1, (1982) 15–19

[12] O.I. Mokhov, *Commuting differential operators of rank 3, and nonlinear differential equations*, Mathematics of the USSR–Izvestiya, **35**:3, (1990) 629–655

[13] E. Previato, G. Wilson, *Differential operators and rank 2 bundles over elliptic curves*, Compositio Math., **81**:1, (1992) 107–119.

[14] G. Latham, E. Previato, *Darboux transformations for higher-rank Kadomtsev–Petviashvili and Krichever–Novikov equations*, Acta Appl. Math., **39**, (1995) 405–433.

[15] A.E. Mironov, *Self–adjoint commuting differential operators.*, Inventiones mathematicae, **197**:2, (2014) 417–431.

[16] A.E. Mironov, *Self–adjoint commuting differential operators of rank two*, Russian Mathematical Surveys, **71**:4, (2016) 751–779.

Gulnara S. Mauleshova
Sobolev Institute of Mathematics,
4 Acad. Koptyug avenue, 630090, Novosibirsk, Russia and
Novosibirsk State University,
Pirogova st 1, 630090, Novosibirsk, Russia
e-mail: mauleshova_gs@mail.ru

Andrey E. Mironov
Sobolev Institute of Mathematics,
4 Acad. Koptyug avenue, 630090, Novosibirsk, Russia and
Novosibirsk State University,
Pirogova st 1, 630090, Novosibirsk, Russia
e-mail: mironov@math.nsc.ru